Evolutionary Computation

Evolutionary Computation

Special Issue Editors

Gai-Ge Wang
Amir H. Alavi

MDPI • Basel • Beijing • Wuhan • Barcelona • Belgrade

MDPI

Special Issue Editors

Gai-Ge Wang
Ocean University of China
China

Amir H. Alavi
University of Pittsburgh
USA

Editorial Office
MDPI
St. Alban-Anlage 66
4052 Basel, Switzerland

This is a reprint of articles from the Special Issue published online in the open access journal *Mathematics* (ISSN 2227-7390) from 2018 to 2019 (available at: https://www.mdpi.com/journal/mathematics/special_issues/Mathematics_Evolutionary_Computation).

For citation purposes, cite each article independently as indicated on the article page online and as indicated below:

LastName, A.A.; LastName, B.B.; LastName, C.C. Article Title. *Journal Name* **Year**, *Article Number*, Page Range.

ISBN 978-3-03921-928-5 (Pbk)
ISBN 978-3-03921-929-2 (PDF)

Contents

About the Special Issue Editors

Gai-Ge Wang is an Associate Professor in Ocean University of China, China. His entire array of publications have been cited over 5000 times (Google Scholar). Twelve and thirty-nine papers are selected as Highly Cited Paper by Web of Science, and Scopus (till November 2019), respectively. One paper is selected as "Top Articles from Outstanding S&T Journals of China-F5000 Frontrunner". The latest Google h-index and i10-index are 40 and 78, respectively. He is a senior member of SAISE, SCIEI, a member of IEEE, IEEE CIS, ISMOST. He served as Editors-in-Chief of OAJRC Computer and Communications, Editorial Advisory Board Member of Communications in Computational and Applied Mathematics (CCAM), Associate Editor of IJCISIM, an Editorial Board Member of IEEE Access, Mathematics, IJBIC, Karbala International Journal of Modern Science, and Journal of Artificial Intelligence and Systems. He served as Guest Editor for many journals including Mathematics, IJBIC, FGCS, Memetic Computing and Operational Research. His research interests are swarm intelligence, evolutionary computation, and big data optimization.

Amir H. Alavi is an Assistant Professor in the Department of Civil and Environmental Engineering, and holds a courtesy appointment in the Department of Bioengineering at the University of Pittsburgh. Prior to joining Pitt, Dr. Alavi was an Assistant Professor of Civil Engineering at the University of Missouri. Dr. Alavi's research interests include structural health monitoring, smart civil infrastructure systems, deployment of advanced sensors, energy harvesting, and engineering information systems. At Pitt, his Intelligent Structural Monitoring and Response Testing (iSMaRT) Laboratory focuses on advancing the knowledge and technology required to create self-sustained and multifunctional sensing and monitoring systems that are enhanced by engineering system informatics. His research activities involve implementation of these smart systems in the fields of civil infrastructure, construction, aerospace, and biomedical engineering. Dr. Alavi has worked on research projects supported by Federal Highway Administration (FHWA), National Institutes of Health (NIH), National Science Foundation (NSF), Missouri DOT, and Michigan DOT. Dr. Alavi has authored five books and over 170 publications in archival journals, book chapters, and conference proceedings. He has received a number of award certificates for his journal articles. He is among the Google Scholar 200 Most Cited Authors in Civil Engineering, as well as Web of Science ESI's World Top 1% Scientific Minds. He has served as the editor/guest editor of several journals such as Sensors, Case Studies in Construction Material, Automation in Construction, Geoscience Frontiers, Smart Cities, ASCE-ASME Journal of Risk and Uncertainty in Engineering Systems, and Advances in Mechanical Engineering.

Preface to "Evolutionary Computation"

Intelligent optimization is based on the mechanism of computational intelligence, refining the appropriate feature model, designing an effective optimization algorithm, and then obtaining the optimal solution or satisfactory solution for complex problems. Intelligent algorithms should try to ensure global optimization quality, fast optimization efficiency and robust optimization performance. Many researchers have different discoveries in the study of intelligent optimization algorithms. Feng et al. presented a novel monarch butterfly optimization with a global position updating operator (GMBO), which can address 0-1 KP known as an NP-complete problem. Ran et al. proposed an accurate approach for identifying rock types in the field based on image analysis using deep convolutional neural networks, which can identify six common rock types with an overall classification accuracy of 97.96%. Kong et al. presented an adaptive multi-swarm particle swarm optimizer, which adaptively divides a swarm into several sub-swarms and a competition mechanism is employed to select exemplars to address large-scale optimization problem. Zhang et al. proposed a memetic algorithm for MLCP based on an improved K-OPT local search and an evolutionary operation to improve the best known results of MLCP. Guo et al. proposed a novel PSO that employs competitive strategy and entropy measurement to manage convergence operator and diversity maintenance respectively, which was applied to the large-scale optimization benchmark suite on CEC 2013 and the results demonstrate the proposed algorithm is feasible and competitive to address large scale optimization problems. Li et al. proposed several new updating strategies for EHO, in which one, two, or three individuals are selected from the previous iterations, and their useful information is incorporated into the updating process. Accordingly, the final individual at this iteration is generated according to the elephant generated by the basic EHO, and the selected previous elephants through a weighted sum. A novel improved whale optimization algorithm (IWOA), based on the integrated approach, is presented by Luan et al. for solving the flexible job shop scheduling problem (FJSP) with the objective of minimizing makespan. Xiao et al. proposed an improved ABC variant based on elite strategy and dimension learning (called ABC-ESDL). The elite strategy selects better solutions to accelerate the search of ABC. The dimension learning uses the differences between two random dimensions to generate a large jump. Balande et al. proposed a hybrid algorithm, namely the Stochastic Ranking with Improved Firefly Algorithm (SRIFA) for solving constrained real-world engineering optimization problems. Yuan et al. proposed a hybrid evolutionary algorithm, which combines local search and genetic algorithm to solve minimum total dominating set (MTDS) problems. Pei et al. proposed a method to accelerate evolutionary multi-objective optimization (EMO) search using an estimated convergence point. Zhang et al. proposed a novel Simple Particle Swarm Optimization based on Random weight and Confidence term (SPSORC). The original two improvements of the algorithm are called Simple Particle Swarm Optimization (SPSO) and Simple Particle Swarm Optimization with Confidence term (SPSOC), respectively. An energy-efficient job shop scheduling problem (EJSP) is investigated by Jiang et al. with the objective of minimizing the sum of the energy consumption cost and the completion-time cost. Jin et al. proposed a developed ABC algorithm based on a cloud model to enhance accuracy of the basic ABC algorithm and avoid getting trapped into local optima by introducing a new select mechanism, replacing the onlooker bees' search formula and changing the scout bees' updating formula.Studying advanced intelligent optimization theory, designing efficient intelligent optimization method and popularizing effective intelligent optimization application not only has important academic value and discipline development significance, but also has very important practical significance for improving enterprise management level, increasing enterprise benefit and promoting enterprise development. Finally, it is hoped that the publication of this book will help beginners to understand the principles and design of intelligent algorithms, help basic readers to carry out the application and promotion of intelligent algorithms, and further promote the development and improvement of intelligent algorithm research, strengthen the research of computational intelligent algorithms, and promote the intersection and integration of related disciplines.

Gai-Ge Wang, Amir H. Alavi
Special Issue Editors

mathematics

MDPI

Article

A Novel Monarch Butterfly Optimization with Global Position Updating Operator for Large-Scale 0-1 Knapsack Problems

Yanhong Feng [1], Xu Yu [2] and Gai-Ge Wang [3,4,5,6,*]

[1] School of Information Engineering, Hebei GEO University, Shijiazhuang 050031, China; qinfyh@163.com or qinfyh@hgu.edu.cn
[2] School of Information Science and Technology, Qingdao University of Science and Technology, Qingdao 266061, China; yuxu0532@163.com
[3] Department of Computer Science and Technology, Ocean University of China, Qingdao 266100, China
[4] Institute of Algorithm and Big Data Analysis, Northeast Normal University, Changchun 130117, China
[5] School of Computer Science and Information Technology, Northeast Normal University, Changchun 130117, China
[6] Key Laboratory of Symbolic Computation and Knowledge Engineering of Ministry of Education, Jilin University, Changchun 130012, China
* Correspondence: gaigewang@163.com or gaigewang@gmail.com or wgg@ouc.edu.cn

Received: 18 September 2019; Accepted: 25 October 2019; Published: 4 November 2019

Abstract: As a significant subset of the family of discrete optimization problems, the 0-1 knapsack problem (0-1 KP) has received considerable attention among the relevant researchers. The monarch butterfly optimization (MBO) is a recent metaheuristic algorithm inspired by the migration behavior of monarch butterflies. The original MBO is proposed to solve continuous optimization problems. This paper presents a novel monarch butterfly optimization with a global position updating operator (GMBO), which can address 0-1 KP known as an NP-complete problem. The global position updating operator is incorporated to help all the monarch butterflies rapidly move towards the global best position. Moreover, a dichotomy encoding scheme is adopted to represent monarch butterflies for solving 0-1 KP. In addition, a specific two-stage repair operator is used to repair the infeasible solutions and further optimize the feasible solutions. Finally, Orthogonal Design (OD) is employed in order to find the most suitable parameters. Two sets of low-dimensional 0-1 KP instances and three kinds of 15 high-dimensional 0-1 KP instances are used to verify the ability of the proposed GMBO. An extensive comparative study of GMBO with five classical and two state-of-the-art algorithms is carried out. The experimental results clearly indicate that GMBO can achieve better solutions on almost all the 0-1 KP instances and significantly outperforms the rest.

Keywords: monarch butterfly optimization; greedy optimization algorithm; global position updating operator; 0-1 knapsack problems

---

1. Introduction

The 0-1 knapsack problem (0-1 KP) is a classical combinatorial optimization task and a challenging *NP*-complete problem as well. That is to say, it can be solved by nondeterministic algorithms in polynomial time. Similar to other *NP*-complete problems, such as vertex cover (VC), hamiltonian circuit (HC), and set cover (SC), the 0-1 KP is intractable. In other words, no polynomial-time exact algorithms have been found for it thus far. This problem was originated from the resource allocation involving financial constraints and since then, has been extensively studied in an array of scientific fields, such as combinatorial theory, computational complexity theory, applied mathematics, and computer science [1]. Additionally, it has been found to have many practical applications, such as project selection [2], investment decision-making [3], and network interdiction problem [4]. Mathematically, we can describe the 0-1 KP as follows:

$$
\begin{aligned}
Maximize \ f(x) &= \sum_{i=1}^{n} p_i x_i \\
subject \ to \ \sum_{i=1}^{n} w_i x_i &\leq C, \\
x_i = 0 \ or \ 1, \ i &= 1, \ldots, n,
\end{aligned}
\tag{1}
$$

where n is the number of items, p_i, and w_i denote the profit and weight of item i, respectively. C represents the total capacity of the knapsack. The 0-1 variable x_i indicates whether the item i is put into the knapsack, i.e., if any item i is selected and belongs to the knapsack, $x_i = 1$, otherwise, $x_i = 0$. The objective of the 0-1 KP is to maximize the total profits of the items placed in the knapsack, subject to the condition that the sum of the weights of the corresponding items does not exceed a given capacity C.

Since the 0-1 KP was reported by Dantzig [5] in 1957, a large number of researchers have focused on addressing it in diverse areas. Some of the main early methods in this field are exact methods, such as the branch and bound method (B&B) [6] and the dynamic programming (DP) method [7]. It is a breakthrough to introduce the concept of the core by Martello et al. [8]. In addition, some effective algorithms have been proposed for 0-1 KP [9], the multidimensional knapsack problem (MKP) [10]. With the rapid development of computational intelligence, some modern metaheuristic algorithms have been proposed for addressing the 0-1 KP. Some of those related algorithms include genetic algorithm (GA) [11], differential evolution (DE) [12], shuffled frog-leaping algorithm (SFLA) [13], cuckoo search (CS) [14,15], artificial bee colony (ABC) [16,17], harmony search (HS) [17–21], and bat algorithm (BA) [22,23]. Many research methods are applied to the 0-1 KP problem. Zhang et al. converted the 0-1 KP problem into a directed graph by the network converting algorithm [24]. Kong et al. proposed an ingenious binary operator to solve the 0-1 KP problem by simplified binary harmony search [20]. Zhou et al. presented a complex-valued encoding scheme for the 0-1 KP problem [22].

In recent years, inspired by natural phenomena, a variety of novel meta-heuristic algorithms have been reported, e.g., bat algorithm (BA) [23], amoeboid organism algorithm [24], animal migration optimization (AMO) [25], artificial plant optimization algorithm (APOA) [26], biogeography-based optimization (BBO) [27,28], human learning optimization (HLO) [29], krill herd (KH) [30–32], monarch butterfly optimization (MBO) [33], elephant herding optimization (EHO) [34], invasive weed optimization (IWO) algorithm [35], earthworm optimization algorithm (EWA) [36], squirrel search algorithm (SSA) [37], butterfly optimization algorithm (BOA) [38], salp swarm algorithm (SSA) [39], whale optimization algorithm (WOA) [40], and others. A review of swarm intelligence algorithms can be referred to [41].

As a novel biologically inspired computing approach, MBO is inspired by the migration behavior of the monarch butterflies with the change of the seasons. The related investigations [42,43] have demonstrated that the advantage of MBO lies in its simplicity, being easy to carry out, and efficiency. In order to address the 0-1 KP, which falls within the domain of the discrete combinatorial optimization problems with constraints, this paper presents a specially designed monarch butterfly optimization with global position updating operator (GMBO). What needs special mention is that GMBO is a supplement and perfection to previous related work, namely, a binary monarch butterfly optimization (BMBO) and a novel chaotic MBO with Gaussian mutation (CMBO) [42]. The main difference and contributions of this paper are as follows, compared with BMBO and CMBO.

Firstly, the original MBO was proposed to address the continuous optimization problems, i.e., it cannot be directly applied in the discrete space. For this reason, in this paper, a dichotomy encoding strategy [44] was employed. More specifically, each monarch butterfly individual is represented as two-tuples consisting of a real-valued vector and a binary vector. Secondly, although BMBO demonstrated excellent performance in solving 0-1 KP, it did not show a prominent advantage [42]. In other words, some techniques can be combined with BMBO for the purpose of improving its global optimization ability. Based on this, an efficient global position updating operator [16] was introduced to enhance the optimization ability and ensure its rapid convergence. Thirdly, a novel two-stage repair operator [45,46] called the greedy modification operator (GMO), and greedy optimization operator (GOO), respectively, was adopted. The former repairs the infeasible solutions while the latter optimizes the feasible solutions during the search process. Fourthly, empirical studies reveal that evolutionary algorithms have certain dependencies on the selection of parameters. Moreover, certain coupling between the parameters still exists. However, suitable parameter combination for a particular problem was not analyzed in BMBO and CMBO. In order to verify the influence degree of four important parameters on the performance of GMBO, an orthogonal design (OD) [47] was applied, and then the appropriate parameter settings were examined and recommended. Fifthly, generally speaking, the approximate solution of an *NP*-hard problem can be obtained by evolutionary algorithms. However, the most important thing is to obtain higher quality approximate solutions, which are closer to the optimal solutions more profitably. In BMBO, the optimal solutions of all the 0-1 KP instances were not provided. It is difficult to judge the quality of an approximate solution obtained by an evolutionary algorithm. In GMBO, the optimal solutions of 0-1 KP instances are calculated by a dynamic programming algorithm. Meanwhile, the approximation ratio based on the best values and the worst values are provided, which clearly reflect the degree of the closeness of the approximate solutions to the optimal solutions. In addition, the application of statistical methods in GMBO is one of the differences between GMBO and BMBO, CMBO, including Wilcoxon's rank-sum tests [48] with a 5% significance level. Moreover, boxplots can visualize the experimental results from the statistical perspective.

The rest of the paper is organized as follows. Section 2 presents a snapshot of the original MBO, while Section 3 introduces the GMBO for large-scale 0-1 KP in detail. Section 4 reports the outcomes of a series of simulation experiments as well as to compare results. Finally, the paper ends with Section 5 after providing some conclusions, along with some directions for further work.

2. Monarch Butterfly Optimization

Animal migration involves mainly long-distance movement, usually in groups, on a regular seasonal basis. MBO [33,43] is a population-based intelligent stochastic optimization algorithm that mimics the seasonal migration behavior of monarch butterflies in nature. It should be noted that the entire population is divided into two subpopulations, named *subpopulation*_1 and *subpopulation*_2, respectively. Based on this, the optimization process consists of two operators, which operate on *subpopulation*_1 and *subpopulation*_2, respectively. The information is interchanged among the individuals of *subpopulation*_1 and *subpopulation*_2 by applying the migration operator. The butterfly adjusting operator delivers the information of the best individual to the next generation. Additionally, Lévy flights [49,50] are introduced into MBO. The main steps of MBO are outlined as follows:

Step 1. Initialize the parameters of MBO. There are five basic parameters to be considered while addressing various optimization problems, including the number of the population (*NP*), the ratio of the number of monarch butterflies in *subpopulation*_1 (*p*), migration period (*peri*), the monarch butterfly adjusting rate (*BAR*), the max walk step of the Lévy flights (*Smax*).

Step 2. Initialize the population with *NP* randomly generated individuals according to a uniform distribution in the search space.

Step 3. Sort the individuals according to their fitness in descending order (Here assumptions for the maximum). The better NP_1 (*p**NP) individuals constitute *subpopulation*_1, and NP_2 (*NP*-NP_1) individuals make up *subpopulation*_2.

Step 4. The position updating of individuals in *subpopulation*_1 is determined by the migration operator. The specific procedure is described in Algorithm 1.

Step 5. The moving direction of the individuals in *subpopulation*_2 depends on the butterfly adjusting operator. The detailed procedure is shown in Algorithm 2.

Step 6. Recombine two subpopulations into one population

Step 7. If the termination criterion is already satisfied, output the best solution found, otherwise, go to Step 3.

Algorithm 1. Migration Operator

Begin
    for i = 1 to NP_1 (for all monarch butterflies in *subpopulation_1*)
        for j = 1 to d (all the elements in ith monarch butterfly)
            $r = rand * peri$, where $rand \sim U(0,1)$
            if $r \leq p$ **then**
                $x_{i,j} = x_{r1,j}$, where $r1 \sim U[1, 2, \ldots, NP_1]$
            else
                $x_{i,j} = x_{r2,j}$, where $r2 \sim U[1, 2, \ldots, NP_2]$
            end if
        end for j
    end for i
End.

---

where dx is calculated by implementing the Lévy flights. It should be noted that the Lévy flights, which originated from the Lévy distribution, are an impactful random walk model, especially on undiscovered, higher-dimensional search space. The step size of Lévy flights refers to Equation (2).

$$StepSize = ceil(exprnd(2 * Maxgen)) \tag{2}$$

where function $exprnd(x)$ returns random numbers of an exponential distribution with mean x and $ceil(x)$ gets a value to the nearest integer greater than or equal to x. $Maxgen$ is the maximum number of iterations.

The parameter ω is the weighting factor which has inverse proportional relationship to the current generation.

Algorithm 2. Butterfly Adjusting Operator

Begin
    for i = 1 to NP_2 (for all monarch butterflies in *subpopulation_2*)
        for j = 1 to d (all the elements in ith monarch butterfly)
            if $rand \leq p$ **then** where $rand \sim U(0,1)$
                $x_{i,j} = x_{best,j}$
            else
                $x_{i,j} = x_{r3,j}$ where $r3 \sim U[1,2,\ldots, NP_2]$
                if $rand > BAR$ **then**
                    $x_{i,j} = x_{i,j} + \omega \times (dx_j - 0.5)$
                end if
            end if
        end for j
    end for i
End.

---

3. A Novel Monarch Butterfly Optimization with Global Position Updating Operator for the 0-1 KP

In this section, we give the detailed design procedure of the GMBO for the 0-1 KP. Firstly, a dichotomy encoding scheme [46] is used to represent each individual. Secondly, a global position updating operator [16] is embedded in GMBO in order to increase the probability of finding the optimal solution. Thirdly, the two-stage individual optimization method is employed, which successively tackles the infeasible solutions and then further improves the existing feasible solutions. Finally, the basic framework of GMBO for 0-1 KP is formed.

3.1. Dichotomy Encoding Scheme

KP belongs to the category of discrete optimization. The solution space is a collection of discrete points rather than a contiguous area. For this reason, we should either redefine the evolutionary operation of MBO or directly apply a continuous algorithm to discrete problems. In this paper, we prefer the latter for its simplicity of operation, comprehensibility, and generality.

As previously mentioned, each monarch butterfly individual is expressed as a two-tuple <**X**, **Y**>. Here, real number vectors **X** still constitute the search space as in the original MBO, which can be regarded as a phenotype similar to the genetic algorithm. Binary vectors, **Y**, form the solution space, which can be seen as a genotype common in the evolutionary algorithm. It should be noted that **Y** may be a valid solution because 0-1KP is a constraint optimization problem. Here we abbreviate the monarch butterfly population to *MBOP*. Then the structure of *MBOP* is given as follows:

$$
MBOP = \begin{bmatrix}
(x_{1,1}, \ y_{1,1})(x_{1,2}, \ y_{1,2}) \cdots (x_{1,d}, \ y_{1,d}) \\
(x_{2,1}, \ y_{2,1})(x_{2,2}, \ y_{2,2}) \cdots (x_{2,d}, \ y_{2,d}) \\
\cdots \quad\quad \cdots \quad\quad \cdots \quad\quad \cdots \\
(x_{NP,1}, y_{NP,1})(x_{NP,2}, y_{NP,2}) \cdots (x_{NP,d}, y_{NP,d})
\end{bmatrix}
\tag{3}
$$

The first step to adopting a dichotomous encoding scheme is to transfer the phenotype to genotype. Therefore, a surjective function g is used to realize the mapping relationship from each element of **X** to the corresponding element of **Y**.

$$
g(x) = \begin{cases} 1 \ if \ sig(x) \geq 0.5 \\ 0 \ else \end{cases}
\tag{4}
$$

where $sig(x) = 1/(1 + e^{-x})$ is the sigmoid function. The sigmoid function is often used as the threshold function of neural networks. It was applied to the binary particle swarm optimization (BPSO) [51] to convert the position of a particle from a real-valued vector to a 0-1 vector. It should be noted that there are other conversion functions [52] can be used.

Now assume a 0-1 KP problem with 10 items, Figure 1 shows the above process, in which each $x_i \in [-5.0, 5.0]$ ($1 \leq i \leq 10$) is randomly chosen based on the uniform distribution.

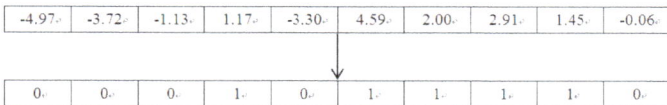

-4.97	-3.72	-1.13	1.17	-3.30	4.59	2.00	2.91	1.45	-0.06
0	0	0	1	0	1	1	1	1	0

Figure 1. The example of the dichotomy encoding scheme.

3.2. Global Position Updating Operator

The main feature of particle swarm optimization (PSO) is that the particle always tends to converge to two extreme positions viz. the best position ever found by itself and the global best position. Inspired by the behavior of swarm intelligence of PSO, a novel position updating operator was recently proposed and successfully embedded in HS for solving 0-1 KP [16]. After that, the position updating operator combines with CS [14] to deal with 0-1 KP.

It is well-known that the evolutionary algorithm can yield strong optimization performance under the condition of the balance between exploitation and exploration, or attraction and diffusion [53]. The original MBO concentrates much on the exploration ability but weak exploitation capability [33,43]. With the aim of enhancing the exploitation capability of MBO, we introduce a global position updating operator mentioned above. The procedure is shown in Algorithm 3, where "best" and "worst" represent the global best individual and the global worst individual, respectively. r_4, r_5, and *rand* are uniform random real numbers in [0, 1]. The p_m parameter is mutation probability.

Algorithm 3. Global Position Updating Operator

Begin
    for $i = 1$ to NP (for all monarch butterflies in the whole population)
        for $j = 1$ to d (all the elements in ith monarch butterfly)
            $step_j = \left| x_{best,j} - x_{worst,j} \right|$
            if $(rand \geq 0.5)$ where $rand \sim U(0,1)$
                $x_j = x_{best,j} + r_4 \times step_j$ where $r_4 \sim U(0,1)$
            else
                $x_j = x_{best,j} - r_4 \times step_j$
            end if
            if $(rand \leq p_m)$
                $x_j = L_j + r_5 \times (U_j - L_j)$ where $r_5 \sim U(0,1)$
                end if
        end for j
    end for i
End.

3.3. Two-Stage Individual Optimization Method Based on Greedy Strategy

Since KP is a constrained optimization problem, it may lead to the occurrence of infeasible solutions. There are usually two major methods: Redefining the objective function by penalty function method (PFM) [54,55] and individual optimization method based on the greedy strategy (IOM) [56,57]. Unfortunately, the former shows poor performance when encountering large-scale KP problems. In this paper, we adopt IOM to address infeasible solutions.

A simple greedy strategy, namely GS [58], is proposed to choose the item with the greatest density p_i/w_i first. Although the feasibility of all individuals can be guaranteed, it is obvious that there are several imperfections. Firstly, for a feasible individual, there is a possibility that the corresponding objective function value may turn to be worse by applying GS. Secondly, the lack of further optimization for all individuals can lead to unsatisfactory solutions.

In order to overcome the shortcomings of GS, the two-stage individual optimization method is proposed by He et al. [45,46]. A greedy modification operator (GMO) is used to repair the infeasible individuals in the first stage. It is followed by the application of the greedy optimization operator (GOO), which further optimizes the feasible individuals. The method proceeds as follows.

Step 1. Quicksort algorithm is used to sort all items in the non-ascending order according to p_i/w_i and the index of items is stored in an array $H[1], H[2]. . . , H[n]$, respectively.

Step 2. For an infeasible individual $\mathbf{X} = \{x_1, x_2, \ldots, x_n\} \in \{0, 1\}^n$, GMO is applied.

Step 3. For a feasible individual $\mathbf{X} = \{x_1, x_2, \ldots, x_n\} \in \{0, 1\}^n$, GOO is performed.

After the above repair process, it is easy to verify that each optimized individual is feasible. The significance of GMO and GOO seems particularly prominent while solving high dimensional KP problems [45,46]. The pseudo-code of GMO and GOO can be shown in Algorithms 4 and 5, respectively.

Algorithm 4. Greedy Modification Operator

Begin
    Input: $X = \{x_1, x_2, \ldots, x_n\} \in \{0, 1\}^n$, $W = \{w_1, w_2, \ldots, w_n\}$, $P = \{p_1, p_2, \ldots, p_n\}$, $H[1 \ldots n]$, C.
    $Weight = 0$
    for $i = 1$ to n
        $weight = weight + x_{H[i]} * w_{H[i]}$
        if $weight > C$
            $weight = weight - x_{H[i]} * w_{H[i]}$
            $x_{H[i]} = 0$
        end if
    end for i
    Output: $X = \{x_1, x_2, \ldots, x_n\} \in \{0, 1\}^n$
End.

Algorithm 5. Greedy Optimization Operator

Begin

   Input: $X = \{x_1, x_2, \ldots, x_n\} \in \{0,1\}^n$, $W = \{w_1, w_2, \ldots, w_n\}$, $P = \{p_1, p_2, \ldots, p_n\}$ $H[1 \ldots n]$, C.

   $Weight = 0$

   for $i = 1$ to n

      $weight = weight + x_i * w_i$

   end for i

   for $i = 1$ to n

      if $x_{H[i]} = 0$ and $weight + w_{H[i]} \leq C$

         $x_{H[i]} = 1$

         $weight = weight + w_{H[i]}$

      end if

   end for i

   Output: $X = \{x_1, x_2, \ldots, x_n\} \in \{0,1\}^n$

End.

Algorithm 6. GMBO for 0-1 KP

Begin

Step 1: **Sorting.** Quicksort is used to sort all items in the non-ascending order by p_i/w_i, $1 \leq i \leq n$ and the index of items is stored in array $H[1 \ldots n]$.

Step 2: **Initialization.** Set the generation counter $g = 1$; set migration period *peri*, the migration ratio p, butterfly adjusting rate *BAR*, and the max walk step of Lévy flights S_{max}; set the maximum generation *MaxGen*. Set the generation interval between recombination *RG*. Generate *NP* monarch butterfly individuals randomly $\{<X_1, Y_1>, <X_2, Y_2>, \ldots, <X_{NP}, Y_{NP}>\}$. Calculate the fitness of each individual, $f(Y_i)$, $1 \leq i \leq NP$.

Step 3: **While** (stopping criterion)

   Divide the whole population (*NP* individuals) into *subpopulation_1* (*NP*$_1$ individuals) and *subpopulation_2* (*NP*$_2$ individuals) according to their fitness;

   Calculate and record the global optimal individual $<X_{gbest}, Y_{gbest}>$.

   Update *subpopulation_1* with migration operator.

   Update *subpopulation_2* with butterfly adjusting operator.

   Update the whole population with Global position updating operator.

   Repair the infeasible solutions by performing GMO.

   Improve the feasible solutions by performing GOO.

   Keep best solutions.

   Find the current best solution $(Y_{gbest}, f(Y_{gbest}))$.

   $g = g + 1$.

   if Mod(g, *RG*) = 0

      Reorganize the two subpopulations into one population.

   end if

Step 4: **end while**

Step 5: Output the best results

End.

3.4. The Procedure of GMBO for the 0-1 KP

In this section, the procedure of GMBO for 0-1 KP is described in Algorithm 6, and the flowchart is illustrated in Figure 2. Apart from the initialization, it is divided into three main processes.

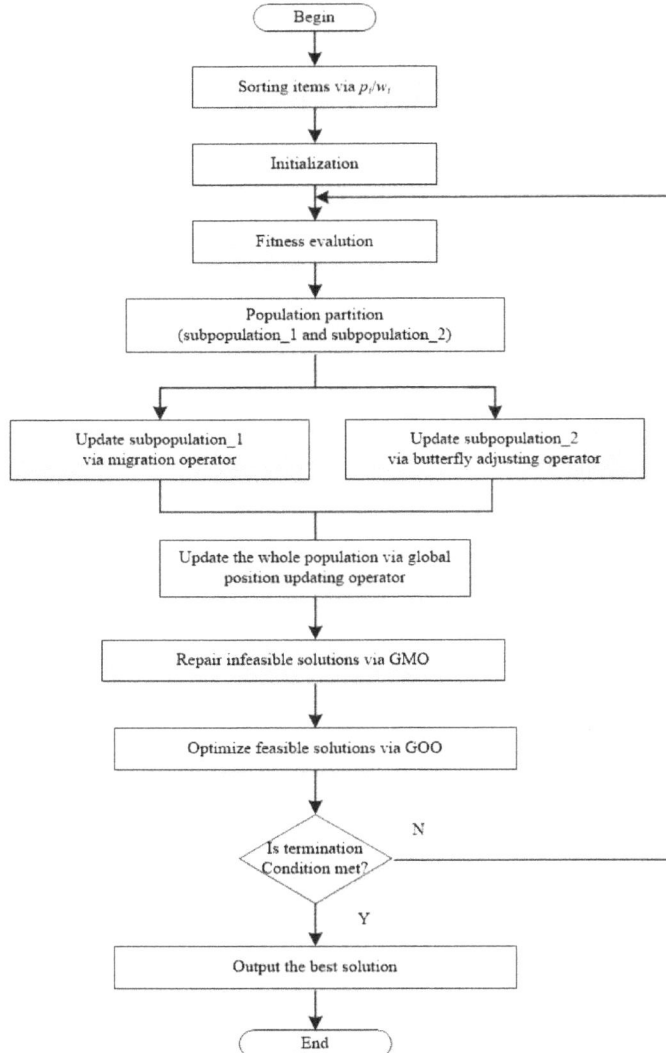

Figure 2. Flowchart of global position updating operator (GMBO) algorithm for 0-1 knapsack (KP) problem.

(1) In the migration process, the position of each monarch butterfly individual in *subpopulation*_1 is updated. We can view this process as exploitation by combining the properties of the currently known individuals in *subpopulation*_1 or *subpopulation*_2.

(2) In the butterfly adjusting process, partial genes of the global best individual are passed on to the next generation. Moreover, Lévy flights come into play owing to longer step length in exploring

the search space. This process can be considered as exploration, which may find new solutions in the unknown domain of the search space.

(3) In the global position updating process, we can define the distance of the global best individual and the global worst individual as the adaptive step. Obviously, the two extreme individuals differ greatly at the early stage of the optimization process. In other words, the adaptive step has a larger value, and the search scope is broader, which is beneficial to the global search over a wide range. With the progress of the evolution, the global worst individual tends to be more similar to the global best individual, and then the difference becomes small at the late stage of the optimization process. Meanwhile, the adaptive step has a smaller value, and the search area narrows, which is useful for performing the local search. In addition, the genetic mutation is applied to preserve the population diversity and avoid premature convergence. It should be noted that, unlike the original MBO, in GMBO, the two newly-generated subpopulations regroup one population at a certain generation rather than each generation, which can reduce time consumption.

3.5. The Time Complexity

In this subsection, the time complexity of GMBO is simply estimated (Algorithm 6). It is not hard to see that the time complexity of GMBO mainly hinges on steps 1–3. In Step 1, Quicksort algorithm costs time $O(n \log n)$. In Step 2, the initialization of NP individuals consumes time $O(NP * n)$. The calculation of fitness has time $O(NP)$. In Step 3, migration operator costs time $O(NP_1 * n)$, and the butterfly adjusting operator has time $O(NP_2 * n)$. Moreover, the global position updating operator consumes $O(NP * n)$. It is noticed that GMO and GOO cost the same time complexity $O(NP * n)$. Thus, the time complexity of GMBO would be $O(n \log n) + O(NP * n) + O(NP) + O(NP_1 * n) + O(NP_2 * n) + O(NP * n) + O(NP * n) = O(n^2)$.

4. Simulation Experiments

We chose 3 different sets of 0-1 KP test instances to verify the feasibility and effectiveness of the proposed GMBO method. The test set 1 and test set 2 were widely used low-dimensional benchmark instances with dimension = 4 to 24. The test set 3 consisted of 15 high-dimensional 0-1 KP test instances generated randomly with dimension = 800 to 2000.

4.1. Experimental Data Set

The generation form of test set 3 was firstly given. Since the difficulty of the knapsack problems was greatly affected by the correlation between the profits and weights [59], 3 typical large scale 0-1 KP instances were randomly generated to demonstrate the performance of the proposed algorithm. Here function *Rand* (*a*, *b*) returned a random integer uniformly distributed in [*a*, *b*]. For each instance, the maximum capacity of the knapsack equaled 0.75 times of the total weights. The procedure is as follows:

- Uncorrelated instances:

$$
\begin{aligned}
w_j &= Rand(10, 100) \\
p_j &= Rand(10, 100)
\end{aligned}
\tag{5}
$$

- Weakly correlated instances:

$$
\begin{aligned}
w_j &= Rand(10, 100) \\
p_j &= Rand(w_j - 10, w_j + 100)
\end{aligned}
\tag{6}
$$

- Strongly correlated instances:

$$
\begin{aligned}
w_j &= Rand(10, 100) \\
p_j &= w_j + 10
\end{aligned}
\tag{7}
$$

In this section, 3 groups of large scale 0-1 KP instances with dimensionality varying from 800 to 2000 were considered. These 15 instances included 5 uncorrelated instances, 5 weakly correlated instances, and 5 strongly correlated instances. The dimension size was 800, 1000, 1200, 1500, and 2000, respectively. We simply denoted these instances by KP1–KP15.

4.2. Parameter Settings

As mentioned earlier, GMBO included 4 important parameters: p, *peri*, *BAR*, and *Smax*. In order to examine the effect of the parameters on the performance of GMBO, Orthogonal Design (OD) [47] was applied with uncorrelated 1000-dimensional 0-1 KP instance. Our experiment contained 4 factors, 4 levels per factor, and 16 combinations of levels. The combinations of different parameter values are given in Table 1.

For each experiment, the average value of the total profits was obtained with 50 independent runs. The results are listed in Table 2.

Table 1. Combinations of different parameter values.

Parameters	Factor Level			
	1	**2**	**3**	**4**
p	1/12	3/12	5/12	10/12
peri	0.8	1	1.2	1.4
BAR	1/12	3/12	5/12	10/12
S_{max}	0.6	0.8	1	1.2

Table 2. Orthogonal array and the experimental results.

Experiment.	Factors				Results
Number	p	*peri*	*BAR*	S_{max}	
1	1	1	1	1	$R_1 = 49{,}542$
2	1	2	2	2	$R_2 = 49{,}538$
3	1	3	3	3	$R_3 = 49{,}503$
4	1	4	4	4	$R_4 = 49{,}528$
5	2	1	2	3	$R_5 = 49{,}745$
6	2	2	1	4	$R_6 = 49{,}739$
7	2	3	4	1	$R_7 = 49{,}763$
8	2	4	3	2	$R_8 = 49{,}739$
9	3	1	3	4	$R_9 = 49{,}704$
10	3	2	4	3	$R_{10} = 49{,}728$
11	3	3	1	2	$R_{11} = 49{,}730$
12	3	4	2	1	$R_{12} = 49{,}714$
13	4	1	4	2	$R_{13} = 49{,}310$
14	4	2	3	1	$R_{14} = 49{,}416$
15	4	3	2	4	$R_{15} = 49{,}460$
16	4	4	1	3	$R_{16} = 49{,}506$

Using data from Table 2, we can carry out factor analysis, rank the 4 parameters according to the degree of influence on the performance of GMBO, and deduce the better level of each factor. The factor analysis results are recorded in Table 3, and the changing trends of all the factor levels are shown in Figure 3.

As we can see from Table 3 and Figure 3, p is the most important parameter and needs a reasonable selection for the 0-1 KP problems. A small p signifies more elements from *subpopulation_2*. Conversely, more elements were selected from *subpopulation_1*. For *peri*, the curve was in a small range in an upward trend. This implied individual elements from *subpopulation_2* had more chance to embody in the newly generated monarch butterfly. For *BAR* and *Smax*, it can be seen from Figure 3 that the effect on the algorithm was not obvious.

According to the above analysis based on OD, the most suitable parameter combination is $p_2 = 3/12$, $peri_4 = 1.4$, $BAR_1 = 1/12$, and $Smax_3 = 1.0$, which will be adopted in the following experiments.

Table 3. Factor analysis with the orthogonal design (OD) method.

Levels	Factor Analysis			
	p	$peri$	BAR	S_{max}
1	$(R_1 + R_1 + R_1 + R_1)/4$ 49,528	$(R_1 + R_5 + R_9 + R_{13})/4$ 49,575	$(R_1 + R_6 + R_{11} + R_{16})/4$ **49,629**	$(R_1 + R_7 + R_{12} + R_{14})/4$ 49,609
2	$(R_5 + R_6 + R_7 + R_8)/4$ **49,746**	$(R_2 + R_6 + R_{10} + R_{14})/4$ 49,605	$(R_2 + R_5 + R_{12} + R_{15})/4$ 49,603	$(R_1 + R_1 + R_1 + R_1)/4$ 49,579
3	$(R_9 + R_{10} + R_{11} + R_{12})/4$ 49,719	$(R_3 + R_7 + R_{11} + R_{15})/4$ 49,614	$(R_3 + R_8 + R_9 + R_{14})/4$ 49,590	$(R_3 + R_5 + R_{10} + R_{16})/4$ **49,620**
4	$(R_{13} + R_{14} + R_{15} + R_{16})/4$ 49,423	$(R_4 + R_8 + R_{12} + R_{16})/4$ **49,622**	$(R_4 + R_7 + R_{10} + R_{13})/4$ 49,582	$(R_4 + R_6 + R_9 + R_{15})/4$ 49,608
Std	134.06	17.72	17.78	15.13
Rank	1	3	2	4
results	p_2	$peri_4$	BAR_1	S_{max3}

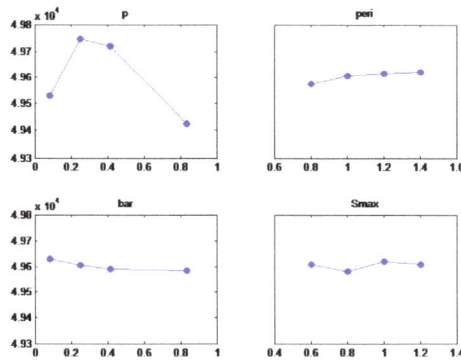

Figure 3. The changing trends of all the factor levels.

4.3. The Comparisons of the GMBO and the Classical Algorithms

In order to investigate the ability of GMBO to find the optimal solutions and to test the convergence speed, 5 representative classical optimization algorithms, including the BMBO [42], ABC [60], CS [61], DE [62] and GA [11], were selected for comparison. GA was an important branch in the field of computational intelligence that has been intensively studied since it was developed by John Holland et al. In addition, GA was representative of the population-based algorithm. DE was a vector-based evolutionary algorithm, and more and more researchers have paid attention to it since it was first proposed. Since then, it has been applied to solve many optimization problems. CS is one of the latest swarm intelligence algorithms. The similarity of CS with GMBO lies in the introduction of Levy flights. ABC is a relatively novel bio-inspired computing method and has the outstanding advantage of the global and local search in each evolution.

There are several points to explain. Firstly, all 5 comparative methods (not including GA) used the previously mentioned dichotomy encoding mechanism. Secondly, all 6 comparative methods used GMO and GOO to carry out the additional repairing and optimization operations. Thirdly, ABC, CS, DE, GA, MBO, and GMBO were short for 6 methods based on binary, respectively.

The parameters were set as follows. For ABC, the number of food sources is set to 25 and maximum search times = 100. CS, DE, and GA are set the same parameters as that of [15]. For MBO, we take the same parameters suggested in Section 4.2. In addition, the 2 subpopulations recombined every 50 generations. GMBO and MBO have identical parameters except for that mutation probability $pm = 0.25$ is included in GMBO.

For the sake of fairness, the population sizes of six methods are set to 50. The maximum run time is set to 8 s for 800, 1000, and 1200 dimensional instances but 10 s for 1500 and 2000 dimensional instances. 50 independent runs are performed to achieve the experimental results.

We use C++ as the programming language and run the codes on a PC with Intel (R) Core(TM) i5-2415M CPU, 2.30GHz, 4GB RAM.

4.3.1. The Experimental Results of GMBO on Solving Two Sets of Low-Dimensional 0-1 Knapsack Problems

In this section, 2 sets of 0-1 KP test instances were considered for testing the efficiency of the GMBO. The maximum number of iterations was set to 50. As mentioned earlier, 50 independent runs were made. The first set, which contained 10 low-dimensional 0-1 knapsack problems [19,20], was adopted with the aim of investigating the basic performance of the GMBO. The standard 10 0-1 KP test instances were studied by many researchers, and detailed information about these instances can be taken from the literature [13,19,20]. Their basic parameters are recorded in Table 4. The experimental results obtained by GMBO are listed in Table 5.

Table 4. The basic information of 10 standard low-dimensional 0-1KP instances.

f	Dim	Opt.value	Opt.solution
$f1$	10	295	(0,1,1,1,0,0,0,1,1,1)
$f2$	20	1024	(1,1,1,1,1,1,1,1,1,1,1,1,1,0,1,0,1,0,1,1)
$f3$	4	35	(1,1,0,1)
$f4$	4	23	(0,1,0,1)
$f5$	15	481.0694	(0,0,1,0,1,0,1,1,0,1,1,1,0,1,1)
$f6$	10	52	(0,0,1,0,1,1,1,1,1,1)
$f7$	7	107	(1,0,0,1,0,0,0)
$f8$	23	9767	(1,1,1,1,1,1,1,1,0,0,1,0,0,0,0,1,1,0,0,0,0,0,0)
$f9$	5	130	(1,1,1,1,0)
$f10$	20	1025	(1,1,1,1,1,1,1,1,1,1,0,1,1,1,1,0,1,0,1,1)

Table 5. The experimental results of 10 standard low-dimensional 0-1 KP instances obtained by GMBO.

f	SR	Time(s)	MinIter	MaxIter	MeanIter	Best	Worst	Mean	Std
$f1$	100%	0.0032	1	1	1	295	295	295	0
$f2$	100%	0.0092	1	52	6.10	1024	1024	1024	0
$f3$	100%	0.0003	1	1	1	35	35	35	0
$f4$	100%	0.0004	1	1	1	23	23	23	0
$f5$	100%	0.0072	1	4	1.30	481.07	481.07	481.07	0
$f6$	100%	0.0023	1	1	1	52	52	52	0
$f7$	100%	0.0000	1	1	1	107	107	107	0
$f8$	100%	0.0024	1	3	1.45	9767	9767	9767	0
$f9$	100%	0.0000	1	1	1	130	130	130	0
$f10$	100%	0.0000	1	1	1	1025	1025	1025	0

Here, "Dim" is the dimension size of test problems; Opt.value is the optimal value obtained by DP method [7]; Opt.solution is the optimal solution; "SR" is success rate; "Time" is the average time to reach the optimal solution among 50 runs; "MinIter", "MaxIter" and "MeanIter" represents the minimum iterations, maximum iterations, and the average iterations to reach the optimal solution among 50 runs, respectively. "Best", "Worst", "Mean" and "Std" are the best value, worst value, mean value, and the standard deviation, respectively.

As can be seen from Table 5, GMBO can achieve the optimal solution for all 10 instances with 100% success rates. Furthermore, the best value, the worst value, and the mean value are all equal to the optimal value for every test problem. Obviously, the efficiency of GMBO is very high for the considered set of instances because GMBO can get the optimal solution in a very short time. The minimum iterations are only 1, and the mean iterations are less than 6 for all the test problems. In particular, for $f6$, HS [18], HIS [63], and NGHS [19] can only achieve the best value 50 while GMBO can get the optimal value 52.

The second set, which includes 25 0-1 KP instances, was taken from references [64,65]. For all we know, the optimal value and the optimal solution of each instance are provided for the first time in this paper. The primary parameters are recorded in Table 6. The experimental results are summarized in Table 7. Compared to Table 5 above, three new evaluation criteria, that is "ARB", "ARW", and "ARM", are used to evaluate the proposed method. "Opt.value" represents the optimal solution value obtained by the DP method. Here, the following definitions are given:

$$ARB = \frac{Opt.value}{Best} \tag{8}$$

$$ARW = \frac{Opt.value}{Worst} \tag{9}$$

$$ARM = \frac{Opt.value}{Mean} \tag{10}$$

Here, "ARB" represents the approximate ratio [66] of the optimal solution value (Opt.value) to the best approximate solution value (Best). Similarly, "ARW" and "ARM" are based on the worst approximate solution value (Worst) and the mean approximate solution value (mean), respectively. ARB, ARW, and ARM indicate the proximity of Best, Worst, and Mean to the Opt.value, respectively. Plainly, ARB, ARW, and ARM are real numbers greater than or equal to 1.0.

Table 6. The basic information of 25 low-dimensional 0-1KP instances.

0-1 KP	Dim	Opt.value	Opt.solution
ks_8a	8	3,924,400	11101100
ks_8b	8	3,813,669	11001001
ks_8c	8	3,347,452	10010100
ks_8d	8	4,187,707	00100111
ks_8e	8	4,955,555	01010011
ks_12a ks_8b	12	5,688,887	100011011010
ks_12b	12	6,498,597	000110101110
ks_12c	12	5,170,626	011010011011
ks_12d	12	6,992,404	110001110100
ks_12e	12	5,337,472	010000001101
ks_16a	16	7,850,983	0100110110100110
ks_16b	16	9,352,998	1000001001111110
ks_16c	16	9,151,147	1101000111010010
ks_16d	16	9,348,889	1011111001000100
ks_16e	16	7,769,117	0011001110111010
ks_20a	20	10,727,049	00100111011100011101
ks_20b	20	9,818,261	10000111010101111110001
ks_20c	20	10,714,023	11111100011001000101
ks_20d	20	8,929,156	00000001111011011100
Ks_20e	20	9,357,969	00101000001001011101
ks_24a	24	13,549,094	110111000110100100000111
ks_24b	24	12,233,713	100010101111110011101011
ks_24c	24	12,448,780	100011010100110001011010
ks_24d	24	11,815,315	101011000010111100110110
ks_24e	24	13,940,099	001011000110111110110010

From Table 7, it was clear that GMBO could obtain the optimal solution value for all the 25 instances. Among them, GMBO could find the optimal solution values of 13 instances with 100% SR, and the success rate of nine instances was more than 80%. In addition, the standard deviation of 13 instances was 0. In particular, ARB can reflect well the proximity between the best approximate solution value and the optimal solution value. ARW and ARM were similar to this. For the three new evaluation criteria, it can be seen that the values were equal to 1.0 or very close to 1.0 for all the 25 instances.

Thus, the conclusion is that GMBO had superior performance in solving low-dimensional 0-1 KP instances.

Table 7. The experimental results of 25 low-dimensional 0-1 KP instances obtained by GMBO.

0-1 KP	SR	Best	Worst	Mean	Std	ARB	ARW	ARM
ks_8a	100%	925,369	925,369	925,369	0	1.0000	1.0000	1.0000
ks_8b	100%	3,813,669	3,813,669	3,813,669	0	1.0000	1.0000	1.0000
ks_8c	100%	3,837,398	3,837,398	3,837,398	0	1.0000	1.0000	1.0000
ks_8d	100%	4,187,707	4,187,707	4,187,707	0	1.0000	1.0000	1.0000
ks_8e	100%	4,955,555	4,955,555	4,955,555	0	1.0000	1.0000	1.0000
ks_12a	88%	5,688,887	5,681,360	5,688,046	2283.52	1.0000	1.0013	1.0001
ks_12b	86%	6,498,597	6,473,019	6,495,016	8875.23	1.0000	1.0040	1.0006
ks_12c	100%	5,170,626	5,170,626	5,170,626	0	1.0000	1.0000	1.0000
ks_12d	100%	6,992,404	6,992,404	6,992,404	0	1.0000	1.0000	1.0000
ks_12e	88%	5,337,472	5,289,570	5,331,724	15,566.30	1.0000	1.0091	1.0011
ks_16a	100%	7,850,983	7,850,983	7,850,983	0	1.0000	1.0000	1.0000
ks_16b	100%	9,352,998	9,352,998	9,352,998	0	1.0000	1.0000	1.0000
ks_16c	100%	9,151,147	9,151,147	9,151,147	0	1.0000	1.0000	1.0000
ks_16d	56%	9,348,889	9,300,041	9,342,056	10,405.28	1.0000	1.0053	1.0007
ks_16e	82%	7,769,117	7,750,491	7,765,991	6713.97	1.0000	1.0024	1.0004
ks_20a	100%	10,727,049	10,727,049	10,727,049	0	1.0000	1.0000	1.0000
ks_20b	98%	9,818,261	9,797,399	9,817,844	2920.68	1.0000	1.0021	1.0000
ks_20c	96%	10,714,023	10,700,635	10,713,487	2623.50	1.0000	1.0013	1.0000
ks_20d	100%	8,929,156	8,929,156	8,929,156	0	1.0000	1.0000	1.0000
Ks_20e	48%	9,357,969	9,357,192	9,357,565	388.18	1.0000	1.0001	1.0000
ks_24a	80%	13,549,094	13,504,878	13,543,476	11,554.74	1.0000	1.0033	1.0004
ks_24b	100%	12,233,713	12,233,713	12,233,713	0	1.0000	1.0000	1.0000
ks_24c	96%	12,448,780	12,445,379	12,448,644	666.45	1.0000	1.0003	1.0000
ks_24d	72%	11,815,315	11,810,051	11,813,841	2363.53	1.0000	1.0004	1.0001
ks_24e	98%	13,940,099	13,929,872	13,939,894	1431.78	1.0000	1.0007	1.0000

4.3.2. Comparisons of Three Kinds of Large-Scale 0-1 KP Instances

In this section, in order to make a comprehensive investigation on the optimization ability of the proposed GMBO, test set 3, which included 5 uncorrelated, 5 weakly correlated, and 5 strongly correlated large-scale 0-1 KP instances, were considered. The experimental results are listed in Tables 8–10 below. The best results on all the statistical criteria of each 0-1 KP instances, i.e., the best values, the mean values, the worst values, the standard deviation, and the approximation ratio, appear in bold. As noted earlier, Opt and Time represent the optimal value and time spending taken by the DP method, respectively.

The performance comparisons of the six methods on the five large-scale uncorrelated 0-1 KP instances are listed in Table 8. It can be seen that GMBO outperformed the other five algorithms on the six and five evaluation criteria for KP1 and KP4, respectively. In addition, GMBO obtained the best values concerning the best and the mean value for KP3 and was superior to the other five algorithms in the worst value for KP2. Unfortunately, GMBO failed to come up with superior performance while encountering 2000-dimensional 0-1 KP instances (KP5). MBO beat the competitors on KP5. Moreover, an apparent phenomenon can be observed, which points out that ABC has better stability. The best value of KP2 was achieved by CS. Obviously, DE and GA showed the worst performance for KP1–KP5.

Meanwhile, the approximation ratio of the best value of GMBO for KP1 equaled approximately 1.0. Additionally, there was little difference between the worst approximation ratio of the best value (1.0242) of GMBO and the best approximation ratio of the best value (1.0237) of MBO for KP5.

Table 8. Performance comparison on large-scale five uncorrelated 0-1KP instances.

		KP1	KP2	KP3	KP4	KP5
DP	Opt	40,686	50,592	61,846	77,033	102,316
	Time(s)	0.952	1.235	1.914	2.521	2.705
ABC	Best	39816	49,374	60,222	74,959	99,353
	ARB	1.0219	1.0247	1.0270	1.0277	1.0298
	Worst	39,542	49,105	59,867	74,571	99,822
	ARW	1.0289	1.0303	1.0331	1.0330	1.0250
	Mean	39,639	49,256	60,059	74,742	99,035
	Std	55.5	**58.56**	**82.28**	**90.07**	130.80
CS	Best	40,445	**50,104**	60,490	75,828	99,248
	ARB	1.0060	1.0097	1.0224	1.0159	1.0309
	Worst	39,411	49,056	59,764	74,472	98,706
	ARW	1.0324	1.0313	1.0348	1.0344	1.0366
	Mean	39,602	49,211	59,938	74,666	98,926
	Std	218.11	205.08	120.76	245.28	**124.58**
DE	Best	39,486	49,303	59,921	74,671	98,943
	ARB	1.0304	1.0261	1.0321	1.0316	1.0341
	Worst	39,154	48,696	59,435	74,077	98,330
	ARW	1.0391	1.0389	1.0406	1.0399	1.0405
	Mean	39,323	48,945	59,645	74,319	98,645
	Std	80.60	111.08	114.17	113.92	154.40
GA	Best	39,190	48,955	59,578	74,372	98,828
	ARB	1.0382	1.0334	1.0381	1.0358	1.0353
	Worst	38,274	47,809	58,106	72,477	96,830
	ARW	1.0630	1.0582	1.0644	1.0629	1.0567
	Mean	38,838	48,384	58,996	73,584	97,765
	Std	196.70	256.69	362.53	414.02	480.15
MBO	Best	40,276	50,023	61,090	75,405	**99,946**
	ARB	1.0102	1.0114	1.0124	1.0216	**1.0237**
	Worst	39,839	49,411	**60,401**	74,815	**99,017**
	ARW	1.0213	1.0239	1.0239	1.0296	**1.0333**
	Mean	40,036	**49,743**	60,732	75,072	**99,512**
	Std	100.34	133.40	163.76	149.57	187.15
GMBO	Best	**40,684**	49,992	**61,764**	**76,929**	99,898
	ARB	**1.0000**	1.0120	**1.0013**	**1.0014**	1.0242
	Worst	**40,527**	49,524	60,225	**75,410**	98,848
	ARW	**1.0039**	1.0216	1.0269	**1.0215**	1.0351
	Mean	**40,641**	49,732	**61,430**	**76,691**	99,424
	Std	**40.09**	116.12	379.76	267.90	200.38

Table 9 records the comparison of the performances of six methods on five large-scale weakly correlated 0-1 KP instances. The experimental results in Table 9 differ from that in Table 8. It is clear that GMBO had a striking advantage in almost all the six statistical standards for KP6–KP9. For KP10, similarly to KP5, GMBO was still not able to win out over MBO. It is worth mentioning that the approximation ratio of the best value of GMBO for KP6–KP7, and KP9 equaled 1.0. Moreover, the standard deviation value of KP6–KP7 and KP9 obtained by GMBO was much smaller than the corresponding value of the other five algorithms.

Table 9. Performance comparison of large-scale five weakly correlated 0-1KP instances.

		KP6	KP7	KP8	KP9	KP10
DP	Opt	35069	43,786	53,553	65,710	118,200
	Time(s)	1.188	1.174	1.413	2.717	2.504
ABC	Best	34706	43,321	52,061	64,864	115,305
	ARB	1.0105	1.0107	1.0287	1.0130	1.0251
	Worst	34,650	43,243	51,711	64,752	114,586
	ARW	1.0121	1.0126	1.0356	1.0148	1.0315
	Mean	34,675	43,275	51,876	64,806	114,922
	Std	16.00	18.74	**79.72**	25.45	**123.59**
CS	Best	34,975	43,708	52,848	65,549	**116,597**
	ARB	1.0027	1.0018	1.0133	1.0025	**1.0137**
	Worst	34,621	43,215	51,617	64,749	114,560
	ARW	1.0129	1.0132	1.0375	1.0148	1.0318
	Mean	34,676	43,326	51,838	64,932	114,879
	Std	65.25	143.50	260.46	245.06	428.93
DE	Best	34,629	43,251	51,900	64,770	114,929
	ARB	1.0127	1.0124	1.0318	1.0145	1.0285
	Worst	34,549	43,140	51,289	64,620	114,199
	ARW	1.0151	1.0150	1.0441	1.0169	1.0350
	Mean	34,588	43,187	51,547	64,692	114,462
	Std	20.93	23.94	123.67	35.66	160.77
GA	Best	34,585	43,172	51,460	64,769	114,539
	ARB	1.0140	1.0142	1.0407	1.0145	1.0320
	Worst	34,361	42,901	50,112	64,315	112,681
	ARW	1.0206	1.0206	1.0687	1.0217	1.0490
	Mean	34,476	43,049	50,945	64,535	113,674
	Std	60.91	74.36	281.41	85.75	405.23
MBO	Best	34,850	43,487	52,720	65,144	116,466
	ARB	1.0063	1.0069	1.0158	1.0087	1.0149
	Worst	34,724	43,349	52,185	64,941	**115,273**
	ARW	1.0099	1.0101	1.0262	1.0118	**1.0254**
	Mean	34,795	43,425	52,449	65,041	**115,998**
	Std	31.41	31.78	111.26	48.66	248.70
GMBO	Best	**35,069**	**43,786**	**53,426**	**65,708**	116,496
	ARB	**1.0000**	**1.0000**	**1.0024**	**1.0000**	1.0146
	Worst	**35,052**	**43,781**	**52,376**	**65,625**	114,761
	ARW	**1.0005**	**1.0001**	**1.0225**	**1.0013**	1.0300
	Mean	**35,064**	**43,784**	**53,167**	**65,666**	115,718
	Std	**4.04**	**1.57**	300.90	**18.48**	492.92

A comparative study of the six methods on five large-scale strongly correlated 0-1 KP instances are recorded in Table 10. Obviously, GMBO outperforms the other five methods for KP11–KP14 on five statistical standards except for Std. ABC obtains the best Std values for KP11–KP15. To KP15, GMBO can get better values on the worst. CS, DE, and GA fail to show outstanding performance for this case. Under these circumstances, the approximation ratio of the worst value of GMBO for KP11–KP15 was less than 1.0019.

Table 10. Performance comparison of large-scale five strongly correlated 0-1KP instances.

		KP11	KP12	KP13	KP14	KP15
DP	Opt	40,167	49,443	60,640	74,932	99,683
	Time(s)	0.793	1.123	1.200	1.971	2.232
ABC	Best	40,127	49,390	60,567	74,822	99,523
	ARB	1.0010	1.0011	1.0012	1.0015	1.0016
	Worst	40,107	49,363	60,540	74,792	99,490
	ARW	1.0015	1.0016	1.0017	1.0019	1.0019
	Mean	40,116	49,376	60,554	74,805	99,506
	Std	**4.52**	**5.61**	**5.54**	**6.85**	**7.29**
CS	Best	40,127	49,393	60,559	74,837	99,517
	ARB	1.0010	1.0010	1.0013	1.0013	1.0017
	Worst	40,096	49,353	60,533	74,779	99,473
	ARW	1.0018	1.0018	1.0018	1.0020	1.0021
	Mean	40,108	49,364	60,543	74,794	99,489
	Std	6.59	6.80	5.39	9.19	8.19
DE	Best	40,137	49,363	60,545	74,778	99,501
	ARB	1.0007	1.0016	1.0016	1.0021	1.0018
	Worst	40,087	49,323	60,498	74737	99,436
	ARW	1.0020	1.0024	1.0023	1.0026	1.0025
	Mean	40,119	49,340	60,518	74,759	99,459
	Std	10.19	8.31	10.16	10.23	14.03
GA	Best	40,069	49,333	60,520	74,766	99,461
	ARB	1.0024	1.0022	1.0020	1.0022	1.0022
	Worst	39,930	49,231	60,391	74,606	99,305
	ARW	1.0059	1.0043	1.0041	1.0044	1.0038
	Mean	40,023	49,287	60,451	74,689	99,382
	Std	31.12	29.76	29.87	37.20	38.42
MBO	Best	40,137	49,393	60,580	74,849	**99,573**
	ARB	1.0007	1.0010	1.0010	1.0011	**1.0011**
	Worst	40,102	49,363	60,539	74,778	99,496
	ARW	1.0016	1.0016	1.0017	1.0021	1.0019
	Mean	40,119	49,379	60,562	74,822	**99,536**
	Std	7.18	9.94	10.77	14.70	15.63
GMBO	Best	**40,167**	**49,442**	**60,630**	**74,852**	99,553
	ARB	**1.0000**	**1.0000**	**1.0002**	**1.0011**	1.0013
	Worst	**40,147**	**49,371**	**60,540**	**74,792**	**99,503**
	ARW	**1.0005**	**1.0015**	**1.0017**	**1.0019**	**1.0018**
	Mean	**40,162**	**49,425**	**60,604**	**74,825**	99,534
	Std	5.11	11.58	20.88	12.00	14.14

For a clearer and more intuitive measure of the similar level of the theoretical optimal value and the actual value obtained by each algorithm, the values of ARB on three types of 0-1 KP instances are illustrated in Figures 4–6. From Figure 4, the ARB of GMBO for KP1, KP3, and KP4 were extremely close to or equal to 1. GMBO had the smallest ARB for KP1, KP3–KP5, except for KP2, for which CS obtained the smallest ARB. Similar to Figure 4, from Figure 5, GMBO still had the smallest ARB values, which are 1.0 (KP6, KP7, and KP9) or less than 1.015 (KP8, KP10). In terms of the strongly correlated 0-1 KP instances, GMBO consistently outperformed the other five methods (see Figure 6), in which GMBO had the smallest ARB values except for KP15. Particularly, the ARB of GMBO was even less than 1.0015 for KP15.

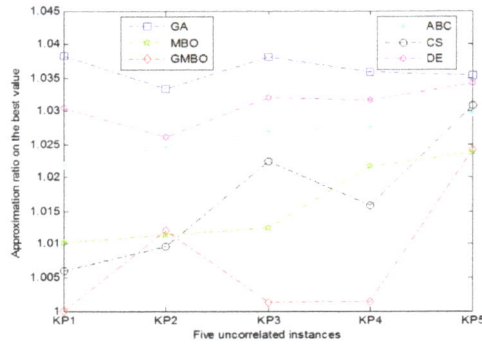

Figure 4. Comparison of ARB for KP1-KP5.

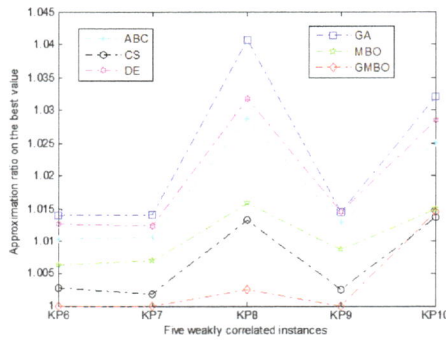

Figure 5. Comparison of ARB for KP6-KP10.

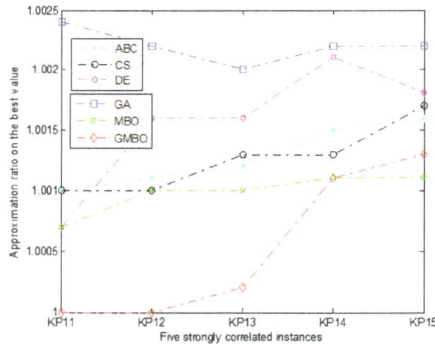

Figure 6. Comparison of ARB for KP11-KP15.

Overall, Tables 8–10 and Figures 4–6 indicate that GMBO was superior to the other five methods when addressing large-scale 0-1 KP problems. In addition, if we look at the worst values achieved by GMBO and the best values obtained by other methods, we can observe that for the majority instances, the former were even far better than the latter.

With regard to the best values, GMBO can gain better values than the others for almost all the instances except KP2, KP5, KP10, and KP15, in which CS and MBO twice achieved the best values, respectively. More specifically, compared to the suboptimal values researched by others, the improvements in KP1–KP15 brought by GMBO were 0.59%, −0.22%, 1.10%, 1.45%, −0.05%, 0.27%, 0.18%, 1.09%, 0.24%, −0.09%, 0.07%, 0.10%, 0.00%, and −0.02%, respectively.

With regard to the mean values, they were very similar to the best values. The improvements in KP1–KP15 were 1.51%, −0.02%, 1.15%, 2.16%, −0.09%, 0.77%, 0.83%, 1.37%, 0.96%, −0.24%, 0.11%, 0.09%, 0.07%, 0.00% and 0.00%, respectively.

With regard to the worst values, GMBO can still reach better values for almost all the 15 instances except KP3, KP5, and KP10 in which MBO was a little better than GMBO. The improvements in KP1–KP15 were 1.73%, 0.23%, −0.29%, 0.80%, −0.17%, 0.94%, 1.00%, 0.37%, 1.05%, −0.44%, 0.10%, 0.02%, 0.00%, 0.00%, and 0.01%, respectively.

In order to test the differences between the proposed GMBO and the other five methods, Wilcoxon's rank-sum tests with the 5% significance level were used. Table 11 records the results of rank-sum tests for KP1–KP15. In Table 11, "1" indicates that GMBO outperforms other methods at 95% confidence. Conversely, "−1". Particularly, "0" represents that the two compared methods possess similar performance. The last three rows summarized the times that GMBO performed better than, similar to, and worse than the corresponding algorithm among 50 runs.

Table 11. Rank sum tests for GMBO with the other five methods on KP1–KP15.

GMBO	ABC	CS	DE	GA	MBO
KP1	1	1	1	1	1
KP2	1	1	1	1	0
KP3	1	1	1	1	1
KP4	1	1	1	1	1
KP5	1	1	1	1	−1
KP6	1	1	1	1	1
KP7	1	1	1	1	1
KP8	1	1	1	1	1
KP9	1	1	1	1	1
KP10	1	1	1	1	−1
KP11	1	1	1	1	−1
KP12	1	1	1	1	1
KP13	1	1	1	1	1
KP14	1	1	1	1	0
KP15	1	1	1	1	0
1	15	15	15	15	9
0	0	0	0	0	3
−1	0	0	0	0	3

From Table 11, GMBO outperformed ABC, CS, DE, and GA on 15 0-1 KP instances. Compared with MBO, GMBO performed better than, similar to, or worse than MBO on 9, 3, 3 0-1KP instances, respectively. Therefore, one conclusion is easy to draw that GMBO was superior to or at least comparable to the other five methods. This conclusion is consistent with the foregoing analysis.

Tables 12–14 illustrate the ranks of six methods for 15 large-scale 0-1 KP instances on the best values, the mean values, and the worst values, respectively. These clearly show the performance of GMBO in comparison with the other five algorithms.

According to Table 12, obviously, the proposed GMBO exhibited superior performance compared with all the other five methods. In addition, CS and MBO can be regarded as the second-best methods, having identical performance. GA consistently showed the worst performance. Overall, the average rank in descending order according to the best values were: GMBO (1.33), MBO (2.33), CS (2.53), ABC (3.80), DE (4.80), and GA (6).

Table 12. Rankings of six algorithms based on the best values.

	ABC	CS	DE	GA	MBO	GMBO
KP1	4	2	5	6	3	1
KP2	4	1	5	6	2	3
KP3	4	3	5	6	2	1
KP4	4	2	5	6	3	1
KP5	3	4	5	6	1	2
KP6	4	2	5	6	3	1
KP7	4	2	5	6	3	1
KP8	4	2	5	6	3	1
KP9	4	2	5	6	3	1
KP10	4	1	5	6	3	2
KP11	4	4	2	6	2	1
KP12	4	2	5	6	2	1
KP13	3	4	5	6	2	1
KP14	4	3	5	6	2	1
KP15	3	4	5	6	1	2
Mean rank	3.80	2.53	4.80	6	2.33	1.33

Table 13. Rankings of six algorithms based on the mean values.

	ABC	CS	DE	GA	MBO	GMBO
KP1	3	4	5	6	2	1
KP2	3	4	5	6	1	2
KP3	3	4	5	6	2	1
KP4	3	4	5	6	2	1
KP5	3	4	5	6	1	2
KP6	4	3	5	6	2	1
KP7	4	3	5	6	2	1
KP8	3	4	5	6	2	1
KP9	4	3	5	6	2	1
KP10	3	4	5	6	1	2
KP11	4	5	3	6	2	1
KP12	3	4	5	6	2	1
KP13	3	4	5	6	2	1
KP14	3	4	5	6	2	1
KP15	3	4	5	6	1	2
Mean rank	3.27	3.87	4.87	6	1.73	1.27

Table 14. Rankings of six algorithms based on the worst values.

	ABC	CS	DE	GA	MBO	GMBO
KP1	3	4	5	6	2	1
KP2	3	4	5	6	2	1
KP3	3	4	5	6	1	2
KP4	3	4	5	6	2	1
KP5	3	4	5	6	1	2
KP6	3	4	5	6	2	1
KP7	3	4	5	6	2	1
KP8	3	4	5	6	2	1
KP9	3	4	5	6	2	1
KP10	3	4	5	6	1	2
KP11	2	4	5	6	3	1
KP12	2	4	5	6	2	1
KP13	1	4	5	6	3	1
KP14	1	3	5	6	4	1
KP15	3	4	5	6	2	1
Mean rank	2.60	3.93	5	6	2.07	1.20

From Table 13, we can see that the average rank of GMBO still occupied the first. MBO consistently outperformed the other four methods. Note that the rank value of ABC was identical to that of CS. The detailed rank was as follows: GMBO (1.27), MBO (1.73), ABC (3.27), CS (3.87), DE (4.87), and GA (6).

Table 14 shows the statistical results of the six methods based on the worst values. The ranking order of the six methods was GMBO (1.20), MBO (2.07), ABC (2.60), CS (3.93), DE (5), and GA (6), which was identical with that in Table 12.

Then, a comparison of the six highest dimensional 0-1 KP instances, i.e., KP4, KP5, KP9, KP10, KP14, and KP15, is illustrated in Figures 7–12, which was based on the best profits achieved by 50 runs.

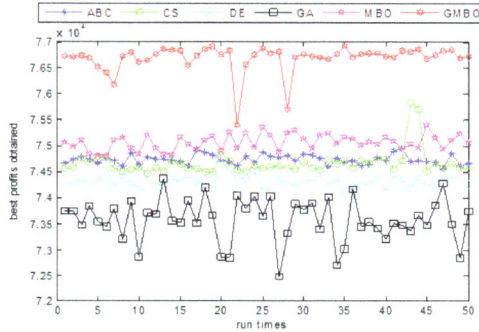

Figure 7. Comparison of the best values on KP4 in 50 runs.

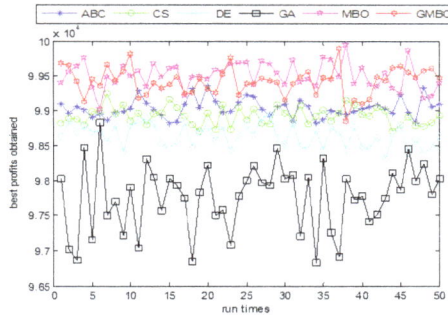

Figure 8. Comparison of the best values on KP5 in 50 runs.

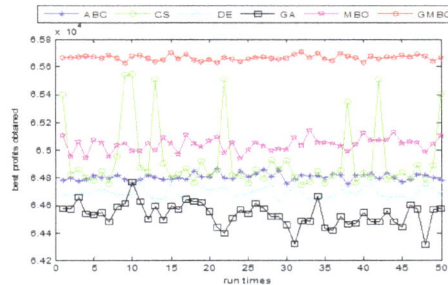

Figure 9. Comparison of the best values on KP9 in 50 runs.

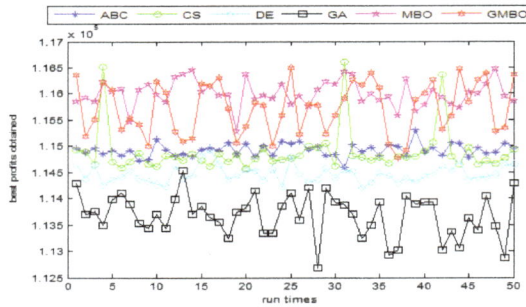

Figure 10. Comparison of the best values on KP10 in 50 runs.

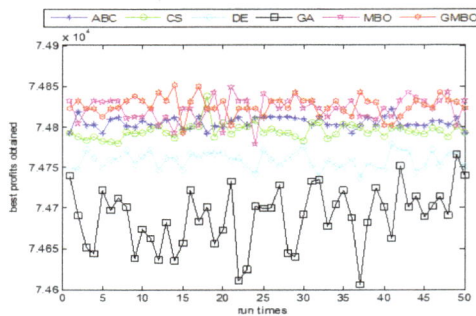

Figure 11. Comparison of the best values on KP14 in 50 runs.

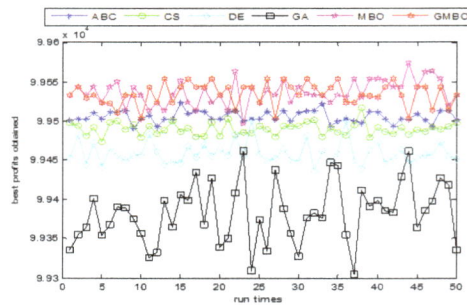

Figure 12. Comparison of the best values on KP15 in 50 runs.

Figures 7, 9 and 11 illustrate the best values achieved by the six methods on 1500-dimensional uncorrelated, weakly correlated, and strongly correlated 0-1 KP instances in 50 runs, respectively. From Figure 7, it can be easily seen that the best values obtained by GMBO far exceed that of the other five methods. Meanwhile, the two best values of CS outstripped the two worst values of GMBO. By looking at Figure 9, we can conclude that GMBO greatly outperformed the other five methods. The distribution of best values of GMBO in 50 times was close to a horizontal line, which pointed towards the excellent stability of GMBO in this case. With regard to numerical stability, CS had the worst performance. From Figure 11, the curve of GMBO still overtopped that of ABC, CS, DE, and GA, as illustrated in Figures 7 and 9. This advantage, however, was not obvious when compared with MBO.

Figures 8, 10 and 12 show the best values obtained by six methods on 2000-dimensional uncorrelated, weakly correlated, and strongly correlated 0-1 KP instances in 50 runs, respectively. As the dimension becomes large, space is expanded dramatically to 2^{2000}, which represents a challenge for any method. It can be said with certainty that almost all the values of GMBO are bigger than that of the other five methods except MBO. Similar to Figure 11, the curves of MBO partially overlaps that of GMBO in Figure 12, which may be interpreted as the ability of GMBO towards competing with MBO.

For the purpose of visualizing the experimental results from the statistical perspective, the corresponding boxplots of six higher dimensional KP4–KP5, KP9–KP10, and KP14-15 are shown in Figures 13–18. On the whole, the boxplot for GMBO has greater value and less height than those of the other five methods, which indicates the stronger optimization ability and stability of GMBO even encountering high-dimensional instances.

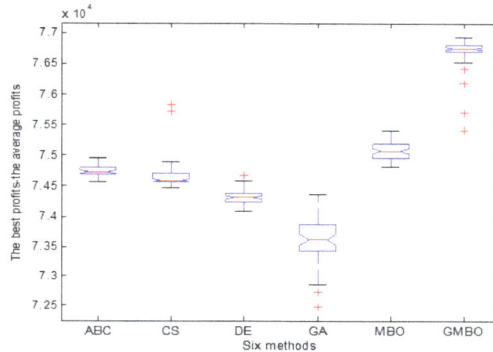

Figure 13. Boxplot of the best values on KP4 in 50 runs.

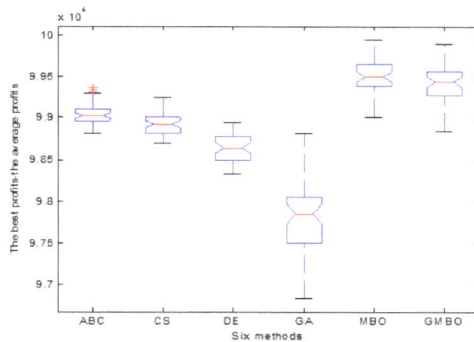

Figure 14. Boxplot of the best values on KP5 in 50 runs.

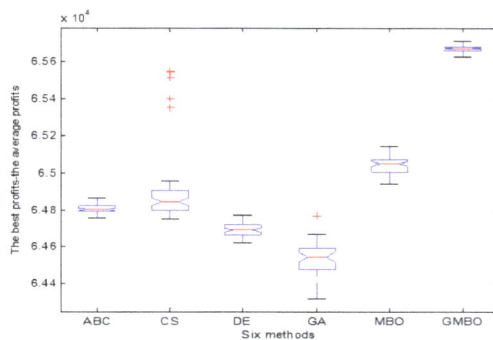

Figure 15. Boxplot of the best values on KP9 in 50 runs.

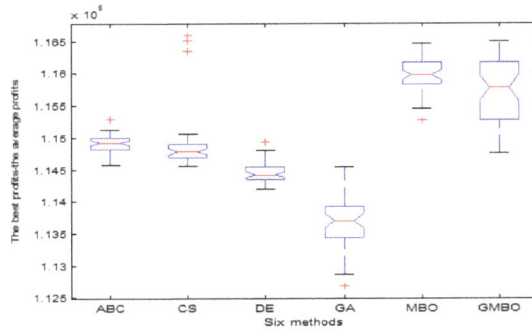

Figure 16. Boxplot of the best values on KP10 in 50 runs.

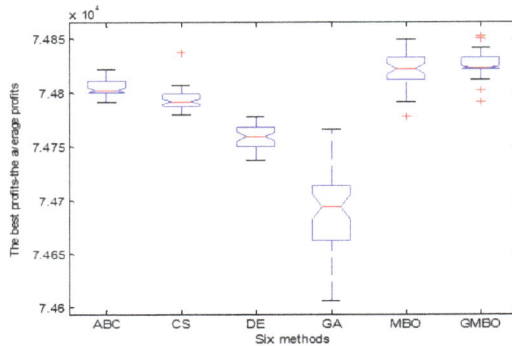

Figure 17. Boxplot of the best values on KP14 in 50 runs.

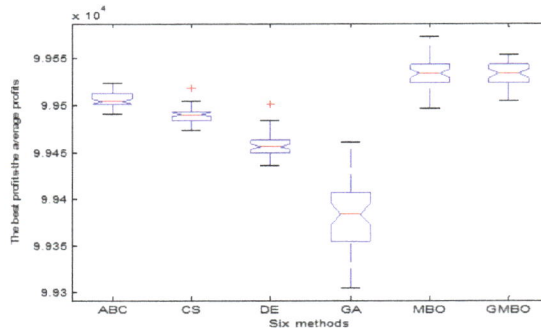

Figure 18. Boxplot of the best values on KP15 in 50 runs.

In order to examine the convergence rate of GMBO, the evolutionary process and convergent trajectories of six methods are illustrated in Figures 19–24. It should be noted that six high dimensional instances, viz., KP4, KP5, KP9, KP10, KP14, and KP15, were chosen. In addition, Figures 19–24 show the average best values with 50 runs, and not one independent experimental result.

From Figure 19, the curves of GMBO and MBO were almost coincident before 6 s, but afterward, GMBO converged rapidly to a better value as compared to the others. From Figure 20, it is indeed interesting to note that MBO has a weak advantage in the average values as compared to GMBO. From Figure 21, MBO and GMBO have identical initial function values, and the average values obtained by MBO were better than that of GMBO before 3 s. However, similar to the trend in Figure 19, 3 s later,

GMBO quickly converged to a higher value. As depicted in Figure 22, unexpectedly, when addressing the 2000-dimensional weakly correlated 0-1 KP instances, GMBO was inferior to MBO. Figures 23 and 24 illustrate the evolutionary process of strongly correlated 0-1 KP instances. By observation of 2 convergence graphs, we can conclude that GMBO and MBO have similar performance. Throughout Figures 19–24, GMBO has a stronger optimization ability and faster convergence speed to reach optimum solutions than the other five methods.

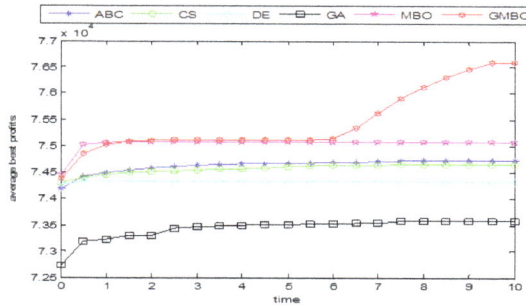

Figure 19. The convergence graph of six methods on KP4 in 10 s.

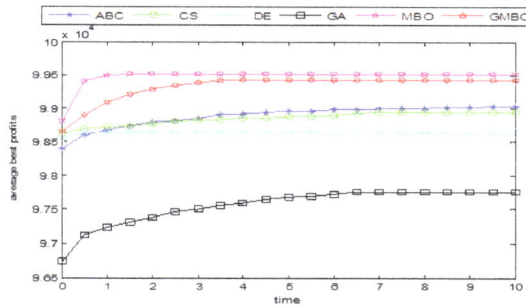

Figure 20. The convergence graph of six methods on KP5 in 10 s.

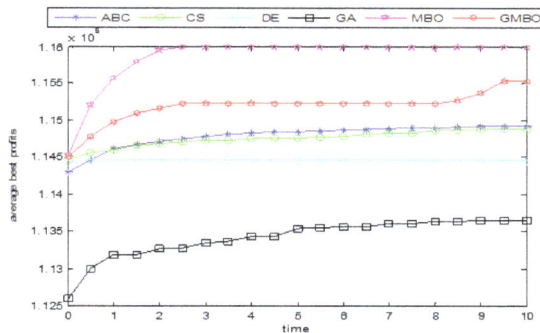

Figure 21. The convergence graph of six methods on KP9 in 10 s.

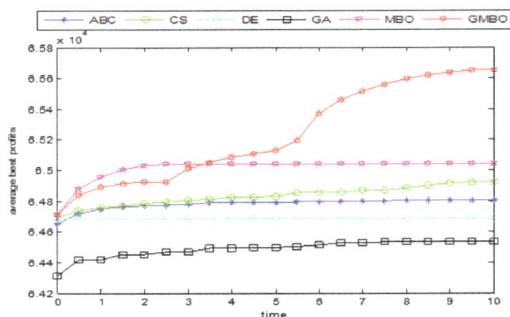

Figure 22. The convergence graph of six methods on KP10 in 10 s.

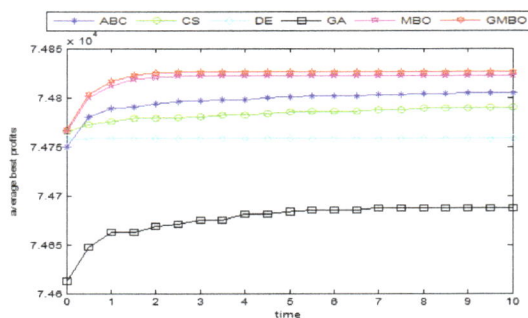

Figure 23. The convergence graph of six methods on KP14 in 10 s.

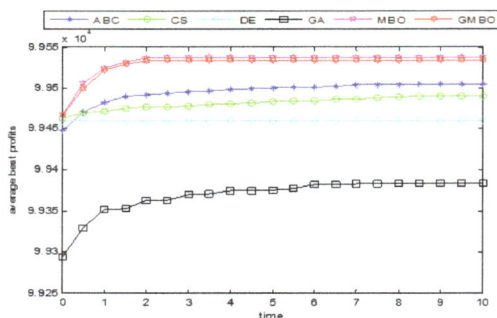

Figure 24. The convergence graph of six methods on KP15 in 10 s.

4.4. The Comparisons of the GMBO and the Latest Algorithms

To evaluate the performance of the proposed GMBO, the two latest algorithms, namely, moth search (MS) [67] and moth-flame optimization (MFO) [68], were especially selected to compare with GMBO. The following factors were mainly considered. (1) The literature on the application of MS and MFO to solve 0-1 KP problem was not found. (2) The GMBO, MS, and MFO were novel nature-inspired swarm intelligence algorithms, which simulated the migration behavior of the monarch butterfly, the Lévy flight mode, or the navigation method of moths.

For MS, the max step $S_{max} = 1.0$, acceleration factor $\varphi = 0.618$, and the index $\beta = 1.5$. For MFO, the maximum number of flames $N = 30$. In order to make a fair comparison, all experiments were conducted in the same experimental environment as described above. The detailed experimental results of GMBO and the other two algorithms on the three kinds of large-scale 0-1 KP instances were presented in Table 15. The best, mean, and standard deviation values in bold indicate superiority. The dominant times of the three algorithms in the three statistical values are given in the last line of Table 15. As the results presented in Table 15, the number of times the GMBO algorithm had priority in the best, the mean, and the standard deviation values were 8, 10, 5, respectively. The simulation results indicated that GMBO generally provided very excellent performance in most instances compared with MFO and MS. The two metrics, namely mean and standard deviation, demonstrated again that GMBO was more stable. The comprehensive performance of MFO was superior to that of MS.

Table 15. Performance comparison of three algorithms on large-scale 0-1 KP instances.

No.	MBO			MFO			MS		
	Best	**Mean**	**Std**	**Best**	**Mean**	**Std**	**Best**	**Mean**	**Std**
KP1	**40,684**	**40,641**	**40.09**	40,538	39,976	309.00	40,242	40,101	56.37
KP2	49,992	49,732	116.12	**50590**	**50200**	384.72	50,056	49,790	79.57
KP3	**61764**	**61430**	379.76	61,836	61,238	608.78	61,059	60,721	101.80
KP4	76,929	**76,691**	267.90	**77,007**	76,353	656.36	75,716	75,505	95.94
KP5	99,898	99,424	200.38	**102,276**	**101,475**	781.66	100,348	100,036	120.08
KP6	**35,069**	**35,064**	**4.04**	**35,069**	34,952	116.84	34,850	34,799	20.64
KP7	**43,786**	**43,784**	**1.57**	43,784	43,630	132.21	43,474	43,424	20.34
KP8	53,426	**53,167**	300.90	**53,552**	53,048	556.60	52,637	52,489	73.73
KP9	**65,708**	**65,666**	18.48	65,692	65,421	253.29	65,093	65,025	27.59
KP10	116,496	115,718	492.92	**118,183**	**117,381**	838.07	116,283	115,937	117.92
KP11	**40,167**	**40,162**	5.11	40,157	40,142	15.48	40,137	40,127	**5.58**
KP12	**49,442**	**49,425**	11.58	49,433	49,411	15.64	49,403	49,390	**7.10**
KP13	60,630	60,604	20.88	60,581	60,557	68.28	60,581	60,571	**9.24**
KP14	74,852	74,852	12.00	**74,910**	**74,874**	32.49	74,852	74,833	**7.71**
KP15	99,553	99,534	14.14	**99,643**	**99,602**	37.10	99,572	99,546	**9.28**
Total	**8**	**10**	**5**	**8**	**5**	0	0	0	**10**

To summarize, by analyzing Tables 4–15 and Figures 4–24, it can be inferred that GMBO had better optimization capability, numerical stability, and higher convergence speed. In other words, it can be claimed that GMBO is an excellent MBO variant, which is capable of addressing large-scale 0-1 KP instances.

5. Conclusions

In order to tackle high-dimensional 0-1 KP problems more efficiently and effectively, as well as to overcome the shortcomings of the original MBO simultaneously, a novel monarch butterfly optimization with the global position updating operator (GMBO) has been proposed in this manuscript. Firstly, a simple and effective dichotomy encoding scheme, without changing the evolutionary formula, is used. Moreover, an ingenious global position updating operator is introduced with the intention of enhancing the optimization capacity and convergence speed. The inspiration behind the new operator lies in creating a balance between intensification and diversification, a very important feature in the field of metaheuristics. Furthermore, a two-stage individual optimization method based on the greedy strategy is employed, which besides guaranteeing the feasibility of the solutions, is able to improve the quality further. In addition, the Orthogonal Design (OD) was applied to find suitable parameters. Finally, GMBO was verified and compared with ABC, CS, DE, GA, and MBO on large-scale 0-1 KP instances. The experimental results demonstrate that GMBO outperforms the other five algorithms on solution precision, convergence speed, and numerical stability.

The introduction of a global position operator coupled with an efficient two-stage repairing operator is instrumental towards the superior performance of GMBO. However, there is room for further enhancing the performance of GMBO. Firstly, the hybridization of the two methods complementing each other is becoming more and more popular, such as the hybridization of HS with CS [69]. Combining MBO with other methods could indeed be very promising and hence worth

Mathematics **2019**, 7, 1056

experimentation. Secondly, in the present work, three groups of high-dimensional 0-1 KP instances were selected. In the future, a multidimensional knapsack problem, quadratic knapsack problem, knapsack sharing problem, and randomized time-varying knapsack problem can be considered to investigate the performance of MBO. Thirdly, some typical combinatorial optimization problems, such as job scheduling problems [70–72], feature selection [73–75], and classification [76], deserve serious investigation and discussion. For these challenging engineering problems, the key issue is how to encode and process constraints. The application of MBO for these problems is another interesting research area. Finally, perturb [77], ensemble [78], learning mechanisms [79,80], or information feedback mechanisms [81] can be effectively combined with MBO to improve performance.

Author Contributions: Investigation, Y.F. and X.Y.; Methodology, Y.F.; Resources, X.Y.; Supervision, G.-G.W.; Validation, G.-G.W.; Visualization, G.-G.W.; Writing—original draft, Y.F. and X.Y.; Writing—review & editing, G.-G.W.

Funding: This research was funded by National Natural Science Foundation of China, grant number 61503165, 61806069.

Conflicts of Interest: The authors declare no conflict of interest.

References

1. Martello, S.; Toth, P. *Knapsack Problems: Algorithms and Computer Implementations*; John Wiley & Sons, Inc.: Hoboken, NJ, USA, 1990.
2. Mavrotas, G.; Diakoulaki, D.; Kourentzis, A. Selection among ranked projects under segmentation, policy and logical constraints. *Eur. J. Oper. Res.* **2008**, *187*, 177–192. [CrossRef]
3. Peeta, S.; Salman, F.S.; Gunnec, D.; Viswanath, K. Pre-disaster investment decisions for strengthening a highway network. *Comput. Oper. Res.* **2010**, *37*, 1708–1719. [CrossRef]
4. Yates, J.; Lakshmanan, K. A constrained binary knapsack approximation for shortest path network interdiction. *Comput. Ind. Eng.* **2011**, *61*, 981–992. [CrossRef]
5. Dantzig, G.B. Discrete-variable extremum problems. *Oper. Res.* **1957**, *5*, 266–288. [CrossRef]
6. Shih, W. A branch and bound method for the multi-constraint zero-one knapsack problem. *J. Oper. Res. Soc.* **1979**, *30*, 369–378. [CrossRef]
7. Toth, P. Dynamic programing algorithms for the zero-one knapsack problem. *Computing* **1980**, *25*, 29–45. [CrossRef]
8. Martello, S.; Toth, P. A new algorithm for the 0-1 knapsack problem. *Manag. Sci.* **1988**, *34*, 633–644. [CrossRef]
9. Pisinger, D. An expanding-core algorithm for the exact 0–1 knapsack problem. *Eur. J. Oper. Res.* **1995**, *87*, 175–187. [CrossRef]
10. Puchinger, J.; Raidl, G.R.; Pferschy, U. The Core Concept for the Multidimensional Knapsack Problem. In Proceedings of the European Conference on Evolutionary Computation in Combinatorial Optimization, Budapest, Hungary, 10–12 April 2006; Gottlieb, J., Raidl, G.R., Eds.; Springer: Berlin, Germany, 2006; pp. 195–208.
11. Thiel, J.; Voss, S. Some experiences on solving multi constraint zero-one knapsack problems with genetic algorithms. *Inf. Syst. Oper. Res.* **1994**, *32*, 226–242.
12. Chen, P.; Li, J.; Liu, Z.M. Solving 0-1 knapsack problems by a discrete binary version of differential evolution. In Proceedings of the Second International Symposium on Intelligent Information Technology Application, Shanghai, China, 21–22 December 2008; Volume 2, pp. 513–516.
13. Bhattacharjee, K.K.; Sarmah, S.P. Shuffled frog leaping algorithm and its application to 0/1 knapsack problem. *Appl. Soft Comput.* **2014**, *19*, 252–263. [CrossRef]
14. Feng, Y.H.; Jia, K.; He, Y.C. An improved hybrid encoding cuckoo search algorithm for 0-1 knapsack problems. *Comput. Intell. Neurosci.* **2014**, *2014*, 1. [CrossRef]
15. Feng, Y.H.; Wang, G.G.; Feng, Q.J.; Zhao, X.J. An effective hybrid cuckoo search algorithm with improved shuffled frog leaping algorithm for 0-1 knapsack problems. *Comput. Intell. Neurosci.* **2014**, *2014*, 36. [CrossRef]
16. Kashan, M.H.; Nahavandi, N.; Kashan, A.H. DisABC: A new artificial bee colony algorithm for binary optimization. *Appl. Soft Comput.* **2012**, *12*, 342–352. [CrossRef]

17. Xue, Y.; Jiang, J.; Zhao, B.; Ma, T. A self-adaptive artificial bee colony algorithm based on global best for global optimization. *Soft Comput.* **2018**, *22*, 2935–2952. [CrossRef]

18. Zong, W.G.; Kim, J.H.; Loganathan, G.V. A New Heuristic Optimization Algorithm: Harmony Search. *Simulation* **2001**, *76*, 60–68. [CrossRef]

19. Zou, D.; Gao, L.; Li, S.; Wu, J. Solving 0-1 knapsack problem by a novel global harmony search algorithm. *Appl. Soft Comput.* **2011**, *11*, 1556–1564. [CrossRef]

20. Kong, X.; Gao, L.; Ouyang, H.; Li, S. A simplified binary harmony search algorithm for large scale 0-1 knapsack problems. *Expert Syst. Appl.* **2015**, *42*, 5337–5355. [CrossRef]

21. Rezoug, A.; Boughaci, D. A self-adaptive harmony search combined with a stochastic local search for the 0-1 multidimensional knapsack problem. *Int. J. Biol. Inspir. Comput.* **2016**, *8*, 234–239. [CrossRef]

22. Zhou, Y.; Li, L.; Ma, M. A complex-valued encoding bat algorithm for solving 0-1 knapsack problem. *Neural Process. Lett.* **2016**, *44*, 407–430. [CrossRef]

23. Cai, X.; Gao, X.-Z.; Xue, Y. Improved bat algorithm with optimal forage strategy and random disturbance strategy. *Int. J. Biol. Inspir. Comput.* **2016**, *8*, 205–214. [CrossRef]

24. Zhang, X.; Huang, S.; Hu, Y.; Zhang, Y.; Mahadevan, S.; Deng, Y. Solving 0-1 knapsack problems based on amoeboid organism algorithm. *Appl. Math. Comput.* **2013**, *219*, 9959–9970. [CrossRef]

25. Li, X.; Zhang, J.; Yin, M. Animal migration optimization: An optimization algorithm inspired by animal migration behavior. *Neural Comput. Appl.* **2014**, *24*, 1867–1877. [CrossRef]

26. Cui, Z.; Fan, S.; Zeng, J.; Shi, Z. Artificial plant optimization algorithm with three-period photosynthesis. *Int. J. Biol. Inspir. Comput.* **2013**, *5*, 133–139. [CrossRef]

27. Simon, D. Biogeography-based optimization. *IEEE Trans. Evol. Comput.* **2008**, *12*, 702–713. [CrossRef]

28. Li, X.; Wang, J.; Zhou, J.; Yin, M. A perturb biogeography based optimization with mutation for global numerical optimization. *Appl. Math. Comput.* **2011**, *218*, 598–609. [CrossRef]

29. Wang, L.; Yang, R.; Ni, H.; Ye, W.; Fei, M.; Pardalos, P.M. A human learning optimization algorithm and its application to multi-dimensional knapsack problems. *Appl. Soft Comput.* **2015**, *34*, 736–743. [CrossRef]

30. Wang, G.-G.; Guo, L.H.; Wang, H.Q.; Duan, H.; Liu, L.; Li, J. Incorporating mutation scheme into krill herd algorithm for global numerical optimization. *Neural Comput. Appl.* **2014**, *24*, 853–871. [CrossRef]

31. Wang, G.-G.; Gandomi, A.H.; Yang, X.-S.; Alavi, H.A. A new hybrid method based on krill herd and cuckoo search for global optimization tasks. *Int. J. Biol. Inspir. Comput.* **2016**, *8*, 286–299. [CrossRef]

32. Wang, G.-G.; Gandomi, A.H.; Alavi, H.A. Stud krill herd algorithm. *Neurocomputing* **2014**, *128*, 363–370. [CrossRef]

33. Wang, G.-G.; Deb, S.; Cui, Z. Monarch butterfly optimization. *Neural Comput. Appl.* **2015**. [CrossRef]

34. Wang, G.-G.; Deb, S.; Gao, X.-Z.; Coelho, L.D.S. A new metaheuristic optimization algorithm motivated by elephant herding behavior. *Int. J. Biol. Inspir. Comput.* **2016**, *8*, 394–409. [CrossRef]

35. Sang, H.-Y.; Duan, Y.P.; Li, J.-Q. An effective invasive weed optimization algorithm for scheduling semiconductor final testing problem. *Swarm Evol. Comput.* **2018**, *38*, 42–53. [CrossRef]

36. Wang, G.-G.; Deb, S.; Coelho, L.D.S. Earthworm optimisation algorithm: A bio-inspired metaheuristic algorithm for global optimisation problems. *Int. J. Biol. Inspir. Comput.* **2018**, *12*, 1–22. [CrossRef]

37. Jain, M.; Singh, V.; Rani, A. A novel nature-inspired algorithm for optimization: Squirrel search algorithm. *Swarm Evol. Comput.* **2019**, *44*, 148–175. [CrossRef]

38. Singh, B.; Anand, P. A novel adaptive butterfly optimization algorithm. *Int. J. Comput. Mater. Sci. Eng.* **2018**, *7*, 69–72. [CrossRef]

39. Sayed, G.I.; Khoriba, G.; Haggag, M.H. A novel chaotic salp swarm algorithm for global optimization and feature selection. *Appl. Intell.* **2018**, *48*, 3462–3481. [CrossRef]

40. Simhadri, K.S.; Mohanty, B. Performance analysis of dual-mode PI controller using quasi-oppositional whale optimization algorithm for load frequency control. *Int. Trans. Electr. Energy Syst.* **2019**. [CrossRef]

41. Brezočnik, L.; Fister, I.; Podgorelec, V. Swarm intelligence algorithms for feature selection: A review. *Appl. Sci.* **2018**, *8*, 1521. [CrossRef]

42. Feng, Y.H.; Wang, G.-G.; Deb, S.; Lu, M.; Zhao, X.-J. Solving 0-1 knapsack problem by a novel binary monarch butterfly optimization. *Neural Comput. Appl.* **2015**. [CrossRef]

43. Wang, G.-G.; Zhao, X.C.; Deb, S. A Novel Monarch Butterfly Optimization with Greedy Strategy and Self-adaptive Crossover Operator. In Proceedings of the 2nd International Conference on Soft Computing & Machine Intelligence (ISCMI 2015), Hong Kong, China, 23–24 November 2015.

44. He, Y.C.; Wang, X.Z.; Kou, Y.Z. A binary differential evolution algorithm with hybrid encoding. *J. Comput. Res. Dev.* **2007**, *44*, 1476–1484. [CrossRef]
45. He, Y.C.; Song, J.M.; Zhang, J.M.; Gou, H.Y. Research on genetic algorithms for solving static and dynamic knapsack problems. *Appl. Res. Comput.* **2015**, *32*, 1011–1015.
46. He, Y.C.; Zhang, X.L.; Li, W.B.; Li, X.; Wu, W.L.; Gao, S.G. Algorithms for randomized time-varying knapsack problems. *J. Comb. Optim.* **2016**, *31*, 95–117. [CrossRef]
47. Fang, K.-T.; Wang, Y. *Number-Theoretic Methods in Statistics*; Chapman & Hall: New York, NY, USA, 1994.
48. Wilcoxon, F.; Katti, S.K.; Wilcox, R.A. Critical values and probability levels for the Wilcoxon rank sum test and the Wilcoxon signed rank test. *Sel. Tables Math. Stat.* **1970**, *1*, 171–259.
49. Gutowski, M. Lévy flights as an underlying mechanism for global optimization algorithms. *arXiv* **2001**, arXiv:math-ph/0106003.
50. Pavlyukevich, I. Levy flights, non-local search and simulated annealing. *Mathematics* **2007**, *226*, 1830–1844. [CrossRef]
51. Kennedy, J.; Eberhart, R.C. A discrete binary version of the particle swarm algorithm. In Proceedings of the 1997 Conference on Systems, Man, and Cybernetics, Orlando, FL, USA, 12–15 October 1997; pp. 4104–4108.
52. Zhu, H.; He, Y.; Wang, X.; Tsang, E.C. Discrete differential evolutions for the discounted {0-1} knapsack problem. *Int. J. Biol. Inspir. Comput.* **2017**, *10*, 219–238. [CrossRef]
53. Yang, X.S.; Deb, S.; Hanne, T.; He, X. Attraction and Diffusion in Nature-Inspired Optimization Algorithms. *Neural Comput. Appl.* **2019**, *31*, 1987–1994. [CrossRef]
54. Joines, J.A.; Houck, C.R. On the use of non-stationary penalty functions to solve nonlinear constrained optimization problems with GA's. Evolutionary Computation. In Proceedings of the First IEEE Conference on Evolutionary Computation, Orlando, FL, USA, 27–29 June 1994; pp. 579–584.
55. Olsen, A.L. Penalty functions and the knapsack problem. Evolutionary Computation. In Proceedings of the First IEEE Conference on Evolutionary Computation, Orlando, FL, USA, 27–29 June 1994; pp. 554–558.
56. Goldberg, D.E. *Genetic Algorithms in Search, Optimization, and Machine Learning*; Addison-Wesley: Boston, MA, USA, 1989.
57. Simon, D. *Evolutionary Optimization Algorithms*; Wiley: New York, NY, USA, 2013.
58. Du, D.Z.; Ko, K.I.; Hu, X. *Design and Analysis of Approximation Algorithms*; Springer Science & Business Media: Berlin, Germany, 2011.
59. Pisinger, D. Where are the hard knapsack problems. *Comput. Oper. Res.* **2005**, *32*, 2271–2284. [CrossRef]
60. Karaboga, D.; Basturk, B. A powerful and efficient algorithm for numerical function optimization: Artificial bee colony (ABC) algorithm. *J. Glob. Optim.* **2007**, *39*, 459–471. [CrossRef]
61. Yang, X.S.; Deb, S. Cuckoo search via Lévy flights. In Proceedings of the World Congress on Nature and Biologically Inspired Computing (NaBIC 2009), Coimbatore, India, 9–11 December 2009; pp. 210–214.
62. Storn, R.; Price, K. Differential evolution–A simple and efficient heuristic for global optimization over continuous spaces. *J. Glob. Optim.* **1997**, *11*, 341–359. [CrossRef]
63. Mahdavi, M.; Fesanghary, M.; Damangir, E. An improved harmony search algorithm for solving optimization problems. *Appl. Math. Comput.* **2007**, *188*, 1567–1579. [CrossRef]
64. Bansal, J.C.; Deep, K. A Modified Binary Particle Swarm Optimization for Knapsack Problems. *Appl. Math. Comput.* **2012**, *218*, 11042–11061. [CrossRef]
65. Lee, C.Y.; Lee, Z.J.; Su, S.F. A New Approach for Solving 0/1 Knapsack Problem. In Proceedings of the IEEE International Conference on Systems, Man and Cybernetics, Montreal, QC, Canada, 7–10 October 2007; pp. 3138–3143.
66. Cormen, T.H. *Introduction to Algorithms*; MIT Press: Cambridge, MA, USA, 2009.
67. Wang, G.-G. Moth search algorithm: A bio-inspired metaheuristic algorithm for global optimization problems. *Memet. Comput.* **2018**, *10*, 151–164. [CrossRef]
68. Mehne, S.H.H.; Mirjalili, S. Moth-Flame Optimization Algorithm: Theory, Literature Review, and Application in Optimal Nonlinear. In *Nature Inspired Optimizers: Theories, Literature Reviews and Applications*; Mirjalili, S., Jin, S.D., Lewis, A., Eds.; Spring: Berlin, Germany, 2020; Volume 811, p. 143.
69. Gandomi, A.H.; Zhao, X.J.; Chu, H.C.E. Hybridizing harmony search algorithm with cuckoo search for global numerical optimization. *Soft Comput.* **2016**, *20*, 273–285.
70. Li, J.-Q.; Pan, Q.-K.; Liang, Y.-C. An effective hybrid tabu search algorithm for multi-objective flexible job-shop scheduling problems. *Comput. Ind. Eng.* **2010**, *59*, 647–662. [CrossRef]

71. Han, Y.-Y.; Gong, D.; Sun, X. A discrete artificial bee colony algorithm incorporating differential evolution for the flow-shop scheduling problem with blocking. *Eng. Optim.* **2014**, *47*, 927–946. [CrossRef]
72. Li, J.-Q.; Pan, Q.-K.; Tasgetiren, M.F. A discrete artificial bee colony algorithm for the multi-objective flexible job-shop scheduling problem with maintenance activities. *Appl. Math. Model.* **2014**, *38*, 1111–1132. [CrossRef]
73. Zhang, W.-Q.; Zhang, Y.; Peng, C. Brain storm optimization for feature selection using new individual clustering and updating mechanism. *Appl. Intell.* **2019**. [CrossRef]
74. Zhang, Y.; Li, H.-G.; Wang, Q.; Peng, C. A filter-based bare-bone particle swarm optimization algorithm for unsupervised feature selection. *Appl. Intell.* **2019**, *49*, 2889–2898. [CrossRef]
75. Zhang, Y.; Wang, Q.; Gong, D.-W.; Song, X.-F. Nonnegative laplacian embedding guided subspace learning for unsupervised feature selection. *Pattern Recognit.* **2019**, *93*, 337–352. [CrossRef]
76. Zhang, Y.; Gong, D.W.; Cheng, J. Multi-objective particle swarm optimization approach for cost-based feature selection in classification. *IEEE ACM Trans. Comput. Biol. Bioinform.* **2017**, *14*, 64–75. [CrossRef]
77. Zhao, X. A perturbed particle swarm algorithm for numerical optimization. *Appl. Soft Comput.* **2010**, *10*, 119–124.
78. Wu, G.; Shen, X.; Li, H.; Chen, H.; Lin, A.; Suganthan, P.N. Ensemble of differential evolution variants. *Inf. Sci.* **2018**, *423*, 172–186. [CrossRef]
79. Wang, G.G.; Deb, S.; Gandomi, A.H.; Alavi, A.H. Opposition-based krill herd algorithm with Cauchy mutation and position clamping. *Neurocomputing* **2016**, *177*, 147–157. [CrossRef]
80. Zhang, Y.; Gong, D.-W.; Gao, X.-Z.; Tian, T.; Sun, X.-Y. Binary differential evolution with self-learning for multi-objective feature selection. *Inf. Sci.* **2020**, *507*, 67–85. [CrossRef]
81. Wang, G.-G.; Tan, Y. Improving metaheuristic algorithms with information feedback models. *IEEE Trans. Cybern.* **2019**, *49*, 542–555. [CrossRef]

![mathematics logo] *mathematics*

MDPI

Article

Rock Classification from Field Image Patches Analyzed Using a Deep Convolutional Neural Network

Xiangjin Ran [1,2], Linfu Xue [1,*], Yanyan Zhang [3], Zeyu Liu [1], Xuejia Sang [4] and Jinxin He [1]

[1]   College of Earth Science, Jilin University, Changchun 130061, China
[2]   College of Applied Technology, Jilin University, Changchun 130012, China
[3]   Jilin Business and Technology College, Changchun 130012, China
[4]   School of Environment Science and Spatial Informatics (CESI), China University of Mining and Technology, Xuzhou 221008, China
*   Correspondence: xuelf@jlu.edu.cn; Tel.: +86-135-008-053-87

Received: 3 July 2019; Accepted: 13 August 2019; Published: 18 August 2019

Abstract: The automatic identification of rock type in the field would aid geological surveying, education, and automatic mapping. Deep learning is receiving significant research attention for pattern recognition and machine learning. Its application here has effectively identified rock types from images captured in the field. This paper proposes an accurate approach for identifying rock types in the field based on image analysis using deep convolutional neural networks. The proposed approach can identify six common rock types with an overall classification accuracy of 97.96%, thus outperforming other established deep-learning models and a linear model. The results show that the proposed approach based on deep learning represents an improvement in intelligent rock-type identification and solves several difficulties facing the automated identification of rock types in the field.

Keywords: deep learning; convolutional neural network; rock types; automatic identification

---

1. Introduction

Rocks are a fundamental component of Earth. They contain the raw materials for virtually all modern construction and manufacturing and are thus indispensable to almost all the endeavors of an advanced society. In addition to the direct use of rocks, mining, drilling, and excavating provide the material sources for metals, plastics, and fuels. Natural rock types have a variety of origins and uses. The three major groups of rocks (igneous, sedimentary, and metamorphic) are further divided into sub-types according to various characteristics. Rock type identification is a basic part of geological surveying and research, and mineral resources exploration. It is an important technical skill that must be mastered by students of geoscience.

Rocks can be identified in a variety of ways, such as visually (by the naked eye or with a magnifying glass), under a microscope, or by chemical analysis. Working conditions in the field generally limit identification to visual methods, including using a magnifying glass for fine-grained rocks. Visual inspection assesses properties such as color, composition, grain size, and structure. The attributes of rocks reflect their mineral and chemical composition, formation environment, and genesis. The color of rock reflects its chemical composition. For example, dark rocks usually contain dark mafic minerals (e.g., pyroxene and hornblende) and are commonly basic, whereas lighter rocks tend to contain felsic minerals (e.g., quartz and feldspar) and are acidic. The sizes of detrital grains provide further information and can help to distinguish between conglomerate, sandstone, and limestone, for example. The textural features of the rock assist in identifying its structure [1] and thus aid classification. The colors, grain sizes, and textural properties of rocks vary markedly between different rock types, allowing a basis for distinguishing them [2]. However, the accurate identification of rock type remains challenging because of the diversity of rock types and the heterogeneity of their properties [3] as well as further limitations imposed by the experience and skill of geologists [4]. The identification of rock type by the naked eye is effectively an image recognition task based on knowledge of rock classification. The rapid development of image acquisition and computer image pattern recognition technology has thus allowed the development of automatic systems to identify rocks from images taken in the field. These systems will greatly assist geologists by improving identification accuracy and efficiency and will also help student and newly qualified geologists practice rock-type identification. Identification systems can be incorporated into automatic remote sensing and geological mapping systems carried by unmanned aerial vehicles (UAVs).

The availability of digital cameras, hand-held devices and the development of computerized image analysis provide technical support for various applications [5], so, they allow several characteristics of rocks to be collected and assessed digitally. Photographs can clearly show the characteristics of color, grain size, and texture of rocks (Figure 1). Although images of rocks do not show homogeneous shapes, textures [1,6], or colors, computer image analysis can be used to classify some types of rock images. Partio et al. [7] used gray-level co-occurrence matrices for texture retrieval from rock images. Lepistö et al. [6] classified rock images based on textural and spectral features.

Advances in satellite and remote sensing technology have encouraged the development of multi-spectral remote sensing technology to classify ground objects of different types [8,9], including rock. However, it is expensive to obtain ultra-high-resolution rock images in the field with the use of remote sensing technology. Therefore, the high cost of data acquisition using hyperspectral technology carried by aircraft and satellites often prevents its use in teaching and the automation of rock type identification.

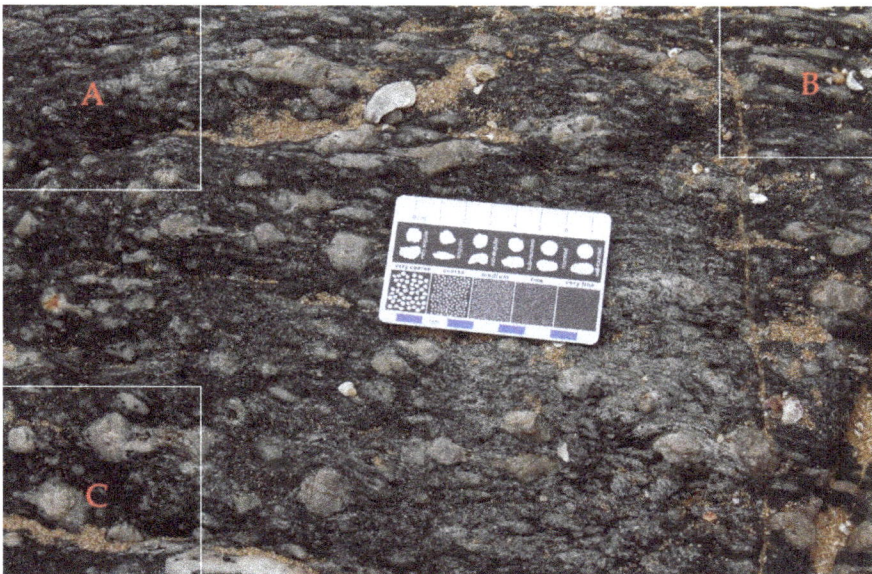

Figure 1. Digital image obtained in the field, allowing the rock type to be identified as mylonite by the naked eye. Partition (**A**) shows the smaller changes in grain size of mylonite; partition (**B**) shows larger tensile deformation of quartz particles; partition (**C**) shows larger grains than partition A and B.

Machine learning algorithms applied to digital image analysis have been used to improve the accuracy and speed of rock identification, and researchers have studied automated rock-type classification based on traditional machine learning algorithms. Lepistö et al. [1] used image analysis to investigated bedrock properties, and Chatterjee [2] tested a genetic algorithm on photographs of samples from a limestone mine to establish a visual rock classification model based on imaging and the Support Vector Machine (SVM) algorithm. Patel and Chatterjee [4] used a probabilistic neural network to classify lithotypes based on image features extracted from the images of limestone. Perez et al. [10] photographed rocks on a conveyor belt and then extracted features of the images to classify their types using the SVM algorithm.

The quality of a digital image used in rock-type identification significantly affects the accuracy of the assessment [2,4]. Traditional machine learning approaches can be effective in analyzing rock lithology, but they are easily disturbed by the selection of artificial features [11]. Moreover, the requirements for image quality and illumination are strict, thus limiting the choice of equipment used and requiring a certain level of expertise on the part of the geologist. In the field, the complex characteristics of weathered rocks and the variable conditions of light and weather, amongst others, can compromise the quality of the obtained images, thus complicating the extraction of rock features from digital images. Therefore, existing available methods are difficult to apply to the automated identification of rock types in the field.

In recent years, deep learning, also known as deep neural networks, has received attention in various research fields [12]. Many methods for deep learning have been proposed [13]. Deep convolutional neural networks (CNNs) are able to automatically learn the features required for image classification from training-image data, thus improving classification accuracy and efficiency without relying on artificial feature selection. Very recent studies have proposed deep learning algorithms to achieve significant empirical improvements in areas such as image classification [14], object detection [15], human behavior recognition [16,17], speech recognition [18,19], traffic signal recognition [20,21], clinical diagnosis [22,23], and plant disease identification [11,24]. The successes of applying CNNs to image recognition have led geologists to investigate their use in identifying rock types [8,9,25], and deep learning has been used in several studies to identify the rock types from images. Zhang et al. [26] used transfer learning to identify granite, phyllite, and breccia based on the GoogLeNet Inception v3 deep CNNs model, achieving an overall accuracy of 85%. Cheng et al. [27] proposed a deep learning model based on CNNs to identify three types of sandstone in image slices with an accuracy of 98.5%. These studies show that CNNs have obtained good results when applied to geological surveying and rock-type recognition. Deep CNNs can identify rock types from images without requiring the manual selection of image features. However, deep CNNs have not yet been applied in the field, and the accuracy of the above results was not sufficient for the identification of rocks.

This paper proposes a new method for automatically classifying field rock images based on deep CNNs. A total of 2290 field rock photographs were first cropped to form a database of 24,315 image patches. The sample patches were then utilized to train and test CNNs, with 14,589 samples being used as the training dataset, 4863 samples being used as the validation dataset and the remaining 4863 samples being used as the testing dataset. The results show that the proposed model achieves higher accuracy than other models. The main contributions of this paper are as follows: (1) the very high resolution of the digital rock images allows them to include interference elements such as grass, soil, and water, which do not aid rock type's identification. This paper proposes a method of training-image generation that can decrease computation and prevent overfitting of the CNNs-based model during training. The method slices the original rock image into patches, selects patches typical of rock images to form a dataset, and removes the interference elements that are irrelevant to rock classification. (2) Rock Types deep CNNs (RTCNNs) model is employed to classify field rock types. Compared with the established SVM, AlexNet, VGGNet-16, and GoogLeNet Inception v3 models, the RTCNNs model has a simpler structure and higher accuracy for identifying rock types in the field. Based on various factors, such as model type, sample size, and model level, a series of comparisons verified the high performance of the RTCNNs model, demonstrating its reliability and yielding an overall identification accuracy of 97.96%.

The remainder of this paper is organized as follows. Section 2 presents details of the modification and customization of the RTCNNs for the automated identification of field rock types. Section 3 describes the techniques of classifying the field rock types (including acquiring images of rock outcrops and generating patched samples) and the software and hardware configurations of the method, followed by a presentation of the results. Section 4 analyzes the factors that affect the identification accuracy, such as the type of model, sample size, and model level, and presents the results. Section 5 provides the conclusions of the study.

2. Architecture of the Rock Types Deep Convolutional Neural Networks Model

Developments in deep learning technology have allowed continuous improvements to be made in the accuracy of CNNs models. Such advances have been gained by models becoming ever deeper, which has meant that such models demand increased computing resources and time. This paper proposes a RTCNNs model for identifying rock types in the field. The computing time of the RTCNNs model is much less than that of a model 10 or more layers. The hardware requirements are quite modest, with computations being carried out with commonly used device CPUs and Graphics Processing Units (GPUs). The RTCNNs model includes six layers (Figure 2).

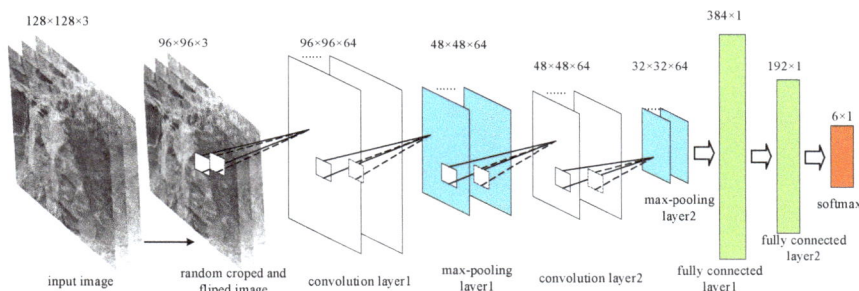

Figure 2. The Rock Types deep CNNs (RTCNNs) model for classifying rock type in the field.

Before feeding the sample images into the model, Random_Clip and Random_Flip operations are applied to the input samples. Each part of the image retains different feature of the target object. Random clipping can reserve the different features of the image. For example, partition A of the image shown in Figure 1 records smaller changes in grain size of mylonite, in which quartz particles do not undergo obvious deformation, while partition B records larger tensile deformation of quartz particles, and the quartz grains in the partition C are generally larger. In addition, in the proposed model, each layers of training have fixed size parameters, such as the input size of convolution layer1 is $96 \times 96 \times 3$, while the output size of feature is $96 \times 96 \times 64$ (Figure 2). The input images are cropped into sub-images with given size, while the given size is less. In the proposed model, the cropped size is $96 \times 96 \times 3$, while the input size is $128 \times 128 \times 3$. Through the random clipping operation of fixed size and different positions, different partitions of the same image are fed into the model during different training epochs. The flipping function can flip the image horizontally randomly. Both clipping and flipping operations are realized through the corresponding functions of TensorFlow deep learning framework [28]. The sample images fed into the model are therefore different in each epoch, which expands the training dataset, improving the accuracy of the model and avoiding overfitting.

Before performing patch-based sampling, the various features of the rock are spread all over the entire original field-captured image. The experiments described in Section 4 show that a smaller convolution kernel can filter the rock features better than the bigger kernel of other models. As a consequence, the first convolutional layer is designed to be 64 kernels of size $5 \times 5 \times 3$, followed by a max-pooling layer (Section 2.2), which can shrink the output feature map by 50%. A Rectified Linear Unit (ReLU, Section 2.3) activation function is then utilized to activate the output neuron. The second convolutional layer has 64 kernels of size $5 \times 5 \times 64$ connected to the outputs of the ReLU function, and it is similarly followed by a max-pooling layer. Below this layer, two fully connected layers are designed to predict six classes of field rock, and the final layer consists of a six-way Softmax layer. Detailed parameters of the model, as obtained by experimental optimization, are listed in Table 1.

Table 1. Parameters and output shapes of the RTCNNs model.

Layer Name	Function	Weight Filter Sizes/Kernels	Padding	Stride	Output Tensor
Input	/	/	/	/	$128 \times 128 \times 3$
Cropped image	random_crop	/	/	/	$96 \times 96 \times 3$
Conv1	conv2d	$5 \times 5 \times 3/64$	SAME	1	$96 \times 96 \times 64$
Pool1	max_pool	3×3	SAME	2	$48 \times 48 \times 64$
Conv2	conv2d	$5 \times 5 \times 64/64$	SAME	1	$48 \times 48 \times 64$
Pool2	max_pool	3×3	SAME	2	$24 \times 24 \times 64$
Output	softmax	/	/	/	6×1

2.1. Convolution Layer

A convolution layer extracts the features of the input images by convolution and outputs the feature maps (Figure 3). It is composed of a series of fixed size filters, known as convolution kernels, which are used to perform convolution operations on image data to produce the feature maps [29]. Generally, the output feature map can be realized by Equation (1):

$$h_{ij}^k = \sum_{i \in M_j} \left(\left(w^k \times x \right)_{ij} + b_k \right) \tag{1}$$

where k represents the kth layer, h represents the value of the feature, (i, j) are coordinates of pixels, w^k represents the convolution kernel of the current layer, and b_k is the bias. The parameters of CNNs, such as the bias (b_k) and convolution kernel (w^k), are usually trained without supervision [11]. Experiments optimized the convolution kernel size by comparing sizes of 3×3, 5×5, and 7×7; the 5×5 size achieves the best classification accuracy. The number of convolution kernels also affects the accuracy rate, so 32, 64, 128, and 256 convolution kernels were experimentally tested here. The highest accuracy is obtained using 64 kernels. Based on these experiments, the RTCNNs model adopts a 5×5 size and 64 kernels to output feature maps.

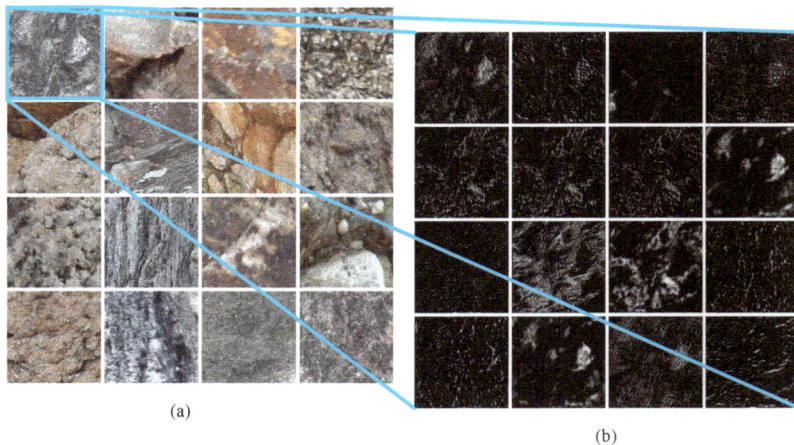

(a)

(b)

Figure 3. Learned rock features after convolution by the RTCNNs model. (**a**) Input patched field rock sample images. (**b**) Outputted feature maps partly after the first convolution of the input image, from the upper left corner in (a).

Figure 3 shows the feature maps outputted from the convolution of the patched field images. Figure 3a depicts the patch images from field photographs inputted to the proposed model during training, and Figure 3b shows the edge features of the sample patches learned by the model after the first layer convolution. The Figure indicates that the RTCNNs model can automatically extract the basic features of the images for learning.

2.2. Max-Pooling Layer

The pooling layer performs nonlinear down-sampling and reduces the size of the feature map, also accelerating convergence and improving computing performance [12]. The RTCNNs model uses max-pooling rather than mean-pooling because the former can obtain more textural features than can the latter [30]. The max-pooling operation maximizes the feature area of a specified size and is formulated by

$$h_j = \max_{i \in R_j} a_i \qquad (2)$$

where R_j is the pooling region j in feature map a, i is the index of each element within the region, and h is the pooled feature map.

2.3. ReLU Activation Function

The ReLU activation function nonlinearly maps the characteristic graph of the convolution layer output to activate neurons while avoiding overfitting and improving learning ability. This function was originally introduced in the AlexNet model [14]. The RTCNNs model uses the ReLU activation function (Equation (3)) for the output feature maps of every convolutional layer:

$$f(x) = max(0, x) \qquad (3)$$

2.4. Fully Connected Layers

Each node of the fully connected layers is connected to all the nodes of the upper layer. The fully connected layers are used to synthesize the features extracted from the image and to transform the two-dimensional feature map into a one-dimensional feature vector [12]. The fully connected layers map the distributed feature representation to the sample label space. The fully connected operation is formulated by Equation (4):

$$a_i = \sum_{j=0}^{m*n*d-1} w_{ij} * x_i + b_i \qquad (4)$$

where i is the index of the output of the fully connected layer; m, n, and d are the width, height, and depth of the feature map outputted from the last layer, respectively; w represents the shared weights; and b is the bias.

Finally, the Softmax layer generates a probability distribution over the six classes using the output from the second fully connected layer as its input. The highest value of the output vector of the Softmax is considered the correct index type for the rock images.

3. Rock-Type Classification Method for Field Images of Rocks

The main steps for classifying field samples are acquiring images, collecting typical rock-type images, establishing databases of rock-type images, setting up deep learning neural networks, and identifying rock types (Figure 4).

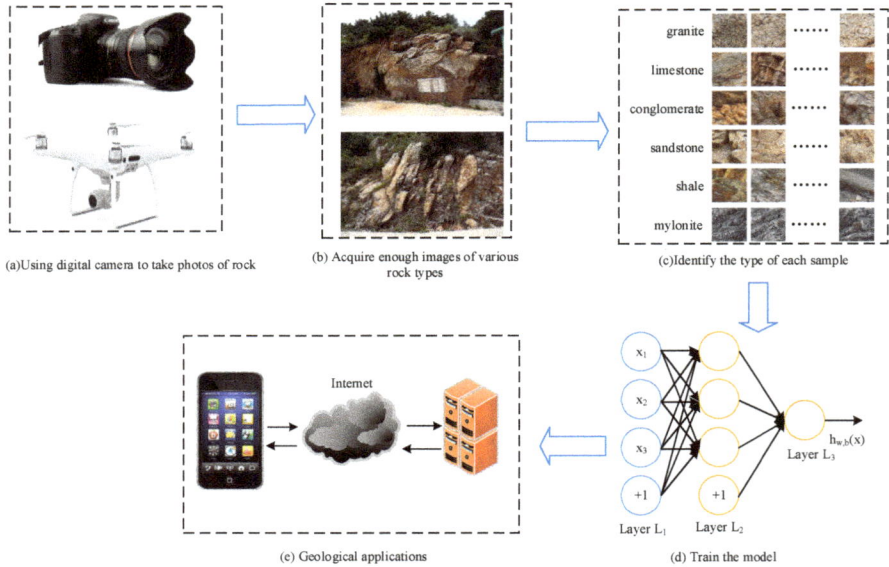

Figure 4. Whole flow chart for the automated identification of field rock types. (**a**) Cameras: Canon EOS 5D Mark III (above) and a Phantum 4 Pro DJi UAV with FC300C camera (below). (**b**) Rock images obtained from outcrops. (**c**) Cutting images (512 × 512 pixels) of marked features from the originals. (**d**) Rock-type identification training using CNNs. (**e**) Application of the trained model to related geological fields.

3.1. Acquisition of Original Field Rock Images

The Xingcheng Practical Teaching Base of Jilin University in Xingcheng (southwest Liaoning Province in NE China) was the field site for the collection of rock images. The site is situated in Liaodong Bay and borders the Bohai Sea. There are various types of rock with good outcrops in this area, mainly granite, tuff and other magmatic rocks, limestone, conglomerate, sandstone, and shale and other sedimentary rocks as well as some mylonite. This diverse geological environment enables the collected images to be used to test the reliability and consistency of the classification method.

The development of UAVs has led to their use in geological research [31–33], as they allow image acquisition to take place in inaccessible areas. As part of this study's objective of obtaining as many photographs of surface rocks as possible, a UAV carrying a camera captured images of many of the better outcrops of rocks on cliffs and in other unapproachable areas. Two cameras were used: a Canon EOS 5D Mark III (EF 24–70 mm F2.8L II USM) was used to take photographs (5760 × 3840 pixels) of outcrops that field geologists could access, and a Phantum 4 Pro DJi UAV with FC300C camera (FOV 84°8.8 mm/24 mm f/2.8–f/11 with autofocus) captured images (4000 × 3000 pixels) of inaccessible outcrops.

Figure 5 shows typical images of the six rock types. There are clear differences in grain size distribution, structure, and color between the rocks, allowing them to be distinguished. However, weathering and other factors in the field can significantly affect the color of sedimentary rocks, for example, which increases the complexity of rock-type identification in the field.

The photographic image capture used different subject distances and focal lengths for different rock types to best capture their particular features. For example, for conglomerates with large grains, the subject distance was 2.0 m, and the focal length was short (e.g., 20 mm), so that the structural characteristics of these rocks could be recorded. For sandstones with smaller grains, the subject distance was 0.8 m with a longer focal length (e.g., 50 mm), allowing the grains to be detectable.

Figure 5. The six types of rock in the field: (**a**) mylonite, (**b**) granite, (**c**) conglomerate, (**d**) sandstone, (**e**) shale, and (**f**) limestone.

A total of 2290 images with typical rock characteristics of six rock types were obtained: 95 of mylonite, 625 of granite, 530 of conglomerate, 355 of sandstone, 210 of shale, and 475 of limestone. These six rock types include four sedimentary rocks (conglomerate, sandstone, shale, and limestone), one metamorphic rock (mylonite), and one igneous rock (granite). After every three samples, one sample was selected as the validation date, and then another sample as selected as the testing data, so 60% of the images of each rock type were selected for the training dataset, 20% for the validation dataset, and leaving 20% for the testing dataset (Table 2).

Table 2. Numbers of original field rock images.

Type	Training Dataset	Validation Dataset	Number of Testing Data
Mylonite	57	19	19
Granite	375	125	125
Conglomerate	318	106	106
Sandstone	213	71	71
Shale	126	42	42
Limestone	285	95	95
Total	1374	458	458

3.2. Preprocessing Field Rock Image Data

In the field, a variety of features may obscure rocks or otherwise detract from the quality of rock images obtained. Grass, water, and soil commonly appear in the collected images (e.g., area A in Figure 6). These features hinder recognition accuracy and consume computing resources. In addition, any image of a three-dimensional rock outcrop will contain some areas that are out of focus and which cannot therefore be seen clearly or properly analyzed (e.g., area B in Figure 6). Furthermore, if the captured image is directly used for training, then the image size of 5760×3840 pixels consumes large amounts of computing resources. Therefore, before training the model, it is necessary to crop the original image into sample patches without the interfering elements, thus reducing the total size of imagery used in the analysis.

The color, mineral composition, and structure of a rock are the basic features for identifying its type. These features have to be identifiable in the cropped images. The original images (of either 5760×3840 pixels or 4000×3000 pixels) are first labeled according to the clarity of the rock and are then cropped into a variable number of sample patches of 512×512 pixels (e.g., boxes 1–7 in Figure 6), before being compressed to 128×128 pixels. Labeling is performed manually and is based on the open-source software "LabelImg" [34], a graphical image annotation tool. Cropping is achieved automatically by a python script based on the QT library. The steps used for processing are as follows:

(1) Open the original field rock image;
(2) Label the areas in the image with typical rock features (Figure 6);
(3) Save the current annotation, after the labeling operation; and
(4) Read all annotated locations and crop the annotated image locations to the specified pixel size for the sample patches.

After the above-mentioned steps, the sample patch images were separated into a training dataset containing 14,589 samples (60% of the total), a validation dataset of 4863 images (20% of the total) and a testing dataset of 4863 images (20% of the total). Table 3 gives the specific distribution of training, validation and testing images across rock types. Using sample patches retains the best representation of rock features and benefits the training of the RTCNNs model.

Figure 6. The extraction of typical rock samples from high-resolution images. Two or more image samples (512 × 512 pixels) are cropped from an original field rock image of 5760 × 3840 pixels. Area A is identified as vegetation cover, and area B is out of focus. Boxes 1–7 are manually labeled as sample patch images.

Table 3. Datasets for image classification of field rocks.

Type	Training Data	Validation Data	Testing Data
Mylonite	1584	528	528
Granite	3753	1251	1251
Conglomerate	3372	1124	1124
Sandstone	2958	986	986
Shale	1686	562	562
Limestone	1236	412	412
Total	14589	4863	4863

3.3. Training the Model

3.3.1. Software and Hardware Configurations

As the RTCNNs model has fewer layers than VGGNet-16 and other models, the computations were carried out on laptops. Table 4 gives the detailed hardware and software specifications. The RTCNNs model was realized under the TensorFlow deep learning framework [28].

Table 4. Software and hardware configurations.

Configuration Item	Value
Type and specification	Dell Inspiron 15-7567-R4645B
CPU	Intel Core i5-7300HQ 2.5 GHz
Graphics Processor Unit	NVIDIA GeForce GTX 1050Ti with 4GB RAM
Memory	8 GB
Hard Disk	1 TB
Solid State Disk	120 GB
Operating System	Windows 10 Home Edition
Python	3.5.2
Tensorflow-gpu	1.2.1

3.3.2. Experimental Results

Training employs random initial weights. After each batch of training is complete, the learning rate changes and the weights are constantly adjusted to find the optimal value, which decreases the loss value of training. After each epoch, the trained parameters are saved in files and used to evaluate the validation dataset and obtain the identification accuracy of each epoch. After 200 epochs, the training loss gradually converged to the minimum. The trained parameters trained after 200 epochs are used to evaluate the testing dataset and obtain the identification accuracy. 10 identical experiments are established totally Figure 7 shows the average loss and accuracy curves for the training and validation datasets from the model using sample patch images of 128 × 128 pixels in the same 10 experiments. The curves show that the model has good convergence after 50 training epochs, with the loss value being below 1.0, and the training accuracy being 95.7%, validation accuracy achieved 95.4%. The highest accuracy of training and validation achieved was 98.6% and 98.2% at 197th epoch. After 200 training epochs, the final training and validation accuracy of the model reached 98.5% and 98.0% respectively. The saved parameters at 197th epoch with the highest validation accuracy was used to test the testing dataset, and the confusion matrix was gained (Table 5). Finally, the testing accuracy achieved was 97.96%.

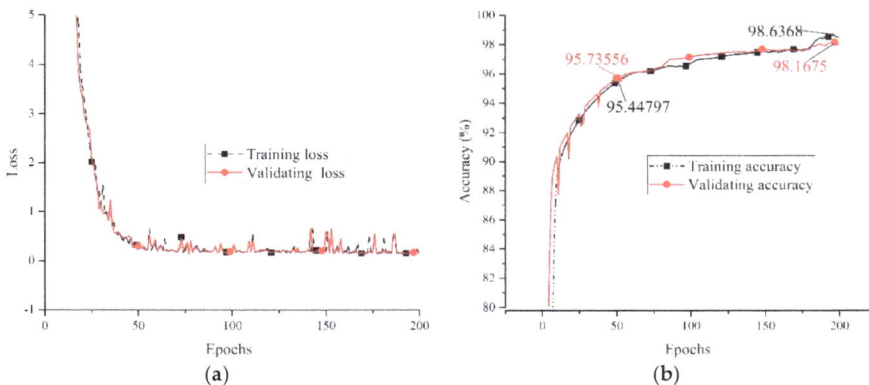

Figure 7. Average loss (**a**) and accuracy curves (**b**) for the training and validation dataset using samples of 128 × 128 pixels in 10 experiments.

The confusion matrix in Table 5 shows that the RTCNNs model can effectively classify mylonite, but is less effective in classifying sandstone and limestone, which yielded error rates of 4.06% and 3.4%, respectively.

Table 5. Confusion matrix of the RTCNNs model based on the testing dataset.

Predicted / Actual	Mylonite	Granite	Conglomerate	Sandstone	Shale	Limestone	Error Rate
mylonite	528	0	0	0	0	0	0.00%
granite	0	1221	6	18	4	2	2.40%
conglomerate	0	0	1114	2	2	6	0.89%
sandstone	5	16	2	946	2	15	4.06%
shale	0	0	2	3	557	0	0.89%
limestone	2	0	4	8	0	398	3.4%

The sample images in Figure 8 show sandstone (a and b) and limestone (c and d) incorrectly classified as granite, limestone, conglomerate, and sandstone, respectively. These samples have similar characteristics to the predicted rock types and are thus misclassified. For example, the grain size, texture, and shape of minerals in the sandstone in (a) are similar to those of minerals in granite.

(a) granite   (b)limestone

(c) conglomerate   (d) sandstone

Figure 8. Samples that were incorrectly classified: (**a**,**b**) sandstone classified as granite and limestone, respectively; (**c**,**d**) limestone classified as conglomerate and sandstone, respectively.

4. Discussion

The identification of rock type from field images is affected by many factors. The choice of model, the size of training images, and the training parameters used will all influence training accuracy. This section reports and discusses various comparative tests and related results.

4.1. Influence of Model Choice on Recognition Accuracy

To test the effectiveness of classification, the RTCNNs model's performance was compared with three other learning models (SVM, AlexNet, GoogLeNet Inception v3, and VGGNet-16) using the same training and testing datasets. All models were trained in 200 epochs using the batch size parameters listed in Table 6. The linear SVM classifier was applied to the datasets to test the performance using the super parameters listed in Table 6. Three other existing models, AlexNet, GoogLeNet Inception v3, and VGGNet-16, were also run using transfer learning, with initial learning rates of 0.01, 0.01, and 0.001, respectively (Table 6). During transfer learning, all the convolution and pooling layers of each model are frozen, and the trainings are conducted only for the fully-connected layers. For AlexNet model, the final FC6, FC7, and FC8 layers are trained. While training the GoogLeNet Inception V3 model, the final FC layer is trained. For VGGNet-16 model, the final FC7 and FC8 layers are trained.

The experimental results show that the RTCNNs model proposed in the present study achieved the highest overall accuracy (97.96%) on the testing dataset. Given that the same training and testing images were used for each model, we ascribe this high accuracy mainly to the proposed CNNs model. The next best performing model was GoogLeNet Inception v3, which obtained an overall accuracy of 97.1% with transfer learning. Although the overall testing accuracy of RTCNNs model is only 0.86% higher than that of GoogLeNet Inception V3 model, it leads to 42 more images identified by RTCNNs model than by GoogLeNet Inception V3 model. When identifying larger dataset, the advantage of RTCNNs model will be more obvious. Meanwhile, the results show that the CNNs model outperforms the linear SVM model in terms of classifying rocks from field images.

In addition, the RTCNNs model has fewer layers than the other models, meaning it is less computationally expensive and can be easily trained on common hardware (see Section 3.3.1). It also requires less time for training than the other deep learning models (Table 6).

Table 6. Recognition performance and related parameters.

Method	Accuracy (%)	Batch Size	Initial Learning Rate	Computing Time
SVM	85.5	200	0.001	3:32:20
AlexNet	92.78	128	0.01	4:49:28
GoogLeNet Inception v3	97.1	100	0.01	7:12:53
VGGNet-16	94.2	100	0.001	5:18:42
Our present study	97.96	16	0.03	4:41:47

4.2. The Effect of Sample Patch Images' Size on Rock-Type Identification

The sample patch images preserve those rock features (e.g., structure, mineral composition, and texture) that are most important to its identification. To test the influence of the size of sample patch images on the accuracy of rock identification, we compressed the sample patches from 512×512 pixels to 32×32, 64×64, 128×128, and 256×256 pixels and compared the results under otherwise identical conditions. The results show that using a training dataset with patches of 128×128 pixels achieved the best performance (Figure 9).

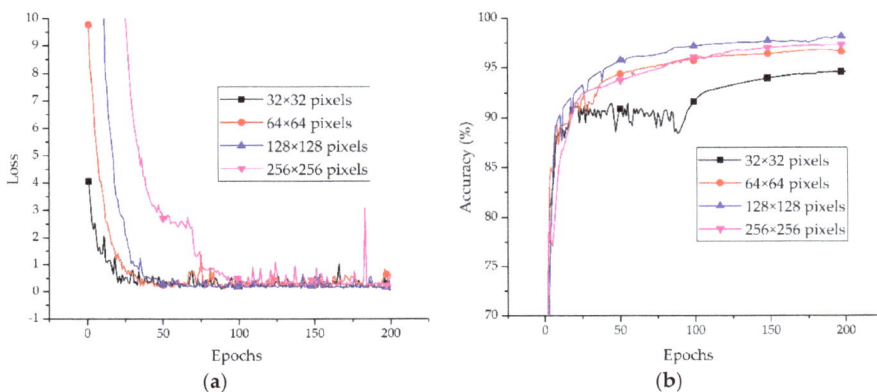

Figure 9. (**a**) Validation loss and (**b**) validation accuracy curves for four sample patch image sizes.

4.3. The Effect of Model Depth on Identification Accuracy

Many previous studies have established that increasing the depth of a model improves its recognition accuracy. Two modifications to the proposed model with different depths are shown in Figure 10; Figure 11 plots the performance accuracy of the two modified models and of the original model.

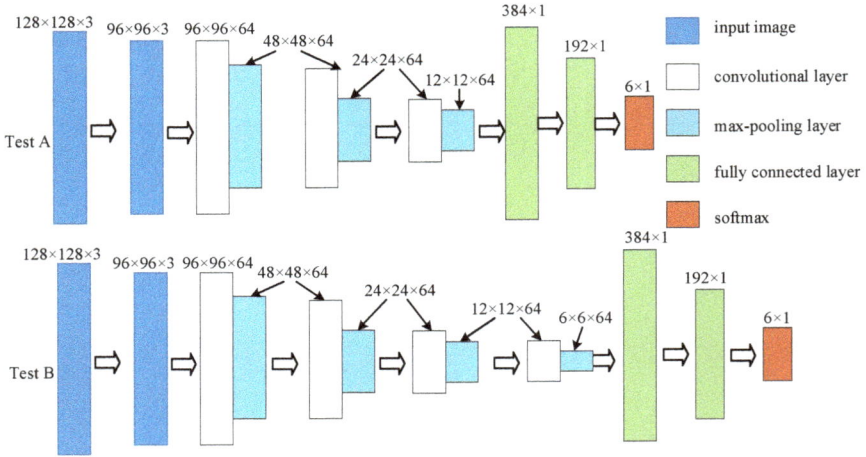

Figure 10. Schematics of two modifications to the proposed model by introducing additional layers. Test A uses one additional convolution layer and one additional pooling layer. Test B has two additional layers of each type.

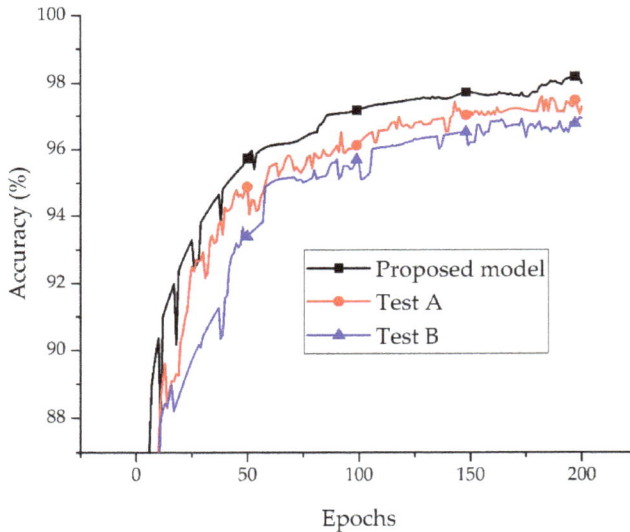

Figure 11. Validation accuracy curves for three models with different depths. The two models Test A and Test B are described in Figure 10 and its caption.

The results of the comparison show that increasing the depth of the model (model Test A and Test B) does not improve the accuracy of recognition/identification in the present case; in fact, increasing the depth reduces such identification (Figure 11). We infer that the feature extraction operation of the proposed CNNs for rock image recognition does not require additional levels, with the convolution operation at a deeper level serving only to lose features and cause classification errors.

5. Conclusions

The continuing development of CNNs has made them suitable for application in many fields. A deep CNNs model with optimized parameters is proposed here for the accurate identification of rock types from images taken in the field. Novelly, we sliced and patched the original obtained photographic images to increase their suitability for training the model. The sliced samples clearly retain the relevant features of the rock and augment the training dataset. Finally, the proposed deep CNNs model was trained and tested using 24,315 sample rock image patches and achieved an overall accuracy of 97.96%. This accuracy level is higher than those of established models (SVM, AlexNet, VGNet-16, and GoogLeNet Inception v3), thereby signifying that the model represents an advance in the automated identification of rock types in the field. The identification of rock type using a deep CNN is quick and easily applied in the field, making this approach useful for geological surveying and for students of geoscience. Meanwhile, the method of identifying rock types proposed in the paper can be applied to the identification of other textures after retraining the corresponding parameters, such as rock thin section images, sporopollen fossil images and so on.

Although CNNs have helped to identify and classify rock types in the field, some challenges remain. First, the recognition accuracy still needs to be improved. The accuracy of 97.96% achieved using the proposed model meant that 99 images were misidentified in the testing dataset. The model attained relatively low identification accuracy for sandstone and limestone, which is attributed to the small grain size and similar colors of these rocks (Table 5; Figure 8). Furthermore, only a narrow range of sample types (six rock types overall) was considered in this study. The three main rock groups (igneous, sedimentary, and metamorphic) can be divided into hundreds of types (and subtypes) according to mineral composition. Therefore, our future work will combine the deep learning model with a knowledge library, containing more rock knowledge and relationships among different rock types, to classify more rock types and improve both the accuracy and the range of rock-type identification in the field. In addition, each field photograph often contains more than one rock type, but the proposed model can classify each image into only one category, stressing the importance of the quality of the original image capture.

Our future work will aim to apply the trained model to field geological surveying using UAVs, which are becoming increasingly important in geological data acquisition and analysis. The geological interpretation of these high-resolution UAV images is currently performed mainly using manual methods, and the workload is enormous. Therefore, the automated identification of rock types will greatly increase the efficiency of large-scale geological mapping in areas with good outcrops. In such areas (e.g., western China), UAVs can collect many high-resolution outcrop images, which could be analyzed using the proposed method to assist in both mapping and geological interpretation while improving efficiency and reducing costs. In order to improve the efficiency of labeling, the feature extraction algorithm [35] will be studied to automatically extract the advantageous factors in the image. We also plan to apply other deep learning models, such as the state-of-art Mask RCNN [36], to identify many types of rock in the same image. In addition, we will study various mature optimization algorithms [37–39] to improve computing efficiency. These efforts should greatly improve large-scale geological mapping and contribute to the automation of mapping.

Author Contributions: Conceptualization, X.R. and L.X.; Data curation, Z.L.; Formal analysis, X.R.; Funding acquisition, L.X.; Investigation, X.R. and L.X.; Methodology, X.R. and L.X.; Project administration, L.X.; Resources, Z.L. and X.S.; Software, X.R. and Y.Z.; Supervision, L.X.; Validation, Y.Z. and J.H.; Visualization, X.R.; Writing—Original draft, X.R.; Writing—Review & Editing, L.X.

Funding: This research was funded by the China Geological Survey, grant number 1212011220247, and Department of Science and Technology of Jilin Province, grant number 20170201001SF, and the Education Department of Jilin Province, grant number JJKH20180161KJ and JJKH20180518KJ.

Acknowledgments: The authors are grateful for anonymous reviewers' hard work and comments that allowed us to improve the quality of this paper. The authors would like to thank Gaige Wang for discussions and suggestions. The authors wish to acknowledge the Xingcheng Practical Teaching Base of Jilin University for providing the data for this project.

Conflicts of Interest: The authors declare no conflict of interest.

References

1. Lepistö, L.; Kunttu, I.; Visa, A. Rock image classification using color features in Gabor space. *J. Electron. Imaging* **2005**, *14*, 040503. [CrossRef]
2. Chatterjee, S. Vision-based rock-type classification of limestone using multi-class support vector machine. *Appl. Intell.* **2013**, *39*, 14–27. [CrossRef]
3. Lepistö, L.; Kunttu, I.; Autio, J.; Visa, A. Rock image retrieval and classification based on granularity. In Proceedings of the 5th International Workshop on Image Analysis for Multimedia Interactive Services, Lisboa, Portugal, 21–23 April 2004.
4. Patel, A.K.; Chatterjee, S. Computer vision-based limestone rock-type classification using probabilistic neural network. *Geosci. Front.* **2016**, *7*, 53–60. [CrossRef]
5. Ke, L.; Gong, D.; Meng, F.; Chen, H.; Wang, G.G. Gesture segmentation based on a two-phase estimation of distribution algorithm. *Inf. Sci.* **2017**, *394*, 88–105.
6. Lepistö, L.; Kunttu, I.; Autio, J.; Visa, A. Rock image classification using non-homogenous textures and spectral imaging. In Proceedings of the 11th International Conference in Central Europe on Computer Graphics, Visualization and Computer Vision 2003 (WSCG 2003), Plzen-Bory, Czech Republic, 3–7 February 2003; pp. 82–86.
7. Partio, M.; Cramariuc, B.; Gabbouj, M.; Visa, A. Rock texture retrieval using gray level co-occurrence matrix. In Proceedings of the 5th Nordic Signal Processing Symposium (NORSIG-2002), Trondheim, Norway, 4–10 October 2002; pp. 4–7.
8. Sharma, A.; Liu, X.; Yang, X.; Shi, D. A patch-based convolutional neural network for remote sensing image classification. *Neural Netw.* **2017**, *95*, 19–28. [CrossRef]
9. Nogueira, K.; Penatti, O.A.B.; dos Santos, J.A. Towards better exploiting convolutional neural networks for remote sensing scene classification. *Pattern Recognit.* **2017**, *61*, 539–556. [CrossRef]
10. Perez, C.A.; Saravia, J.A.; Navarro, C.F.; Schulz, D.A.; Aravena, C.M.; Galdames, F.J. Rock lithological classification using multi-scale Gabor features from sub-images, and voting with rock contour information. *Int. J. Miner. Process.* **2015**, *144*, 56–64. [CrossRef]
11. Liu, B.; Zhang, Y.; He, D.; Li, Y. Identification of Apple Leaf Diseases Based on Deep Convolutional Neural Networks. *Symmetry* **2017**, *10*, 11. [CrossRef]
12. Guo, Y.; Liu, Y.; Oerlemans, A.; Lao, S.; Wu, S.; Lew, M.S. Deep learning for visual understanding: A review. *Neurocomputing* **2016**, *187*, 27–48. [CrossRef]
13. Lecun, Y.; Bengio, Y.; Hinton, G. Deep learning. *Nature* **2015**, *521*, 436. [CrossRef]
14. Krizhevsky, A.; Sutskever, I.; Hinton, G.E. ImageNet classification with deep convolutional neural networks. *Commun. ACM* **2012**, *60*, 2012. [CrossRef]
15. Yang, C.; Li, W.; Lin, Z. Vehicle Object Detection in Remote Sensing Imagery Based on Multi-Perspective Convolutional Neural Network. *Int. J. Geo-Inf.* **2018**, *7*, 249. [CrossRef]
16. Han, S.; Ren, F.; Wu, C.; Chen, Y.; Du, Q.; Ye, X. Using the TensorFlow Deep Neural Network to Classify Mainland China Visitor Behaviours in Hong Kong from Check-in Data. *Int. J. Geo-Inf.* **2018**, *7*, 158. [CrossRef]
17. Sargano, A.; Angelov, P.; Habib, Z. A Comprehensive Review on Handcrafted and Learning-Based Action Representation Approaches for Human Activity Recognition. *Appl. Sci.* **2017**, *7*, 110. [CrossRef]
18. Sainath, T.N.; Kingsbury, B.; Saon, G.; Soltau, H.; Mohamed, A.R.; Dahl, G.; Ramabhadran, B. Deep Convolutional Neural Networks for large-scale speech tasks. *Neural Netw. Off. J. Int. Neural Netw. Soc.* **2015**, *64*, 39. [CrossRef] [PubMed]
19. Noda, K.; Yamaguchi, Y.; Nakadai, K.; Okuno, H.G.; Ogata, T. Audio-visual speech recognition using deep learning. *Appl. Intell.* **2015**, *42*, 722–737. [CrossRef]
20. Lv, Y.; Duan, Y.; Kang, W.; Li, Z.; Wang, F.Y. Traffic Flow Prediction with Big Data: A Deep Learning Approach. *IEEE Trans. Intell. Transp. Syst.* **2015**, *16*, 865–873. [CrossRef]
21. Sermanet, P.; Lecun, Y. Traffic sign recognition with multi-scale Convolutional Networks. *Int. Jt. Conf. Neural Netw.* **2011**, *7*, 2809–2813.
22. Zhang, Y.C.; Kagen, A.C. Machine Learning Interface for Medical Image Analysis. *J. Digit. Imaging* **2017**, *30*, 615–621. [CrossRef]

23. Alipanahi, B.; Delong, A.; Weirauch, M.T.; Frey, B.J. Predicting the sequence specificities of DNA- and RNA-binding proteins by deep learning. *Nat. Biotechnol.* **2015**, *33*, 831–838. [CrossRef]

24. Lu, Y.; Yi, S.; Zeng, N.; Liu, Y.; Zhang, Y. Identification of rice diseases using deep convolutional neural networks. *Neurocomputing* **2017**, *267*, 378–384. [CrossRef]

25. Perol, T.; Gharbi, M.; Denolle, M. Convolutional Neural Network for Earthquake Detection and Location. *Sci. Adv.* **2017**, *4*, 1700578. [CrossRef]

26. Zhang, Y.; Li, M.; Han, S. Automatic identification and classification in lithology based on deep learning in rock images. *Acta Petrol. Sin.* **2018**, *34*, 333–342.

27. Cheng, G.; Guo, W.; Fan, P. Study on Rock Image Classification Based on Convolution Neural Network. *J. Xi'an Shiyou Univ. (Nat. Sci.)* **2017**, *4*, 116–122. [CrossRef]

28. Inc, G. TensorFlow. Available online: https://www.tensorflow.org/ (accessed on 5 August 2018).

29. Ferreira, A.; Giraldi, G. Convolutional Neural Network approaches to granite tiles classification. *Expert Syst. Appl.* **2017**, *84*, 1–11. [CrossRef]

30. Boureau, Y.L.; Ponce, J.; Lecun, Y. A Theoretical Analysis of Feature Pooling in Visual Recognition. In Proceedings of the International Conference on Machine Learning, Haifa, Israel, 21–24 June 2010; pp. 111–118.

31. Blistan, P.; Kovanič, Ľ.; Zelizňaková, V.; Palková, J. Using UAV photogrammetry to document rock outcrops. *Acta Montan. Slovaca* **2016**, *21*, 154–161.

32. Vasuki, Y.; Holden, E.J.; Kovesi, P.; Micklethwaite, S. Semi-automatic mapping of geological Structures using UAV-based photogrammetric data: An image analysis approach. *Comput. Geosci.* **2014**, *69*, 22–32. [CrossRef]

33. Zheng, C.G.; Yuan, D.X.; Yang, Q.Y.; Zhang, X.C.; Li, S.C. UAVRS Technique Applied to Emergency Response Management of Geological Hazard at Mountainous Area. *Appl. Mech. Mater.* **2013**, *239*, 516–520. [CrossRef]

34. Tzutalin LabelImg. Git code (2015). Available online: https://github.com/tzutalin/labelImg (accessed on 5 August 2018).

35. Zhang, Y.; Song, X.F.; Gong, D.W. A Return-Cost-based Binary Firefly Algorithm for Feature Selection. *Inf. Sci.* **2017**, *418*, 567–574. [CrossRef]

36. He, K.; Gkioxari, G.; Dollar, P.; Girshick, R. Mask R-CNN. In Proceedings of the IEEE International Conference on Computer Vision, Venice, Italy, 22–29 October 2017; p. 1.

37. Rizk-Allah, R.M.; El-Sehiemy, R.A.; Wang, G.G. A novel parallel hurricane optimization algorithm for secure emission/economic load dispatch solution. *Appl. Soft Comput.* **2018**, *63*, 206–222. [CrossRef]

38. Zhang, Y.; Gong, D.W.; Cheng, J. Multi-Objective Particle Swarm Optimization Approach for Cost-Based Feature Selection in Classification. *IEEE/ACM Trans. Comput. Biol. Bioinform.* **2017**, *14*, 64–75. [CrossRef]

39. Wang, G.G.; Gandomi, A.H.; Alavi, A.H. An effective krill herd algorithm with migration operator in biogeography-based optimization. *Appl. Math. Model.* **2014**, *38*, 2454–2462. [CrossRef]

Article

An Adaptive Multi-Swarm Competition Particle Swarm Optimizer for Large-Scale Optimization

Fanrong Kong [1,2,3], Jianhui Jiang [1,*] and Yan Huang [2,3]

[1] School of Software Engineering, Tongji University, Shanghai 201804, China; kfr@ssc.stn.sh.cn
[2] Shanghai Development Center of Computer Software Technology, Shanghai 201112, China; huangy@ssc.stn.sh.cn
[3] Shanghai Industrial Technology Institute, Shanghai 201206, China
[*] Correspondence: jhjiang@tongji.edu.cn

Received: 25 March 2019; Accepted: 4 June 2019; Published: 6 June 2019

Abstract: As a powerful tool in optimization, particle swarm optimizers have been widely applied to many different optimization areas and drawn much attention. However, for large-scale optimization problems, the algorithms exhibit poor ability to pursue satisfactory results due to the lack of ability in diversity maintenance. In this paper, an adaptive multi-swarm particle swarm optimizer is proposed, which adaptively divides a swarm into several sub-swarms and a competition mechanism is employed to select exemplars. In this way, on the one hand, the diversity of exemplars increases, which helps the swarm preserve the exploitation ability. On the other hand, the number of sub-swarms adaptively changes from a large value to a small value, which helps the algorithm make a suitable balance between exploitation and exploration. By employing several peer algorithms, we conducted comparisons to validate the proposed algorithm on a large-scale optimization benchmark suite of CEC 2013. The experiments results demonstrate the proposed algorithm is effective and competitive to address large-scale optimization problems.

Keywords: particle swarm optimization; large-scale optimization; adaptive multi-swarm; diversity maintenance

---

1. Introduction

Particle swarm optimization (PSO), as an active tool in dealing with optimization problems, have been widely applied to various kinds of optimization problems [1,2], such as industry, transportation, economics and so forth. Especially, in current years, with the rising development of the industrial network, many optimization problems with complex properties of nonlinearity, multi-modularity, large-scale and so on are proposed. To well address these problems, PSO exhibits powerful abilities to meet the requirement of practical demands and thus gains much attention in deep research and further applications [3,4]. However, according to the current research, there exists a notorious problem in PSO: the algorithm lacks the ability in diversity maintenance and therefore the solutions trend to local optima, especially for complex optimization problems. In the algorithm design, a large weight on diversity maintenance will break down the convergence process, while a small weight on diversity maintenance will cause a poor ability to get rid of local optima. Therefore, as a crucial aspect in the design of PSO, population diversity maintenance has been a hot issue and drawn much attention globally.

To improve PSO's ability in population diversity maintenance, many works are conducted, which can be generally categorized into the following aspects: (1) mutation strategy [5–8]; (2) distance-based indicator [9–11]; (3) adaptive control strategy [12–14]; (4) multi-swarm strategy [15–21]; and (5) hybridization technology [22–24]. However, in the five categories, there exists many parameters and laws in the design of PSO. Improper parameters or laws will negatively affect an algorithm's performance. To provide an easy tool for readers in the design of PSO, Spolaor [25] and Nobile [26] proposed reboot strategies based PSO and fuzzy self-tuning PSO, respectively, to provide a simple way to tube PSO parameters. However, for large-scale optimization, it is also a challenge for PSO implementation. To address this issue, in this paper, we propose a novel algorithm named adaptive multi-swarm competition PSO (AMCPSO), which randomly divides the whole swarm into several sub-swarms. In each sub-swarm, a competitive mechanism is adopted to select a winner particle who attracts the poor particles in the same sub-swarm. The number of winners, namely exemplars, is equal to the number of sub-swarms. In multi-swarm strategy, each exemplar is selected from each sub-swarm. In this way, the diversity of exemplars increases, which is helpful for algorithms to eliminate the effects of local optima.

The contributions of this paper are listed as follows. First, in this proposed algorithm, the exemplars are all selected from the current swarm, rather than personal historical best positions. In this way, no historical information is needed in the design of the algorithm. Second, we consider different optimization stages and propose a law to adaptively adjust the size of sub-swarms. For an optimization process, in the beginning, exploration ability is crucial for algorithms to explore searching space. To enhance algorithm's ability in exploration, we set a large number of sub-swarms, which means a large number of exemplars will be obtained. On the other hand, in the late stage, algorithms should pay more attention to exploitation for enhancing the accuracy of optimization results. Hence, a small number of exemplars is preferred. Third, to increase the ability of diversity maintenance, we do not employ the global best solution, but the mean value of each sub-swarm, which is inspired by Arabas and Biedrzycki [27]. According to Arabas and Biedrzycki [27], the quality of mean value has a higher probability to be better than that of a global best solution. In addition, because the mean value is calculated by the whole swarm, it has a high probability to be updated, unlike the global best solution, which is consistent for several generations.

The rest of this paper is organized as follows. Section 2 introduces related work on diversity maintenance for PSO and some comments and discussions on the works are presented. In Section 3, we design an adaptive multi-swarm competition PSO. The algorithm structure and pseudo-codes are presented in the section. We employed several state-of-the-art algorithms to compare the proposed algorithm on large-scale optimization problems and analyzed the performance, which is presented in Section 4. Finally, we end this paper with conclusions and present our future work in Section 5.

2. Related Work

In standard PSO, each particle has two attributes: velocity and position. The update mechanism for the two attributes is given in Equation (1).

$$v_i(t+1) = \omega v_i(t) + r_1 * c_1(pbest_i - p_i(t)) + r_2 * c_2(gbest - p_i(t)) \tag{1}$$
$$p_i(t+1) = p_i(t) + v_i(t) \tag{2}$$

where v_i and p_i are the velocity and position of ith particle, t is the index of generation, $pbest_i$ is the best position that particle i found thus far, and $gbest$ is the global best position. r_1 and $r_2 \in [0,1]^D$ are two random values and c_1 and c_2 are the acceleration coefficients. On the right side of the velocity update equation, there are three components. The first one is called inertia term, which retains the particle's own property, e.g. its current velocity. The weight of the property is controlled by ω named inertia weight. The second component and the third component are called cognitive component and social component, respectively. The two components guide the ith particle to move towards better positions. Since $gbest$ guides the whole swarm, if it is in local optimal position, the standard PSO has a poor ability to get rid of it. Moreover, the algorithm gets premature convergence.

To improve PSO's ability in diversity maintenance, many works are conducted. As mentioned in Section 1, there are generally five ways. The first is about the mutation operator. In most canonical swarm intelligence, there is no mutation operator. To pursue a better performance, several kinds of mutation, such as Gaussian mutation, wavelet mutation, etc., are implemented to SI [6,7] so that the population diversity can be regained to some extent. In [5], Sun et al. defined a lower distance-based limit. Once the diversity is below the limit, a mutation happens on the current global best individual. Wang proposed a Gauss mutation PSO in [8] to help the algorithm retain population diversity. Nevertheless, mutation rate and mutation degree are difficult to predefine. A small mutation rate/degree plays a weak role to increase population diversity, while a large rate/degree is harmful to the algorithm's convergence, as mentioned in [28].

The second way is based on the distance-based indicator. The diversity maintenance is considered as a compensation term. During the optimization process, the distances among particles are continuously monitored. By predefining a limit, if the diversity situation is worse than the threshold, some strategies are activated to increase the distance. In [9], the authors used criticality to depict the diversity status of a swarm. During the optimization process, if the criticality value is larger than a predefined threshold, a relocating strategy is activated to disperse the particles. In this way, particles will not be too close. Inspired from the electrostatics, Blackwell and Bentley endowed particles with a new attribute, called charging status [10]. For two charged particles, an electrostatic reaction is launched to modulate their velocities. The same authors presented a similar idea in a follow up work [11]. The big challenge in this kind of method is the difficulty to predefine a suitable threshold as for the requirement of diversity is different for different problems or different optimization phases.

The third method to increase swarm diversity is based on an adaptive way. According to the authors of [12,13], the parameters in the velocity and position update mechanism play different roles during the whole optimization process. At the beginning phase, exploration helps the swarm explore the searching space, while exploitation occupies a priority position in the later phase of an optimization process. According to the update mechanism in standard PSO, the values of ω, c_1 and c_2 are influential to the weights of exploitation and exploration. Hence, time-varying weights are employed in [12,13]. In [12], the value of ω decreases from 0.9 to 0.4, which changes the focus from exploration to exploitation. In [13,14], some mathematical functions are empirically established to dynamically and adaptively adjust parameters during the optimization process. However, the performances of these proposed algorithms are very sensitive to the adaptive rules which depend much on authors' experiences. Therefore, it is not easy to apply these methods in real applications.

Hybridization is also an effective way to help integrate the advantages of different meta-heuristics. To improve PSO's ability in diversity maintenance, some works are conducted to prevent particles from searching discovered areas, which is similar to the idea of tabu searching. The prevention methods including deflecting, stretching, repulsion and so forth [22]. A cooperative strategy is employed to improve PSO in [23]. In the proposed algorithm, the essence of the algorithm divides the problem into several subproblems. By solving the sub-problems, final solutions can be integrated by the sub-solutions. Similar ideas are also investigated in [24].

As a feasible and effective way, the multi-swarm strategy is well applied to PSO. Generally, there are two kinds of design in multi-swarm strategy. The first one is the niching strategy. By defining a niche radius, the particles in one niche are considered as similar individuals and only the particles in the different niche will be selected for information recombination [16]. In the standard niching strategy, the algorithm's performance is sensitive to the niche radius [17]. To overcome this problem, Li proposed a parameter-free method in [15]. The author constructed a localized ring topology instead of the niche radius. In addition, in [21], the authors proposed a multi-swarm particle swarm optimizer for large-scale optimization problems. However, the size of each sub-swarm is fixed, which means that the strategy does not consider the swarm size to dynamically manage exploitation and exploration. The second method is to assign different tasks to different sub-swarms. In [18], the authors proposed to regulate population diversity according to predefined value named decreasing rate of the number of sub-swarms. In [20], the authors defined a frequency to switch exploitation and exploration for different sub-swarms. In this way, the algorithm can give considerations to convergence speed and diversity maintenance. In [19], the authors divided the swarm into pairs. By comparing the fitness in pairs, the loser will learn from the winner, which increases the diversity of exemplars. However, the learning efficiency decreases since the learn may happen between two good particles or two bad particles.

However, in the above methods, there is still no research focusing on large-scale optimization problems. To address large-scale optimization problems, there are generally two ways, which are cooperation co-evolution and balancing of exploration and exploitation. For the first kind of methods, a cooperative co-evolution (CC) framework is proposed, which divides a whole dimension into several segments. For the segments, conventional evolutionary algorithms are employed to address them to access sub-solutions. By integrating the sub-solutions, the final solution will be obtained. The framework now has been well applied to various kinds of evolutionary algorithms. Cooperative coevolution differential evolution (DECC), proposed by Omidvar et al., is a series of DE algorithm for large-scale optimization problems. DECC-DG2 first groups the variables considering the interrelationship between them and then optimizes the sub-components using DECC [29]. CBCC3 was proposed by Omidvar for large-scale optimization based on Contribution-Based Cooperative Co-evolution (CBCC). It includes an exploration phase that is controlled by a random method and an exploitation phase controlled by a contribution information based mechanism where the contribution information of a given component is computed based on the last non-zero difference in the objective value of two consecutive iterations [30]. In [31], the authors applied the CC-framework to PSO and obtained satisfactory performance. However, the CC framework will cost huge computational resources in the sub-space division. In the second way to address large-scale optimization problems, researchers propose novel operators of exploitation and exploration. In addition, the balance between the two abilities is also crucial to the final performance. Cheng and Jin proposed SL-PSO (social learning PSO) i [32]. In the algorithm, the authors proposed novel ideas in dealing with the selection of poor particles. However, the algorithm's performance is sensitive to parameter setting.

To summarize, in the current research, the algorithms either lack parameter tuning during the optimization process or aggravate a huge burden to computational resources. To address the problems, in Section 3, we propose an adaptive multi-swarm competition PSO.

3. Adaptive Multi-Swarm Competition PSO

As mentioned in Section 1, the global best position always attracts all particles until it is updated. When the global best position gets trapped into local optima, it is very difficult for the whole swarm to get rid of it. Hence, the optimization progress stagnates or a premature convergence occurs. To overcome this problem, in this paper, we employ an adaptive multi-swarm strategy and propose a novel swarm optimization algorithm termed adaptive multi-swarm competition PSO (AMCPSO). From the beginning to the end of the optimization process, we uniformly and randomly divide the whole swarm into several sub-swarms. In each sub-swarm, the particles compare with each other. The losers will learn from the winner in the same sub-swarm, while the winners do nothing in the generation. The number of the winners is equal to the number of sub-swarms and the number is adaptive to the optimization process. At the beginning stages, a large number of sub-swarms will help algorithm increase the diversity of exemplars and explore the searching space. With the optimization process, the number of sub-swarms decreases in the algorithm's exploitation. On the one hand, for each generation, there are several exemplars, i.e., winners, for losers' updating. Even some exemplars are in local optimal positions, they will not affect the whole swarm. On the other hand, since we divide the swarm randomly, both winners and losers may be different, which means the exemplars are not consistent for many generations. In this way, the ability of diversity maintenance for the whole swarm is improved. Compared with the standard PSO, we abandon the historical information, e.g., the personal historical best position, etc., which is easier for users in implementations. Inspired by Arabas and Biedrzycki [27], the quality of mean value has a higher probability to be better than that of a global best solution. Therefore, we employ the mean position of the whole swarm to attract all particles. On the one hand, the global information can be achieved in particles update. On the other hand, the mean position generally changes for every generation, which has a large probability to avoid local optima. For the proposed algorithm, the velocity and position update mechanisms are given in Equation (3).

$$v_{l\_i}(t+1) = \omega v_{l\_i}(t) + r_1 * c_1(p_w(t) - p_{l\_i}(t)) + r_2 * c_2(p_{mean} - p_{l\_i}(t)) \tag{3}$$

$$p_{l\_i}(t+1) = p_{l\_i}(t) + v_{l\_i}(t) \tag{4}$$

where t records the index of generation, i is the index of losers in each sub-swarm, and v and p are velocity and position, respectively, for each particle. The subscript l means the loser particles, while the subscript w means the winner particle in the same sub-swarm. p_{mean} is employed as global information to depict the mean position of the whole swarm. ω, called the inertia coefficient, is used to control the weight of particle's own velocity property. c_1 and c_2 are cognitive coefficient and social coefficient, respectively. r_1 and $r_2 \in [0, 1]$ are parameters to control the weights of the cognitive component and social component, respectively. In the proposed algorithm, a number of sub-swarms is involved, which also defines the number of exemplars in advance. To help algorithm change the focus from exploration to exploitation, in this paper, we employ an adaptive strategy for the division strategy, which means that the number of sub-swarms are not consistent, but a varying number from a large number of sub-swarms to a small number. The number of the sub-swarms is set according to experimental experience and presented as follows.

$$m = round\left(gs_{ini} - 0.9gs_{ini} * \left(1 - exp\left(-\frac{FEs}{Max\_Gen * Dr}\right)\right)\right); \tag{5}$$

where $round(\Omega)$ is the symbol to round a real value Ω, gs_{ini} is the initial number of sub-swarms, Max_Gen is maximum generation limit, and Dr is used to control the decreasing rate. A small value of Dr causes a sharp decreasing for the number of sub-swarms, while a large value of Dr provides a gentle slope for number decreasing. In Equation (5), the population size of each sub-swarm increases from 5‰ to 5%. For Equation (5), it provides an adaptive way to set the number of sub-swarms, namely m. Meanwhile, the number of exemplars is also adaptive. Considering that, for some generations, the whole population size cannot exact divide m, we randomly select particles with size of the residual and do nothing for them in the generation, which means at most 5% particles are not involved in the update. However, in the algorithm's implementation, we use maximum fitness evaluations as termination condition to guarantee adequate runs for the algorithm. The pseudo-codes of AMCPSO are given in Algorithm 1.

Algorithm 1: Pseudo-codes of adaptive multi-swarm particle swarm optimization.

Input: Number of particles ps, parameters $omega$, c_1, c_2, Maximum number of fitness evaluations Max_FEs

Output: The current global best particle

1 Randomly generate a swarm P and evaluate their fitness f.;
2 *Loop*: Calculate the mean position according to the whole swarm's positions;
3 Calculate m (the size of each sub-swarm) by Equation (5);
4 Randomly select $ps - mod(ps, m)$ particles for update;
5 In each sub-swarm, compare the particles according to their fitness, and select the local best particle by Equation (3);
6 Up date FEs;
7 If $FEs \geq Max\_FEs$, output the current global best particle; Otherwise, goto *Loop*;

For the proposed algorithm, the analysis on the time complexity is presented as follows. According to Algorithm 1, the time complexity for the sub-swarm division is $O(m)$, while the calculation of the mean value in each sub-swarm is also $O(m)$. Since the two operations do not aggravate the burden of computational cost, the main computational cost in the proposed algorithm is the update of each particle, which is $O(mn)$. It is an inevitable cost in most swarm-based evolutionary optimizers. Therefore, the whole time cost of the proposed algorithm is $O(mn)$, where m is the swarm size and n is the search dimensionality.

4. Experiments and Discussions

Experimental Settings

As explained above, large-scale optimization problems demand an algorithm much ability in balancing of diversity maintenance and convergence. To validate the performance of the proposed algorithm, we employed a benchmark suite in CEC 2013 on large-scale optimization problems [33]. In the performance comparisons, four popular algorithms were adopted: CBCC3 [30], DECC-dg [24], DECC-dg2 [29] and SL-PSO [34]. The four algorithms are proposed to address large-scale optimization problems in the corresponding papers. DECC-dg and DECC-dg2 are two improved DECC algorithms with differential evolution strategy [24,29]. SL-PSO was proposed by Cheng [34] to address large-scale optimization problems and achieve competitive performance. For the comparison, we ran each algorithm 25 times to pursue an average performance. The termination condition for each run was set by the maximum number of fitness evaluations (FEs) predefined as 3×10^6. The parameters of the peer algorithms were set as the same as in their reported paper. For the proposed algorithm AMCPSO, we set the parameter as follows. For the gs_{ini}, we set the value as 20, which means the number of sub-swarms increases from 2 to 20. c_1 and c_2 were set as 1 and 0.001. The performances of the five algorithms are presented in Table 1.

Table 1. The experimental results of 1000-dimensional IEEE CEC' 2013 benchmark functions with fitness evaluations of 3×10^6.

Function	Quality	SLPSO	CBCC-3	DECC-dg	DECC2	AMCPSO
F_1	Mean	3.70×10^{-14}	8.65×10^5	2.79×10^6	8.65×10^5	4.55×10^{-2}
	Std	1.44×10^{-15}	2.27×10^4	6.70×10^5	2.27×10^4	4.92×10^{-3}
F_2	Mean	6.70×10^3	1.41×10^4	1.41×10^4	1.41×10^4	2.51×10^3
	Std	4.98×10^1	3.02×10^2	3.03×10^2	3.02×10^2	2.61×10^2
F_3	Mean	2.16×10^1	2.06×10^1	2.07×10^1	2.06×10^1	2.15×10^1
	Std	1.14×10^{-3}	1.69×10^{-3}	2.19×10^{-3}	1.69×10^{-3}	2.16×10^{-3}
F_4	Mean	1.20×10^{10}	3.39×10^7	6.72×10^{10}	2.51×10^8	1.12×10^{10}
	Std	5.54×10^8	3.48×10^6	5.76×10^9	1.89×10^7	1.49×10^9
F_5	Mean	7.58×10^5	2.14×10^6	3.13×10^6	2.74×10^6	4.54×10^5
	Std	2.14×10^4	8.31×10^4	1.23×10^5	5.66×10^4	3.14×10^4
F_6	Mean	1.06×10^6	1.05×10^6	1.06×10^6	1.06×10^6	1.06×10^6
	Std	1.64×10^2	6.61×10^2	3.70×10^2	4.50×10^2	2.19×10^2
F_7	Mean	1.73×10^7	2.95×10^7	3.45×10^8	8.93×10^7	4.40×10^6
	Std	1.49×10^6	5.44×10^6	7.60×10^7	7.16×10^6	6.71×10^5
F_8	Mean	2.89×10^{14}	6.74×10^{10}	1.73×10^{15}	1.01×10^{14}	9.76×10^{13}
	Std	1.75×10^{13}	2.08×10^9	2.78×10^{14}	1.31×10^{13}	1.69×10^{13}
F_9	Mean	4.44×10^7	1.70×10^8	2.79×10^8	3.08×10^8	2.63×10^7
	Std	1.47×10^6	6.20×10^6	1.32×10^7	1.39×10^7	2.58×10^6
F_{10}	Mean	9.43×10^7	9.28×10^7	9.43×10^7	9.44×10^7	9.11×10^7
	Std	3.99×10^4	1.40×10^5	6.45×10^4	5.82×10^4	6.41×10^4
F_{11}	Mean	9.98×10^9	7.70×10^8	1.26×10^{11}	9.93×10^9	4.08×10^8
	Std	1.82×10^9	6.80×10^7	2.44×10^{10}	3.26×10^8	2.83×10^7
F_{12}	Mean	1.13×10^3	5.81×10^7	5.89×10^7	5.81×10^7	1.60×10^3
	Std	2.12×10^1	1.53×10^7	2.75×10^6	1.53×10^6	1.72×10^2
F_{13}	Mean	2.05×10^9	6.03×10^8	1.06×10^{10}	6.03×10^8	3.88×10^8
	Std	2.13×10^8	2.69×10^7	7.94×10^8	2.69×10^7	4.21×10^7
F_{14}	Mean	1.60×10^{10}	1.11×10^9	3.69×10^{10}	1.11×10^9	9.36×10^8
	Std	1.62×10^9	2.10×10^8	6.58×10^9	2.10×10^8	8.14×10^7
F_{15}	Mean	6.68×10^7	7.11×10^6	6.32×10^6	7.11×10^6	6.27×10^7
	Std	1.01×10^6	2.70×10^5	2.69×10^5	2.70×10^5	7.04×10^6

In Table 1, "Mean" is the mean performance of 25 runs for each algorithm, while "Std" is the standard deviation of the algorithms' performance. The best performance of "Mean" for each algorithm is marked by bold font. For the 15 benchmark functions, AMCPSO won eight times. For the five other benchmark functions, AMCPSO also had very competitive performance. Therefore, according to the comparison results, the proposed algorithm AMCPSO exhibits powerful ability in addressing large-scale optimization problems, which also demonstrates that the proposed strategy is feasible and effective to help PSO enhance the ability in balancing diversity maintenance and convergence. The convergence figures are presented in Figure 1.

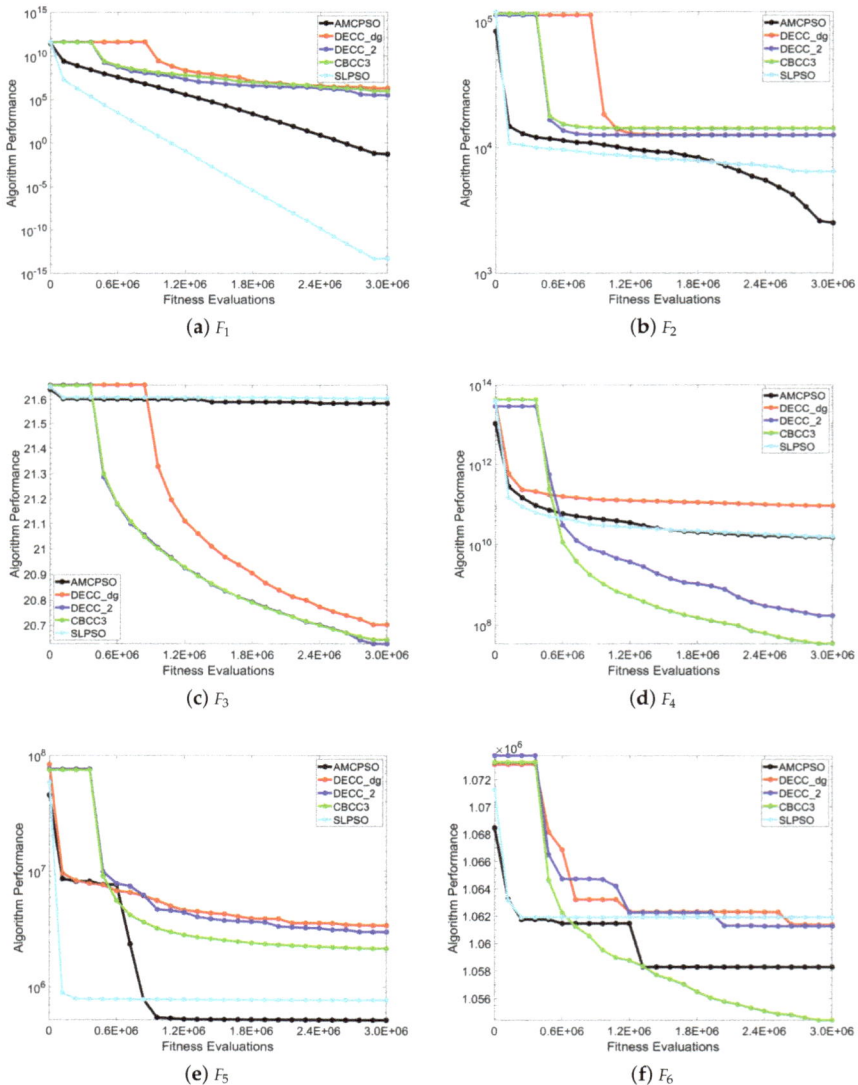

Figure 1. *Cont.*

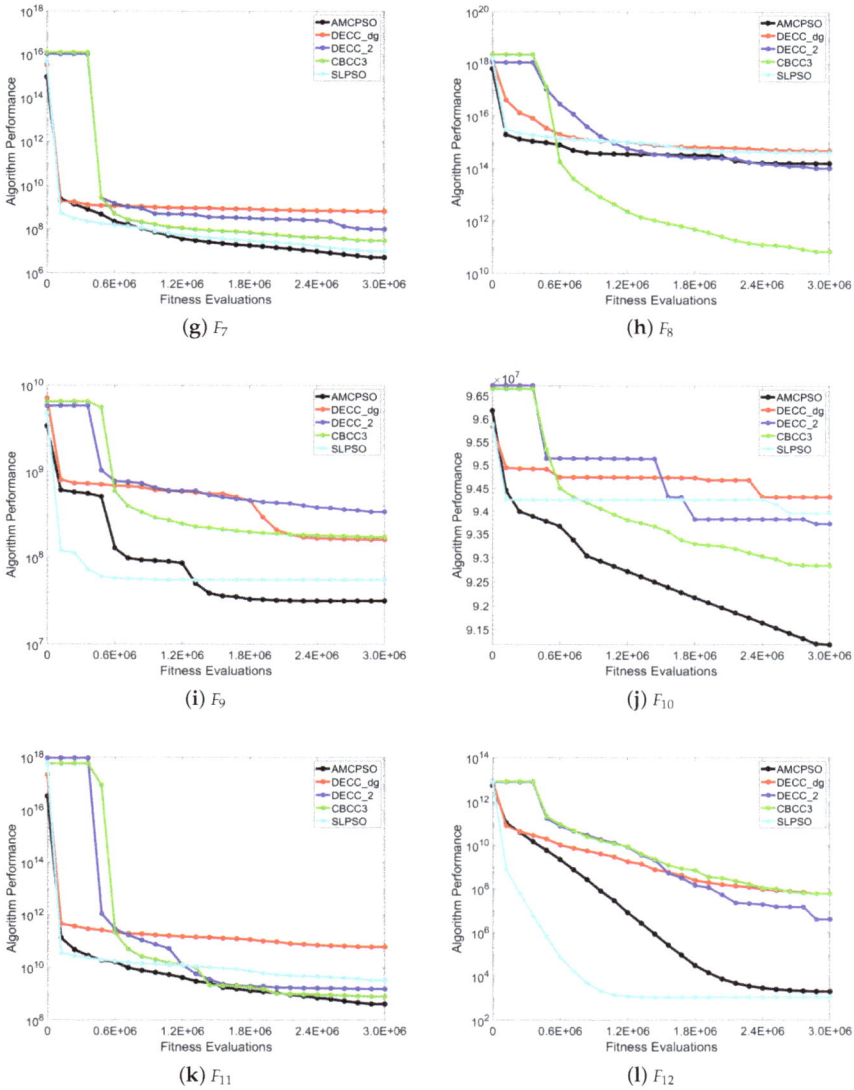

(**g**) F_7

(**h**) F_8

(**i**) F_9

(**j**) F_{10}

(**k**) F_{11}

(**l**) F_{12}

Figure 1. *Cont.*

(m) F_{13}

(n) F_{14}

(o) F_{15}

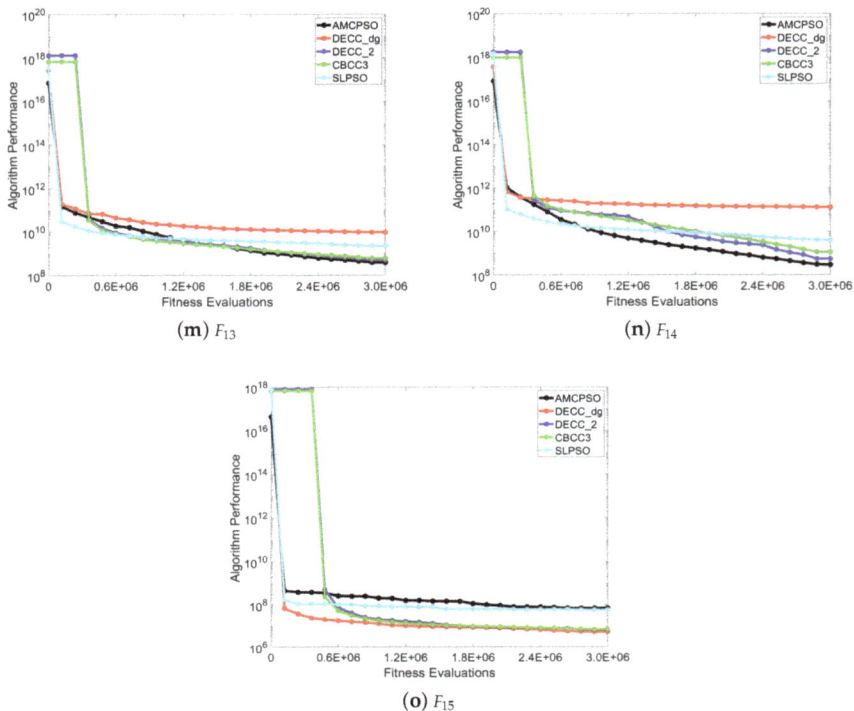

Figure 1. Convergence profiles of different algorithms obtained on the CEC'2013 test suite with 1000 dimensions.

According to the figures, for many benchmark functions, the final optimization performances of AMCPSO was better than the performances of the other algorithms, which demonstrates the proposed algorithm is competitive to address large-scale optimization problems. For some benchmarks, such as F_1, F_{12}, and F_{15}, even though AMCPSO was not the best, its performance kept getting better with the increase of FES, as shown in Figure 1a,m,p, which demonstrates the algorithm has a competitive performance in optimization.

As shown in Equation (3), c_1 and c_2 are employed to balance the abilities of exploration and exploitation. In this study, we fixed the value of c_1 as 1 and conducted experiments on tuning the value of c_2 to investigate the sensitivity. For the value of c_2, we tested 0.001, 0.002, 0.005, 0.008 and 0.01, respectively, and present the results in Table 2. In the table, for the first line, the value of c_2 decreased from 0.01 to 0.001. For other parameter settings, we still employed the same values as in Table 1. According to the results, there were not significant differences in the algorithm's performance, which demonstrates that the value of c_2 is not too sensitive to the algorithm's performance.

Table 2. The sensitivity of c_2 to AMCPSO's performance.

Function	Quality	0.01	0.008	0.005	0.002	0.001
F_1	Mean	5.39×10^{-2}	4.83×10^{-2}	3.90×10^{-2}	3.31×10^{-2}	4.55×10^{-2}
	Std	6.55×10^{-3}	3.35×10^{-3}	4.49×10^{-3}	3.13×10^{-3}	6.78×10^{-3}
F_2	Mean	1.74×10^{3}	1.51×10^{3}	2.18×10^{3}	2.10×10^{3}	2.51×10^{3}
	Std	5.67×10^{2}	7.43×10^{2}	3.92×10^{2}	6.51×10^{2}	4.37×10^{2}
F_3	Mean	2.14×10^{1}	2.15×10^{1}	2.15×10^{1}	2.15×10^{1}	2.15×10^{1}
	Std	1.57×10^{-3}	4.31×10^{-3}	3.92×10^{-3}	6.71×10^{-3}	3.92×10^{-3}
F_4	Mean	1.41×10^{10}	1.30×10^{10}	1.41×10^{10}	1.63×10^{10}	1.12×10^{10}
	Std	6.05×10^{9}	3.18×10^{9}	2.77×10^{9}	4.17×10^{9}	4.37×10^{9}
F_5	Mean	4.61×10^{5}	4.62×10^{5}	4.53×10^{5}	4.44×10^{5}	4.54×10^{5}
	Std	2.94×10^{4}	4.83×10^{4}	3.17×10^{4}	2.05×10^{4}	6.78×10^{4}
F_6	Mean	1.06×10^{6}	1.06×10^{6}	1.06×10^{6}	1.06×10^{6}	1.06×10^{6}
	Std	4.38×10^{2}	3.81×10^{2}	2.55×10^{2}	1.38×10^{2}	5.97×10^{2}
F_7	Mean	4.98×10^{6}	3.26×10^{6}	4.45×10^{6}	5.46×10^{6}	4.40×10^{6}
	Std	6.79×10^{5}	4.45×10^{5}	4.63×10^{5}	5.47×10^{5}	6.18×10^{5}
F_8	Mean	1.31×10^{14}	1.47×10^{14}	2.01×10^{14}	1.39×10^{14}	9.76×10^{13}
	Std	2.76×10^{13}	4.97×10^{13}	5.51×10^{13}	1.62×10^{13}	1.17×10^{13}
F_9	Mean	3.71×10^{7}	3.68×10^{7}	3.61×10^{7}	4.21×10^{7}	2.63×10^{7}
	Std	3.24×10^{6}	2.14×10^{6}	1.31×10^{6}	6.97×10^{6}	5.41×10^{6}
F_{10}	Mean	9.22×10^{7}	9.23×10^{7}	9.17×10^{7}	9.14×10^{7}	9.11×10^{7}
	Std	8.18×10^{4}	6.59×10^{4}	4.37×10^{4}	2.37×10^{4}	5.22×10^{4}
F_{11}	Mean	4.41×10^{8}	4.33×10^{8}	2.08×10^{9}	7.66×10^{8}	4.08×10^{8}
	Std	5.24×10^{7}	5.91×10^{7}	4.39×10^{7}	2.28×10^{7}	2.11×10^{7}
F_{12}	Mean	1.56×10^{3}	2.07×10^{3}	1.67×10^{3}	1.81×10^{3}	1.60×10^{3}
	Std	2.27×10^{2}	4.51×10^{2}	3.41×10^{2}	3.97×10^{2}	5.22×10^{2}
F_{13}	Mean	4.85×10^{8}	5.27×10^{8}	3.49×10^{8}	4.98×10^{8}	3.88×10^{8}
	Std	3.17×10^{7}	2.25×10^{7}	2.94×10^{7}	8.57×10^{7}	5.19×10^{7}
F_{14}	Mean	7.76×10^{8}	1.43×10^{9}	2.25×10^{9}	7.21×10^{8}	9.36×10^{8}
	Std	2.58×10^{7}	6.97×10^{7}	4.55×10^{7}	3.93×10^{7}	2.88×10^{7}
F_{15}	Mean	5.19×10^{7}	5.98×10^{7}	6.16×10^{7}	6.05×10^{7}	6.27×10^{7}
	Std	2.81×10^{6}	3.08×10^{6}	7.01×10^{6}	4.78×10^{6}	8.53×10^{6}

5. Conclusions and Future Work

In this paper, to enhance PSO's ability in dealing with large-scale optimization problems, we propose a novel PSO named adaptive multi-swarm competition PSO (AMCPSO), which adaptively divides a swarm into several sub-swarms. In each sub-swarm, a local winner is selected, while the local losers will learn from the local winner. In this way, for the whole swarm, not only one position is selected to attract all others and therefore the diversity of exemplars increases. On the other hand, with the process of optimization, the number of sub-swarms decreases, which helps the algorithm adaptively change the swarm's focus from exploration to exploitation. At the beginning of optimization process, many sub-swarms are adopted, which makes the algorithm focus on exploration, while the number of sub-swarms decreases with the optimization process to help the algorithm enhance the ability in exploitation. By employing benchmark functions in CEC 2013, the performance of the proposed algorithm AMCPSO was validated and the comparison results demonstrate AMCPSO has a competitive ability in dealing with large-scale optimization problems.

In our future work, on the one hand, the proposed algorithm's structure will be investigated on several different kinds of algorithms [26,35–37] to improve their performances in addressing large-scale optimization problems. On the other hand, we will apply the proposed algorithm to real applications, such as optimization problems in an industrial network, traffic network, location problems and so forth [38,39].

Author Contributions: Conceptualization, F.K.; methodology, F.K.; software, F.K.; validation, F.K.; formal analysis, F.K.; investigation, F.K.; resources, J.J.; data curation, Y.H.; writing—original draft preparation, F.K.; writing—review and editing, F.K.; visualization, F.K.; supervision, F.K.; project administration, F.K.; funding acquisition, F.K. and J.J.

Funding: This work was funded by National Key Research and Development Program of China (2018YFB1702300), the National Natural Science Foundation of China under Grant Nos. 61432017 and 61772199.

Conflicts of Interest: The authors declare no conflict of interest.

References

1. Shi, Y.; Eberhart, R. A modified particle swarm optimizer. In Proceedings of the 1998 IEEE International Conference on Evolutionary Computation Proceedings, IEEE World Congress on Computational Intelligence, Anchorage, AK, USA, 4–9 May 1998; pp. 69–73.
2. Shi, Y.; Eberhart, R. Fuzzy adaptive particle swarm optimization. In Proceedings of the 2001 Congress on Evolutionary Computation, Seoul, Korea, 27–30 May 2001; Volume 1, pp. 101–106.
3. Cao, B.; Zhao, J.; Lv, Z.; Liu, X.; Kang, X.; Yang, S. Deployment optimization for 3D industrial wireless sensor networks based on particle swarm optimizers with distributed parallelism. *J. Netw. Comput. Appl.* **2018**, *103*, 225–238. [CrossRef]
4. Wang, L.; Ye, W.; Wang, H.; Fu, X.; Fei, M.; Menhas, M.I. Optimal Node Placement of Industrial Wireless Sensor Networks Based on Adaptive Mutation Probability Binary Particle Swarm Optimization Algorithm. *Comput. Sci. Inf. Syst.* **2012**, *9*, 1553–1576. [CrossRef]
5. Sun, J.; Xu, W.; Fang, W. A diversity guided quantum behaved particle swarm optimization algorithm. In *Simulated Evolution and Learning*; Wang, T.D., Li, X., Chen, S.H., Wang, X., Abbass, H., Iba, H., Chen, G., Yao, X., Eds.; Volume 4247 of Lecture Notes in Computer Science; Springer: Berlin/Heidelberg, Germany, 2006; pp. 497–504.
6. Higashi, N.; Iba, H. Particle swarm optimization with Gaussian mutation. In Proceedings of the 2003 IEEE Swarm Intelligence Symposium, SIS '03, Indianapolis, IN, USA, 26 April 2003; pp. 72–79.
7. Ling, S.H.; Iu, H.H.C.; Chan, K.Y.; Lam, H.K.; Yeung, B.C.W.; Leung, F.H. Hybrid Particle Swarm Optimization with Wavelet Mutation and Its Industrial Applications. *IEEE Trans. Syst. Man Cybern. Part B (Cybern.)* **2008**, *38*, 743–763. [CrossRef] [PubMed]
8. Wang, H.; Sun, H.; Li, C.; Rahnamayan, S.; Pan, J.S. Diversity enhanced particle swarm optimization with neighborhood search. *Inf. Sci.* **2013**, *223*, 119–135. [CrossRef]
9. Lovbjerg, M.; Krink, T. Extending particle swarm optimisers with self-organized criticality. In Proceedings of the 2002 Congress on Evolutionary Computation, CEC '02, Honolulu, HI, USA, 12–17 May 2002; Volume 2, pp. 1588–1593.
10. Blackwell, T.M.; Bentley, P.J. Dynamic Search With Charged Swarms. In Proceedings of the Genetic and Evolutionary Computation Conference, New York, NY, USA, 9–13 July 2002; pp. 19–26.
11. Blackwell, T. Particle swarms and population diversity. *Soft Comput.* **2005**, *9*, 793–802. [CrossRef]
12. Zhan, Z.H.; Zhang, J.; Li, Y.; Chung, H.S.H. Adaptive Particle Swarm Optimization. *IEEE Trans. Syst. Man Cybern. Part B Cybern.* **2009**, *39*, 1362–1381. [CrossRef] [PubMed]
13. Hu, M.; Wu, T.; Weir, J.D. An Adaptive Particle Swarm Optimization With Multiple Adaptive Methods. *IEEE Trans. Evol. Comput.* **2013**, *17*, 705–720. [CrossRef]
14. Liang, J.J.; Qin, A.K.; Suganthan, P.N.; Baskar, S. Comprehensive learning particle swarm optimizer for global optimization of multimodal functions. *IEEE Trans. Evol. Comput.* **2006**, *10*, 281–295. [CrossRef]
15. Li, X. Niching Without Niching Parameters: Particle Swarm Optimization Using a Ring Topology. *IEEE Trans. Evol. Comput.* **2010**, *14*, 150–169. [CrossRef]
16. Cioppa, A.D.; Stefano, C.D.; Marcelli, A. Where Are the Niches? Dynamic Fitness Sharing. *IEEE Trans. Evol. Comput.* **2007**, *11*, 453–465. [CrossRef]

17. Bird, S.; Li, X. Adaptively Choosing Niching Parameters in a PSO. In Proceedings of the Annual Conference on Genetic and Evolutionary Computation, Seattle, WA, USA, 8–12 July 2006; pp. 3–10.

18. Li, C.; Yang, S.; Yang, M. An Adaptive Multi-Swarm Optimizer for Dynamic Optimization Problems. *Evol. Comput.* **2014**, *22*, 559–594. [CrossRef]

19. Cheng, R.; Jin, Y. A Competitive Swarm Optimizer for Large Scale Optimization. *IEEE Trans. Cybern.* **2015**, *45*, 191–204. [CrossRef] [PubMed]

20. Siarry, P.; Pétrowski, A.; Bessaou, M. A multipopulation genetic algorithm aimed at multimodal optimization. *Adv. Eng. Softw.* **2002**, *33*, 207–213. [CrossRef]

21. Zhao, S.; Liang, J.; Suganthan, P.; Tasgetiren, M. Dynamic multi-swarm particle swarm optimizer with local search for Large Scale Global Optimization. In Proceedings of the IEEE Congress on Evolutionary Computation, CEC 2008 (IEEE World Congress on Computational Intelligence), Hong Kong, China, 1–6 June 2008; pp. 3845–3852.

22. Parsopoulos, K.; Vrahatis, M. On the computation of all global minimizers through particle swarm optimization. *IEEE Trans. Evol. Comput.* **2004**, *8*, 211–224. [CrossRef]

23. van den Bergh, F.; Engelbrecht, A. A cooperative approach to particle swarm optimization. *IEEE Trans. Evol. Comput.* **2004**, *8*, 225–239. [CrossRef]

24. Omidvar, M.N.; Li, X.; Mei, Y.; Yao, X. Cooperative Co-Evolution With Differential Grouping for Large Scale Optimization. *IEEE Trans. Evol. Comput.* **2014**, *18*, 378–393. [CrossRef]

25. Spolaor, S.; Tangherloni, A.; Rundo, L.; Nobile, M.S.; Cazzaniga, P. Reboot strategies in particle swarm optimization and their impact on parameter estimation of biochemical systems. In Proceedings of the 2017 IEEE Conference on Computational Intelligence in Bioinformatics and Computational Biology (CIBCB), Manchester, UK, 23–25 August 2017; pp. 1–8.

26. Nobile, M.S.; Cazzaniga, P.; Besozzi, D.; Colombo, R.; Mauri, G.; Pasi, G. Fuzzy Self-Tuning PSO: A settings-free algorithm for global optimization. *Swarm Evol. Comput.* **2018**, *39*, 70–85. [CrossRef]

27. Arabas, J.; Biedrzycki, R. Improving Evolutionary Algorithms in a Continuous Domain by Monitoring the Population Midpoint. *IEEE Trans. Evol. Comput.* **2017**, *21*, 807–812. [CrossRef]

28. Jin, Y.; Branke, J. Evolutionary Optimization in Uncertain Environments—A Survey. *IEEE Trans. Evol. Comput.* **2005**, *9*, 303–317. [CrossRef]

29. Omidvar, M.N.; Yang, M.; Mei, Y.; Li, X.; Yao, X. DG2: A Faster and More Accurate Differential Grouping for Large-Scale Black-Box Optimization. *IEEE Trans. Evol. Comput.* **2017**, *21*, 929–942. [CrossRef]

30. Omidvar, M.N.; Kazimipour, B.; Li, X.; Yao, X. CBCC3—A Contribution-Based Cooperative Co-evolutionary Algorithm with Improved Exploration/Exploitation Balance. In Proceedings of the IEEE Congress on Evolutionary Computation (CEC) Held as part of IEEE World Congress on Computational Intelligence (IEEE WCCI), Vancouver, BC, Canada, 24–29 July 2016; pp. 3541–3548.

31. Li, X.; Yao, X. Cooperatively Coevolving Particle Swarms for Large Scale Optimization. *IEEE Trans. Evol. Comput.* **2012**, *16*, 210–224.

32. Cheng, J.; Zhang, G.; Neri, F. Enhancing distributed differential evolution with multicultural migration for global numerical optimization. *Inf. Sci.* **2013**, *247*, 72–93. [CrossRef]

33. Li, X.; Tang, K.; Omidvar, M.N.; Yang, Z.; Qin, K. *Benchmark Functions for the CEC'2013 Special Session and Competition on Large-Scale Global Optimization*; Technical Report; School of Computer Science and Information Technology, RMIT University: Melbourne, Australia, 2013.

34. Cheng, R.; Jin, Y. A social learning particle swarm optimization algorithm for scalable optimization. *Inf. Sci.* **2015**, *291*, 43–60. [CrossRef]

35. Wang, Y.; Wang, P.; Zhang, J.; Cui, Z.; Cai, X.; Zhang, W.; Chen, J. A Novel Bat Algorithm with Multiple Strategies Coupling for Numerical Optimization. *Mathematics* **2019**, *7*, 135. [CrossRef]

36. Cui, Z.; Zhang, J.; Wang, Y.; Cao, Y.; Cai, X.; Zhang, W.; Chen, J. A pigeon-inspired optimization algorithm for many-objective optimization problems. *Sci. China Inf. Sci.* **2019**, *62*, 070212. [CrossRef]

37. Cai, X.; Gao, X.Z.; Xue, Y. Improved bat algorithm with optimal forage strategy and random disturbance strategy. *Int. J. Bio-Inpired Comput.* **2016**, *8*, 205–214. [CrossRef]

38. Cui, Z.; Du, L.; Wang, P.; Cai, X.; Zhang, W. Malicious code detection based on CNNs and multi-objective algorithm. *J. Parallel Distrib. Comput.* **2019**, *129*, 50–58. [CrossRef]

39. Cui, Z.; Xue, F.; Cai, X.; Cao, Y.; Wang, G.; Chen, J. Detection of Malicious Code Variants Based on Deep Learning. *IEEE Trans. Ind. Inform.* **2018**, *14*, 3187–3196. [CrossRef]

mathematics

MDPI

Article

An Efficient Memetic Algorithm for the Minimum Load Coloring Problem

Zhiqiang Zhang [1,2,*], Zhongwen Li [1,2,*], Xiaobing Qiao [3] and Weijun Wang [2]

[1] Key Laboratory of Pattern Recognition and Intelligent Information Processing, Institutions of Higher
    Education of Sichuan Province, Chengdu University, Chengdu 610106, China
[2] School of Information Science and Engineering, Chengdu University, Chengdu 610106, China;
    wangweijun@cdu.edu.cn
[3] College of Teachers, Chengdu University, Chengdu 610106, China; qiaoxiaobing@cdu.edu.cn
*   Correspondence: zqzhang@cdu.edu.cn (Z.Z.); lizw@cdu.edu.cn (Z.L.)

Received: 29 March 2019; Accepted: 21 May 2019; Published: 25 May 2019

Abstract: Given a graph G with n vertices and l edges, the load distribution of a coloring $q: V \rightarrow \{$red, blue$\}$ is defined as $d_q = (r_q, b_q)$, in which r_q is the number of edges with at least one end-vertex colored red and b_q is the number of edges with at least one end-vertex colored blue. The minimum load coloring problem (MLCP) is to find a coloring q such that the maximum load, $l_q = 1/l \times \max\{r_q, b_q\}$, is minimized. This problem has been proved to be NP-complete. This paper proposes a memetic algorithm for MLCP based on an improved K-OPT local search and an evolutionary operation. Furthermore, a data splitting operation is executed to expand the data amount of global search, and a disturbance operation is employed to improve the search ability of the algorithm. Experiments are carried out on the benchmark DIMACS to compare the searching results from memetic algorithm and the proposed algorithms. The experimental results show that a greater number of best results for the graphs can be found by the memetic algorithm, which can improve the best known results of MLCP.

Keywords: minimum load coloring; memetic algorithm; evolutionary; local search

1. Introduction

The minimum load coloring problem (MLCP) of the graph, discussed in this paper, was introduced by Nitin Ahuja et al. [1]. This problem is described as follows: a graph $G = (V, E)$ is given, in which V is a set of n vertices, and E is a set of l edges. The load of a k-coloring $\varphi: V \rightarrow \{1, 2, 3, \ldots, k\}$ is defined as

$$1/l \times \max_{i \in \{1,2,3\ldots,k\}} |\{e \in E | \varphi^{-1}(i) \cap e \neq \varnothing\}|,$$

the maximum fraction of edges with at least one end-point in color i, where the maximum is taken over all $i \in \{1,2,3, \ldots, k\}$. The aim of the minimum load coloring problem is to minimize the load over all k-colorings.

This paper is dedicated to the NP-complete minimum load coloring problem [1]. We focus on coloring the vertices with the colors of red and blue. A graph $G = (V, E)$ is given, in which V is a set of n vertices, and E is a set of l edges. The load distribution of a coloring $q: V \rightarrow \{$red, blue$\}$ is defined as $d_q = (r_q, b_q)$, in which r_q is the number of edges with at least one end-vertex colored red, and b_q is the number of edges with at least one end-vertex colored blue. The objective of MLCP is to find a coloring q such that the maximum load, $l_q = 1/l \times \max\{r_q, b_q\}$, is minimized. MLCP can be applied to solve the wavelength division multiplexing (WDM) problem of network communication, and build the WDM network and complex power network [1–3].

This paper proposes an effective memetic algorithm for the minimum load coloring problem, which relies on four key components. Firstly, an improved K-OPT local search procedure, combining a tabu search strategy and a vertices addition strategy, is especially designed for MLCP to explore the search space and escape from the local optima. Secondly, a data splitting operation is used to expand the amount of data in the search space, which enables the memetic algorithm to explore in a larger search space. Thirdly, to find better global results, through randomly changing the current search patterns a disturbance operation is employed to improve the probability of escaping from the local optima. Finally, a population evolution mechanism is devised to determine how the better solution is inserted into the population.

We evaluate the performance of memetic algorithm on 59 well-known graphs from benchmark DIMACS coloring competitions. The computational results show that the search ability of memetic algorithm is better than those of simulated annealing algorithm, greedy algorithm, artificial bee colony algorithm [4] and variable neighborhood search algorithm [5]. In particular, it improves the best known results of 16 graphs in known literature algorithms.

The paper is organized as follows. Section 2 describes the related work of heuristic algorithms. Section 3 describes the general framework and the components of memetic algorithm, including the population initialization, the data splitting operation, the improved K-OPT local search procedure of individuals, the evolutionary operation and the disturbance operation. Section 4 describes the design process of simulated annealing algorithm. Section 5 describes the design process of greedy algorithm. Section 6 describes the experimental results. Section 7 describes the conclusion of the paper.

2. Related Work

In [6,7], the parameterized and approximation algorithms are proposed to solve the load coloring problem, and theoretically prove their capability in finding the best solution. On the other hand, considering the theoretical intractability of MLCP, several heuristic algorithms are proposed to find the best solutions. Heuristic algorithms use rules based on previous experience to solve a combinatorial optimization problem at the cost of acceptable time and space, and, at the same time, comparatively better results can be obtained. The heuristic algorithms used here include an artificial bee colony algorithm [4], a tabu search algorithm [5] and a variable neighborhood search algorithm [5] to solve MLCP.

Furthermore, to find the best solutions of the other combinatorial optimization problems, several heuristic algorithms are employed, such as a variable neighborhood search algorithm [8,9], a tabu search algorithm [10–13], a simulated annealing algorithm [14–17], and a greedy algorithm [18].

Local search algorithm, as an important heuristic algorithm, has been improved and evolved into many updated forms, such as a variable depth search algorithm [19], a reactive local search algorithm [20], an iterated local search algorithm [21], and a phased local search algorithm [22].

Memetic algorithm [23,24] is an optimization algorithm which combines population-based global search and individual-based local heuristic search, whose application is found in solving combinatorial optimization problems. Memetic algorithm is also proposed to solve the minimum sum coloring problem of graphs [24].

3. A Memetic Algorithm for MLCP

In this paper, we propose an efficient memetic algorithm to solve MLCP of graphs. In our algorithm, there are several important design parts.

(1) Construct the population for the global search.
(2) Search heuristically the individuals to find better solutions.
(3) Evolve the population to find better solutions.

Memetic algorithm is summarized in Memetic_D_O_MLCP (Algorithm 1). After population initialization, the algorithm randomly generates a population X consisting of p individuals (Algorithm 1, Line 2, Section 3.2). Then, the memetic algorithm repeats a series of generations (limited to a stop condition) to explore the search space defined by the set of all proper 2-colorings (Section 3.1). For each generation, by data splitting operation, the population X is expanded to population Z with twice as much as the data amount (Algorithm 1, Line 5, Section 3.3). An improved K- OPT local search is carried out for each individual Z_j ($0 \le j < |Z|$) of the population Z to find the best solution of MLCP (Algorithm 1, Line 8, Section 3.4). If the improved solution has a better value, it is then used to update the best solution found so far (Algorithm 1, Lines 9-10). Finally, an evolutionary operation is conducted in population Z to get a replaced one instead of population X (Algorithm 1, Line 14, Section 3.5). To further improve the search ability of the algorithm and find better solutions, we add a disturbance operation into the memetic algorithm (Algorithm 1, Line 15, Section 3.6).

Algorithm 1 Memetic_D_O_MLCP (G, m, p, b, k, X, Z).

Require:
G: $G = (V, E)$, $|V| = n$, $|E| = l$
m: initial number of red vertices in graph G
p: number of individuals in the population
b, k: control parameters of disturbance operation
X: set that stores the population
Z: set that stores the extended population of X, $|Z| = 2 \times |X|$
Output: s_1, the best solution found by the algorithm
$f(s_1)$, the value of the objective function
begin
1 $d1 \leftarrow 0, d2 \leftarrow 0$; /* control variables of disturbance operation, Section 3.6 */
2 Init_population(X, m, p); /* generates population X consisting of p individuals, Section 3.2 */
3 $W_{best} \leftarrow 0$;
4 repeat
5 $\quad Z \leftarrow$ Data_splitting(X); /* population X is extended to population Z with twice as much the data amount, Section 3.3 */
6 $\quad j \leftarrow 0$;
7 $\quad$ while $j < 2 \times p$ do
8 $\quad\quad W \leftarrow$ New_K-OPT_MLCP (G, Z_j, T, L);
$\quad\quad\quad\quad$ /* a heuristic search is carried out for individual Z_j, (T is tabu table and L is tabu tenure value, Section 3.4) */
9 $\quad\quad$ if $f(W) > W_{best}$ then
10 $\quad\quad\quad W_{best} \leftarrow f(W), s_1 \leftarrow W$;
11 $\quad\quad$ end if
12 $\quad\quad j \leftarrow j + 1$;
13 $\quad$ end while
14 $\quad X \leftarrow$ Evolution_population (Z, X); /* Section 3.5 */
15 $\quad (s_1, W_{best}, d1, d2) \leftarrow$ Disturbance_operation($X, b, k, d1, d2, W_{best}$); /* Section 3.6 */
16 until stop condition is met;
17 return s_1, W_{best};
end

3.1. Search Space and Objective Function

In [1], the following description is considered to be MLCP's equivalent problem. A graph $G = (V, E)$ is given, in which V is a set of n vertices, and E is a set of l edges. (V_{red}, V_{blue}) is a two-color load coloring bipartition scheme of V, in which V_{red} is the set of vertices which are red, and V_{blue} is the set of vertices which are blue, here $V = V_{red} \cup V_{blue}$. The aim is to find the maximum value of min{$|Er(V_{red})|$, $|Er(V_{blue})|$} from all bipartition schemes of (V_{red}, V_{blue}) such that l_q can minimize. The maximum value is the minimum two-color load problem solution of graph G. Here, $Er(V_{red})$ is the set of edges with both end-points in V_{red}, and $Er(V_{blue})$ is the set of edges with both end-points in V_{blue}.

The algorithm conducts a searching within the bipartition scheme (V_{red}, V_{blue}), here $|V_{red}|{\subset}V$, $V_{blue} = V\backslash V_{red}$, when $|Er(V_{red})| \approx |Er(V_{blue})|$, (V_{red}, V_{blue}) is the solution of the MLCP found by the algorithm. The search space S of the algorithm is defined as follows:

$$S = \{(V_{red}, V_{blue})|V_{red} \subset V, V_{blue} = V\backslash V_{red}|\}. \tag{1}$$

The objective function is as follows:

$$\begin{cases} f((V_{red}, V_{blue})) = \min\{|Er(V_{red})|, |Er(V_{blue})|\} \\ Er(V_{red}) = \{(v, w)|\forall (v, w) \in E, v \in V_{red}, w \in V_{red}\} \\ Er(V_{blue}) = \{(v, w)|\forall (v, w) \in E, v \in V_{blue}, w \in V_{blue}\} \end{cases}. \tag{2}$$

We define the best solution of the MLCP as follows:

$$f_b((V_{red}, V_{blue})) = \max_{1 \le j \le t}\{f((V_{red}, V_{blue})_j)\}. \tag{3}$$

Here, t is the number of all solutions that can be found by the algorithm in graph G, and $(V_{red}, V_{blue})_j$ is the jth solution of the MLCP found by the algorithm.

Suppose a graph $G = (V, E)$ is given in Figure 1. Let $|V| = 6$, $|E| = 8$, and then the best solution for MLCP of graph G is shown in Figure 2, and its best value is 2.

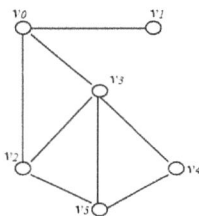

Figure 1. An instance of undirected graph G.

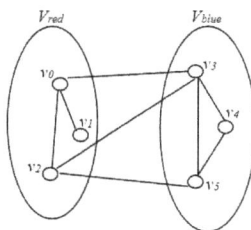

Figure 2. A best solution for MLCP of graph G.

3.2. Initial Population

The algorithm randomly generates population X consisting of p individuals. For the given graph $G = (V, E)$, in which V is a set containing n vertices, and E is a set containing l edges, m vertices are chosen at random from V to construct set V_{red} (m is the initial number of the red vertices); and the remaining vertices are used to construct set V_{blue}, that is, $|V_{red}| = m$, $V_{blue} = V\backslash V_{red}$. Sets V_{red} and V_{blue} are seen as a bipartition scheme (V_{red}, V_{blue}), which is also treated as an individual in population X. In this way, p individuals are generated at random initially, and population X is thus constructed, $|X| = p$.

3.3. Data Splitting Operation

To avoid the defect of the local optima, we expand the data amount of population X, hence we get an expanded scope of data search. We use two data splitting strategies to split a bipartition scheme into two. Thus, by using the first data splitting strategy each individual X_i ($0 \leq i < p$) in population X generates an individual $Z_{2 \times i}$, and by using the second data splitting strategy each individual X_i ($0 \leq i < p$) generates an individual $Z_{2 \times i+1}$. By doing this, p individuals in population X are divided into $2 \times p$ individuals, and the enlarged population Z is constructed ($|X| = p$, $|Z| = 2 \times p$). Figure 3 shows the population expansion, where the red arrow indicates the effects of the first data splitting strategy and the blue arrow the effects of the second data splitting strategy.

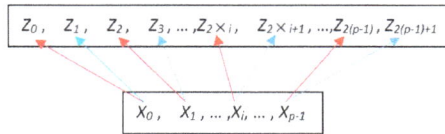

Figure 3. Expanding population X to population Z.

The first data splitting strategy of bipartition scheme (V_{red}, V_{blue}) is an important part of memetic algorithm, which consists of five steps.

First step: Degree set $Degree_{red}$ of all vertices in sub-graph $G_1(V_{red})$ is calculated.

Second step: Find the minimum degree vertex, v, from $Degree_{red}$. If there is more than one vertex with the same minimum degree, randomly select a vertex among them.

Third step: Degree set $Degree_{blue}$ of all vertices in sub-graph $G_2(V_{blue})$ is calculated.

Fourth step: Find the minimum degree vertex, w, from $Degree_{blue}$. If there is more than one vertex with the same minimum degree, randomly select a vertex among them.

Fifth step: A new bipartition scheme (V'_{red}, V'_{blue}) is generated by exchanging the vertices v and w in sets V_{red} and V_{blue}.

Suppose the number of red vertices is 4 in the given graph $G = (V, E)$, $V = \{v_0, v_1, \ldots, v_9\}$. We obtain a bipartition scheme (V_{red}, V_{blue}), as shown in Figure 4a, in which set $V_{red} = \{v_0, v_3, v_8, v_9\}$, $V_{blue} = \{v_1, v_2, v_4, v_5, v_6, v_7\}$, where the degree of vertex v_9 in set V_{red} is the smallest, and that of vertex v_4 in set V_{blue} is the smallest. After exchanging the two vertices, a new bipartition scheme (V'_{red}, V'_{blue}) is generated. The new bipartition scheme (V'_{red}, V'_{blue}) after splitting is: $V'_{red} = \{v_0, v_3, v_4, v_8\}$, $V'_{blue} = \{v_1, v_2, v_5, v_6, v_7, v_9\}$, as shown in Figure 4b.

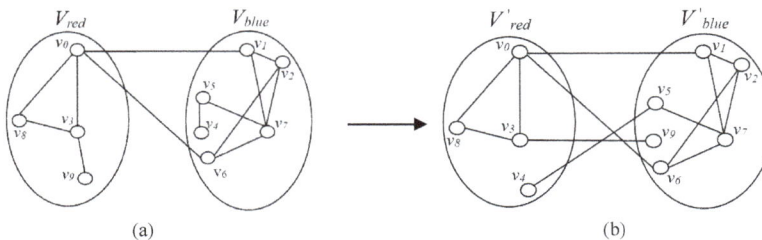

Figure 4. Bipartition scheme splitting: (**a**) bipartition scheme before the splitting; and (**b**) bipartition scheme after the splitting.

The second data splitting strategy is described as follows: for a given bipartition scheme (V_{red}, V_{blue}), in which $|V_{red}| = m$, $V_{blue} = V \backslash V_{red}$, randomly a vertex v in set V_{red} is chosen, and a vertex w in set V_{blue} is randomly chosen. Then, vertices v and w in set V_{red} and set V_{blue} are exchanged to generate a new bipartition scheme (V''_{red}, V''_{blue}).

3.4. Search for the Individuals

Memetic algorithm needs to carry out a heuristic search for each individual in the population by an effective and improved K-OPT local search algorithm designed.

We first obtain an individual Z_j which is a $(V_{red}, V_{blue})_j$, $((V_{red})_j \subset V$, $(V_{blue})_j = V \setminus (V_{red})_j$, $0 \leq j < |Z|)$. The local search algorithm is implemented to add as many selected vertices, acquired through our vertex adding strategy, as possible in set $(V_{blue})_j$ to set $(V_{red})_j$, until the stop condition set by the algorithm is met. Thus, a new bipartition scheme $(V_{red}, V_{blue})'_j$ is constructed. Generally speaking, in $(V_{red}, V_{blue})'_j$, $|Er((V'_{red})_j)|$ is approximately equal to $|Er((V'_{blue})_j)|$. Then, the objective function value f $((V_{red}, V_{blue})'_j)$ is calculated using Equation (2). If $f((V_{red}, V_{blue})'_j) > f(s_1)$, the memetic algorithm accepts the constructed bipartition as the new best solution. The improved K-OPT local search algorithm is implemented by the New_K-OPT_MLCP (Algorithm 2).

Our vertex adding strategy is described as follows:

We first need to define the following three vectors as the foundation on which the vertex adding strategy is constructed.

- CC_{red}: The current set of red vertices in graph G.
- PAV_{red}: The vertex set of possible additions, i.e., each vertex is connected to at least one vertex of CC_{red}.

$$PAV_{red} = \{v | v \in V_{blue}, \exists w \in CC_{red}, (v, w) \in E, V_{blue} = V \setminus CC_{red}\}. \tag{4}$$

- $GPAV_{red}$: The degree set of vertices $v_i \in PAV_{red}$ in sub-graph $G'(PAV_{red})$, where $PAV_{red} \subseteq V_{blue}$.

$$\begin{cases} GPAV_{red}[i] = degree_{G'(PAV_{red})}(v_i) = |\{a | \forall a \in PAV_{red}, (v_i, a) \in E\}| \\ v_i \in PAV_{red} \\ 0 \leq i \leq |PAV_{red}| - 1 \end{cases} \tag{5}$$

To avoid the local optima defect, the vertex adding strategy is employed in two phases: vertex addition phase (Algorithm 2, Lines 8–12) and vertex deletion phase (Algorithm 2, Lines 14–18).

In the vertex addition phase of CC_{red}, we obtain PAV_{red} from the current CC_{red}, then select a vertex w from PAV_{red} and move it from V_{blue} to CC_{red}, and finally update PAV_{red}. The vertex addition phase is repeatedly executed until $PAV_{red} = \varnothing$ or $|Er(CC_{red})| > |Er(V_{blue})|$.

In the vertex deletion phase of CC_{red}, we select a vertex u from CC_{red}, then delete the vertex u from CC_{red}, and add it to V_{blue}. Go back to the vertex addition phase again to continue the execution until the set ending conditions are met.

The approach to select vertex w is first to obtain a $GPAV_{red}$ in sub-graph $G'(PAV_{red})$, then to calculate the vertex selection probability value $\rho(w_i)$ of each vertex w_i $(0 \leq i < |PAV_{red}|)$ in PAV_{red}, and finally to select vertex w_i to maximize $\rho(w_i)$. If there are more than one vertex with the maximum value of $\rho(w_i)$, randomly select one.

$$\begin{cases} maxd = \max_{w_i \in PAV_{red}, 0 \leq i < |PAV_{red}|} (degree_{G'(PAV_{red})}(w_i)) \\ \rho(w_i) = \frac{maxd + 1 - GPAV_{red}[i]}{maxd + 1} \\ w_i \in PAV_{red} \\ 0 \leq i < |PAV_{red}| \end{cases} \tag{6}$$

A vertex w is selected according to the following criterion:

$$f_1(w) = \max_{0 \leq i < |PAV_{red}|} (\rho(w_i)). \tag{7}$$

We found that the larger the probability value $\rho(w_i)$ of vertex w_i is, the smaller the degree value of vertex w_i becomes.

The approach of vertex u selection is as follows: we assume that $(CC_{red}, V_{blue})^{(j)}$ is the bipartition scheme with no possible additions. We successively take the value of i from the range of $0 - (|CC_{red}| - 1)$, and then in turn execute $CC_{red}^{(j)} \setminus \{u_i\}$ as follows: delete vertex u_i from $CC_{red}^{(j)}$ successively to generate new bipartition schemes $(CC'_{red}, V'_{blue})_i$, i.e., $(CC'_{red})_i \leftarrow CC_{red}^{(j)} \setminus \{u_i\}$, $(V'_{blue})_i \leftarrow V_{blue}^{(j)} \cup \{u_i\}$, $u_i \in CC_{red}$, $0 \le i < |CC_{red}|$, and finally obtain $Ed((CC'_{red}, V'_{blue})_i)$.

$$\begin{cases} Ed((CC'_{red}, V'_{blue})_i) = \left| \{(x, y) | \forall (x, y) \in E, x \in (CC'_{red})_i, y \in (V'_{blue})_i \} \right| \\ (V'_{blue})_i = V \setminus (CC'_{red})_i \\ 0 \le i < \left| CC_{red}^{(j)} \right| \end{cases} \quad (8)$$

The maximum value *maxdd* is found from all the values of Ed, and the vertex selection probability value $\rho_2(u_i)$ of vertex u_i can be calculated:

$$\begin{cases} maxdd = \max\limits_{0 \le i < |CC_{red}^{(j)}|} \left(Ed((CC'_{red}, V'_{blue})_i) \right) \\ \rho_2(u_i) = \frac{maxdd + 1 - Ed((CC'_{red}, V'_{blue})_i)}{maxdd + 1} \\ u_i \in CC_{red}^{(j)} \\ 0 \le i < \left| CC_{red}^{(j)} \right| \end{cases} \quad (9)$$

A vertex u is selected according to the following criterion:

$$f_2(u) = \max\limits_{u_i \in CC_{red}^{(j)}, 0 \le i < |CC_{red}^{(j)}|} (\rho_2(u_i)). \quad (10)$$

We found that the larger the probability value $\rho_2(u_i)$ of vertex u_i is, the smaller the corresponding $Ed((CC'_{red}, V'_{blue})_i)$ becomes. If there are more than one vertices with the maximum value of $\rho_2(u_i)$, randomly select one.

At each generation, the variable gA stores the value of vertices number successfully added to the CC_{red} for now, and the variable $gmaxA$ stores the value of vertices number successfully added to the CC_{red} during the previous generations. If $gA > gmaxA$, the incumbent CC_{red} has more red vertices than the previous ones found by the local search algorithm. Then, $gmaxA$ is updated with the value of gA and the incumbent CC_{red} is stored to the set *Abest* (Algorithm 2, Line 12). In the vertex addition phase, the value of $gA + 1$ replaces that of gA ($gA \leftarrow gA + 1$) after a vertex is added. In the vertex deletion phase, the value of $gA - 1$ replaces that of gA ($gA \leftarrow gA - 1$) after a vertex is deleted.

At the completion of the inner loop statements, when $gmaxA > 0$, CC_{red}, which has the greatest number of vertices, is stored in set *Abest*, then the incumbent CC_{red} is updated with *Abest*. When $gmaxA = 0$, CC_{red}, which has the greatest number of vertices, is stored in set *Aprev*; if the execution of the inner loop does not find any new set CC_{red} that has more vertices, *Aprev* is adopted as CC_{red} generated by the previous execution of the inner loop and will replace the incumbent CC_{red} (Algorithm 2, Lines 22–28).

The algorithm's search efficiency may be reduced because of the roundabout searching characteristics. To solve this problem, a restricting tabu table is added to the local search algorithm.

The tabu table can be presented by two-dimensional array or one-dimensional array. We adopt the one-dimensional array T, set the tabu tenure value as L, and store the iteration numbers of running the local search algorithm into the tabu table. When the algorithm runs reach iteration value c, and if $(c - T[w]) < L$ or if $(c - T[u]) < L$, it means vertex w or u has been processed and the vertex should be re-selected. Otherwise, the current value c is stored in the tabu table, i.e., $T[w] \leftarrow c$ or $T[u] \leftarrow c$.

Algorithm 2 New_K-OPT_MLCP (G, Z_j, T, L).

Require:

Z_j: the jth individual $(V_{red}, V_{blue})_j$, $(V_{red})_j \subset V$, $(V_{blue})_j = V \backslash (V_{red})_j$

T: tabu table

L: tabu tenure value

Output: s_2, the solution found by local search algorithm

begin

1 $(CC_{red}, V_{blue}) \leftarrow (V_{red}, V_{blue})_j$;

2 according to CC_{red} and V_{blue}, PAV_{red} and $GPAV_{red}$ are obtained;

3 repeat

4     $Aprev \leftarrow CC_{red}, DA \leftarrow Aprev, PL \leftarrow \{v_0, v_1, \ldots \ldots v_{n\text{-}1}\}, gA \leftarrow 0, gmaxA \leftarrow 0, c \leftarrow 0$;

5   repeat

6         $c \leftarrow c + 1$;

7   if $|PAV_{red} \cap PL| > 0$ and $|Er(CC_{red})| < |Er(V_{blue})|$ then

8   select vertex w according to $f_1(w)$, if there are multiple vertices, select a vertex w randomly;

9   if $c - T[w] < L$ then select a non-tabu vertex w according to $f_1(w)$;

10             $T[w] \leftarrow c$;

11             $CC_{red} \leftarrow CC_{red} \cup \{w\}, V_{blue} \leftarrow V_{blue} \backslash \{w\}, gA \leftarrow gA + 1, PL \leftarrow PL \backslash \{w\}$;

12   if $gA > gmaxA$ then $gmaxA \leftarrow gA, Abest \leftarrow CC_{red}$;

13   else

14   select vertex u according to $f_2(u)$, if there are multiple vertices, select a vertex u randomly;

15   if $c - T[u] < L$ then select a non-tabu vertex u according to $f_2(u)$;

16             $T[u] \leftarrow c$;

17             $CC_{red} \leftarrow CC_{red} \backslash \{u\}, V_{blue} \leftarrow V_{blue} \cup \{u\}, gA \leftarrow gA - 1, PL \leftarrow PL \backslash \{u\}$;

18   if $u \in Aprev$ then $DA \leftarrow DA \backslash \{u\}$;

19   end if

20   based on CC_{red} and V_{blue}, PAV_{red} and $GPAV_{red}$ are updated;

21   until $|DA| = 0$ or the cut-off time condition for CPU running is met;

22   if $gmaxA > 0$ then

23         $CC_{red} \leftarrow Abest$;

24         $V_{blue} \leftarrow V \backslash CC_{red}$;

25   else

26         $CC_{red} \leftarrow Aprev$;

27         $V_{blue} \leftarrow V \backslash CC_{red}$;

28   end if

29 until $gmaxA \leq 0$ or the cut-off time condition for CPU running is met;

30 $(V_{red}, V_{blue})_j \leftarrow (CC_{red}, V_{blue})$;

31 $s_2 \leftarrow (V_{red}, V_{blue})_j$;

32 return s_2;

end

3.5. Evolutionary Operation of Population

An evolutionary operation in the population X is needed to quickly find the best solution of MLCP. We sort the individuals Z_j $(0 \leq j < 2 \times p)$ in population Z in ascending order according to the calculated value of objective function f_* in Equation (11). Then, we replace the individuals $X_0 - X_{p-1}$ of population X with the individuals $Z_0 - Z_{p-1}$ to complete the evolutionary operation.

$$f_*((V_{red}, V_{blue})_j) = |\{(a, b) | \forall (a, b) \in E, a \in (V_{red})_j, b \in (V_{blue})_j\}|, 0 \leq j < 2 \times p. \tag{11}$$

Evolution operation of the population is represented by Evolution_population (Algorithm 3).

Algorithm 3 Evolution_population (Z, X).

Require:

Z: population, $|Z| = 2 \times p$

Output: population X

begin

1 $j \leftarrow 0$;

2 while $j < 2 \times p$ do

3    $R[j] \leftarrow f^*((V_{red}, V_{blue})_j)$;

4    $j \leftarrow j + 1$;

5 end while

6 Individuals $Z_0 - Z_{2 \times p - 1}$ of population Z are sorted in ascending order according to the value of set R;

7 $(X_0 - X_{p-1}) \leftarrow (Z_0 - Z_{p-1})$;

8 return X;

end

3.6. Disturbance Operation

To further improve the search ability of the algorithm and find better values, we add a disturbance operation into the memetic algorithm. This disturbance operation is executed k times.

Algorithm 4 Disturbance_operation$(X, b, k, d1, d2, W_{best})$.

Require:

X: set that stores the population

b, k: control parameters of disturbance operation

$d1, d2$: control variables of disturbance operation

W_{best}: variable that stores the value of the objective function $f(s_t)$, in which s_t is the current best solution of MLCP

Output: s_1, a better solution found by the local search algorithm

    $f(s_1)$, the value of the objective function

    $d1, d2$, values of the control variables

begin

1 $d1 \leftarrow d1 + 1$;

2 if $d1 = b$ then

3   $d2 \leftarrow d2 + 1$;

4   if $d2 \leq k$ then

5   if $d2 = 1$ then

6   randomly choose X_j from X and start disturbance to generate a new X'_j;

7     $W \leftarrow$ New_K-OPT_MLCP (G, X'_j, T, L);

8   if $f(W) > W_{best}$ then $s_1 \leftarrow W$, $W_{best} \leftarrow f(W)$;

9     $t \leftarrow X'_j$;

10   else

11   start disturbance t to generate a new t';

12     $W \leftarrow$ New_K-OPT_MLCP (G, t', T, L);

13   if $f(W) > W_{best}$ then $s_1 \leftarrow W$, $W_{best} \leftarrow f(W)$;

14     $t \leftarrow t'$;

15   end if

16   $d1 \leftarrow d1 - 1$

17 else

18   $d1 \leftarrow 0$, $d2 \leftarrow 0$;

19 end if

20 end if

21 return s_1, W_{best}, $d1$, $d2$;

end

When the number of iterations is b, disturbance operation begins and randomly selects an individual $(V_{red}, V_{blue})_j$ $(0 \leq j < |X|)$ from population X, and chooses at random a vertex from set V_{red} and a vertex from set V_{blue}, then the two vertices are exchanged to generate a new $(V_{red}, V_{blue})'$. Then, employ the New_K-OPT_MLCP algorithm to search in $(V_{red}, V_{blue})'$; if a better solution of MLCP is found, the memetic algorithm will accept it.

The disturbance operation is represented by Disturbance_operation (Algorithm 4).

In Memetic_D_O_MLCP algorithm, setting the value of b and k will determine the disturbance operation's starting condition and the number of times of its execution. In Disturbance_operation algorithm, Lines 1, 3, 16, and 18 store the modified values of variables $d1$ and $d2$, which are the threshold values needed to start off a new disturbance operation.

4. Simulated Annealing Algorithm

Simulated annealing algorithm, a classical heuristic algorithm to solve combinatorial optimization problems, starts off from a higher initial temperature. With the decreasing of temperature parameters, the algorithm can randomly find the global best solution of problems instead of the local optima by combining the perturbations triggered by the probabilities.

For a given graph G, simulated annealing algorithm finds a coloring bipartition scheme (V_{red}, V_{blue}) of V which maximizes $\min\{|Er(V_{red})|, |Er(V_{blue})|\}$. With parameters T_0 (initial temperature value), α (cooling coefficient) and T_{end} (the end temperature value), first, the algorithm divides the vertex set V into two sets, i.e., V_{red} and V_{blue} $(V_{red} = \varnothing, V_{blue} = V)$ and the initial value of the best solution of MLCP C_{best} is set to 0. Next, a vertex is randomly selected in V_{blue} and moved from V_{blue} to V_{red}; here, $|V_{red}| = 1$, $V_{blue} = V \backslash V_{red}$. Then, the algorithm repeats a series of generations to explore the search space defined by the set of all 2-colorings. At each generation, a vertex is randomly selected in V_{blue} and moved from V_{blue} to V_{red}. The additions will take place in the following three forms:

When $2 > |V_{red}|$ and $2 \leq |V_{blue}|$, a vertex is randomly selected in V_{blue} and moved from V_{blue} to V_{red} to generate a new coloring bipartition scheme (V'_{red}, V'_{blue}) and the new status is accepted.

When $2 > |V_{blue}|$ and $2 \leq |V_{red}|$, a vertex is randomly selected in V_{red} and moved from V_{red} to V_{blue} to generate a new coloring bipartition scheme (V'_{red}, V'_{blue}) and accepted as the new status.

When $2 \leq |V_{red}|$ and $2 \leq |V_{blue}|$, a vertex is randomly selected from V_{red} and one randomly from V_{blue}, then the two vertices are exchanged to generate a new coloring bipartition scheme (V'_{red}, V'_{blue}), only when $R_1((V'_{red}, V'_{blue})) \geq R_1((V_{red}, V_{blue}))$, the scheme is accepted as a new status. Otherwise, the probability will decide whether to accept it as a new status or not.

Once the new status is accepted, if $C_{best} < R_1((V'_{red}, V'_{blue}))$, then the bipartition scheme (V'_{red}, V'_{blue}) is accepted as the best solution of MLCP.

At the end of each generation, the temperature T cools down until $T \leq T_{end}$ according to $T = T \times \alpha$, where $\alpha \in (0,1)$. The algorithm runs iteratively as per the above steps until the stop condition is met.

The best solution found by the algorithm is $R_b((V_{red}, V_{blue}))$, i.e.,

$$
\begin{cases}
R_1((V_{red}, V_{blue})_j) = \min\{|Er((V_{red})_j)|, |Er((V_{blue})_j)|\} \\
R_b((V_{red}, V_{blue})) = \max_{0 \leq j < t}\{R_1((V_{red}, V_{blue})_j)\}
\end{cases}
\tag{12}
$$

Here, t is the number of all solutions that can be found by the simulated annealing algorithm in graph G, and $(V_{red}, V_{blue})_j$ is the jth solution of MLCP.

The simulated annealing algorithm is represented by SA (Algorithm 5).

Algorithm 5 SA(G, V_{red}, V_{blue}, T_0, α, T_{end}).

Require: G: $G = (V, E)$, $|V| = $ n, $|E| = l$
V_{red}: a set of red vertices in graph G
V_{blue}: a set of blue vertices in graph G
T_0: initial temperature value
α: cooling coefficient
T_{end}: end temperature value
Output: s_3, the best solution found by SA algorithm
$R_1(s_3)$, value of the objective function
begin
1 $C_{best} \leftarrow 0$;
2 repeat
3   $T \leftarrow T_0$;
4   initialize V_{red} and V_{blue}, randomly select a vertex in V_{blue} and moved from V_{blue} to V_{red};
5   while $T > T_{end}$ do
6   if $2 > |V_{red}|$ and $2 \leq |V_{blue}|$ then
7   a vertex is randomly selected in V_{blue} and moved from V_{blue} to V_{red} to generate a new bipartition scheme
(V'_{red}, V'_{blue});
8   (V_{red}, V_{blue}) $\leftarrow$ (V'_{red}, V'_{blue});
9   if $C_{best} < R_1((V'_{red}, V'_{blue}))$ then $C_{best} \leftarrow R_1((V'_{red}, V'_{blue}))$, $s_3 \leftarrow (V'_{red}, V'_{blue})$;
10   else if $2 > |V_{blue}|$ and $2 \leq |V_{red}|$ then
11   a vertex is randomly selected in V_{red} and moved from V_{red} to V_{blue} to generate a new bipartition scheme
(V'_{red}, V'_{blue});
12   (V_{red}, V_{blue}) $\leftarrow$ (V'_{red}, V'_{blue});
13   if $C_{best} < R_1((V'_{red}, V'_{blue}))$ then $C_{best} \leftarrow R_1((V'_{red}, V'_{blue}))$, $s_3 \leftarrow (V'_{red}, V'_{blue})$;
14   else if $2 \leq |V_{red}|$ and $2 \leq |V_{blue}|$ then
15   according to (V_{red}, V_{blue}), a vertex is randomly selected from V_{red} and a vertex randomly selected from
V_{blue};
16   the two vertices are exchanged to generate a new bipartition scheme (V'_{red}, V'_{blue});
17   if $R_1((V_{red}, V_{blue})) \leq R_1((V'_{red}, V'_{blue}))$ then
18   (V_{red}, V_{blue}) $\leftarrow$ (V'_{red}, V'_{blue});
19   if $C_{best} < R_1((V'_{red}, V'_{blue}))$ then $C_{best} \leftarrow R_1((V'_{red}, V'_{blue}))$, $s_3 \leftarrow (V'_{red}, V'_{blue})$;
20   else if random number in $(0, 1) < e^{\frac{R_1((V'_{red}, V'_{blue})) - R_1((V_{red}, V_{blue}))}{T}}$ then
21   (V_{red}, V_{blue}) $\leftarrow$ (V'_{red}, V'_{blue});
22   end if
23   end if
24   $T \leftarrow T \times \alpha$;
25 end while
26 until stop condition is met;
27 return s_3, C_{best}; /*C_{best} is the value of the objective function $R_1(s_3)$ */
end

---

5. Greedy Algorithm

    Greedy algorithm aims at making the optimal choice at each stage with the hope of finding a global best solution. For a given graph G, greedy algorithm finds a coloring bipartition scheme (V_{red}, V_{blue}) of V which maximizes min{$|Er(V_{red})|$, $|Er(V_{blue})|$}.

    When a graph $G = (V, E)$ is given, the algorithm divides vertex set V into two sets, i.e., V_{red} and V_{blue} ($V_{red} = \varnothing$, $V_{blue} = V$), and the initial value of the best solution of MLCP C_{best} is set to 0. Next, a vertex is randomly selected in V_{blue} and moved from V_{blue} to V_{red}, here $|V_{red}| = 1$, $V_{blue} = V \backslash V_{red}$. Then, the algorithm repeats a series of generations to explore the search space defined by the set of all 2-colorings. At each generation, based on sub-graph $G'(V_{blue})$, choose a vertex w of the minimum degree ($w \in V_{blue}$); if there are more than one vertex with the same minimum degree, randomly select a vertex among them. Then, add the vertex from V_{blue} to V_{red}, that is: $V_{red} \leftarrow V_{red} \cup \{w\}$, $V_{blue} \leftarrow V_{blue} \setminus \{w\}$, thus a new bipartition scheme (V'_{red}, V'_{blue}) is generated, and, when $R_2((V'_{red}, V'_{blue})) > C_{best}$, the scheme is accepted as the best solution. The generation will be repeated until $|Er(V_{red})| > |Er(V_{blue})|$.

The algorithm runs iteratively as per the above steps until the stop condition is met.

The best solution found by the algorithm is $R_g((V_{red}, V_{blue}))$, i.e.,

$$\begin{cases} R_2((V_{red}, V_{blue})_j) = \min\{|Er((V_{red})_j)|, |Er((V_{blue})_j)|\} \\ R_g((V_{red}, V_{blue})) = \max_{0 \le j < t}\{R_2((V_{red}, V_{blue})_j)\} \end{cases} \tag{13}$$

Here, t is the number of all solutions that can be found by the greedy algorithm in graph G, and $(V_{red}, V_{blue})_j$ is the jth solution of MLCP.

The greedy algorithm is represented by Greedy (Algorithm 6).

Algorithm 6 Greedy (G, V_{red}, V_{blue}).

Require: G: $G = (V, E)$, $|V| = n$, $|E| = l$
V_{red}: a set of red vertices in graph G
V_{blue}: a set of blue vertices in graph G
Output: s_4, the best solution found by greedy algorithm
$R_2(s_4)$, value of the objective function
begin
1 $C_{best} \leftarrow 0$;
2 repeat
3    $V_{red} \leftarrow \varnothing$, $V_{blue} \leftarrow V$;
4   randomly select a vertex in V_{blue} and moved from V_{blue} to V_{red};
5   repeat
6       $(V'_{red}, V'_{blue}) \leftarrow (V_{red}, V_{blue})$;
7   if $R_2((V'_{red}, V'_{blue})) > C_{best}$ then $s_4 \leftarrow (V'_{red}, V'_{blue})$, $C_{best} \leftarrow R_2((V'_{red}, V'_{blue}))$;
8   select a vertex w with the minimum degree from sub-graph $G'(V_{blue})$, if there are multiple vertices, select
a vertex w randomly;
9       $V_{red} \leftarrow V_{red} \cup \{w\}$, $V_{blue} \leftarrow V_{blue} \setminus \{w\}$;
10 until $|Er(V_{red})| > |Er(V_{blue})|$;
11 until stop condition is met;
12 return s_4, C_{best}; /* C_{best} is the value of the objective function $R_2(s_4)$ */
end

---

6. Experimental Results

All algorithms were programmed in C++, and run on a PC with Intel Pentium(R) G630 processor 2.70 GHz and 4 GB memory under Windows 7 (64 bits), and the test graphs adopted were the benchmark DIMACS proposed in [5]. We compared the search results by using memetic algorithm, simulated annealing algorithm, and greedy algorithm. Then, the results of memetic algorithm were compared with those obtained from using artificial bee colony algorithm [4], tabu search algorithm [5] and variable neighborhood search algorithm [5].

The first group of experiments was performed to adjust the key parameters and analyze their influence on Memetic_D_O_MLCP. As is known to all, the most important parameters in Memetic_D_O_MLCP implementations are the values of p and L, which determine the number of the individuals of the population and the tabu tenure value during the search process. To find the most suitable values of p and L for Memetic_D_O_MLCP approach to MLCP, we performed experiments with different values of p and L. Memetic_D_O_MLCP was run 10 times for each benchmark instance, and each test lasted 30 min.

The results of experiments are summarized in Table 1, organized as follows: in the first column, *Inst* the benchmark instance name is given, containing the vertices set V; and, in the second column, *m* is the initial number of red vertices in the benchmark graph. For each $p \in \{4, 12, 20\}$ and $L \in \{10, 60, 90\}$, column *Best* contains the best values of MLCP solution found by the algorithm, while column *Avg* represents the average values of MLCP solution found by the algorithm. For each instance, the best values of *Best* and *Avg* are shown in italics. The analysis of the obtained results shows that values of *p* and *L* influence the solution quality. For example, the number of best values of *Best* is 5 for combination $p = 12$ and $L = 90$; *Best* 3 for $p = 4$, $L = 90$ and $p = 20$, $L = 60$; *Best* 2 for $p = 4$ and $L = 10$, $p = 4$ and $L = 60$, $p = 12$ and $L = 60$; *Best* 1 for $p = 20$ and $L = 90$; *Best* 0 for $p = 12$ and $L = 10$, $p = 20$ and $L = 10$. Meanwhile, the number of best values of *Avg* is 2 for combinations $p = 12$ and $L = 90$, $p = 4$ and $L = 90$; *Avg* 1 for $p = 4$ and $L = 10$, $p = 4$ and $L = 60$, $p = 12$ and $L = 60$, $p = 20$ and $L = 10$, $p = 20$ and $L = 60$, $p = 20$ and $L = 90$.

In Table 1, one observes that, for combination $p = 12$ and $L = 90$, the number of instances where the Memetic_D_O_MLCP achieved the best value for *Best* and *Avg* is 5 and 2, respectively. For all other combinations, these numbers are the biggest. Therefore, we used the combination in all other experiments.

The second groups of tests compared the search results of Memetic_D_O_MLCP, SA algorithm and Greedy algorithm, each having been run 20 times for each benchmark instance with the cut-off time of 30 min. In simulated annealing algorithm, the initial temperature T_0 is set at 1000, the cooling coefficient α at 0.9999 and the end temperature T_{end} at 0.0001. The results of experiments are summarized in Table 2, organized as follows: in the second column, $|V|$ is the number of vertices; and, in the third column, $|E|$ is the number of edges. For each instance the best values of *Best* are shown in italics. Among 59 instances, the search results of Memetic_D_O_MLCP, SA algorithm and Greedy algorithm were the same in the instances *myciel3.col*, *myciel4.col*, *queen5_5.col* and *queen6_6.col*. Memetic_D_O_MLCP and Greedy algorithm could find equivalent best value of four instances (i.e., *queen7_7.col*, *queen8_8.col*, *queen8_12.col*, and *queen9_9.col*). In the remaining 51 instances, Memetic_D_O_MLCP could find the best results of 38 instances (accounting for 75%), and Greedy algorithm could find the best results of 13 instances (accounting for 25%). The experiments showed that Memetic_D_O_MLCP could find more instances of best values.

The third group of tests aimed at comparing the search results after each algorithm was run on four benchmark instances, namely *myciel6.col*, *homer.col*, *mulsol.i.5.col* and *inithx.i.1.col*, for the first one 100 s. The results that algorithms found were collected at an interval of 10 s. The running time was regarded as the *X* coordinate on the axis and the value of MLCP solution as the *Y* coordinate.

Figure 5 illustrates that Memetic_D_O_MLCP can find the best result at each time node.

Table 1. Experiments with parameters *p* and *L*.

Inst	m	p = 4						p = 12						p = 20					
		L = 10		L = 60		L = 90		L = 10		L = 60		L = 90		L = 10		L = 60		L = 90	
		Best	Avg	Best	Avg	Best	Avg	Best	Avg	Best	Avg	Best	Avg	Best	Avg	Best	Avg	Best	Avg
fpsol2.i.1.col	|V|/5	3035	2844	3033	2857	3582	2928	3015	2840	3002	2845	3029	2837	3071	2929	2942	2727	2998	2843
fpsol2.i.2.col	|V|/5	2120	1965	2375	1953	2183	1860	2272	1972	2176	1944	2450	1969	2324	2016	2169	1932	2310	2038
fpsol2.i.3.col	|V|/5	2141	1930	2331	1931	2266	2048	2333	1960	2330	1930	2115	1900	1981	1817	2397	1986	2281	1951
DSJC125.1.col	|V|/5	255	252	254	252	254	251	254	252	255	252	255	252	254	252	255	253	254	252
DSJC125.5.col	|V|/5	1081	1072	1082	1067	1088	1078	1084	1076	1087	1080	1089	1080	1084	1078	1087	1072	1084	1074
queen15_15.col	|V|/5	1716	1699	1721	1678	1721	1694	1693	1650	1705	1692	1716	1681	1659	1632	1704	1638	1674	1641
queen16_16.col	|V|/5	2090	2050	2087	2049	2087	2055	2040	1976	2062	1994	2098	2026	2036	1990	2032	1995	2040	1996
mulsol.i.4.col	|V|/5	1704	1694	1704	1698	1704	1694	1701	1695	1704	1696	1704	1698	1700	1697	1704	1697	1704	1694

Table 2. Test results of the Memetic_D_O_MLCP, SA, and Greedy on benchmark instances.

Inst	|V|	|E|	Memetic_D_O_MLCP			SA		Greedy	
			m	Best	Avg	Best	Avg	Best	Avg
anna.col	138	986	|V|/5	*200*	198	160	154	131	131
david.col	87	812	|V|/5	*158*	157	133	130	140	140
DSJC125.1.col	125	736	|V|/5	*255*	252	222	217	240	238
DSJC125.5.col	125	3891	|V|/5	*1091*	1083	1025	1021	1075	1067
DSJC125.9.col	125	6961	|V|/5	*1798*	1789	1761	1756	1776	1772
fpsol2.i.1.col	496	11654	|V|/5	*3091*	2896	3016	2982	2510	2502
fpsol2.i.2.col	451	8691	|V|/5	*2290*	2046	2267	2242	1707	1703
fpsol2.i.3.col	425	8688	|V|/5	*2387*	1996	2291	2247	1664	1664
games120.col	120	1276	|V|/5	*288*	281	215	209	284	284
homer.col	561	3258	10	*662*	655	450	441	492	489
huck.col	74	602	|V|/5	*130*	129	111	109	113	113
inithx.i.1.col	864	18707	10	*6644*	6153	4861	4773	6167	6050
inithx.i.2.col	645	13979	10	*5622*	5104	3641	3597	3571	3481
inithx.i.3.col	621	13969	10	*5589*	4756	3643	3593	3131	3111
jean.col	80	508	|V|/5	*111*	110	98	95	106	106
latin_square_10.col	900	307350	10	85161	85072	77006	75770	*85185*	85185
le450_5a.col	450	5714	10	1824	1801	1516	1495	*1834*	1827
le450_5b.col	450	5734	10	1820	1801	1512	1498	*1843*	1831
le450_5c.col	450	9803	10	2985	2951	2541	2530	*3014*	2995
le450_5d.col	450	9757	10	2943	2913	2542	2519	*2972*	2958
le450_15b.col	450	8169	10	2395	2355	2138	2120	*2409*	2398
le450_15c.col	450	16680	10	4530	4476	4294	4267	*4560*	4539
le450_15d.col	450	16750	10	4586	4542	4320	4289	*4626*	4609
le450_25a.col	450	8260	10	2467	2434	2183	2148	*2466*	2454
le450_25b.col	450	8263	10	2664	2606	2172	2149	*2682*	2658
le450_25c.col	450	17343	10	4711	4652	4457	4438	*4728*	4714
le450_25d.col	450	17425	10	4872	4807	4470	4449	*4883*	4875
miles250.col	128	774	|V|/5	*185*	184	145	137	183	183
miles500.col	128	2340	|V|/5	*522*	522	393	381	518	518
miles750.col	128	4226	|V|/5	*870*	870	673	638	849	849
miles1000.col	128	6432	|V|/5	*1183*	1180	954	921	1156	1156
miles1500.col	128	10396	|V|/5	*1645*	1616	1461	1421	1485	1484
mulsol.i.1.col	197	3925	|V|/5	*1697*	1690	1193	1152	1624	1624
mulsol.i.2.col	188	3885	|V|/5	*1685*	1682	1153	1117	1202	1189
mulsol.i.3.col	184	3916	|V|/5	*1695*	1692	1174	1131	1211	1174
mulsol.i.4.col	185	2946	|V|/5	*1704*	1701	1172	1134	1218	1195
mulsol.i.5.col	186	3973	|V|/5	*1714*	1713	1189	1144	1216	1210
myciel3.col	11	20	|V|/5	5	5	5	5	5	5
myciel4.col	23	71	|V|/5	21	21	21	21	21	21
myciel6.col	95	755	|V|/5	*233*	231	215	212	194	193
myciel7.col	191	2360	|V|/5	*723*	717	643	634	574	574
queen5_5.col	25	320	|V|/5	46	46	46	46	46	46
queen6_6.col	36	580	|V|/5	91	91	91	88	91	91
queen7_7.col	49	952	|V|/5	148	147	145	141	*148*	148
queen8_8.col	64	1456	|V|/5	*236*	232	219	214	*236*	228
queen8_12.col	96	2736	|V|/5	*458*	453	400	391	*458*	458
queen9_9.col	81	2112	|V|/5	*340*	336	308	304	*340*	334
queen10_10.col	100	2940	|V|/5	*485*	479	419	415	468	466
queen11_11.col	121	3960	|V|/5	*644*	643	563	556	633	633
queen12_12.col	144	5192	|V|/5	*866*	853	725	717	833	833
queen13_13.col	169	6656	|V|/5	*1097*	1093	918	909	1067	1067
queen14_14.col	196	8372	|V|/5	*1407*	1385	1148	1131	1346	1345
queen15_15.col	225	10360	|V|/5	*1721*	1706	1402	1376	1676	1675
queen16_16.col	256	12640	|V|/5	*2107*	2075	1668	1652	2051	2048
school1.col	385	19095	10	6633	6553	4951	4886	*6644*	6644
school1_nsh.col	352	14612	10	5545	5450	3838	3780	*5548*	5548
zeroin.i.1.col	211	4100	10	*1210*	1198	1113	1095	924	923
zeroin.i.2.col	211	3541	10	*1135*	1126	975	959	803	800
zeroin.i.3.col	206	3540	10	*1134*	1126	981	964	800	799

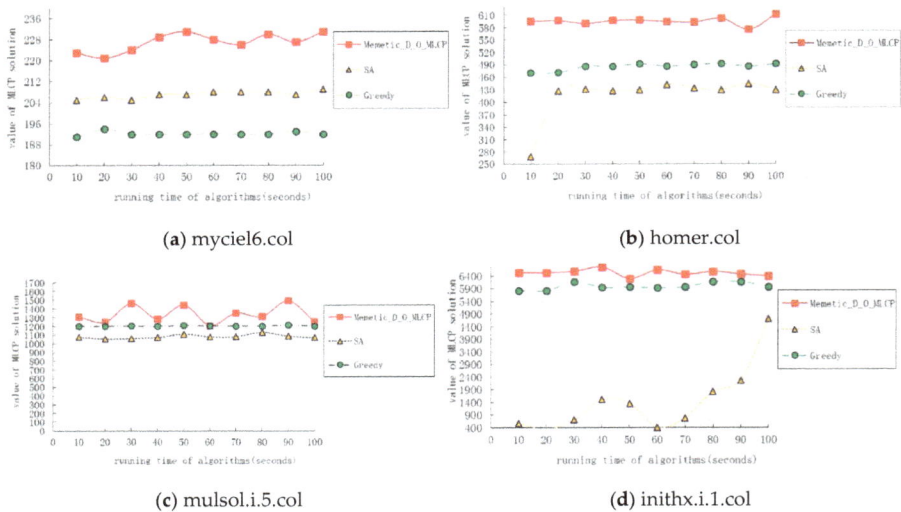

(**a**) myciel6.col

(**b**) homer.col

(**c**) mulsol.i.5.col

(**d**) inithx.i.1.col

Figure 5. Running curves of the Memetic_D_O_MLCP, SA and Greedy on benchmark instances.

The fourth group of tests compared the time each algorithm took to find the best results, each being run 20 times for 32 instances with the cut-off time of 30 min.

The results are summarized in Table 3. Compared with SA algorithm and Greedy algorithm, it took less time for Memetic_D_O_MLCP to find the best results for the 11 instances (shown in italics). Accounting for 34% in the total, these 11 instances were: *fpsol2.i.2.col*, *huck.col*, *mulsol.i.3.col*, *mulsol.i.4.col*, *mulsol.i.5.col*, *myciel6.col*, *queen10_10.col*, *queen11_11.col*, *queen15_15.col*, *inithx.i.3.col*, and *zeroin.i.2.col*. For six instances, namely *david.col*, *DSJC125.9.col*, *games120.col*, *miles250.col*, *miles750.col*, and *jean.col*, which accounted for 19% in the total, the time spent by Memetic_D_O_MLCP and Greedy algorithm showed little difference. Additionally, the former found better results than the latter. For the remaining 15 instances, although the time taken by Memetic_D_O_MLCP was longer than that by Greedy algorithm, as it consumed more time for executing the operations of data splitting, searching, evolution and disturbance, the results found by the former were better than those by the latter. Of all 32 instances, comparing with Memetic_D_O_MLCP, SA algorithm spent more time to find the best results; besides, the *Best* SA algorithm results were inferior.

The comparison between Memetic_D_O_MLCP and artificial bee colony (ABC) algorithm [4] is summarized in Table 4. For each instance, the best values of *Best* are shown in italics. Of all 21 instances proposed in [4], except that the search results of instances *myciel3.col* and *myciel4.col* were equivalent to that of artificial bee colony algorithm, Memetic_D_O_MLCP found better results (accounting for 90%) and improved the best-known result of instance *myciel5.col*.

Table 3. Running time of the Memetic_D_O_MLCP, SA and Greedy on benchmark instances.

Inst		Memetic_D_O_MLCP			SA			Greedy		
	m	*Best*	*Avg*	*Time(min)*	*Best*	*Avg*	*Time(min)*	*Best*	*Avg*	*Time(min)*
anna.col	|V|/5	200	199	5.30	159	150	17.29	131	131	0.09
david.col	|V|/5	158	157	0.47	140	130	21.97	140	140	0.02
DSJC125.1.col	|V|/5	254	252	27.92	220	214	29.76	239	237	19.24
DSJC125.5.col	|V|/5	1086	1078	22.82	1026	1012	28.20	1075	1067	14.07
DSJC125.9.col	|V|/5	1797	1785	18.86	1757	1752	14.04	1782	1773	17.81
fpsol2.i.1.col	|V|/5	3073	2871	23.52	2984	2966	13.42	2510	2504	4.78
fpsol2.i.2.col	|V|/5	*2274*	1882	*17.20*	2250	2226	25.58	1707	1706	22.30
games120.col	|V|/5	288	280	1.05	216	205	29.74	284	284	0.05
huck.col	|V|/5	*130*	129	*0.03*	113	109	6.86	113	113	0.09
miles250.col	|V|/5	185	184	5.92	140	134	29.67	183	183	4.05
miles500.col	|V|/5	522	522	1.11	389	375	25.71	518	518	< 0.01
miles750.col	|V|/5	870	870	1.72	644	618	17.17	849	849	0.07
miles1000.col	|V|/5	1186	1178	18.53	938	892	24.83	1156	1156	0.22
miles1500.col	|V|/5	1619	1613	27.56	1453	1411	15.90	1485	1485	2.36
mulsol.i.1.col	|V|/5	1695	1689	20.94	1164	1089	28.06	1624	1624	0.29
mulsol.i.2.col	|V|/5	1685	1680	26.63	1157	1065	22.97	1202	1188	8.10
mulsol.i.3.col	|V|/5	*1693*	1687	*22.76*	1147	1112	23.56	1209	1183	25.14
mulsol.i.4.col	|V|/5	*1704*	1694	*25.22*	1139	1091	29.25	1218	1186	28.08
mulsol.i.5.col	|V|/5	*1714*	1705	*23.77*	1165	1093	20.23	1214	1207	28.68
jean.col	|V|/5	111	110	0.13	97	93	18.93	106	106	0.02
myciel6.col	|V|/5	*232*	231	*20.01*	213	211	19.19	194	192	26.07
myciel7.col	|V|/5	719	710	18.73	631	624	27.23	574	574	3.68
queen10_10.col	|V|/5	*485*	478	*1.42*	418	410	10.08	468	466	21.61
queen11_11.col	|V|/5	*644*	643	*2.56*	554	539	21.94	640	633	12.95
queen12_12.col	|V|/5	866	853	5.32	721	703	22.64	833	833	0.07
queen13_13.col	|V|/5	1097	1093	25.21	907	884	21.66	1067	1067	2.98
queen15_15.col	|V|/5	*1697*	1675	*21.65*	1377	1357	29.27	1676	1675	26.07
inithx.i.1.col	10	6622	5982	23.36	4774	4696	11.55	6169	6044	1.45
inithx.i.3.col	10	*5362*	4123	*22.89*	3616	3569	9.58	3151	3117	22.93
zeroin.i.1.col	10	1207	1189	12.76	1111	1083	19.77	924	923	5.78
zeroin.i.2.col	10	*1131*	1124	*18.46*	967	939	21.43	802	799	26.64
zeroin.i.3.col	10	1131	1125	28.96	959	937	28.32	800	798	10.53

Table 4. Comparison results on Memetic_D_O_MLCP and ABC.

Inst	Memetic_D_O_MLCP		ABC
	m	*Best*	*Best*
DSJC125.1.col	|V|/5	*255*	209
DSJC125.5.col	|V|/5	*1091*	1005
DSJC125.9.col	|V|/5	*1798*	1746
fpsol2.i.1.col	|V|/5	*3091*	2956
fpsol2.i.2.col	|V|/5	*2290*	2231
fpsol2.i.3.col	|V|/5	*2387*	2207
inithx.i.1.col	10	*6644*	1295
inithx.i.2.col	10	*5622*	3574
inithx.i.3.col	10	*5589*	3548
myciel3.col	|V|/5	*5*	5
myciel4.col	|V|/5	*21*	21
myciel5.col	|V|/5	*73*	68
myciel6.col	|V|/5	*233*	207
myciel7.col	|V|/5	*723*	621
le450_5a.col	10	*1824*	1475
le450_5b.col	10	*1820*	1490
le450_5c.col	10	*2985*	2505
le450_5d.col	10	*2943*	2493
le450_15b.col	10	*2395*	2110
le450_15c.col	10	*4530*	4217
le450_15d.col	10	*4586*	4227

Furthermore, we compared the search results from Memetic_D_O_MLCP, tabu search (Tabu) algorithm [5] and variable neighborhood search (VNS) algorithm [5]; the results are shown in Table 5 (the algorithms in the literature were run 20 times, each lasting 30 min for each benchmark instance). Memetic_D_O_MLCP could find the best results of 26 instances (shown in italics), in which the best results of 11 instances equaled those found by Tabu algorithm. Hence, Memetic_D_O_MLCP could improve the best-known results of the remaining 15 instances. Besides, of the 53 instances in Table 5, the best results of 22 instances found by Memetic_D_O_MLCP were better than those by Tabu algorithm, and the best results of 42 instances found by Memetic_D_O_MLCP were better than that of VNS algorithm.

Table 5. Comparison results on Memetic_D_O_MLCP, Tabu and VNS.

Inst	m	Memetic_D_O_MLCP		Tabu		VNS	
		Best	Avg	Best	Avg	Best	Avg
anna.col	|V|/5	200	198	195	182	218	189
david.col	|V|/5	158	157	153	142	164	152
DSJC125.1.col	|V|/5	*255*	252	248	238	227	215
DSJC125.5.col	|V|/5	*1091*	1083	1078	1073	1047	1033
DSJC125.9.col	|V|/5	*1798*	1789	1794	1787	1793	1785
games120.col	|V|/5	*288*	281	282	269	192	181
homer.col	10	*662*	655	651	625	603	541
huck.col	|V|/5	*130*	129	*130*	126	123	110
inithx.i.1.col	10	6644	6153	7412	6272	6215	5838
inithx.i.2.col	10	5622	5104	5956	5831	4771	4478
inithx.i.3.col	10	5589	4756	5943	5818	4804	4464
jean.col	|V|/5	*111*	110	110	104	110	95
latin_square_10.col	10	*85161*	85072	76925	76925	77031	76956
le450_5a.col	10	1824	1801	1977	1923	1545	1520
le450_5b.col	10	1820	1801	1969	1923	1550	1522
le450_5c.col	10	2985	2951	3154	3124	2578	2553
le450_5d.col	10	2943	2913	3140	3108	2583	2546
le450_15b.col	10	2395	2355	2795	2719	2338	2268
le450_25b.col	10	2664	2606	2903	2863	2382	2337
le450_25d.col	10	4872	4807	5420	5376	4844	4747
miles250.col	|V|/5	*185*	184	183	172	134	126
miles500.col	|V|/5	*522*	522	502	483	402	367
miles750.col	|V|/5	*870*	870	836	833	708	648
miles1000.col	|V|/5	*1183*	1180	1114	1108	1035	963
miles1500.col	|V|/5	*1645*	1616	1517	1513	1565	1490
mulsol.i.1.col	|V|/5	*1697*	1690	1649	1649	1313	1240
mulsol.i.2.col	|V|/5	*1685*	1682	*1685*	1668	1319	1211
mulsol.i.3.col	|V|/5	*1695*	1692	*1695*	1669	1260	1217
mulsol.i.4.col	|V|/5	*1704*	1701	*1704*	1693	1276	1214
mulsol.i.5.col	|V|/5	*1714*	1713	1697	1686	1296	1233
myciel3.col	|V|/5	5	5	5	5	7	7
myciel4.col	|V|/5	21	21	21	20	25	24
myciel6.col	|V|/5	233	231	231	223	247	237
myciel7.col	|V|/5	723	717	714	701	737	711
queen5_5.col	|V|/5	46	46	46	45	50	48
queen6_6.col	|V|/5	*91*	91	*91*	90	86	82
queen7_7.col	|V|/5	*148*	147	*148*	145	142	136
queen8_8.col	|V|/5	*236*	232	*236*	233	208	201
queen8_12.col	|V|/5	*458*	453	*458*	457	380	369
queen9_9.col	|V|/5	*340*	336	336	332	306	293
queen10_10.col	|V|/5	*485*	479	*485*	483	403	394
queen11_11.col	|V|/5	644	643	650	637	546	536
queen12_12.col	|V|/5	*866*	853	*866*	858	703	689
queen13_13.col	|V|/5	1097	1093	1106	1066	910	887
queen14_14.col	|V|/5	*1407*	1385	*1407*	1403	1127	1101
queen15_15.col	|V|/5	1721	1707	1722	1703	1388	1366
queen16_16.col	|V|/5	2107	2075	2136	2125	1682	1650
school1.col	10	6633	6553	6975	6752	5628	5398
school1_nsh.col	10	5545	5450	5721	5622	4169	4066
zeroin.i.1.col	10	1210	1198	1185	1166	1454	1358
zeroin.i.2.col	10	1135	1126	1105	1079	1294	1201
zeroin.i.3.col	10	1134	1126	1107	1082	1221	1158

7. Conclusions

In this paper, we propose a memetic algorithm (Memetic_D_O_MLCP) to deal with the minimum load coloring problem. The algorithm employs an improved K-OPT local search procedure with a combination of data splitting operation, disturbance operation and a population evolutionary operation to assure the quality of the search results and intensify the searching ability.

We assessed the performance of our algorithm on 59 well-known graphs from the benchmark DIMACS competitions. The algorithm could find the best results of 46 graphs. Compared with simulated annealing algorithm and greedy algorithm, which cover the best results for the tested instances, our algorithm was more competent.

In addition, we investigated the artificial bee colony algorithm, variable neighborhood search algorithm and tabu search algorithm proposed in the literature. We carried out comparative experiments between our algorithm and artificial bee colony algorithm using 21 benchmark graphs, and the experiments showed that the algorithm's best results of 19 benchmark graphs were better than those of artificial bee colony algorithm, and the best-known result of one benchmark graph was improved by our algorithm. More experiments were conducted to compare our algorithm with tabu search algorithm and variable neighborhood search algorithm, and proved that the best-known results of 15 benchmark graphs were improved by our algorithm.

Finally, we showed that the proposed Memetic_D_O_MLCP approach significantly improved the classical heuristic search approach for the minimum load coloring problem.

Author Contributions: Writing and methodology, Z.Z.; Software, Z.Z.; Review and editing, Z.L.; Validation, X.Q.; and Supervision, W.W.

Funding: This research was supported by the Scientific Research Fund of Key Laboratory of Pattern Recognition and Intelligent Information Processing of Chengdu University (No. MSSB-2018-08), Chengdu Science and Technology Program (No. 2018-YF05-00731-SN), Sichuan Science and Technology Program (No. 2018GZ0247), and the Application Fundamental Foundation of Sichuan Provincial Science and Technology Department (No. 2018JY0320).

Conflicts of Interest: The authors declare no conflict of interest.

References

1. Ahuja, N.; Baltz, A.; Doerr, B.; Přívětivý, A.; Srivastav, A. On the minimum load coloring problem. *J. Discret. Algorithms* **2007**, *5*, 533–545. [CrossRef]

2. Baldine, I.; Rouskas, G.N. Reconfiguration and dynamic load balancing in broadcast WDM Networks. *Photonic Netw. Commun. J.* **1999**, *1*, 49–64. [CrossRef]

3. Rouskas, G.N.; Thaker, D. Multi-destination communication in broadcast WDM networks: A Survey. *Opt. Netw.* **2002**, *3*, 34–44.

4. Fei, T.; Bo, W.; Jin, W.; Liu, D. Artificial Bee Colony Algorithm for the Minimum Load Coloring Problem. *J. Comput. Theor. Nanosci.* **2013**, *10*, 1968–1971. [CrossRef]

5. Ye, A.; Zhang, Z.; Zhou, X.; Miao, F. Tabu Assisted Local Search for the Minimum Load Coloring Problem. *J. Comput. Theor. Nanosci.* **2014**, *11*, 2476–2480. [CrossRef]

6. Gutin, G.; Jones, M. Parameterized algorithms for load coloring problem. *Inf. Process. Lett.* **2014**, *114*, 446–449. [CrossRef]

7. Barbero, F.; Gutin, G.; Jones, M.; Sheng, B. Parameterized and Approximation Algorithms for the Load Coloring Problem. *Algorithmica* **2017**, *79*, 211–229. [CrossRef]

8. Hansen, P.; Mladenović, N.; Urošević, D. Variable neighborhood search for the maximum clique. *Discret. Appl. Math.* **2004**, *145*, 117–125. [CrossRef]

9. Dražić, Z.; Čangalović, M.; Kovačević-Vujčić, V. A metaheuristic approach to the dominating tree problem. *Optim. Lett.* **2017**, *11*, 1155–1167. [CrossRef]

10. Fadlaoui, K.; Galinier, P. A tabu search algorithm for the covering design problem. *J. Heuristics* **2011**, *17*, 659–674. [CrossRef]

11. Li, X.; Yue, C.; Aneja, Y.P.; Chen, S.; Cui, Y. An Iterated Tabu Search Metaheuristic for the Regenerator Location Problem. *Appl. Soft Comput.* **2018**, *70*, 182–194. [CrossRef]

12. Ho, S.C. An iterated tabu search heuristic for the Single Source Capacitated Facility Location Problem. *Appl. Soft Comput.* **2015**, *27*, 169–178. [CrossRef]

13. Palubeckis, G.; Ostreika, A.; Rubliauskas, D. Maximally diverse grouping: An iterated tabu search approach. *J. Oper. Res. Soc.* **2015**, *66*, 579–592. [CrossRef]
14. Tang, Z.; Feng, Q.; Zhong, P. Nonuniform Neighborhood Sampling based Simulated Annealing for the Directed Feedback Vertex Set Problem. *IEEE Access* **2017**, *5*, 12353–12363. [CrossRef]
15. Palubeckis, G. A variable neighborhood search and simulated annealing hybrid for the profile minimization problem. *Comput. Oper. Res.* **2017**, *87*, 83–97. [CrossRef]
16. Zhao, D.; Shu, Z. A Simulated Annealing Algorithm with Effective Local Search for Solving the Sum Coloring Problem. *J. Comput. Theor. Nanosci.* **2016**, *13*, 945–949. [CrossRef]
17. Li, X.; Li, S.-J.; Li, H. Simulated annealing with large-neighborhood search for two-echelon location routing problem. *Chin. J. Eng.* **2017**, *39*, 953–961.
18. Parekh, A.K. Analysis of a greedy heuristic for finding small dominating sets in graphs. *Inf. Process. Lett.* **1991**, *39*, 237–240. [CrossRef]
19. Katayama, K.; Hamamoto, A.; Narihisa, H. An effective local search for the maximum clique problem. *Inf. Process. Lett.* **2005**, *95*, 503–511. [CrossRef]
20. Battiti, R.; Protasi, M. Reactive local search for maximum clique. *Algorithmica* **2001**, *29*, 610–637. [CrossRef]
21. Xu, J.; Wu, C.C.; Yin, Y.; Lin, W.C. An iterated local search for the multi-objective permutation flowshop scheduling problem with sequence-dependent setup times. *Appl. Soft Comput.* **2017**, *52*, 39–47. [CrossRef]
22. Pullan, W. Phased local search for the maximum clique problem. *J. Comb. Optim.* **2006**, *12*, 303–323. [CrossRef]
23. Moscato, P.; Cotta, C. A gentle introduction to memetic algorithms. In *Handbook of Metaheuristics*; International series in operations research and management science; Kluwer Academic Publishers: Dordrecht, The Netherlands, 2003; Volume 57, pp. 105–144, Chapter 5.
24. Jin, Y.; Hao, J.K.; Hamiez, J.P. A memetic algorithm for the Minimum Sum Coloring Problem. *Comput. Oper. Res.* **2014**, *43*, 318–327. [CrossRef]

Article

An Entropy-Assisted Particle Swarm Optimizer for Large-Scale Optimization Problem

Weian Guo [1,2,*], Lei Zhu [3,*], Lei Wang [4], Qidi Wu [4] and Fanrong Kong [5]

[1] Key Laboratory of Intelligent Computing & Signal Processing (Ministry of Education), Anhui University, Hefei 230039, China
[2] Sino-German College of Applied Sciences, Tongji University, Shanghai 201804, China
[3] Key Lab of Information Network Security Ministry of Public Security, Shanghai 201112, China
[4] School of Electronics and Information Engineering, Tongji University, Shanghai 201804, China; wanglei@tongji.edu.cn (L.W.); wuqidi@tongji.edu.cn (Q.W.)
[5] School of Software Engineering, Tongji University, Shanghai 201804, China; ahshicr@163.com
* Correspondence: guoweian@163.com (W.G.); zhulei_shanghai@163.com or zhulei@stars.org.cn (L.Z.)

Received: 7 April 2019; Accepted: 5 May 2019; Published: 9 May 2019

Abstract: Diversity maintenance is crucial for particle swarm optimizer's (PSO) performance. However, the update mechanism for particles in the conventional PSO is poor in the performance of diversity maintenance, which usually results in a premature convergence or a stagnation of exploration in the searching space. To help particle swarm optimization enhance the ability in diversity maintenance, many works have proposed to adjust the distances among particles. However, such operators will result in a situation where the diversity maintenance and fitness evaluation are conducted in the same distance-based space. Therefore, it also brings a new challenge in trade-off between convergence speed and diversity preserving. In this paper, a novel PSO is proposed that employs competitive strategy and entropy measurement to manage convergence operator and diversity maintenance respectively. The proposed algorithm was applied to the large-scale optimization benchmark suite on CEC 2013 and the results demonstrate the proposed algorithm is feasible and competitive to address large scale optimization problems.

Keywords: diversity maintenance; particle swarm optimizer; entropy; large scale optimization

1. Introduction

Swarm intelligence plays a very active role in optimization areas. As a powerful tool in swarm optimizers, particles swarm optimizer (PSO) has been widely and successfully applied to many different areas, including electronics [1], communication technique [2], energy forecasting [3], job-shop scheduling [4], economic dispatch problems [5], and many others [6]. In the design PSO, each particle has two properties that are velocity and position, respectively. For each generation in the algorithm, particles' properties update according to the mechanisms presented in Equations (1) and (2).

$$
\begin{aligned}
V_i(t+1) &= \omega V_i(t) + c_1 R_1 (P_{i,pbest}(t) - P_i(t)) \\
&+ c_2 R_2 (P_{gbest}(t) - P_i(t)) \tag{1} \\
P_i(t+1) &= P_i(t) + V_i(t+1) \tag{2}
\end{aligned}
$$

where $V_i(t)$ and $P_i(t)$ are used to represent the velocity and position of the ith particle in the tth generation. $\omega \in [0,1]$ is an inertia weight and $c_1, c_2 \in [0,1]$ are acceleration coefficients. $R_1, R_2 \in [0,1]^n$ are two random vectors, where n is the dimension of the problem. $P_{i,pbest}(t)$ is the best position where the ith particle ever arrived, while P_{gbest} is the current global best position found by the whole swarm so far.

According to the update mechanism of PSO, the current global best particle P_{gbest} attracts the whole swarm. However, if P_{gbest} is a local optimal position, it is very difficult for the whole swarm to get rid of it. Therefore, for PSO, it is a notorious problem that the algorithm lacks competitive ability in diversity maintenance, which usually causes a premature convergence or a stagnation in convergence. To overcome this issue, many works are proposed in current decades, which are presented in Section 2 in detail. However, since diversity maintenance and fitness evaluation are conducted in the same distance-based space, it is difficult to distinguish the role of an operator in exploration and exploitation, respectively. It is also a big challenge to explicitly balance the two abilities. Hence, in current research, the proposed methods usually encounter problems, such as structure design, parameter tuning and so on. To overcome the problem, in this paper, on one the hand, we propose a novel method to maintain swarm diversity by an entropy measurement, while, on the other hand, a competitive strategy is employed for swarm convergence. Since entropy is a frequency measurement, while competitive strategy is based on the Euclidean space, the proposed method eliminates the coupling in traditional way to balance exploration and exploitation.

The rest of this paper is organized as follows. In Section 2, the related work to enhance PSO's ability in diversity maintenance is introduced. In Section 3, we propose a novel algorithm named entropy-assisted PSO, which considers convergence and diversity maintenance simultaneously and independently. The experiments on the proposed algorithm are presented in Section 4. We also select several peer algorithms in the comparisons to validate the optimization ability. The conclusions and future works are proposed in Section 5.

2. Related Work

Considering that the standard PSO has the weakness in diversity maintenance, many researchers focused on this topic to improve PSO. By mimicking genetic algorithms, mutation operators are adopted in PSO's design. In [7–9], the authors applied different kinds of mutation operators including Gaussian mutation operator, wavelet mutation operator and so forth to swarm optimizers. In this way, the elements in a particle will be changed according to probabilities and therefore the particle's position changes. However, the change will causes a break down of the convergence process, which is harmful to algorithm's performance. To address the issue, some researchers predefined a threshold to activate mutation operator, which means that mutation operator does not always work, but only happens when the swarm diversity worsens. In [10], a distance-based limit is predefined to activate mutation operator so that the method preserves swarm diversity. A similar idea is adopted in [9], where a Gauss mutation operator is employed. However, as mentioned in [11], for the design of mutation operator, it is difficult to well preset a suitable mutation rate. A large value of mutation rate will result in a loss for the swarm in convergence, while a small value of mutation rate is helpless to preserve swarm diversity.

Besides mutation operator, several other strategies will be activated when the swarm diversity is worse than a predefined limit. Since many distance-indicators, such variance of particles' positions, are employed to evaluate swarm diversity, a natural idea is to increase the distances among particles. In [12], the authors defined a criticality to evaluate the current state of swarm diversity is suitable or not. A crowded swarm has a high value of criticality, while a sparse swarm's criticality is small. A relocating strategy will be activated to disperse the swarm if the criticality is larger than a preset limit. Inspired from the electrostatics, in [13,14], the authors endowed particles with a new property named charging status. For any two charged particles, an electrostatic reaction is launched to regulate their velocities and therefore the charged particles will not be too close. Nevertheless, in the threshold-based design, it is a big challenge to preset the suitable threshold for different optimization problems. In addition, even in one optimization process, the weights of exploitation and exploration are different, it is very difficult to suitably regulate the threshold.

To avoid presetting a threshold, many researchers proposed adaptive way to maintain swarm diversity. The focus is on the parameters setting in PSO's update mechanism. For PSO, there are three components involved in velocity update. The first is inertia component which plays the role to retain each particle's own property [15,16]. As shown in Equation (1), the value of ω is used to control the weight of this component. A large value of ω helps swarm explore the searching space, while a small value of ω assists a swarm on the exploitation. To help a swarm shift the role from exploration to exploitation, an adaptive strategy is proposed in [15]; by the authors' experience, the value of *omega* decreasing from 0.9 to 0.4 is helpful for a swarm to properly explore and exploit the searching space. For the cognitive component and social component, which are the second term and third term in Equation (1), they focused more on exploration and exploitation, respectively. To provide an adaptive way to tune their weights, Hu et al constructed several mathematical functions empirically [16,17], which can dynamically regulate the weights of the two components. Besides parameter setting, researchers also provided novel structures for swarm searching. A common way is multi-swarm strategy, which means a whole swarm is divided into several sub-swarms. Each sub-swarm has different roles. On the one hand, to increase the diversity of exemplars, niching strategies are proposed. The particles in the same niche are considered as similar ones, and no information sharing occurs between similar particles. In this way, the searching efficiency is improved. However, the strategy provides a sensitive parameter, e.g. niching radius, in algorithm design. To address this problem, Li used a localized ring topology to propose a parameter-free niching strategy, which improves algorithm design [18]. On the other hand, in multi-swarm strategy, sub-swarms have different tasks. In [19], the authors defined a frequency to switch exploitation and exploration for different sub-swarms, which assists the whole swarm converge and maintain diversity in different optimization phases.

However, in the current research, the diversity measurement and management are conducted in distance-based space, where fitness evaluations are also done. In this way, both particles' quality evaluation and diversity assessment have a heavy coupling. It is very hard to tell the focus on exploitation and exploration of a learning operator. Hence, the algorithms' performances are very sensitive to the design of algorithm's structure and the parameters tuning, which brings a big challenge for users' implementation. To address the issues, in this paper, the contributions are listed as follows. First, we propose a novel way to measure population diversity by entropy, which is from the view of frequency. Second, based on the maximum entropy principle, we propose a novel idea in diversity management. In this way, the exploitation and exploration are conducted independently and simultaneously, which eliminates the coupling the convergence and diversity maintenance and provides a flexible algorithm's structure for users in real implementations.

3. Algorithm

Iin traditional PSO, both diversity maintenance operator and fitness evaluation operator are conducted in distance-based measurement space. This will result in a heavy coupling in particles' update for exploitation and exploration, which brings a big challenge in balance the weights of the two abilities. To overcome this problem, in this paper, we propose a novel idea to improve PSO which is termed entropy-assisted particle swarm optimization (EAPSO). In the proposed algorithm, we consider both diversity maintenance and fitness evaluation independently and simultaneously. Diversity maintenance and fitness evaluation are conducted by frequency-based space and distance-based space, respectively. To reduce the computation load in large scale optimization problem, in this paper, we only consider the phonotypic diversity, which is depicted by fitness domain, rather than genetic diversity. In each generation, the fitness domain is divided into several segments. We account the number of particles in each segment, as shown in Figure 1.

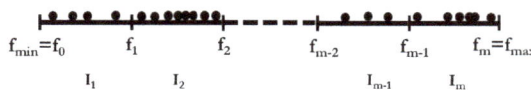

Figure 1. The illustration for the entropy diversity measurement.

The maximum fitness and minimal fitness are set as the fitness landscape boundaries. For the landscape, it is uniformly divided into several segments. For each segment, we account for the number of particles, namely number of fitness values. For the entropy calculation, we use the following formulas, which are inspired by Shannon entropy.

$$H = -\sum_i^m p_i \log p_i \tag{3}$$

where H is used to depict the entropy of a swarm and p_i is the probability that fitness values are located in the ith segment, which can be obtained by Equation (4).

$$p_i = \frac{num_i}{m} \tag{4}$$

where m is the swarm size, and num_i is the number of fitness values that appear in the ith segment. Inspired by the maximum entropy principle, the value of H is maximized if and only if $p_i = p_j$, where $i, j \in [1, n]$. Hence, to gain a large value of entropy, the fitness values are supposed to be distributed uniformly in all segments. To pursue this goal, we define a novel measurement to select global best particle, which considers fitness and entropy simultaneously. All particles are evaluated by Equation (5).

$$Q_i = \alpha fitness_{rank} + \beta entropy_{rank} \tag{5}$$

where $fitness_{rank}$ is the fitness value rank of a particle, while $entropy_{rank}$ is the entropy value of a particle. α and β are employed to manage the weights of the two ranks. However, in real applications, the tuning on two parameters will increase the difficulty. Considering that the two parameters are used to adjust the weights of exploration and exploitation respectively, in real applications, we fix the value of one of them, while tuning the other one. In this paper, we set the weight of β as 1, and therefore, by regulating the value of α, the weights of exploration and exploitation can be adjusted. To calculate the value of $fitness_{rank}$, all particles are ranked according to their fitness values. For a particle's $entropy_{rank}$, it is defined as the segment rank where the particle is. A segment has a top rank if there is few particles in the segment, while a segment ranks behind if there are crowded particles in the segment. According to Equation (5), a small value of Q_i means a good performance of particle i.

In the proposed algorithm, we propose a novel learning mechanism as shown in Equation (6). We randomly and uniformly divide a swarm into several groups. Namely, the numbers of particles in each group are equal. The particle with high quality of Q in a group is considered as an exemplar, which means that the exemplars are selected according to both fitness evaluation and entropy selection. In this paper, we abandon the historical information, but only use the information of the current swarm, which reduces the spacing complexity of the algorithm. The update mechanism of the proposed algorithm is given in Equation (6).

$$
\begin{aligned}
V_i(t+1) &= \omega V_i(t) \\
&+ r_1 * c_1 * (P_{lw}(t) - P_{ll}(t)) \\
&+ r_2 * c_2 * (P_g - P_{ll}(t)) \tag{6} \\
P_{ll}(t+1) &= P_{ll}(t) + V_i(t+1) \tag{7}
\end{aligned}
$$

where V_i is the velocity of the ith particle, ω is the inertia parameter, P_{lw} is the position of the local winner in a group, P_{lw} is the position of local loser in the same group, P_g is the current best position found, c_1 and c_2 are the cognitive coefficient and social coefficient respectively, r_1 and r_2 are random values that belong to $[0, 1]$, and t is used to present the index of generation. On the one hand, the fitness is evaluated according to the objective function. On the other hand, the divergence situation of a particle is evaluated by the entropy measurement. By assigning weights to the divergence and convergence, the update mechanism involves both exploration and exploitation. The pseudo-codes of the proposed algorithms are given in Algorithm 1.

Algorithm 1: Pseudo-codes of entropy-assisted PSO.

Input: Swarm size n, Group size g, Number of segments m, Weight value α
Output: The current global best particle

1 Loop 1: Evaluate the fitness for all particles, and for the ith particle, its fitness is f_i;
2 Set the maximum value and minimal value of fitness as f_{max} and f_{min} respectively;
3 Divide the interval $[f_{min}, f_{max}]$ into m segments;
4 Calculate the number of fitness values in each segment, for the ith segment, the number of fitness values is recorded as num_i;
5 Sort the number of fitness values, and record the fitness rank fr_i for each particle;
6 Sort the segments according to the number of fitness values, and record the segment rank sr_i for each particle;
7 Evaluate each particle' quality Q by Equation (5);
8 Divide the swarm into g groups, and compare the particles by their performances Q;
9 Select the global best particle according to Q;
10 Update particles according to Equation (6);
11 If the termination condition is satisfied, output the current global best particle; Otherwise, goto Loop 1;

In the proposed algorithm, for each particle, it has two exemplars, which are local best exemplar and global best particle respectively. The ability to maintain diversity is improved from two aspects. First, in the evaluation of particles, we consider both fitness and diversity by objective function and entropy respectively. Second, we divide a swarm into several sub-swarms, the number of local best exemplars are equal to the number of sub-swarms. In this way, even some exemplars are located in local optimal positions, and they will not affect the whole swarm so that the diversity of exemplars is maintained. Finally, in this paper, the value of α in Equation (5) provides an explicit way to manage the weights of exploration ability and exploitation ability and therefore eliminate the coupling of the two abilities.

4. Experiments and Discussions

We applied the proposed algorithm to large scale optimization problems (LSOPs). In general, LSOPs have hundreds or thousands of variables. For meta-heuristic algorithms, they usually suffer from "the curse of dimensionality", which means the performance deteriorates dramatically as the problem dimension increases [20]. Due to a large number of variables, the searching space is complex, which brings challenges for meta-heuristic algorithms. First, the searching space is huge and wide, which demands of a high searching efficiency [21,22]. Second, the large scale causes capacious attraction domains of local optimum and exacerbates the difficulty for algorithms to get rid of local optimal positions [23]. Hence, in the optimization process, both convergence ability and diversity maintenance of a swarm are crucial to algorithm's performance. We employed LSOPs in CEC 2013 as benchmark suits to test the proposed algorithm. The details of the benchmarks are listed in [24]. In comparisons, several peer algorithms, including DECC-dg (Cooperative Co-Evolution with Differential Grouping), MMOCC(Multimodal optimization enhanced cooperative coevolutio), SLPSO (Social Learning Particle Swarm Optimization), and CSO (Competitive Swarm Optimizer), were selected. DECC-dg is an improved version of DECC, which is reported in [25]. CSO was proposed by Cheng and Jin, which exhibits a powerful ability in dealing with the large scale optimization problems of IEEE CEC 2008 [21]. SLPSO was proposed by the same authors in [22], where a social learning concept is employed. For MMOCC, which is currently proposed by Peng etc, which adopts the idea of CC framework and the techniques of multi-modular optimization [26].

For each algorithm, we present a mean performance of 25 independent runs. The termination condition was limited by the maximum number of Fitness Evaluations (FEs), i.e., 3×10^6, as recommended in [24]. For EA-PSO, the population size was 1200. The reasons to employ a large size of population are presented as follows: First, a large size of population enhances the parallel computation ability for the algorithm. Second, the grouping strategy will be more efficient when a large size of population is employed. If the population size is too small, the size of groups will also be small and therefore the learning efficiency in each group decreases. Third, in EA-PSO, the diversity management is conducted by entropy control, which is a frequency based approach. As mentioned in [27], a large population size is recommended when using FBMs. Fourth, a large population size is helpful to avoid empty segments. Although a large size of population was employed, we used the number of fitness evaluations (FEs) to limit the computational resources in comparisons to guarantee a fair comparison. The number of intervals (m) was 30. The group size and α were set as 20 and 0.1, respectively. The experimental results are presented in Table 1. The best results of mean performance for each benchmark function are marked by bold font. To provide a statistical analysis, the p values were obtained by Wilcoxon signed ranks test. According to the statistical analysis, most of the p values were smaller than 0.05, which demonstrates the differences were effective. However, for benchmark F6, the comparisons "EA-PSO vs. CSO" and "EA-PSO vs. DECC-DG", the p values were larger than 0.05, which means that there was no significant differences between the algorithms' performance for the benchmark. The same was found for "EA-PSO vs. MMO-CC" on benchmark F8, "EA-PSO vs. SLPSO" on benchmark F12.

According to Table 1, EA-PSO outperformed the other algorithms for 10 benchmark functions. For F2, F4, F12, and F13, EA-PSO took the second or third ranking in the comparisons. The comparison results demonstrate that the proposed algorithm is very competitive to address large scale optimization problems. We present the convergence profiles of different algorithms in Figure 2.

Table 1. The experimental results of 1000-dimensional IEEE CEC' 2013 benchmark functions with fitness evaluations of 3×10^6. The best performance is marked by bold font in each line.

Function	Quality	CSO	SLPSO	DECC-DG	MMO-CC	EA-PSO
F_1	mean	3.68×10^{-17}	3.70×10^{-14}	2.79×10^6	$\mathbf{4.82 \times 10^{-20}}$	3.53×10^{-16}
	std	3.70×10^{-19}	1.44×10^{-15}	6.70×10^5	1.30×10^{-21}	1.14×10^4
	p-value	1.22×10^{-18}	2.31×10^{-18}	3.40×10^{-4}	9.76×10^{-21}	-
F_2	mean	$\mathbf{7.08 \times 10^2}$	6.70×10^3	1.41×10^4	1.51×10^3	1.45×10^3
	std	7.04×10^0	4.98×10^1	3.03×10^2	8.43×10^0	4.21×10^1
	p-value	1.57×10^{-5}	9.70×10^{-27}	9.57×10^{-27}	4.85×10^{-24}	-
F_3	mean	2.16×10^1	2.16×10^1	2.07×10^1	$\mathbf{2.06 \times 10^1}$	2.15×10^1
	std	1.39×10^{-3}	1.14×10^{-3}	2.19×10^{-3}	2.36×10^{-3}	2.26×10^{-3}
	p-value	1.41×10^{-5}	2.01×10^{-2}	1.44×10^{-36}	1.53×10^{-36}	-
F_4	mean	1.14×10^{10}	1.20×10^{10}	6.72×10^{10}	5.15×10^{11}	$\mathbf{4.36 \times 10^9}$
	std	2.66×10^8	5.54×10^8	5.76×10^9	9.71×10^{10}	7.72×10^8
	p-value	7.92×10^{-12}	4.94×10^{-12}	6.55×10^{-11}	2.57×10^{-11}	-
F_5	mean	7.44×10^5	7.58×10^5	3.13×10^6	2.42×10^6	$\mathbf{6.68 \times 10^5}$
	std	2.49×10^4	2.14×10^4	1.23×10^5	1.14×10^5	3.57×10^5
	p-value	7.57E-09	7.43E-09	3.92×10^{-15}	5.48×10^{-15}	-
F_6	mean	1.06×10^6	1.06×10^6	1.06×10^6	1.06×10^6	$\mathbf{1.05 \times 10^6}$
	std	1.90×10^6	1.64×10^2	3.70×10^2	6.41×10^2	4.91×10^2
	p-value	1.71×10^{-1}	7.06×10^{-3}	3.18×10^{-1}	2.70×10^{-3}	-
F_7	mean	8.19×10^6	1.73×10^7	3.45×10^8	1.28×10^{10}	$\mathbf{1.43 \times 10^6}$
	std	4.85×10^5	1.49×10^6	7.60×10^7	1.07×10^9	4.87×10^6
	p-value	7.06×10^{-14}	3.18×10^{-11}	2.76×10^{-4}	4.61×10^{-12}	-
F_8	mean	3.14×10^{14}	2.89×10^{14}	1.73×10^{15}	1.54×10^{14}	$\mathbf{1.47 \times 10^{14}}$
	std	1.09×10^{13}	1.75×10^{13}	2.78×10^{14}	4.45×10^{13}	6.17×10^{12}
	p-value	9.71×10^{-15}	8.23×10^{-11}	3.17×10^{-6}	9.50×10^{-1}	-
F_9	mean	$\mathbf{4.42 \times 10^7}$	4.44×10^7	2.79×10^8	1.76×10^8	5.05×10^7
	std	1.59×10^6	1.47×10^6	1.32×10^7	7.03×10^6	1.17×10^7
	p-value	4.38×10^{-6}	3.81×10^{-6}	7.65×10^{-12}	1.86×10^{-12}	-
F_{10}	mean	9.40×10^7	9.43×10^7	9.43×10^7	9.38×10^7	$\mathbf{9.35 \times 10^7}$
	std	4.28×10^4	3.99×10^4	6.45×10^4	1.02×10^5	7.92×10^4
	p-value	4.89×10^{-1}	3.81×10^{-4}	7.65×10^{-3}	1.86×10^{-5}	-

Table 1. *Cont.*

Function	Quality	CSO	SLPSO	DECC-DG	MMO-CC	EA-PSO
F_{11}	mean	$\mathbf{3.56 \times 10^8}$	9.98×10^9	1.26×10^{11}	5.66×10^{12}	5.00×10^8
	std	1.47×10^7	1.82×10^9	2.44×10^{10}	1.09×10^{12}	1.92×10^7
	p-value	6.46×10^{-15}	7.09×10^{-5}	7.54×10^{-5}	2.05×10^{-5}	-
F_{12}	mean	1.39×10^3	$\mathbf{1.13 \times 10^3}$	5.89×10^7	1.14×10^{11}	1.40×10^3
	std	2.19×10^1	2.12×10^1	2.75×10^6	6.32×10^{10}	2.23×10^1
	p-value	2.76×10^{-10}	6.79×10^{-1}	6.55×10^{-17}	1.62×10^{-3}	-
F_{13}	mean	1.75×10^9	2.05×10^9	1.06×10^{10}	1.32×10^{12}	$\mathbf{1.66 \times 10^9}$
	std	6.47×10^7	2.13×10^8	7.94×10^8	2.88×10^{11}	5.54×10^7
	p-value	1.19×10^{-11}	4.98×10^{-9}	5.97×10^{-12}	5.85×10^{-15}	-
F_{14}	mean	6.95×10^9	1.60×10^{10}	3.69×10^{10}	4.12×10^{11}	$\mathbf{1.40 \times 10^8}$
	std	9.22×10^8	1.62×10^9	6.58×10^9	1.21×10^{11}	2.79×10^7
	p-value	7.51×10^{-7}	2.55×10^{-10}	5.05×10^{-5}	6.99×10^{-5}	-
F_{15}	mean	1.65×10^7	6.68×10^7	6.32×10^6	4.05×10^8	$\mathbf{7.69 \times 10^6}$
	std	2.21×10^5	1.01×10^6	2.69×10^5	1.91×10^7	3.39×10^5
	p-value	8.91×10^{-23}	5.47×10^{-25}	1.49×10^{-25}	2.57×10^{-4}	-

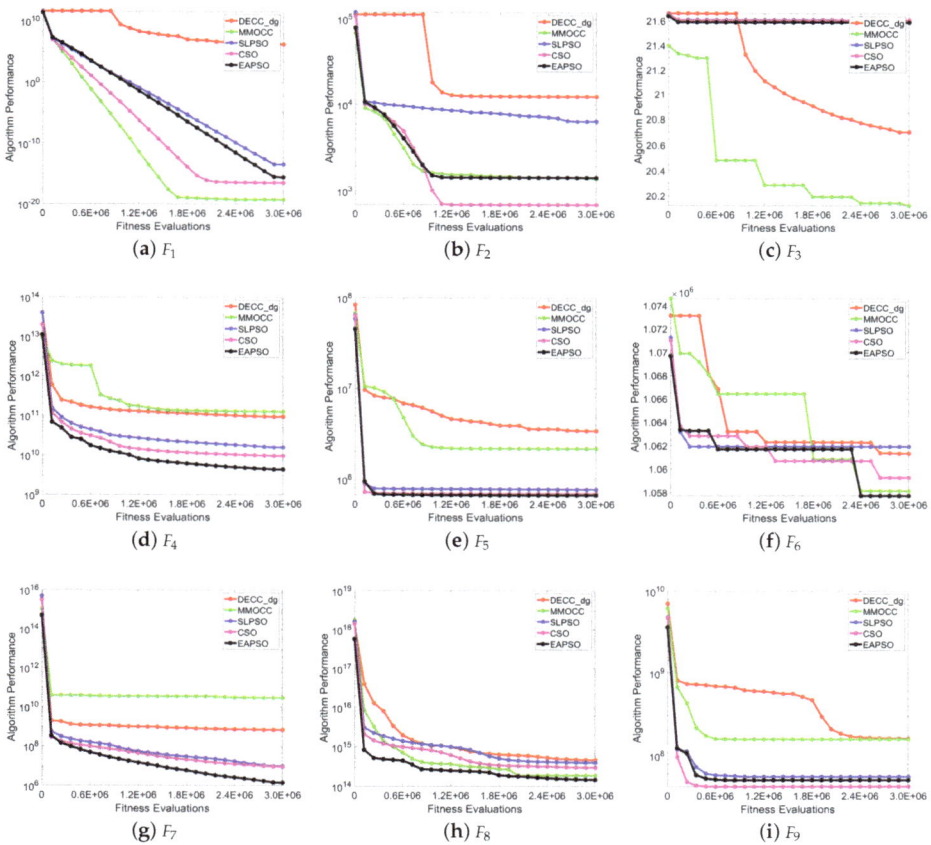

(**a**) F_1 (**b**) F_2 (**c**) F_3

(**d**) F_4 (**e**) F_5 (**f**) F_6

(**g**) F_7 (**h**) F_8 (**i**) F_9

Figure 2. *Cont.*

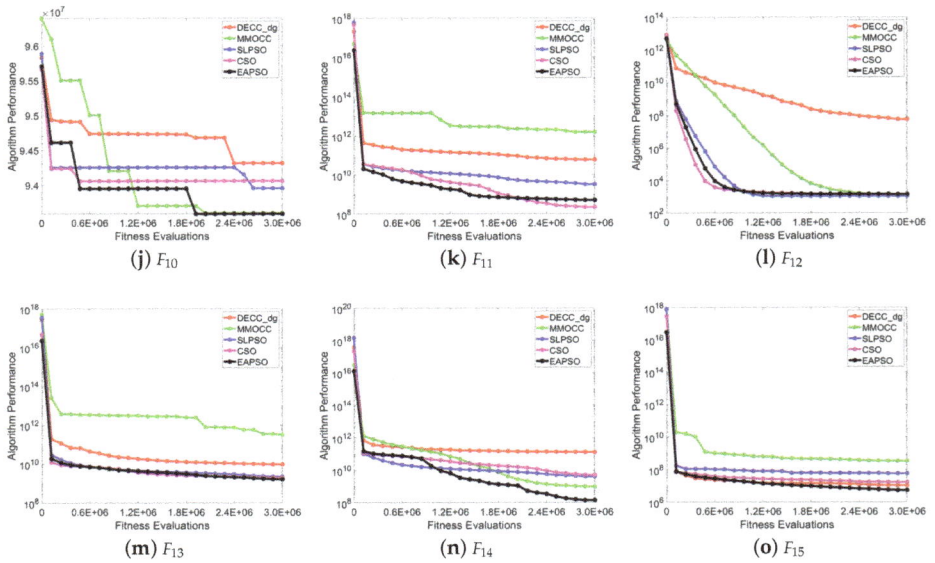

(j) F_{10} **(k)** F_{11} **(l)** F_{12}

(m) F_{13} **(n)** F_{14} **(o)** F_{15}

Figure 2. Convergence profiles of different algorithms obtained on the CEC'2013 test suite with 1000 dimensions.

In this study, the value of α was used to balance the abilities of exploration and exploitation. Hence, we investigated the influence of α to algorithm's performance. In this test, we set α as 0.2, 0.3 and 0.4. For other parameters, we used the same setting as in Table 1. For each value of α, we ran the algorithm 25 times and present the mean optimization results in Table 2. According to the results, there was no significant difference in the order of magnitude. On the other hand, for the four values, $\alpha = 0.1$ and $\alpha = 0.2$, both won six times, which demonstrates that a small value of α would help the algorithm achieve a more competitive optimization performance. The convergence profiles for algorithm's performances with different values of α are presented in Figure 3.

Table 2. The different values of α to EA-PSO's performances on IEEE CEC 2013 large scale optimization problems with 1000 dimensions (fitness evaluations = 3×10^6).

Function	$\alpha = 0.1$	$\alpha = 0.2$	$\alpha = 0.3$	$\alpha = 0.4$
F1	3.53×10^{-16}	2.97×10^{-16}	5.07×10^{-16}	9.43×10^{-16}
F2	1.45×10^3	1.45×10^3	1.58×10^3	1.45×10^3
F3	2.15×10^1	2.15×10^1	2.15×10^1	2.15×10^1
F4	4.36×10^9	6.37×10^9	6.97×10^9	9.02×10^9
F5	6.68×10^5	5.48×10^5	8.72×10^5	6.87×10^5
F6	1.06×10^6	1.06×10^6	1.06×10^6	1.06×10^6
F7	1.43×10^6	2.02×10^6	2.51×10^7	9.86×10^6
F8	1.47×10^{14}	3.11×10^{13}	1.29×10^{14}	8.66×10^{13}
F9	5.05×10^7	4.59×10^7	5.79×10^7	7.02×10^7
F10	9.35×10^7	9.40×10^7	9.41×10^7	9.42×10^7
F11	5.00×10^8	4.98×10^8	3.74×10^8	4.23×10^8
F12	1.40×10^3	1.30×10^3	1.33×10^3	1.51×10^3
F13	1.66×10^9	7.38×10^8	1.61×10^9	5.86×10^8
F14	1.40×10^8	1.44×10^8	4.21×10^8	4.87×10^8
F15	7.69×10^6	7.42×10^6	8.04×10^6	7.65×10^6

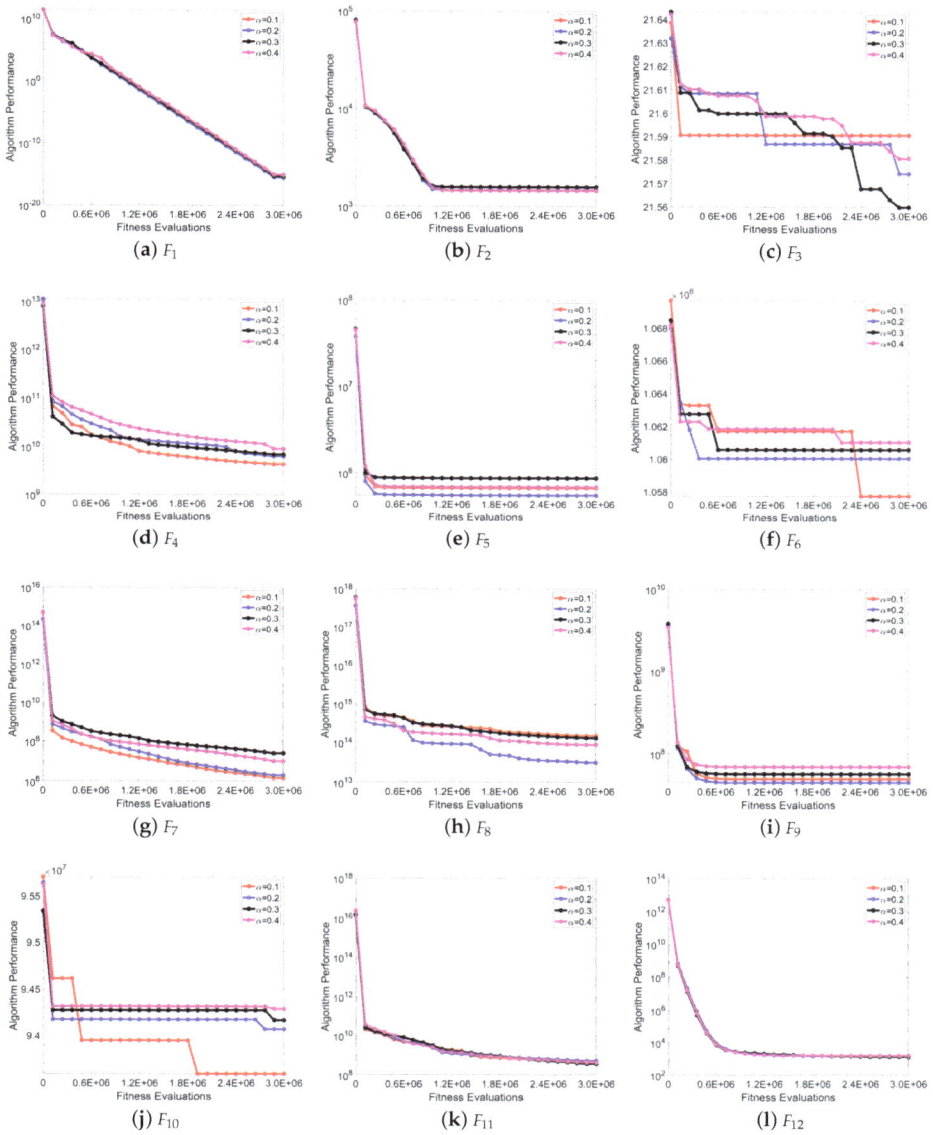

(**a**) F_1

(**b**) F_2

(**c**) F_3

(**d**) F_4

(**e**) F_5

(**f**) F_6

(**g**) F_7

(**h**) F_8

(**i**) F_9

(**j**) F_{10}

(**k**) F_{11}

(**l**) F_{12}

Figure 3. *Cont.*

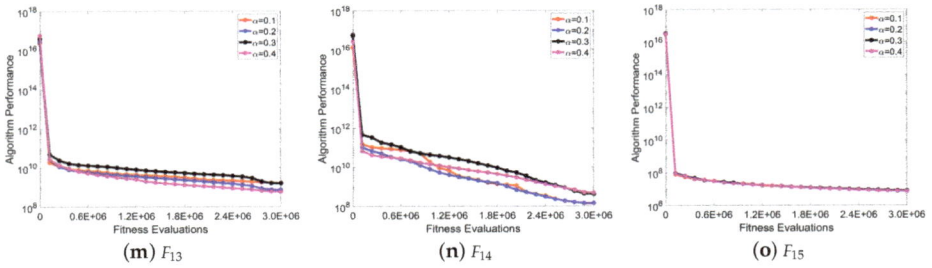

Figure 3. Convergence profiles of different algorithms obtained on the CEC'2013 test suite with 1000 dimensions.

5. Conclusions

In this paper, a novel particle swarm optimizer named entropy-assisted PSO is proposed. All particles are evaluated by fitness and diversity simultaneously and the historical information of the particles are no longer needed in particle update. The optimization experiments were conducted on the benchmarks suite of CEC 2013 with the topic of large scale optimization problems. The comparison results demonstrate the proposed structure helped enhance the ability of PSO in addressing large scale optimization and the proposed algorithm EA-PSO achieved competitive performance in the comparisons. Moreover, since the exploration and exploitation are conducted independently and simultaneously in the proposed structure, the proposed algorithm's structure is flexible to many different kinds of optimization problems.

In the future, the mathematical mechanism of the proposed algorithm will be further investigated and discussed. Considering that, for many other kinds of optimization problems, such as multi-modular optimization problems, dynamic optimization problems, and multi-objective optimization, the population divergence is also crucial to algorithms' performances, we will apply the entropy idea to such problems and investigate the roles in divergence maintenance.

Author Contributions: Conceptualization, W.G., L.W. and Q.W.; Methodology, W.G.; Software, W.G. and L.Z.; Validation, W.G. and L.Z.; Formal analysis, W.G. and L.W.; Investigation, W.G. and F.K.; Resources, W.G. and L.Z.; Data curation, W.G.; Writing original draft preparation, W.G. and F.K.; Writing review and editing, W.G., L.W. and Q.W.; Visualization, W.G. and L.Z.; Supervision, L.W. and Q.W.; Project administration, L.W. and Q.W.; Funding acquisition, W.G. and L.Z.

Funding: This work was sponsored by the National Natural Science Foundation of China under Grant Nos. 71771176 and 61503287, and supported by Key Lab of Information Network Security, Ministry of Public Security and Key Laboratory of Intelligent Computing & Signal Processing, Ministry of Education.

Conflicts of Interest: The authors declare no conflict of interest.

References

1. Shi, H.; Wen, H.; Hu, Y.; Jiang, L. Reactive Power Minimization in Bidirectional DC-DC Converters Using a Unified-Phasor-Based Particle Swarm Optimization. *IEEE Trans. Power Electron.* **2018**, *33*, 10990–11006. [CrossRef]
2. Bera, R.; Mandal, D.; Kar, R.; Ghoshal, S.P. Non-uniform single-ring antenna array design using wavelet mutation based novel particle swarm optimization technique. *Comput. Electr. Eng.* **2017**, *61*, 151–172. [CrossRef]
3. Osorio, G.J.; Matias, J.C.O.; Catalao, J.P.S. Short-term wind power forecasting using adaptive neuro-fuzzy inference system combined with evolutionary particle swarm optimization, wavelet transform and mutual information. *Renew. Energy* **2015**, *75*, 301–307. [CrossRef]

4.   Nouiri, M.; Bekrar, A.; Jemai, A.; Niar, S.; Ammari, A.C. An effective and distributed particle swarm optimization algorithm for flexible job-shop scheduling problem. *J. Intell. Manuf.* **2018**, *29*, 603–615. [CrossRef]

5.   Aliyari, H.; Effatnejad, R.; Izadi, M.; Hosseinian, S.H. Economic Dispatch with Particle Swarm Optimization for Large Scale System with Non-smooth Cost Functions Combine with Genetic Algorithm. *J. Appl. Sci. Eng.* **2017**, *20*, 141–148. [CrossRef]

6.   Bonyadi, M.R.; Michalewicz, Z. Particle swarm optimization for single objective continuous space problems: A review. *Evol. Comput.* **2017**, *25*, 1–54. [CrossRef]

7.   Higashi, N.; Iba, H. Particle swarm optimization with gaussian mutation. In Proceedings of the 2003 IEEE Swarm Intelligence Symposium, Indianapolis, IN, USA, 26 April 2003; pp. 72–79.

8.   Ling, S.H.; Iu, H.H.C.; Chan, K.Y.; Lam, H.K.; Yeung, B.C.W.; Leung, F.H. Hybrid particle swarm optimization with wavelet mutation and its industrial applications. *IEEE Trans. Syst. Man Cybern. Part B Cybern.* **2008**, *38*, 743–763. [CrossRef]

9.   Wang, H.; Sun, H.; Li, C.; Rahnamayan, S.; Pan, J.-S. Diversity enhanced particle swarm optimization with neighborhood search. *Inf. Sci.* **2013**, *223*, 119–135. [CrossRef]

10. Sun, J.; Xu, W.; Fang, W. A diversity guided quantum behaved particle swarm optimization algorithm. In *Simulated Evolution and Learning*; Wang, T.D., Li, X., Chen, S.H., Wang, X., Abbass, H., Iba, H., Chen, G., Yao, X., Eds.; Lecture Notes in Computer Science; Springer: Berlin/Heidelberg, Germany, 2006; Volume 4247, pp. 497–504.

11. Jin, Y.; Branke, J. Evolutionary optimization in uncertain environments—A survey. *IEEE Trans. Evol. Comput.* **2005**, *9*, 303–317. [CrossRef]

12. Lovbjerg, M.; Krink, T. Extending particle swarm optimisers with self-organized criticality. In Proceedings of the 2002 Congress on Evolutionary Computation, Honolulu, HI, USA, 12–17 May 2002; Volume 2, pp. 1588–1593.

13. Blackwell, T.M.; Bentley, P.J. Dynamic search with charged swarms. In Proceedings of the Genetic and Evolutionary Computation Conference, New York, NY, USA, 9–13 July 2002; pp. 19–26.

14. Blackwell, T. Particle swarms and population diversity. *Soft Comput.* **2005**, *9*, 793–802. [CrossRef]

15. Zhan, Z.; Zhang, J.; Li, Y.; Chung, H.S. Adaptive particle swarm optimization. *IEEE Trans. Syst. Man Cybern. Part B Cybern.* **2009**, *39*, 1362–1381. [CrossRef] [PubMed]

16. Hu, M.; Wu, T.; Weir, J.D. An adaptive particle swarm optimization with multiple adaptive methods. *IEEE Trans. Evol. Comput.* **2013**, *17*, 705–720. [CrossRef]

17. Liang, J.J.; Qin, A.K.; Suganthan, P.N.; Baskar, S. Comprehensive learning particle swarm optimizer for global optimization of multimodal functions. *IEEE Trans. Evol. Comput.* **2006**, *10*, 281–295. [CrossRef]

18. Li, X. Niching without niching parameters: Particle swarm optimization using a ring topology. *IEEE Trans. Evol. Comput.* **2010**, *14*, 150–169. [CrossRef]

19. Siarry, P.; Pétrowski, A.; Bessaou, M. A multipopulation genetic algorithm aimed at multimodal optimization. *Adv. Eng. Softw.* **2002**, *33*, 207–213. [CrossRef]

20. Yang, Z.; Tanga, K.; Yao, X. Large scale evolutionary optimization using cooperative coevolution. *Inf. Sci.* **2008**, *178*, 2985–2999. [CrossRef]

21. Cheng, R.; Jin, Y. A competitive swarm optimizer for large scale optimization. *IEEE Trans. Cybern.* **2015**, *45*, 191–204. [CrossRef]

22. Cheng, R.; Jin, Y. A social learning particle swarm optimization algorithm for scalable optimization. *Inf. Sci.* **2015**, *291*, 43–60. [CrossRef]

23. Yang, Q.; Chen, W.-N.; Deng, J.D.; Li, Y.; Gu, T.; Zhang, J. A Level-Based Learning Swarm Optimizer for Large-Scale Optimization. *IEEE Trans. Evol. Comput.* **2018**, *22*, 578–594. [CrossRef]

24. Li, X.; Tang, K.; Omidvar, M.N.; Yang, Z.; Qin, K. *Benchmark Functions for the CEC 2013 Special Session And Competition on Large-Scale Global Optimization*; Tech. Rep.; School of Computer Science and Information Technology, RMIT University: Melbourne, Australia, 2013.

25. Omidvar, M.N.; Li, X.; Mei, Y.; Yao, X. Cooperative Co-Evolution with Differential Grouping for Large Scale Optimization. *IEEE Trans. Evol. Comput.* **2014**, *18*, 378–393. [CrossRef]

26. Peng, X.; Jin, Y.; Wang, H. Multimodal optimization enhanced cooperative coevolution for large-scale optimization. *IEEE Trans. Cybern.* **2018**, [CrossRef]

27. Corriveau, G.; Guilbault, R.; Tahan, A.; Sabourin, R. Review and Study of Genotypic Diversity Measures for Real-Coded Representations. *IEEE Trans. Evol. Comput.* **2012**, *16*, 695–710. [CrossRef]

Article

Enhancing Elephant Herding Optimization with Novel Individual Updating Strategies for Large-Scale Optimization Problems

Jiang Li, Lihong Guo *, Yan Li and Chang Liu

Changchun Institute of Optics, Fine Mechanics and Physics, Chinese Academy of Sciences, Changchun 130033, China; cclijiang@163.com (J.L.); ly2455@sina.com (Y.L.); lc1120112964@163.com (C.L.)
* Correspondence: guolh@ciomp.ac.cn

Received: 16 February 2019; Accepted: 27 April 2019; Published: 30 April 2019

Abstract: Inspired by the behavior of elephants in nature, elephant herd optimization (EHO) was proposed recently for global optimization. Like most other metaheuristic algorithms, EHO does not use the previous individuals in the later updating process. If the useful information in the previous individuals were fully exploited and used in the later optimization process, the quality of solutions may be improved significantly. In this paper, we propose several new updating strategies for EHO, in which one, two, or three individuals are selected from the previous iterations, and their useful information is incorporated into the updating process. Accordingly, the final individual at this iteration is generated according to the elephant generated by the basic EHO, and the selected previous elephants through a weighted sum. The weights are determined by a random number and the fitness of the elephant individuals at the previous iteration. We incorporated each of the six individual updating strategies individually into the basic EHO, creating six improved variants of EHO. We benchmarked these proposed methods using sixteen test functions. Our experimental results demonstrated that the proposed improved methods significantly outperformed the basic EHO.

Keywords: elephant herding optimization; EHO; swarm intelligence; individual updating strategy; large-scale; benchmark

1. Introduction

Inspired by nature, a large variety of metaheuristic algorithms [1] have been proposed that provide optimal or near-optimal solutions to various complex large-scale problems that are difficult to solve using traditional techniques. Some of the many successful metaheuristic approaches include particle swarm optimization (PSO) [2,3], cooperative coevolution [4–6], seagull optimization algorithm [7], GRASP [8], clustering algorithm [9], and differential evolution (DE) [10,11], among others.

In 2015, Wang et al. [12,13] proposed a new metaheuristic algorithm called elephant herd optimization (EHO), for finding the optimal or near-optimal function values. Although EHO exhibits a good performance on benchmark evaluations [12,13], like most other metaheuristic methods, it does not utilize the best information from the previous elephant individuals to guide current and future searches. This gap will be addressed, because previous individuals can provide a variety of useful information. If such information could be fully exploited and applied in the later updating process, the performance of EHO may be improved significantly, without adding unnecessary operations and fitness evaluations.

In the research presented in this paper, we extended and improved the performance of the original EHO (which we call "the basic EHO") by fully investigating the information in the previous elephant individuals. Then, we designed six updating strategies to update the individuals. For each of the six individual updating strategies, first, we selected a certain number of elephants from the previous iterations. This selection could be made in either a fixed or random way, with one, two, or three individuals selected from previous iterations. Next, we used the information from these selected previous individual elephants to update the individuals. In this way, the information from the previous individuals could be reused fully. The final elephant individual at this iteration was generated according to the elephant individual generated by the basic EHO at the current iteration, along with the selected previous elephants using a weighted sum. While there are many ways to determine the weights, in our current work, they were determined by random numbers. Last, by combining the six individual updating strategies with EHO, we developed the improved variants of EHO. To verify our work, we benchmarked these variants using sixteen cases involving large-scale complex functions. Our experimental results showed that the proposed variants of EHO significantly outperformed the originally described EHO.

The organization of the remainder of this paper is as follows. Section 2 reviews the main steps of the basic EHO. In Section 3, we describe the proposed method for incorporating useful information from previous elephants into the EHO. Section 4 provides details of our various experiments on sixteen large-scale functions. Lastly, Section 5 offers our conclusion and suggestions for future work.

2. Related Work

As EHO [12,13] is a newly-proposed swarm intelligence-based algorithm, in this section, some of the most representative work regarding swarm intelligence, including EHO, are summarized and reviewed.

Meena et al. [14] proposed an improved EHO algorithm, which is used to solve the multi-objective distributed energy resources (DER) accommodation problem of distribution systems by combining a technique for order of preference by similarity to ideal solution (TOPSIS). The proposed technique is productively implemented on three small- to large-scale benchmark test distribution systems of 33-bus, 118-bus, and 880-bus.

When the spectral resolution of the satellite imagery is increased, the higher within-class variability reduces the statistical separability between the LU/LC classes in spectral space and tends to continue diminishing the classification accuracy of the traditional classifiers. These are mostly per pixel and parametric in nature. Jayanth et al. [15] used EHO to solve the problems. The experimental results revealed that EHO shows an improvement of 10.7% on Arsikere Taluk and 6.63% on the NITK campus over the support vector machine.

Rashwan et al. [16] carried out a series of experiments on a standard test bench, as well as engineering problems and real-world problems, in order to understand the impact of the control parameters. On top of that, the main aim of this paper is to propose different approaches to enhance the performance of EHO. Case studies ranging from the recent test bench problems of Congress on Evolutionary Computation (CEC) 2016, to the popular engineering problems of the gear train, welded beam, three-bar truss design problem, continuous stirred tank reactor, and fed-batch fermentor, are used to validate and test the performances of the proposed EHOs against existing techniques.

Correia et al. [17] firstly used a metaheuristic algorithm, namely EHO, to address the energy-based source localization problem in wireless sensor networks. Through extensive simulations, the key parameters of the EHO algorithm are optimized, such that they match the energy decay model between two sensor nodes. The simulation results show that the new approach significantly outperforms the existing solutions in noisy environments, encouraging further improvement and testing of metaheuristic methods.

Jafari et al. [18] proposed a new hybrid algorithm that was based on EHO and cultural algorithm (CA), known as the elephant herding optimization cultural (EHOC) algorithm. In this process, the belief space defined by the cultural algorithm was used to improve the standard EHO. In EHOC, based on the belief space, the separating operator is defined, which can create new local optimums in the search space, so as to improve the algorithm search ability and to create an algorithm with an optimal exploration–exploitation balance. The CA, EHO, and EHOC algorithms are applied to eight mathematical optimization problems and four truss weight minimization problems, and to assess the performance of the proposed algorithm, the results are compared. The results clearly indicate that EHOC can accelerate the convergence rate effectively and can develop better solutions compared with CA and EHO.

Hassanien et al. [19] combined support vector regression (SVR) with EHO in order to predict the values of the three emotional scales as continuous variables. Multiple experiments are applied to evaluate the prediction performance. EHO was applied in two stages of the optimization. Firstly, to fine-tune the regression parameters of the SVR. Secondly, to select the most relevant features extracted from all 40 EEG channels, and to eliminate the ineffective and redundant features. To verify the proposed approach, the results proved EHO-SVR's ability to gain a relatively enhanced performance, measured by a regression accuracy of 98.64%.

Besides EHO, many other swarm intelligence-based algorithms have been proposed, and some of the most representative ones are summarized and reviewed as follows.

Monarch butterfly optimization (MBO) [20] is proposed for global optimization problems, inspired by the migration behavior of monarch butterflies. Yi et al. [21] proposed a novel quantum inspired MBO methodology, called QMBO, by incorporating quantum computation into MBO, which is further used to solve uninhabited combat air vehicles (UCAV) path planning navigation problem [22,23]. In addition, Feng et al. proposed various variants of MBO algorithms to solve the knapsack problem [24–28]. In addition, Wang et al. also improved on the performance of the MBO algorithm from various aspects [29–32]; a variant of the MBO method in combination with two optimization strategies, namely GCMBO, was also put forward.

Inspired by the phototaxis and Lévy flights of the moths, Wang developed a new kind of metaheuristic algorithm, called the moth search (MS) algorithm [33]. Feng et al. [34] divided twelve transfer functions into three families, and combined them with MS, and then twelve discrete versions of MS algorithms are proposed for solving the set-union knapsack problem (SUKP). Based on the improvement of the moth search algorithm (MSA) using differential evolution (DE), Elaziz et al. [35] proposed a new method for the cloud task scheduling problem. In addition, Feng and Wang [36] verified the influence of the Lévy flights operator and fly straightly operator in MS. Nine types of new mutation operators based on the global harmony search have been specially devised to replace the Lévy flights operator.

Inspired by the herding behavior of krill, Gandomi and Alavi proposed a krill herd (KH) [37]. After that, Wang et al. improved the KH algorithms through different optimization strategies [38–46]. More literature regarding the KH algorithm can be found in the literature [47].

The artificial bee colony (ABC) algorithm [48] is a swarm-based meta-heuristic algorithm that was introduced by Karaboga in 2005 for optimizing numerical problems. Wang and Yi [49] presented a robust optimization algorithm, namely KHABC, based on hybridization of KH and ABC methods and the information exchange concept. In addition, Liu et al. [50] presented an ABC algorithm based on the dynamic penalty function and Lévy flight (DPLABC) for constrained optimization problems.

Also, many other researchers have proposed other state-of-the-art metaheuristic algorithms, such as particle swarm optimization (PSO) [51–56], cuckoo search (CS) [57–61], probability-based incremental learning (PBIL) [62], differential evolution (DE) [63–66], evolutionary strategy (ES) [67,68], monarch butterfly optimization (MBO) [20], firefly algorithm (FA) [69–72], earthworm optimization algorithm (EWA) [73], genetic algorithms (GAs) [74–76], ant colony optimization (ACO) [77–79], krill herd (KH) [37,80,81], invasive weed optimization [82–84], stud GA (SGA) [85], biogeography-based optimization (BBO) [86,87], harmony search (HS) [88–90], and bat algorithm (BA) [91,92], among others

Besides benchmark evaluations [93,94], these proposed state-of-the-art metaheuristic algorithms are also used to solve various practical engineering problems, like test-sheet composition [95], scheduling [96,97], clustering [98–100], cyber-physical social systems [101], economic load dispatch [102,103], fault diagnosis [104], flowshop [105], big data optimization [106,107], gesture segmentation [108], target recognition [109,110], prediction of pupylation sites [111], system identification [112], shape design [113], multi-objective optimization [114], and many-objective optimization [115–117].

3. Elephant Herd Optimization

The basic EHO can be described using the following simplified rules [12]:

(1)   Elephants belonging to different clans live together led by a matriarch. Each clan has a fixed number of elephants. For the purposes of modelling, we assume that each clan consists of an equal, unchanging number of elephants.

(2)   The positions of the elephants in a clan are updated based on their relationship to the matriarch. EHO models this behavior through an updating operator.

(3)   Mature male elephants leave their family groups to live alone. We assume that during each generation, a fixed number of male elephants leave their clans. Accordingly, EHO models the updating process using a separating operator.

(4)   Generally, the matriarch in each clan is the eldest female elephant. For the purposes of modelling and solving the optimization problems, the matriarch is considered the fittest elephant individual in the clan.

As this paper is focused on improving the EHO updating process, in the following subsection, we provide further details of the EHO updating operator as it was originally presented. For details regarding the EHO separating operator, see the literature [12].

3.1. Clan Updating Operator

The following updating strategy of the basic EHO was described by the authors of [12], as follows. Assume that an elephant clan is denoted as ci. The next position of any elephant, j, in the clan is updated using (1), as follows:

$$x_{new,ci,j} = x_{ci,j} + \alpha \times \left(x_{best,ci} - x_{ci,j}\right) \times r, \tag{1}$$

where $x_{new,ci,j}$ is the updated position, and $x_{ci,j}$ is the prior position of elephant j in clan ci. $x_{best,ci}$ denotes the matriarch of clan ci; she is the fittest elephant individual in the clan. The scale factor $\alpha \in [0,1]$ determines the influence of the matriarch of ci on $x_{ci,j}$. $r \in [0, 1]$, which is a type of stochastic distribution, can provide a significant improvement for the diversity of the population in the later search phase. For the present work, a uniform distribution was used.

It should be noted that $x_{ci,j} = x_{best,ci,}$ which means that the matriarch (fittest elephant) in the clan cannot be updated by (1). To avoid this situation, we can update the fittest elephant using the following equation:

$$x_{new,ci,j} = \beta \times x_{center,ci}, \tag{2}$$

where the influence of $x_{center,ci}$ on $x_{new,ci}$, is regulated by $\beta \in [0,1]$.

In Equation (2), the information from all of the individuals in clan ci is used to create the new individual $x_{new,ci,j}$. The centre of clan ci, $x_{center,ci}$, can be calculated for the d-th dimension through D calculations, where D is the total dimension, as follows:

$$x_{center,ci,d} = \frac{1}{n_{ci}} \times \sum_{j=1}^{n_{ci}} x_{ci,j,d} \tag{3}$$

Here, $1 \leq d \leq D$ represents the d-th dimension, n_{ci} is the number of individuals in ci, and $x_{ci,j,d}$ is the d-th dimension of the individual $x_{ci,j}$.

Algorithm 1 provides the pseudocode for the updating operator.

Algorithm 1: Clan updating operator [12]

Begin

    for $ci = 1$ to $nClan$ (for all clans in elephant population) **do**

        for $j = 1$ to n_{ci} (for all elephant individuals in clan ci) **do**

            Update $x_{ci,j}$ and generate $x_{new,ci,j}$ according to (1).

            if $x_{ci,j} = x_{best,ci}$ **then**

                Update $x_{ci,j}$ and generate $x_{new,ci,j}$ according to (2).

            end if

        end for j

    end for ci

End.

3.2. Separating Operator

In groups of elephants, male elephants leave their family group and live alone upon reaching puberty. This process of separation can be modeled into a separating operator when solving optimization problems. In order to further improve the search ability of the EHO method, let us assume that the individual elephants with the worst fitness will implement the separating operator for each generation, as shown in (4).

$$x_{worst,ci} = x_{min} + (x_{max} - x_{min} + 1) \times rand \qquad (4)$$

where x_{max} and x_{min} are the upper and lower bound, respectively, of the position of the individual elephant. $x_{worst,ci}$ is the worst individual elephant in clan ci. $rand \in [0, 1]$ is a kind of stochastic distribution, and the uniform distribution in the range [0, 1] is used in our current work.

Accordingly, the separating operator can be formed, as shown in Algorithm 2.

Algorithm 2: Separating operator

Begin

    for $ci =1$ to $nClan$ (all of the clans in the elephant population) **do**

        Replace the worst elephant individual in clan ci using (4).

    end for ci

End.

3.3. Schematic Presentation of the Basic EHO Algorithm

For EHO, like the other metaheuristic algorithms, a kind of elitism strategy is used with the aim of protecting the best elephant individuals from being ruined by the clan updating and separating operators. In the beginning, the best elephant individuals are saved, and the worst ones are replaced by the saved best elephant individuals at the end of the search process. This elitism ensures that the later elephant population is not always worse than the former one. The schematic description can be summarized as shown in Algorithm 3.

As described before, the basic EHO algorithm does not take the best available information in the previous group of individual elephants to guide the current and later searches. This may lead to a slow convergence during the solution of certain complex, large-scale optimization problems. In our current work, some of the information used for the previous individual elephants was reused, with the aim of improving the search ability of the basic EHO algorithm.

Algorithm 3: Elephant Herd Optimization (EHO) [12]

Begin

  Step 1: Initialization.

        Set the generation counter $t = 1$.

    Initialize the population P of NP elephant individuals randomly, with uniform distribution in the search space.

    Set the number of the kept elephants $nKEL$, the maximum generation $MaxGen$, the scale factor α and β, the number of clan $nClan$, and the number of elephants for the ci-th clan n_{ci}.

  Step 2: Fitness evaluation.

        Evaluate each elephant individual according to its position.

  Step 3: While $t < MaxGen$ **do the following:**

        Sort all of the elephant individuals according to their fitness.

        Save the $nKEL$ elephant individuals.

        Implement the clan updating operator as shown in Algorithm 1.

        Implement the separating operator as shown in Algorithm 2.

        Evaluate the population according to the newly updated positions.

        Replace the worst elephant individuals with the $nKEL$ saved ones.

        Update the generation counter, $t = t + 1$.

  Step 4: End while

  Step 5: Output the best solution.

End.

---

4. Improving EHO with Individual Updating Strategies

In this research, we propose six new versions of EHO based on individual updating strategies. In theory, k ($k \geq 1$) previous elephant individuals can be selected, but as more individuals ($k \geq 4$) are chosen, the calculations of the weights become more complex. Therefore, for this paper, we investigate $k \in \{1, 2, 3\}$.

Suppose that x_i^t is the ith individual at iteration t, and x_i and f_i^t are its position and fitness values, respectively. Here, t is the current iteration, $1 \leq i \leq N_P$ is an integer number, and N_P is the population size. y_i^{t+1} is the individual generated by the basic EHO, and f_i^{t+1} is its fitness. The framework of our proposed method is given through the individuals at the $(t-2)$th, $(t-1)$th, tth, and $(t+1)$th iterations.

4.1. Case of $k = 1$

The simplest case is when $k = 1$. The ith individual x_i^{t+1} can be generated as follows:

$$x_i^{t+1} = \theta y_i^{t+1} + \omega x_j^t, \tag{5}$$

where x_j^t is the position for individual j ($j \in \{1, 2, \cdots, N_P\}$) at iteration t, and f_j^{t+1} is its fitness. θ and ω are weighting factors satisfying $\theta + \omega = 1$. They can be given as follows:

$$\theta = r, \ \omega = 1 - r \tag{6}$$

Here, r is a random number that is drawn from the uniform distribution in $[0, 1]$. The individual j can be determined in the following ways:

(1) $j = i$;

(2) $j = r_1$, where r_1 is an integer between 1 and N_P that is selected randomly.

The individual generated by the second method has more population diversity than the individual generated the first way. We refer to these updating strategies as R1 and RR1, respectively. Their incorporation into the basic EHO results in EHOR1 and EHORR1, respectively.

4.2. Case of k = 2

Two individuals at two previous iterations are collected and used to generate elephant i. For this case, the ith individual x_i^{t+1} can be generated as follows:

$$x_i^{t+1} = \theta y_i^{t+1} + \omega_1 x_{j_1}^t + \omega_2 x_{j_2}^{t-1}, \tag{7}$$

where $x_{j_1}^t$ and $x_{j_2}^{t-1}$ are the positions for individuals j_1 and j_2 ($j_1, j_2 \in \{1, 2, \cdots, N_P\}$) at iterations t and $t-1$, and $f_{j_1}^t$ and $f_{j_2}^{t-1}$ are their fitness values, respectively. θ, ω_1, and ω_2 are weighting factors satisfying $\theta + \omega_1 + \omega_2 = 1$. They can be calculated as follows:

$$\theta = r,$$
$$\omega_1 = (1-r) \times \frac{f_{j_2}^{t-1}}{f_{j_2}^{t-1} + f_{j_1}^t},$$
$$\omega_2 = (1-r) \times \frac{f_{j_1}^t}{f_{j_2}^{t-1} + f_{j_1}^t}. \tag{8}$$

Here, r is a random number that is drawn from the uniform distribution in $[0, 1]$. Individuals j_1 and j_2 in (8) can be determined in several different ways, but in this paper, we focus on the following two approaches:

(1) $j_1 = j_2 = i$;
(2) $j_1 = r_1$, and $j_2 = r_2$, where r_1 and r_2 are integers between 1 and NP selected randomly.

As in the previous case, the individuals generated by the second method have more population diversity than the individuals generated the first way. We refer to these updating strategies as R2 and RR2, respectively. Their incorporation into EHO yields EHOR2 and EHORR2, respectively.

4.3. Case of k = 3

Three individuals at three previous iterations are collected and used to generate individual i. For this case, the ith individual x_i^{t+1} can be generated as follows:

$$x_i^{t+1} = \theta y_i^{t+1} + \omega_1 x_{j_1}^t + \omega_2 x_{j_2}^{t-1} + \omega_3 x_{j_3}^{t-2}, \tag{9}$$

where $x_{j_1}^t$, $x_{j_2}^{t-1}$ and $x_{j_3}^{t-2}$ are the positions of individuals j_1, j_2, and j_3 ($j_1, j_2, j_3 \in \{1, 2, \cdots, N_P\}$) at iterations t, $t-1$, and $t-2$, and $f_{j_1}^t$, $f_{j_2}^{t-1}$, and $f_{j_3}^{t-2}$ are their fitness values, respectively. Their weighting factors are θ, ω_1, ω_2, and ω_3 with $\theta + \omega_1 + \omega_2 + \omega_3 = 1$. The calculation can be given as follows:

$$\theta = r,$$
$$\omega_1 = \tfrac{1}{2} \times (1-r) \times \frac{f_{j_2}^{t-1} + f_{j_3}^{t-2}}{f_{j_1}^t + f_{j_2}^{t-1} + f_{j_3}^{t-2}},$$
$$\omega_2 = \tfrac{1}{2} \times (1-r) \times \frac{f_{j_1}^t + f_{j_3}^{t-2}}{f_{j_1}^t + f_{j_2}^{t-1} + f_{j_3}^{t-2}},$$
$$\omega_3 = \tfrac{1}{2} \times (1-r) \times \frac{f_{j_1}^t + f_{j_2}^{t-1}}{f_{j_1}^t + f_{j_2}^{t-1} + f_{j_3}^{t-2}}. \tag{10}$$

Although $j_1 \sim j_3$ can be determined in several ways, in this work, we adopt the following two methods:

(1) $j_1 = j_2 = j_3 = i$;
(2) $j_1 = r_1, j_2 = r_2$, and $j_3 = r_3$, where $r_1 \sim r_3$ are integer numbers between 1 and N_P selected at random.

As in the previous two cases, the individuals generated using the second method have more population diversity. We refer to these updating strategies as R3 and RR3, respectively. Their incorporation into EHO leads to EHOR3 and EHORR3, respectively.

5. Simulation Results

As discussed in Section 4, in the experimental part of our work, the six individual updating strategies (R1, RR1, R2, RR2, R3, and RR3) were incorporated separately into the basic EHO. Accordingly, we proposed six improved versions of EHO, namely: EHOR1, EHORR1, EHOR2, EHORR2, EHOR3, and EHORR3. For the sake of clarity, the basic EHO can also be identified as EHOR0, and we can call the updating strategies R0, R1, RR1, R2, RR2, R3, and RR3 for short. To provide a full assessment of the performance of each of the proposed individual updating strategies, we compared the six improved EHOs with each other and with the basic EHO. Through this comparison, we could look at the performance of the six updating strategies in order to determine whether these strategies were able to improve the performance of the EHO.

The six variants of the EHO were investigated fully from various respects through a series of experiments, using sixteen large-scale benchmarks with dimensions $D = 50, 100, 200, 500,$ and 1000. These complicated large-scale benchmarks can be found in Table 1. More information about all the benchmarks can be found in the literature [86,118,119].

Table 1. Sixteen benchmark functions.

No.	Name	No.	Name
F01	Ackley	F09	Rastrigin
F02	Alpine	F10	Schwefel 2.26
F03	Brown	F11	Schwefel 1.2
F04	Holzman 2 function	F12	Schwefel 2.22
F05	Levy	F13	Schwefel 2.21
F06	Penalty #1	F14	Sphere
F07	Powell	F15	Sum function
F08	Quartic *with noise*	F16	Zakharov

As all metaheuristic algorithms are based on a certain distribution, different runs will generate different results. With the aim of getting the most representative statistical results, we performed 30 independent runs under the same conditions, as shown in the literature [120].

For all of the methods studied in this paper, their parameters were set as follows: the scale factor $\alpha = 0.5, \beta = 0.1$, the number of the kept elephants $nKEL = 2$, and the number of clans $nClan = 5$. In the simplest form, all of the clans have an equal number of elephants. In our current work, all of the clans have the same number of elephants (i.e., $n_{ci} = 20$). Except for the number of elephants in each clan, the other parameters are the same as in the basic EHO, which can be found in the literature [12,13]. The best function values found by a certain intelligent algorithm are shown in bold font.

5.1. Unconstrained Optimization

5.1.1. D = 50

In this section of our work, seven kinds of EHOs (the basic EHO plus the six proposed improved variants) were evaluated using the 16 benchmarks mentioned previously, with dimension $D = 50$. The obtained mean function values and standard values from thirty runs are recorded in Tables 2 and 3.

From Table 2, we can see that in terms of the mean function values, R2 performed the best, at a level far better than the other methods. As for the other methods, R1 and RR1 provided a similar performance to each other, and they could find the smallest fitness values successfully on only one of the complex functions used for benchmarking. From Table 3, obviously, R2 performed in the most stable way, while for the other algorithms, EHO has a significant advantage over the other algorithms.

Table 2. Mean function values obtained by elephant herd optimization (EHO) and six improved methods with $D = 50$.

	EHO	R1	RR1	R2	RR2	R3	RR3
F01	2.57×10^{-4}	7.11×10^{-4}	0.05	$\mathbf{8.38 \times 10^{-5}}$	1.57	1.53×10^{-4}	0.01
F02	1.04×10^{-4}	2.69×10^{-4}	0.01	$\mathbf{2.74 \times 10^{-5}}$	0.23	5.13×10^{-5}	2.38×10^{-3}
F03	4.41×10^{-7}	9.16×10^{-6}	3.37×10^{-3}	$\mathbf{6.14 \times 10^{-9}}$	0.76	4.32×10^{-8}	8.25×10^{-5}
F04	1.50×10^{-15}	4.97×10^{-11}	3.58×10^{-6}	$\mathbf{2.27 \times 10^{-16}}$	0.03	3.38×10^{-16}	1.82×10^{-9}
F05	4.49	4.25	$\mathbf{3.95}$	4.43	4.89	4.44	4.50
F06	1.22	$\mathbf{1.06}$	1.62	1.72	2.01	1.76	1.79
F07	5.13×10^{-7}	2.55×10^{-6}	0.02	$\mathbf{3.50 \times 10^{-8}}$	2.25	1.51×10^{-7}	4.85×10^{-4}
F08	2.57×10^{-16}	1.24×10^{-15}	7.23×10^{-9}	$\mathbf{2.21 \times 10^{-16}}$	1.72×10^{-5}	2.21×10^{-16}	4.03×10^{-13}
F09	2.59×10^{-6}	6.86×10^{-5}	0.03	$\mathbf{9.83 \times 10^{-8}}$	9.28	5.06×10^{-7}	3.92×10^{-3}
F10	1.65×10^4	1.64×10^4	1.63×10^4	$\mathbf{1.65 \times 10^4}$	1.61×10^4	1.64×10^4	1.64×10^4
F11	1.44×10^{-5}	4.47×10^{-4}	0.36	$\mathbf{1.02 \times 10^{-6}}$	49.04	3.52×10^{-6}	2.18
F12	1.07×10^{-3}	3.81×10^{-3}	0.14	$\mathbf{2.96 \times 10^{-4}}$	2.37	5.31×10^{-4}	0.02
F13	6.69×10^{-4}	1.34×10^{-3}	0.07	$\mathbf{1.52 \times 10^{-4}}$	1.21	3.05×10^{-4}	0.01
F14	1.27×10^{-8}	1.63×10^{-7}	3.85×10^{-4}	$\mathbf{7.00 \times 10^{-10}}$	0.04	2.50×10^{-9}	3.02×10^{-6}
F15	6.70×10^{-7}	1.40×10^{-5}	7.03×10^{-3}	$\mathbf{4.76 \times 10^{-8}}$	3.61	1.77×10^{-7}	9.87×10^{-4}
F16	1.28×10^{-3}	0.30	512.90	$\mathbf{3.00 \times 10^{-5}}$	3.87×10^7	1.71×10^{-4}	0.56
TOTAL	0	1	1	$\mathbf{14}$	0	0	0

Table 3. Standard values obtained by EHO and six improved methods with $D = 50$.

	EHO	R1	RR1	R2	RR2	R3	RR3
F01	$\mathbf{2.39 \times 10^{-5}}$	9.89×10^{-4}	0.06	3.60×10^{-5}	0.22	4.89×10^{-5}	0.01
F02	$\mathbf{1.37 \times 10^{-5}}$	3.32×10^{-4}	0.01	$\mathbf{1.14 \times 10^{-5}}$	0.04	1.38×10^{-5}	2.17×10^{-3}
F03	$\mathbf{1.55 \times 10^{-8}}$	3.34×10^{-5}	6.16×10^{-3}	$\mathbf{3.72 \times 10^{-9}}$	0.07	3.52×10^{-8}	7.80×10^{-5}
F04	4.50×10^{-16}	2.49×10^{-10}	1.08×10^{-5}	$\mathbf{6.72 \times 10^{-18}}$	0.01	2.92×10^{-16}	5.64×10^{-9}
F05	$\mathbf{0.16}$	0.26	0.52	0.30	0.20	0.30	0.33
F06	0.23	$\mathbf{0.18}$	0.41	0.28	0.32	0.25	0.27
F07	1.40×10^{-7}	5.25×10^{-6}	0.04	$\mathbf{2.71 \times 10^{-8}}$	0.76	7.78×10^{-8}	1.58×10^{-3}
F08	6.86×10^{-18}	3.99×10^{-15}	1.96×10^{-8}	$\mathbf{1.68 \times 10^{-20}}$	6.10×10^{-6}	1.26×10^{-19}	1.17×10^{-12}
F09	4.61×10^{-7}	1.75×10^{-4}	0.09	$\mathbf{6.36 \times 10^{-8}}$	1.60	3.11×10^{-7}	0.02
F10	444.40	591.40	502.50	506.70	486.40	454.70	$\mathbf{349.00}$
F11	3.73×10^{-6}	1.98×10^{-3}	0.71	$\mathbf{8.92 \times 10^{-7}}$	29.12	2.30×10^{-6}	9.49
F12	$\mathbf{1.03 \times 10^{-4}}$	5.54×10^{-3}	0.15	1.03×10^{-4}	0.28	1.21×10^{-4}	0.01
F13	9.58×10^{-5}	1.97×10^{-3}	0.07	$\mathbf{5.39 \times 10^{-5}}$	0.15	6.84×10^{-5}	6.05×10^{-3}
F14	2.07×10^{-9}	5.02×10^{-7}	1.09×10^{-3}	$\mathbf{6.10 \times 10^{-10}}$	0.01	2.04×10^{-9}	3.90×10^{-6}
F15	$\mathbf{1.00 \times 10^{-7}}$	4.76×10^{-5}	0.02	3.41×10^{-8}	0.73	8.84×10^{-8}	3.40×10^{-3}
F16	8.21×10^{-4}	1.43	1.23×10^3	$\mathbf{2.06 \times 10^{-5}}$	1.92×10^7	1.21×10^{-4}	0.44
TOTAL	3	1	0	11	0	0	1

5.1.2. $D = 100$

As above, the same seven kinds of EHOs were evaluated using the sixteen benchmarks mentioned previously, with dimension $D = 100$. The obtained mean function values and standard values from 30 runs are recorded in Tables 4 and 5.

Regarding the mean function values, Table 4 shows that R2 performed much better than the other algorithms, providing the smallest function values on 13 out of 16 functions. As for the other algorithms, R0, R1, and RR1 gave a similar performance to each other, performing the best only on one function each (F10, F06, and F05, respectively). From Table 5, obviously, R2 performed in the most stable way, while for the other algorithms, EHO has significant advantage over other algorithms.

Table 4. Mean function values obtained by EHO and six improved methods with $D = 100$.

	EHO	R1	RR1	R2	RR2	R3	RR3
F01	3.22×10^{-4}	6.27×10^{-4}	0.12	$\mathbf{6.92 \times 10^{-5}}$	1.70	1.78×10^{-4}	0.01
F02	2.55×10^{-4}	6.86×10^{-4}	0.04	$\mathbf{6.28 \times 10^{-5}}$	0.48	1.08×10^{-4}	4.78×10^{-3}
F03	9.83×10^{-7}	2.65×10^{-5}	1.27×10^{-3}	$\mathbf{1.55 \times 10^{-8}}$	1.50	9.96×10^{-8}	2.01×10^{-4}
F04	1.18×10^{-14}	1.85×10^{-9}	7.43×10^{-6}	$\mathbf{2.55 \times 10^{-16}}$	0.13	7.25×10^{-16}	1.04×10^{-9}
F05	9.19	9.01	**8.41**	9.22	9.51	9.07	9.26
F06	3.11	**2.91**	3.89	3.74	4.30	3.80	3.80
F07	2.56×10^{-6}	4.65×10^{-5}	0.02	$\mathbf{7.34 \times 10^{-8}}$	5.33	3.31×10^{-7}	3.35×10^{-3}
F08	4.18×10^{-16}	1.13×10^{-13}	1.38×10^{-7}	$\mathbf{2.21 \times 10^{-16}}$	8.16×10^{-5}	2.24×10^{-16}	3.03×10^{-12}
F09	8.20×10^{-6}	4.58×10^{-5}	0.08	$\mathbf{2.47 \times 10^{-7}}$	19.19	1.05×10^{-6}	1.30×10^{-3}
F10	$\mathbf{3.58 \times 10^4}$	3.56×10^4	3.50×10^4	3.55×10^4	3.59×10^4	3.56×10^4	3.68×10^4
F11	6.15×10^{-5}	4.45×10^{-3}	3.54	$\mathbf{5.51 \times 10^{-6}}$	200.20	1.74×10^{-5}	0.07
F12	2.53×10^{-3}	5.37×10^{-3}	0.26	$\mathbf{5.94 \times 10^{-4}}$	5.27	1.04×10^{-3}	0.05
F13	8.12×10^{-4}	1.50×10^{-3}	0.06	$\mathbf{1.72 \times 10^{-4}}$	1.46	3.52×10^{-4}	0.02
F14	3.99×10^{-8}	1.09×10^{-6}	5.59×10^{-4}	$\mathbf{1.23 \times 10^{-9}}$	0.09	4.99×10^{-9}	8.68×10^{-6}
F15	4.45×10^{-6}	5.90×10^{-4}	0.11	$\mathbf{2.52 \times 10^{-7}}$	15.00	8.77×10^{-7}	1.83×10^{-3}
F16	0.04	2.20	7.85×10^5	$\mathbf{7.09 \times 10^{-4}}$	1.58×10^{10}	3.31×10^{-3}	987.40
TOTAL	1	1	1	**13**	0	0	0

Table 5. Standard values obtained by EHO and six improved methods with $D = 100$.

	EHO	R1	RR1	R2	RR2	R3	RR3
F01	$\mathbf{1.85 \times 10^{-5}}$	8.00×10^{-4}	0.38	2.47×10^{-5}	0.12	4.83×10^{-5}	0.01
F02	$\mathbf{1.24 \times 10^{-5}}$	1.10×10^{-3}	0.08	2.65×10^{-5}	0.08	3.99×10^{-5}	3.78×10^{-3}
F03	2.22×10^{-8}	9.04×10^{-5}	1.77×10^{-3}	$\mathbf{8.47 \times 10^{-9}}$	0.21	4.81×10^{-8}	1.77×10^{-4}
F04	2.03×10^{-15}	1.02×10^{-8}	2.98×10^{-5}	$\mathbf{3.90 \times 10^{-17}}$	0.05	6.76×10^{-16}	4.03×10^{-9}
F05	0.11	0.29	0.76	0.25	**0.10**	0.31	0.25
F06	0.31	0.33	0.38	0.29	0.41	**0.25**	0.29
F07	3.85×10^{-7}	1.44×10^{-4}	0.04	$\mathbf{6.16 \times 10^{-8}}$	0.95	1.70×10^{-7}	0.01
F08	2.21×10^{-17}	5.63×10^{-13}	4.37×10^{-7}	$\mathbf{4.23 \times 10^{-20}}$	3.09×10^{-5}	3.81×10^{-19}	9.87×10^{-12}
F09	8.05×10^{-7}	1.53×10^{-4}	0.18	$\mathbf{1.46 \times 10^{-7}}$	3.21	6.43×10^{-7}	5.79×10^{-4}
F10	733.60	769.20	834.60	606.00	**547.10**	725.00	621.20
F11	1.18×10^{-5}	0.01	13.78	$\mathbf{4.34 \times 10^{-6}}$	114.20	9.54×10^{-6}	0.25
F12	$\mathbf{1.23 \times 10^{-4}}$	0.01	0.32	2.23×10^{-4}	0.50	2.95×10^{-4}	0.03
F13	6.13×10^{-5}	3.10×10^{-3}	0.08	$\mathbf{4.52 \times 10^{-5}}$	0.17	1.02×10^{-4}	0.01
F14	3.74×10^{-9}	3.58×10^{-6}	1.84×10^{-3}	$\mathbf{8.08 \times 10^{-10}}$	0.01	2.98×10^{-9}	8.97×10^{-6}
F15	5.46×10^{-7}	2.88×10^{-3}	0.23	$\mathbf{2.59 \times 10^{-7}}$	2.76	6.25×10^{-7}	4.04×10^{-3}
F16	3.18×10^{-3}	8.63	3.03×10^6	$\mathbf{4.19 \times 10^{-4}}$	3.52×10^9	1.74×10^{-3}	4.76×10^3
TOTAL	3	0	0	**10**	2	1	0

5.1.3. $D = 200$

Next, the seven types of EHOs were evaluated using the 16 benchmarks mentioned previously, with dimension $D = 200$. The obtained mean function values and standrd values from 30 runs are recorded in Tables 6 and 7.

From Table 6, we can see that in terms of the mean function values, R2 performed much better than the other algorithms, providing the smallest function values on 13 out of 16 of the benchmark functions. As for the other methods, R1 ranked second, having performed the best on two of the benchmark functions. RR1 ranked third, giving the best result on one of the functions. From Table 7, obviously, R2 performed in the most stable way, while for the other algorithms, EHO has a significant advantage over the other algorithms.

Table 6. Mean function values obtained by EHO and six improved methods with $D = 200$.

	EHO	R1	RR1	R2	RR2	R3	RR3
F01	3.68×10^{-4}	8.01×10^{-4}	0.03	$\mathbf{8.94 \times 10^{-5}}$	1.71	1.78×10^{-4}	9.60×10^{-3}
F02	5.65×10^{-4}	7.10×10^{-4}	0.06	$\mathbf{1.08 \times 10^{-4}}$	1.06	2.19×10^{-4}	0.01
F03	2.17×10^{-6}	2.23×10^{-5}	8.24×10^{-3}	$\mathbf{2.72 \times 10^{-8}}$	3.12	1.83×10^{-7}	7.55×10^{-4}
F04	7.43×10^{-14}	1.46×10^{-9}	1.82×10^{-5}	$\mathbf{3.25 \times 10^{-16}}$	0.56	3.46×10^{-15}	2.37×10^{-8}
F05	18.45	18.39	**17.57**	18.53	18.70	18.50	18.50
F06	6.90	**6.89**	8.03	7.60	8.67	7.72	7.82
F07	7.24×10^{-6}	2.87×10^{-5}	0.04	$\mathbf{1.76 \times 10^{-7}}$	11.34	6.97×10^{-7}	2.21×10^{-3}
F08	1.20×10^{-15}	2.20×10^{-12}	1.05×10^{-8}	$\mathbf{2.22 \times 10^{-16}}$	3.94×10^{-4}	2.21×10^{-16}	8.55×10^{-11}
F09	2.16×10^{-5}	1.46×10^{-4}	0.14	$\mathbf{6.14 \times 10^{-7}}$	40.87	2.38×10^{-6}	0.01
F10	7.58×10^4	$\mathbf{7.43 \times 10^4}$	7.59×10^4	7.58×10^4	7.50×10^4	7.58×10^4	7.58×10^4
F11	2.41×10^{-4}	2.46×10^{-3}	7.16	$\mathbf{2.72 \times 10^{-5}}$	699.60	7.04×10^{-5}	0.14
F12	5.71×10^{-3}	0.02	0.37	$\mathbf{1.29 \times 10^{-3}}$	11.26	2.41×10^{-3}	0.12
F13	9.65×10^{-4}	2.89×10^{-3}	0.06	$\mathbf{1.93 \times 10^{-4}}$	1.57	3.64×10^{-4}	0.02
F14	1.08×10^{-7}	1.82×10^{-6}	6.24×10^{-4}	$\mathbf{2.74 \times 10^{-9}}$	0.19	1.24×10^{-8}	1.46×10^{-5}
F15	2.25×10^{-5}	1.90×10^{-4}	0.96	$\mathbf{1.02 \times 10^{-6}}$	63.15	4.19×10^{-6}	0.01
F16	1.52	1.53×10^3	4.00×10^8	**0.01**	5.17×10^{12}	0.08	4.61×10^5
TOTAL	0	2	1	**13**	0	0	0

Table 7. Standard values obtained by EHO and six improved methods with $D = 200$.

	EHO	R1	RR1	R2	RR2	R3	RR3
F01	$\mathbf{1.10 \times 10^{-5}}$	1.35×10^{-3}	0.04	4.38×10^{-5}	0.13	5.87×10^{-5}	4.78×10^{-3}
F02	$\mathbf{2.70 \times 10^{-5}}$	8.25×10^{-4}	0.10	3.76×10^{-5}	0.12	4.61×10^{-5}	0.01
F03	3.50×10^{-8}	7.44×10^{-5}	0.02	$\mathbf{1.65 \times 10^{-8}}$	0.27	1.10×10^{-7}	1.55×10^{-3}
F04	1.35×10^{-14}	5.94×10^{-9}	4.19×10^{-5}	$\mathbf{1.21 \times 10^{-16}}$	0.14	9.76×10^{-15}	1.22×10^{-7}
F05	0.12	0.16	0.76	0.17	**0.05**	0.18	0.18
F06	0.34	0.40	0.28	0.30	0.58	**0.22**	0.29
F07	6.49×10^{-7}	6.58×10^{-5}	0.04	$\mathbf{1.11 \times 10^{-7}}$	2.00	4.21×10^{-7}	5.56×10^{-3}
F08	7.08×10^{-17}	1.13×10^{-11}	1.80×10^{-8}	$\mathbf{1.61 \times 10^{-19}}$	1.28×10^{-4}	1.05×10^{-18}	4.33×10^{-10}
F09	1.44×10^{-6}	3.78×10^{-4}	0.23	$\mathbf{5.66 \times 10^{-7}}$	3.79	1.43×10^{-6}	0.05
F10	897.80	853.40	971.60	1.12×10^3	**833.10**	1.02×10^3	907.60
F11	4.63×10^{-5}	5.96×10^{-3}	14.39	$\mathbf{2.64 \times 10^{-5}}$	353.00	5.11×10^{-5}	0.27
F12	$\mathbf{2.72 \times 10^{-4}}$	0.03	0.36	4.03×10^{-4}	1.09	8.39×10^{-4}	0.13
F13	$\mathbf{7.38 \times 10^{-5}}$	4.15×10^{-3}	0.08	7.99×10^{-5}	0.16	9.51×10^{-5}	0.01
F14	7.41×10^{-9}	7.42×10^{-6}	1.10×10^{-3}	$\mathbf{1.96 \times 10^{-9}}$	0.02	1.03×10^{-8}	1.03×10^{-5}
F15	1.88×10^{-6}	3.50×10^{-4}	2.13	$\mathbf{8.43 \times 10^{-7}}$	9.40	2.12×10^{-6}	0.04
F16	0.14	4.85×10^3	1.74×10^9	$\mathbf{8.45 \times 10^{-3}}$	7.73×10^{11}	0.04	2.00×10^6
TOTAL	4	0	0	9	2	1	0

5.1.4. $D = 500$

The seven kinds of EHOs also were evaluated using the same 16 benchmarks mentioned previously, with dimension $D = 500$. The obtained mean function values and standard values from 30 runs are recorded in Tables 8 and 9.

In terms of the mean function values, Table 8 shows that R2 performed much better than the other methods, providing the smallest function values on 13 out of 16 functions. In comparison, R0, RR1, and RR3 gave similar performances to each other, performing the best on only one function each. From Table 9, obviously, R2 performed in the most stable way, while for the other algorithms, EHO has a significant advantage over the other algorithms.

Table 8. Mean function values obtained by EHO and six improved methods with $D = 500$.

	EHO	R1	RR1	R2	RR2	R3	RR3
F01	3.92×10^{-4}	7.03×10^{-4}	0.03	$\mathbf{8.75 \times 10^{-5}}$	1.76	1.72×10^{-4}	8.52×10^{-3}
F02	1.53×10^{-3}	3.40×10^{-3}	0.11	$\mathbf{2.99 \times 10^{-4}}$	2.58	5.47×10^{-4}	0.02
F03	5.60×10^{-6}	8.70×10^{-5}	0.04	$\mathbf{5.67 \times 10^{-8}}$	7.91	4.15×10^{-7}	2.14×10^{-3}
F04	6.51×10^{-13}	4.40×10^{-9}	7.69×10^{-3}	$\mathbf{1.48 \times 10^{-15}}$	3.94	1.45×10^{-14}	7.70×10^{-8}
F05	45.80	45.90	**45.44**	45.92	45.99	45.91	45.93
F06	**18.55**	19.23	19.93	19.44	21.13	19.43	19.62
F07	2.35×10^{-5}	6.78×10^{-5}	0.24	$\mathbf{4.42 \times 10^{-7}}$	27.03	1.59×10^{-6}	8.63×10^{-3}
F08	8.52×10^{-15}	3.89×10^{-13}	1.60×10^{-7}	$\mathbf{2.23 \times 10^{-16}}$	2.42×10^{-3}	2.39×10^{-16}	1.51×10^{-10}
F09	6.13×10^{-5}	2.15×10^{-3}	0.87	$\mathbf{1.31 \times 10^{-6}}$	107.70	6.53×10^{-6}	0.02
F10	1.96×10^{5}	1.97×10^{5}	1.97×10^{5}	1.99×10^{5}	1.99×10^{5}	1.92×10^{5}	$\mathbf{1.95 \times 10^{5}}$
F11	1.61×10^{-3}	0.17	10.43	$\mathbf{1.40 \times 10^{-4}}$	5.30×10^{3}	4.16×10^{-4}	0.71
F12	0.02	0.03	1.30	$\mathbf{3.15 \times 10^{-3}}$	30.31	6.04×10^{-3}	0.57
F13	1.07×10^{-3}	3.01×10^{-3}	0.06	$\mathbf{2.07 \times 10^{-4}}$	1.72	4.22×10^{-4}	0.02
F14	3.12×10^{-7}	4.56×10^{-6}	5.26×10^{-3}	$\mathbf{6.15 \times 10^{-9}}$	0.48	2.74×10^{-8}	6.08×10^{-5}
F15	1.71×10^{-4}	7.41×10^{-4}	1.85	$\mathbf{7.59 \times 10^{-6}}$	427.30	2.17×10^{-5}	0.07
F16	1.60×10^{3}	3.60×10^{7}	1.33×10^{12}	**1.20**	8.83×10^{15}	16.25	2.37×10^{9}
TOTAL	1	0	1	**13**	0	0	1

Table 9. Standard function values obtained by EHO and six improved methods with $D = 500$.

	EHO	R1	RR1	R2	RR2	R3	RR3
F01	$\mathbf{7.22 \times 10^{-6}}$	8.01×10^{-4}	0.02	2.66×10^{-5}	0.08	4.32×10^{-5}	4.04×10^{-3}
F02	$\mathbf{4.67 \times 10^{-5}}$	4.09×10^{-3}	0.11	1.19×10^{-4}	0.20	1.49×10^{-4}	0.01
F03	5.65×10^{-8}	2.10×10^{-4}	0.13	$\mathbf{2.62 \times 10^{-8}}$	0.51	2.91×10^{-7}	5.94×10^{-3}
F04	6.17×10^{-14}	1.83×10^{-8}	0.04	$\mathbf{3.12 \times 10^{-15}}$	0.98	2.53×10^{-14}	2.84×10^{-7}
F05	0.09	0.02	0.75	0.03	0.05	0.04	**0.02**
F06	0.31	0.32	0.18	0.21	0.85	**0.15**	0.25
F07	1.05×10^{-6}	1.35×10^{-4}	0.78	$\mathbf{3.05 \times 10^{-7}}$	3.86	9.25×10^{-7}	0.02
F08	3.61×10^{-16}	1.72×10^{-12}	4.03×10^{-7}	$\mathbf{2.01 \times 10^{-18}}$	5.40×10^{-4}	1.31×10^{-17}	6.30×10^{-10}
F09	2.15×10^{-6}	0.01	1.71	$\mathbf{9.91 \times 10^{-7}}$	12.30	4.69×10^{-6}	0.04
F10	1.54×10^{3}	1.43×10^{3}	1.30×10^{3}	$\mathbf{1.28 \times 10^{3}}$	1.38×10^{3}	1.45×10^{3}	1.65×10^{3}
F11	3.44×10^{-4}	0.70	12.21	$\mathbf{9.50 \times 10^{-5}}$	2.00×10^{3}	2.38×10^{-4}	1.51
F12	$\mathbf{3.52 \times 10^{-4}}$	0.04	1.93	1.23×10^{-3}	2.11	2.09×10^{-3}	1.24
F13	$\mathbf{5.90 \times 10^{-5}}$	4.79×10^{-3}	0.06	6.75×10^{-5}	0.12	1.69×10^{-4}	0.01
F14	1.75×10^{-8}	1.60×10^{-5}	0.02	$\mathbf{3.39 \times 10^{-9}}$	0.05	1.32×10^{-8}	1.16×10^{-4}
F15	8.50×10^{-6}	1.04×10^{-3}	3.96	$\mathbf{6.20 \times 10^{-6}}$	50.28	1.15×10^{-5}	0.19
F16	113.00	1.90×10^{8}	4.50×10^{12}	**2.37**	1.61×10^{15}	17.62	1.27×10^{10}
TOTAL	4	0	0	9	0	1	2

5.1.5. $D = 1000$

Finally, the same seven types of EHOs were evaluated using the 16 benchmarks mentioned previously, with dimension $D = 1000$. The obtained mean function values and standard values from 30 runs are recorded in Tables 10 and 11.

In terms of the mean function values, Table 10 shows that R2 had the absolute advantage over the other metaheuristic algorithms, succeeding in finding function values on 12 out of 16 functions. Among the other metaheuristic algorithms, R0 ranked second, having performed the best on three of the benchmark functions. In addition, RR1 was successful in finding the best function value. From Table 11, obviously, R2 performed in the most stable way, while for the other algorithms, EHO has a significant advantage over the other algorithms.

Table 10. Mean function values obtained by EHO and six improved methods with $D = 1000$.

	EHO	R1	RR1	R2	RR2	R3	RR3
F01	4.04×10^{-4}	1.29×10^{-3}	0.03	**8.66×10^{-5}**	1.79	1.90×10^{-4}	0.01
F02	3.14×10^{-3}	5.68×10^{-3}	0.20	**6.20×10^{-4}**	5.45	1.29×10^{-3}	0.04
F03	1.11×10^{-5}	5.10×10^{-5}	0.07	**1.41×10^{-7}**	16.18	8.78×10^{-7}	1.97×10^{-3}
F04	3.05×10^{-12}	1.17×10^{-9}	7.48×10^{-4}	**4.48×10^{-15}**	16.77	3.01×10^{-14}	4.97×10^{-7}
F05	91.29	91.36	**91.19**	91.36	91.53	91.36	91.38
F06	**38.23**	38.90	39.79	39.11	42.71	39.08	39.26
F07	5.15×10^{-5}	1.20×10^{-3}	0.39	**1.01×10^{-6}**	55.16	3.93×10^{-6}	0.01
F08	3.73×10^{-14}	1.00×10^{-11}	3.77×10^{-6}	**2.28×10^{-16}**	0.01	2.73×10^{-16}	1.66×10^{-8}
F09	1.33×10^{-4}	1.91×10^{-4}	0.91	**3.21×10^{-6}**	221.00	1.00×10^{-5}	0.05
F10	**3.94×10^{5}**	3.94×10^{5}	4.01×10^{5}	4.00×10^{5}	3.98×10^{5}	3.97×10^{5}	3.98×10^{5}
F11	5.88×10^{-3}	0.39	1.18×10^{3}	**4.79×10^{-4}**	2.06×10^{4}	1.45×10^{-3}	7.41
F12	**0.03**	0.09	1.94	66.74	72.28	54.85	56.84
F13	1.16×10^{-3}	1.77×10^{-3}	0.12	**2.40×10^{-4}**	1.87	4.74×10^{-4}	0.02
F14	6.68×10^{-7}	4.59×10^{-6}	0.03	**1.79×10^{-8}**	1.00	5.22×10^{-8}	1.46×10^{-4}
F15	7.43×10^{-4}	0.02	14.34	**2.45×10^{-5}**	1.73×10^{3}	9.52×10^{-5}	0.57
F16	4.71×10^{5}	8.97×10^{8}	1.28×10^{14}	**77.47**	2.51×10^{18}	3.89×10^{3}	4.76×10^{12}
TOTAL	3	0	1	**12**	0	0	0

Table 11. Standard function values obtained by EHO and six improved methods with $D = 1000$.

	EHO	R1	RR1	R2	RR2	R3	RR3
F01	**8.50×10^{-6}**	2.69×10^{-3}	0.04	2.81×10^{-5}	0.07	6.05×10^{-5}	0.01
F02	**6.45×10^{-5}**	6.76×10^{-3}	0.24	3.09×10^{-4}	0.39	4.28×10^{-4}	0.03
F03	**7.82×10^{-8}**	1.24×10^{-4}	0.21	1.02×10^{-7}	1.70	4.10×10^{-7}	2.21×10^{-3}
F04	2.22×10^{-13}	3.61×10^{-9}	1.86×10^{-3}	**8.75×10^{-15}**	2.65	3.01×10^{-14}	1.80×10^{-6}
F05	0.06	0.02	0.49	0.01	0.06	**0.01**	0.01
F06	0.22	0.39	**0.15**	0.22	1.08	0.18	0.22
F07	1.06×10^{-6}	4.96×10^{-3}	0.58	**6.86×10^{-7}**	5.68	1.98×10^{-6}	0.01
F08	9.68×10^{-16}	2.77×10^{-11}	1.55×10^{-5}	**3.36×10^{-18}**	2.65×10^{-3}	6.47×10^{-17}	9.11×10^{-8}
F09	4.43×10^{-6}	3.58×10^{-4}	1.64	**2.11×10^{-6}**	20.66	7.28×10^{-6}	0.09
F10	2.19×10^{3}	2.04×10^{3}	2.11×10^{3}	1.98×10^{3}	**1.60×10^{3}**	1.93×10^{3}	2.75×10^{3}
F11	9.97×10^{-4}	1.46	4.01×10^{3}	**4.03×10^{-4}**	9.58×10^{3}	8.04×10^{-4}	24.87
F12	**5.39×10^{-4}**	0.15	1.76	4.90	1.61	2.31	1.48
F13	**6.61×10^{-5}**	2.68×10^{-3}	0.15	9.41×10^{-5}	0.15	1.45×10^{-4}	0.01
F14	2.21×10^{-8}	1.05×10^{-5}	0.10	**1.05×10^{-8}**	0.08	2.83×10^{-8}	2.89×10^{-4}
F15	2.62×10^{-5}	0.04	52.24	**2.14×10^{-5}**	213.10	4.96×10^{-5}	1.43
F16	2.34×10^{4}	3.39×10^{9}	4.12×10^{14}	**93.37**	4.52×10^{17}	3.42×10^{3}	2.67×10^{13}
TOTAL	5	0	1	8	1	1	0

5.1.6. Summary of Function Values Obtained by Seven Variants of EHOs

In Section 4.1, the mean function values obtained from 30 runs were collected and analyzed. In addition, the best, mean, worst, and standard (STD) function values obtained from 30 implementations were summarized and are recorded, as shown in Table 12.

From Table 12, we can see that, in general, R2 performed far better than the six other algorithms. R1 and R0 were second and third in performance, respectively, among all of the seven tested methods. Except for R0, R1, and R2, the other metaheuristic algorithms provided similar performances, which were highly inferior to R0, R1, and R2. Looking carefully at Table 12 for the best function values, R1 provided the best performance among the seven metaheuristic algorithms, which was far better than R2. This indicates that finding a means to improve the best performance of R2 further is a challenging question in EHO studies.

Table 12. Optimization results values obtained by EHO and six improved methods for 16 benchmark functions. STD—standard.

D		EHO	R1	RR1	R2	RR2	R3	RR3
	BEST	0	**13**	2	1	0	0	0
	MEAN	0	1	1	**14**	0	0	0
50	WORST	0	2	0	**13**	0	0	1
	STD	3	1	0	**11**	0	0	1
	TOTAL	3	17	3	**39**	0	0	2
	BEST	0	**13**	1	2	0	0	0
	MEAN	1	1	1	**13**	0	0	0
100	WORST	1	2	0	**13**	0	0	0
	STD	3	0	0	**10**	2	1	0
	TOTAL	5	16	2	**38**	2	1	0
	BEST	1	**10**	1	3	0	1	0
	MEAN	0	2	1	**13**	0	0	0
200	WORST	1	2	0	**13**	0	0	0
	STD	4	0	0	**9**	2	1	0
	TOTAL	6	14	2	**38**	2	2	0
	BEST	1	**11**	1	2	0	0	1
	MEAN	1	0	1	**13**	0	0	1
500	WORST	2	0	0	**14**	0	0	0
	STD	4	0	0	**9**	0	1	2
	TOTAL	8	11	2	**38**	0	1	4
	BEST	0	**11**	1	3	0	0	1
	MEAN	3	0	1	**12**	0	0	0
1000	WORST	3	1	0	**12**	0	0	0
	STD	5	0	1	**8**	1	1	0
	TOTAL	11	12	3	**35**	1	1	1

To provide a clear demonstration of the effectiveness of the different individual updating strategies, in this part of our work, we selected five functions randomly from the 16 large-scale complex functions, and their convergence histories with dimension $D = 50, 100, 200, 500$, and 1000. From Figures 1–5, we can see that of the seven metaheuristic algorithms, R2 succeeded in finding the best function values at the end of the search in each of these five large-scale complicated functions. This trend coincided with our previous analysis.

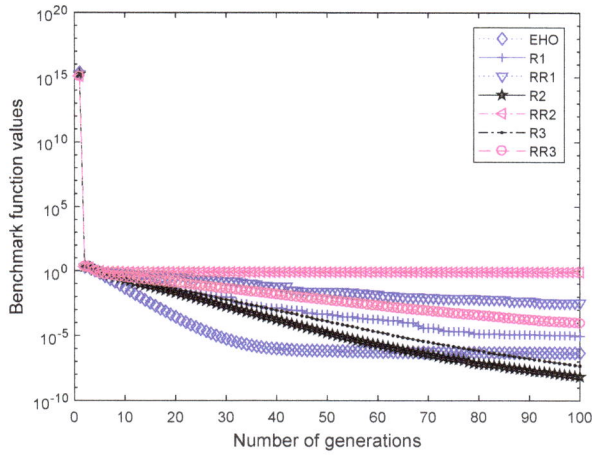

(a) F03

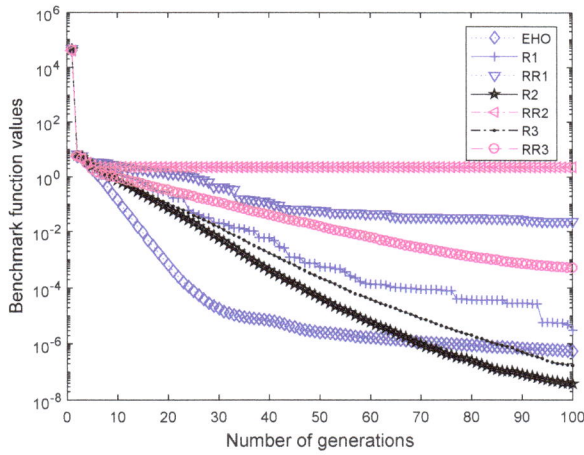

(b) F07

Figure 1. *Cont.*

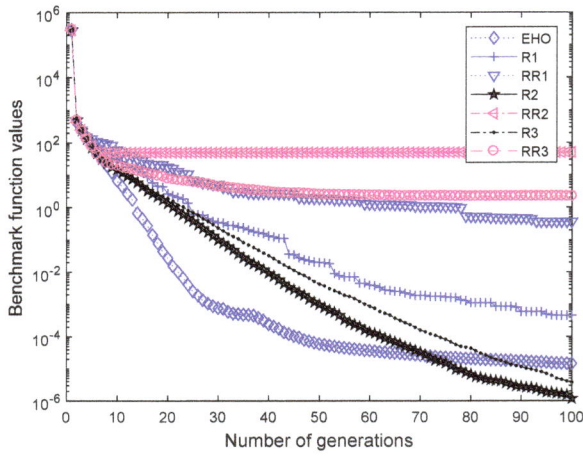

(c) F11

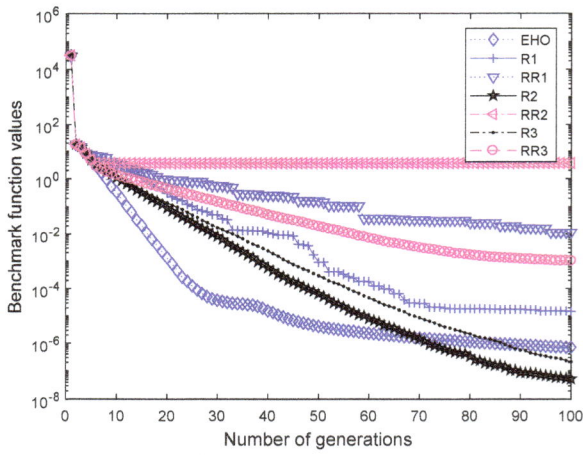

(d) F15

Figure 1. *Cont.*

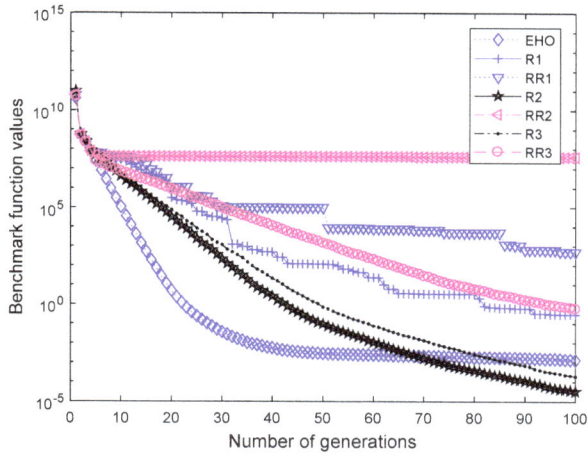

(e) F16

Figure 1. Optimization process of seven algorithms on five functions with $D = 50$. EHO—elephant herd optimization.

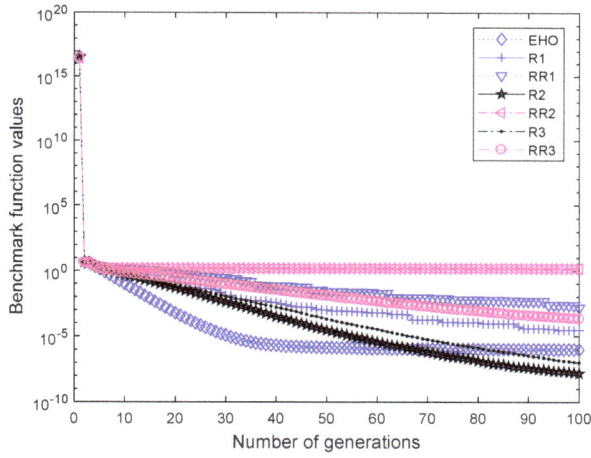

(a) F03

Figure 2. *Cont.*

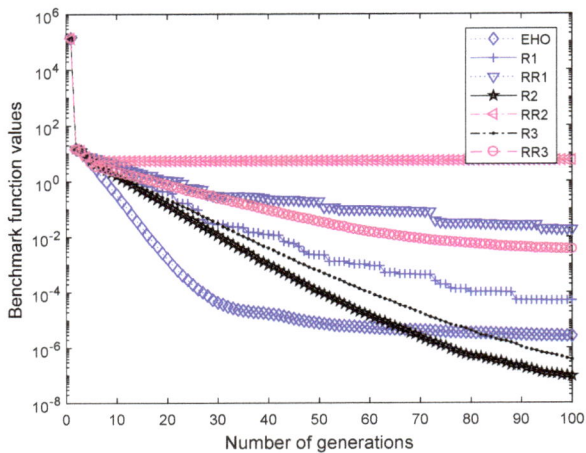

(b) F07

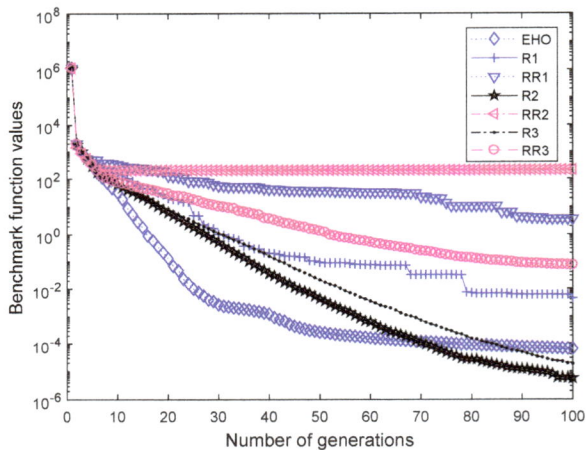

(c) F11

Figure 2. *Cont.*

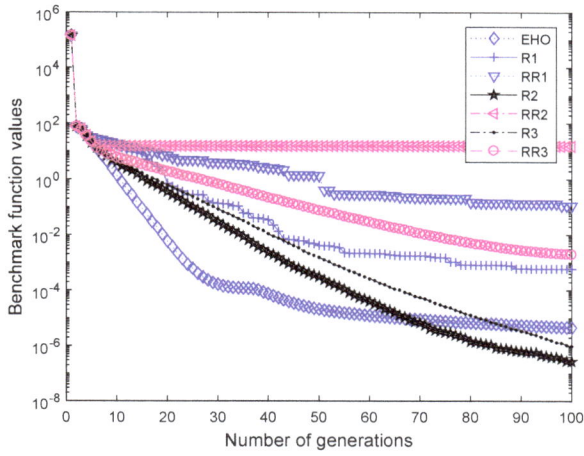

(d) F15

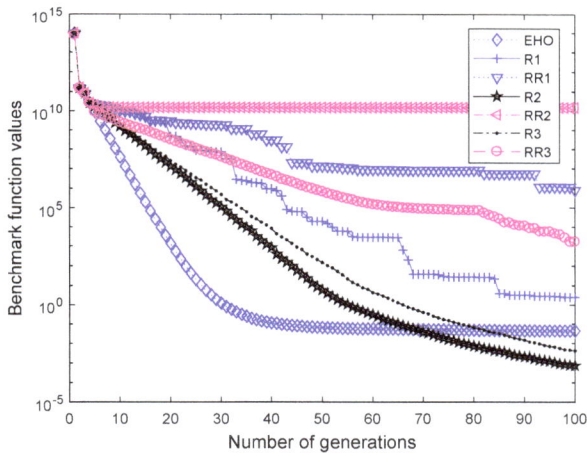

(e) F16

Figure 2. Optimization process of seven algorithms on five functions with $D = 100$.

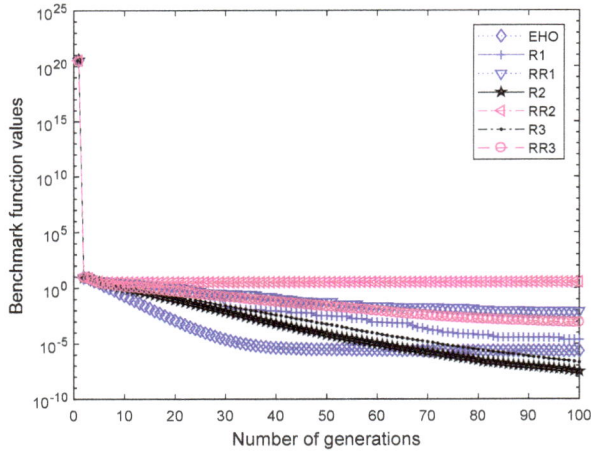

(a) F03

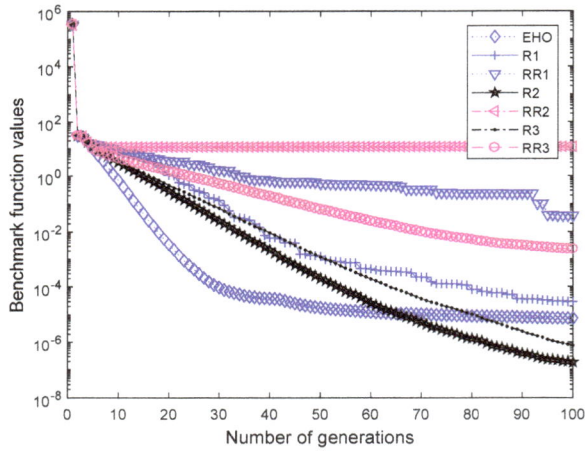

(b) F07

Figure 3. *Cont.*

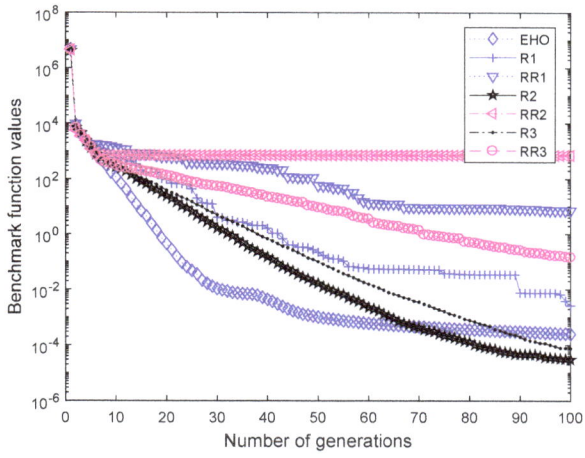

(c) F11

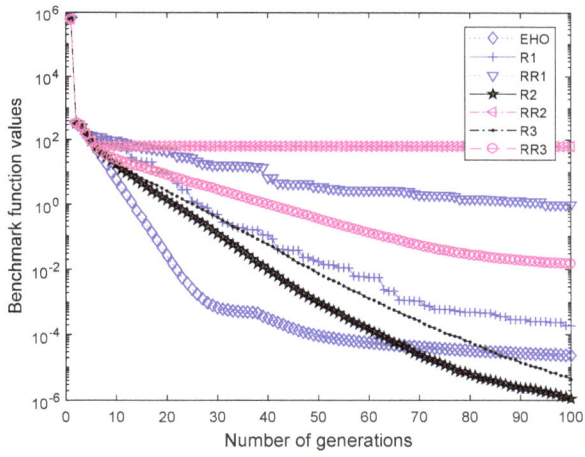

(d) F15

Figure 3. *Cont.*

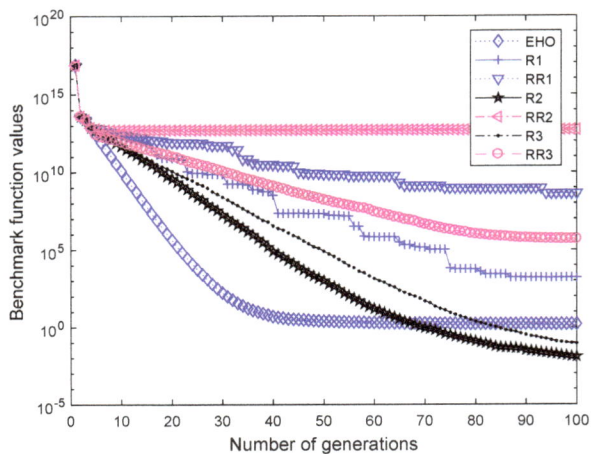

(e) F16

Figure 3. Optimization process of seven algorithms on five functions with $D = 200$.

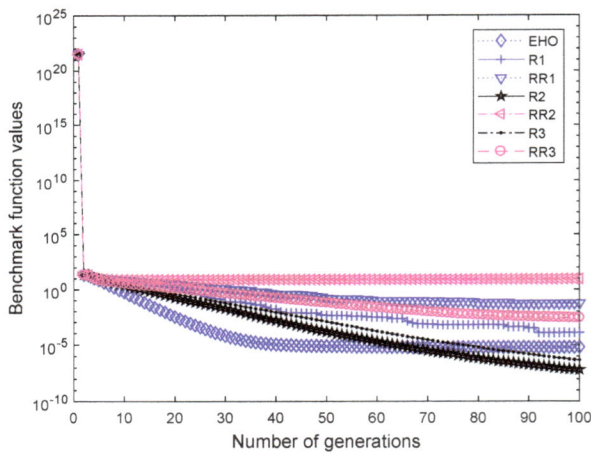

(a) F03

Figure 4. *Cont.*

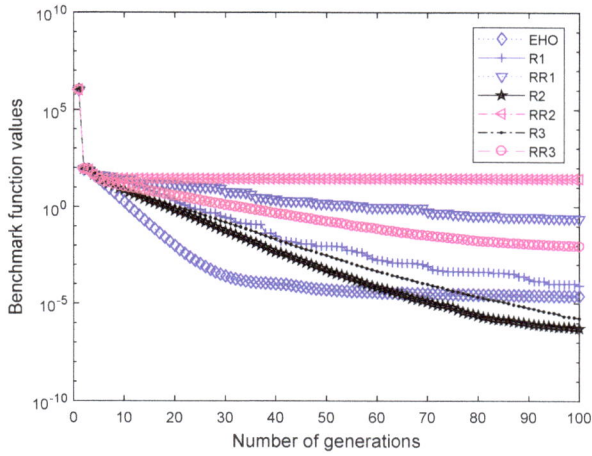

(b) F07

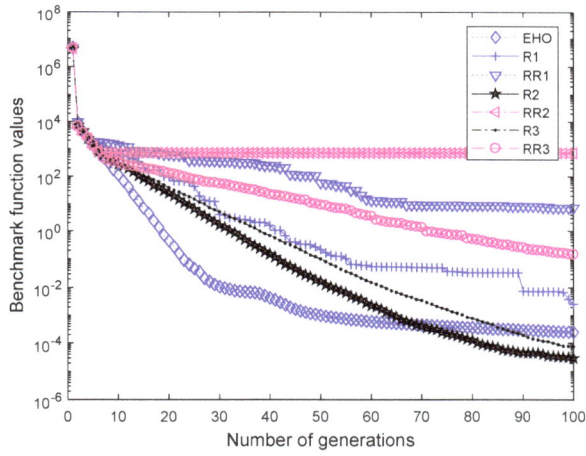

(c) F11

Figure 4. *Cont.*

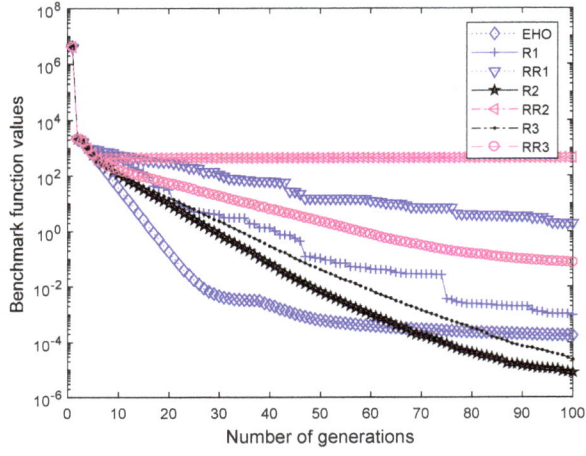

(d) F15

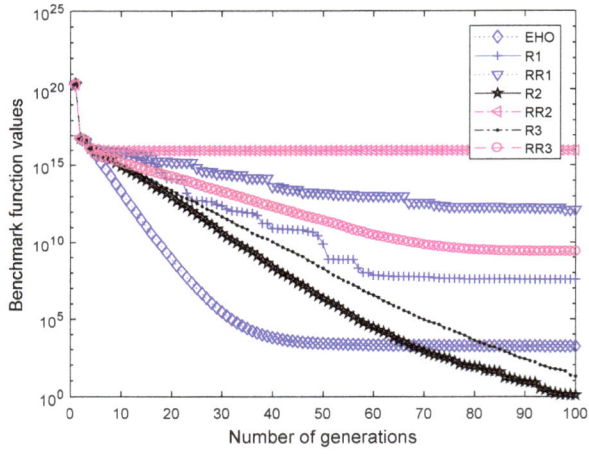

(e) F16

Figure 4. Optimization process of seven algorithms on five functions with $D = 500$.

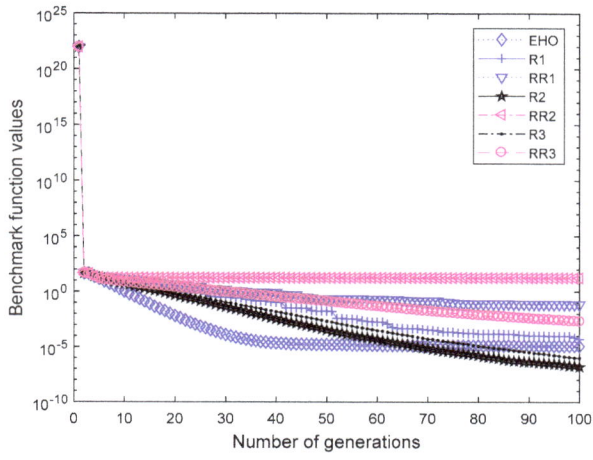

(a) F03

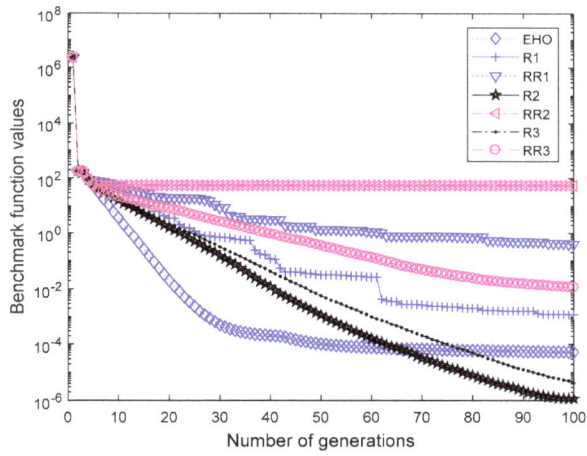

(b) F07

Figure 5. *Cont.*

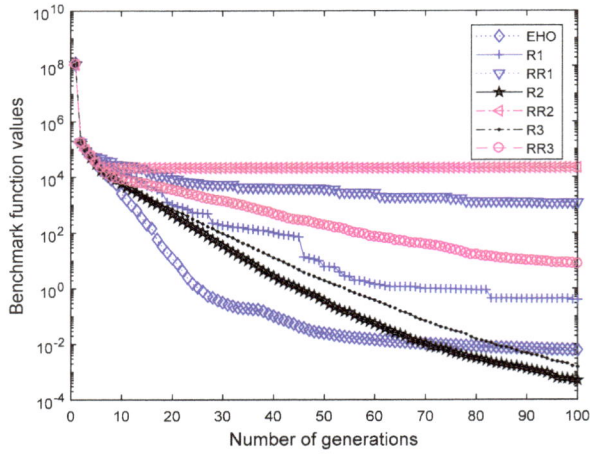

(c) F11

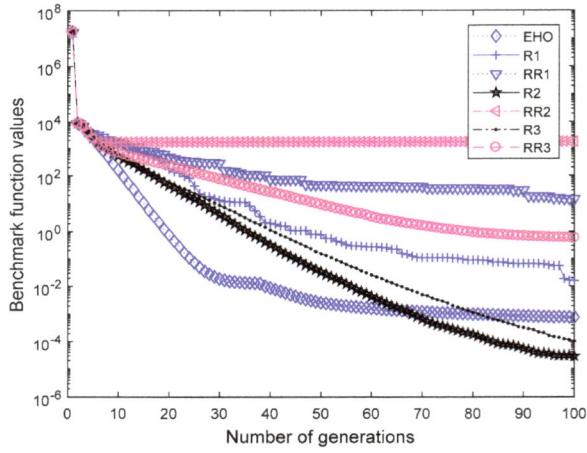

(d) F15

Figure 5. *Cont.*

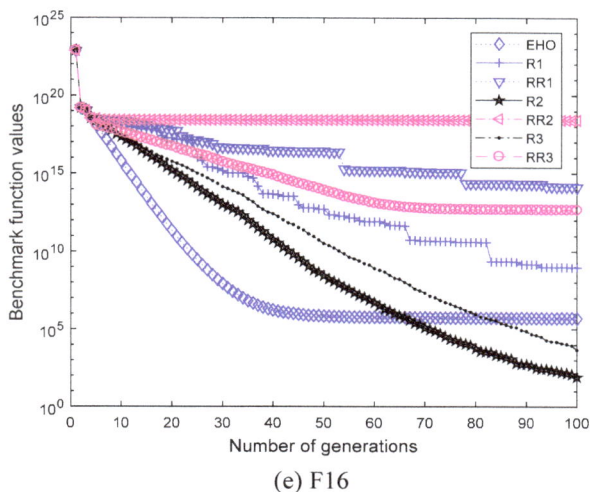

(e) F16

Figure 5. Optimization process of seven algorithms on five functions with $D = 1000$.

5.2. Constrained Optimization

Besides the standard benchmark evaluation, in this section, fourteen constrained optimization problems originated from CEC 2017 [121] are selected in order to further verify the performance of six improved versions of EHO, namely: EHOR1, EHORR1, EHOR2, EHORR2, EHOR3, and EHORR3. The six variants of the EHO were investigated fully through a series of experiments, using fourteen large-scale constrained benchmarks with dimensions $D = 50$, and 100. These complicated large-scale benchmarks can be found in Table 13. More information about all of the benchmarks can be found in [86,118,119]. As before, we performed 30 independent runs under the same conditions as shown in the literature [120]. For all of the parameters, they are the same as before.

Table 13. Details of 14 Congress on Evolutionary Computation (CEC) 2017 constrained functions. D is the number of decision variables, I is the number of inequality constraints, and E is the number of equality constraints.

No.	Problem	Search Range	Type of Objective	Number of Constraints	
				E	I
F01	**C05**	$[-10, 10]^D$	Non-Separable	0	2 Non-Separable, Rotated
F02	C06	$[-20, 20]^D$	Separable	6	0 Separable
F03	C07	$[-50, 50]^D$	Separable	2 Separable	0
F04	C08	$[-100, 100]^D$	Separable	2 Non-Separable	0
F05	C09	$[-10, 10]^D$	Separable	2 Non-Separable	0
F06	C10	$[-100, 100]^D$	Separable	2 Non-Separable	0
F07	C12	$[-100, 100]^D$	Separable	0	2 Separable
F08	C13	$[-100, 100]^D$	Non-Separable	0	3 Separable

Table 13. *Cont.*

No.	Problem	Search Range	Type of Objective	Number of Constraints E	I
F09	C15	$[-100, 100]^D$	Separable	1	1
F10	C16	$[-100, 100]^D$	Separable	1 Non-Separable	1 Separable
F11	C17	$[-100, 100]^D$	Non-Separable	1 Non-Separable	1 Separable
F12	C18	$[-100, 100]^D$	Separable	1	2 Non-Separable
F13	C25	$[-100, 100]^D$	Rotated	1 Rotated	1 Rotated
F14	C26	$[-100, 100]^D$	Rotated	1 Rotated	1 Rotated

5.2.1. $D = 50$

In this section of our work, seven kinds of EHOs were evaluated using the 14 constrained benchmarks mentioned previously, with dimension $D = 50$. The obtained mean function values and standard values from thirty runs are recorded in Tables 14 and 15.

Table 14. Mean function values obtained by EHO and six improved methods on fourteen CEC 2017 constrained optimization functions with $D = 50$.

	EHO	R1	RR1	R2	RR2	R3	RR3
F01	**8.50×10^{-6}**	2.69×10^{-3}	0.04	2.81×10^{-5}	0.07	6.05×10^{-5}	0.01
F02	**6.45×10^{-5}**	6.76×10^{-3}	0.24	3.09×10^{-4}	0.39	4.28×10^{-4}	0.03
F03	**7.82×10^{-8}**	1.24×10^{-4}	0.21	1.02×10^{-7}	1.70	4.10×10^{-7}	2.21×10^{-3}
F04	2.22×10^{-13}	3.61×10^{-9}	1.86×10^{-3}	**8.75×10^{-15}**	2.65	3.01×10^{-14}	1.80×10^{-6}
F05	0.06	0.02	0.49	0.01	0.06	**0.01**	0.01
F06	0.22	0.39	**0.15**	0.22	1.08	0.18	0.22
F07	1.06×10^{-6}	4.96×10^{-3}	0.58	**6.86×10^{-7}**	5.68	1.98×10^{-6}	0.01
F08	9.68×10^{-16}	2.77×10^{-11}	1.55×10^{-5}	**3.36×10^{-18}**	2.65×10^{-3}	6.47×10^{-17}	9.11×10^{-8}
F09	4.43×10^{-6}	3.58×10^{-4}	1.64	**2.11×10^{-6}**	20.66	7.28×10^{-6}	0.09
F10	2.19×10^3	2.04×10^3	2.11×10^3	1.98×10^3	**1.60×10^3**	1.93×10^3	2.75×10^3
F11	9.97×10^{-4}	1.46	4.01×10^3	**4.03×10^{-4}**	9.58×10^3	8.04×10^{-4}	24.87
F12	**5.39×10^{-4}**	0.15	1.76	4.90	1.61	2.31	1.48
F13	**6.61×10^{-5}**	2.68×10^{-3}	0.15	9.41×10^{-5}	0.15	1.45×10^{-4}	0.01
F14	2.21×10^{-8}	1.05×10^{-5}	0.10	**1.05×10^{-8}**	0.08	2.83×10^{-8}	2.89×10^{-4}
TOTAL	1	2	9	2	0	0	0

Table 15. Standard values obtained by EHO and six improved methods on fourteen CEC 2017 constrained optimization functions with $D = 50$.

	EHO	R1	RR1	R2	RR2	R3	RR3
F01	4.06×10^4	**3.58×10^4**	4.08×10^4	5.75×10^4	6.26×10^4	5.65×10^4	4.16×10^4
F02	127.50	123.10	100.70	86.83	102.00	122.10	**59.11**
F03	65.68	46.15	48.94	50.84	**38.43**	42.04	51.71
F04	0.37	0.61	0.72	0.72	**0.34**	0.70	0.76
F05	0.36	0.35	0.31	0.37	0.27	0.29	**0.22**
F06	2.75	2.56	1.91	2.73	2.66	2.15	**1.19**
F07	1.73×10^3	1.42×10^3	1.44×10^3	1.79×10^3	1.67×10^3	1.61×10^3	**1.37×10^3**
F08	2.51×10^8	**2.07×10^8**	2.54×10^8	2.49×10^8	3.87×10^8	2.38×10^8	2.63×10^8
F09	**1.39**	1.77	1.44	2.32	2.31	2.01	1.44
F10	41.32	40.76	**33.77**	39.29	41.69	36.24	34.50
F11	0.45	0.37	0.49	0.52	**0.36**	0.43	0.37
F12	1.81×10^3	**1.48×10^3**	1.65×10^3	1.84×10^3	1.85×10^3	1.75×10^3	1.63×10^3
F13	80.20	67.86	75.43	75.82	72.32	69.12	**61.46**
F14	**1.39**	1.54	2.28	1.66	2.11	1.88	1.76
TOTAL	2	3	1	1	2	0	5

From Table 14, we can see that in terms of the mean function values, RR1 performed the best, at a level far better than the other methods. As for the other methods, R1 and R2 provided a similar performance to each other, and they could find the smallest fitness values successfully on two constrained functions. From Table 15, it can be observed that RR3 performed in the most stable way, while for the other algorithms, they have a similar stable performance.

5.2.2. $D = 100$

As above, the same seven kinds of EHOs were evaluated using the fourteen constrained benchmarks mentioned previously, with dimension $D = 100$. The obtained mean function values and standard values from 30 runs are recorded in Tables 16 and 17.

Table 16. Mean function values obtained by EHO and six improved methods on fourteen CEC 2017 constrained optimization functions with $D = 100$.

	EHO	R1	RR1	R2	RR2	R3	RR3
F01	1.09×10^6	9.71×10^5	9.48×10^5	$\mathbf{9.49 \times 10^5}$	1.13×10^6	9.71×10^5	1.01×10^6
F02	3.88×10^3	$\mathbf{3.76 \times 10^3}$	3.80×10^3	3.77×10^3	4.00×10^3	3.77×10^3	3.88×10^3
F03	9.35×10^3	9.35×10^3	9.35×10^3	9.35×10^3	$\mathbf{9.34 \times 10^3}$	9.35×10^3	9.37×10^3
F04	$\mathbf{1.01 \times 10^3}$	1.01×10^3	1.01×10^3	1.01×10^3	1.01×10^3	1.01×10^3	1.01×10^3
F05	9.39	9.37	9.37	9.36	9.76	**9.35**	9.43
F06	1.04×10^3	1.04×10^3	$\mathbf{1.04 \times 10^3}$	1.04×10^3	1.05×10^3	1.04×10^3	1.04×10^3
F07	6.99×10^4	6.79×10^4	6.58×10^4	$\mathbf{6.50 \times 10^4}$	7.20×10^4	6.68×10^4	6.96×10^4
F08	8.77×10^9	8.02×10^9	$\mathbf{7.80 \times 10^9}$	8.19×10^9	8.95×10^9	8.14×10^9	8.62×10^9
F09	47.52	47.20	47.73	47.98	49.02	47.56	**46.86**
F10	2.18×10^3	2.16×10^3	$\mathbf{2.13 \times 10^3}$	2.14×10^3	2.22×10^3	2.17×10^3	2.16×10^3
F11	17.99	17.63	**17.18**	17.22	18.60	17.51	17.96
F12	6.82×10^4	6.60×10^4	6.50×10^4	$\mathbf{6.52 \times 10^4}$	7.01×10^4	6.63×10^4	6.87×10^4
F13	3.60×10^3	$\mathbf{3.55 \times 10^3}$	3.58×10^3	3.57×10^3	3.78×10^3	3.59×10^3	3.66×10^3
F14	55.89	55.19	53.99	**53.88**	60.16	54.61	56.37
TOTAL	1	2	4	4	1	1	1

Table 17. Standard values obtained by EHO and six improved methods on fourteen CEC 2017 constrained optimization functions with $D = 100$.

	EHO	R1	RR1	R2	RR2	R3	RR3
F01	7.04×10^4	$\mathbf{6.49 \times 10^4}$	7.85×10^4	8.34×10^4	9.27×10^4	6.51×10^4	7.03×10^4
F02	122.00	125.20	106.10	111.90	137.30	113.80	**102.50**
F03	73.82	74.87	72.61	66.75	96.82	80.98	**65.38**
F04	0.25	**0.23**	0.39	0.25	0.31	0.24	0.36
F05	0.14	0.14	0.14	0.19	0.13	0.14	**0.09**
F06	**1.20**	1.23	1.39	1.46	1.73	1.22	1.27
F07	2.93×10^3	$\mathbf{2.42 \times 10^3}$	3.14×10^3	3.02×10^3	3.25×10^3	2.73×10^3	2.52×10^3
F08	5.94×10^8	6.11×10^8	7.07×10^8	7.24×10^8	1.01×10^9	6.43×10^8	$\mathbf{5.77 \times 10^8}$
F09	0.77	0.64	1.25	0.93	**0.57**	0.95	1.06
F10	52.32	60.00	53.11	63.71	52.42	56.61	**51.64**
F11	0.67	0.72	0.73	0.72	0.84	**0.57**	0.59
F12	2.35×10^3	2.77×10^3	3.13×10^3	2.71×10^3	3.00×10^3	3.00×10^3	$\mathbf{2.34 \times 10^3}$
F13	80.54	77.54	125.50	116.70	119.80	108.50	**72.75**
F14	3.02	2.93	3.44	3.27	3.05	3.38	**2.56**
TOTAL	1	3	0	0	1	1	8

Regarding the mean function values, Table 16 shows that RR1 and R2 have the same performance, which performed much better than the other algorithms, providing the smallest function values on 4 out of 14 functions. As for the other algorithms, R1 ranks R2, EHO, RR2, R3, and RR3 gave a similar performance to each other, performing the best only on one function each. From Table 17, it can be observed that RR3 performed in the most stable way, while for the other algorithms, R1 has a significant advantage over the other algorithms.

6. Conclusions

In optimization research, few metaheuristic algorithms reuse previous information to guide the later updating process. In our proposed improvement for basic elephant herd optimization, the previous information in the population is extracted to guide the later search process. We select one, two, or three elephant individuals from the previous iterations in either a fixed or random manner. Using the information from the selected previous elephant individuals, we offer six individual updating strategies (R1, RR1, R2, RR2, R3, and RR3) that are then incorporated into the basic EHO in order to generate six variants of EHO. The final EHO individual at this iteration is generated according to the individual generated by the basic EHO at the current iteration, along with the selected previous individuals using a weighted sum. The weights are determined by a random number and the fitness of the elephant individuals at the previous iteration.

We tested our six proposed algorithms against 16 large-scale test cases. Among the six individual updating strategies, R2 performed much better than the others on most benchmarks. The experimental results demonstrated that the proposed EHO variations significantly outperformed the basic EHO.

In future research, we will propose more individual updating strategies to further improve the performance of EHO. In addition, the proposed individual updating strategies will be incorporated into other metaheuristic algorithms. We believe they will generate promising results on large-scale test functions and practical engineering cases.

Author Contributions: Methodology, J.L. and Y.L.; software, L.G. and Y.L.; validation, Y.L. and C.L.; formal analysis, J.L., Y.L., and C.L.; supervision, L.G.

Acknowledgments: This work was supported by the Overall Design and Application Framework Technology of Photoelectric System (no. 315090501) and the Early Warning and Laser Jamming System for Low and Slow Targets (no. 20170203015GX).

Conflicts of Interest: The authors declare no conflict of interest.

References

1. Wang, G.-G.; Tan, Y. Improving metaheuristic algorithms with information feedback models. *IEEE Trans. Cybern.* **2019**, *49*, 542–555. [CrossRef] [PubMed]
2. Saleh, H.; Nashaat, H.; Saber, W.; Harb, H.M. IPSO task scheduling algorithm for large scale data in cloud computing environment. *IEEE Access* **2019**, *7*, 5412–5420. [CrossRef]
3. Zhang, Y.F.; Chiang, H.D. A novel consensus-based particle swarm optimization-assisted trust-tech methodology for large-scale global optimization. *IEEE Trans. Cybern.* **2017**, *47*, 2717–2729. [PubMed]
4. Kazimipour, B.; Omidvar, M.N.; Qin, A.K.; Li, X.; Yao, X. Bandit-based cooperative coevolution for tackling contribution imbalance in large-scale optimization problems. *Appl. Soft Comput.* **2019**, *76*, 265–281. [CrossRef]
5. Jia, Y.-H.; Zhou, Y.-R.; Lin, Y.; Yu, W.-J.; Gao, Y.; Lu, L. A Distributed Cooperative Co-evolutionary CMA Evolution Strategy for Global Optimization of Large-Scale Overlapping Problems. *IEEE Access* **2019**, *7*, 19821–19834. [CrossRef]
6. De Falco, I.; Della Cioppa, A.; Trunfio, G.A. Investigating surrogate-assisted cooperative coevolution for large-Scale global optimization. *Inf. Sci.* **2019**, *482*, 1–26. [CrossRef]
7. Dhiman, G.; Kumar, V. Seagull optimization algorithm: Theory and its applications for large-scale industrial engineering problems. *Knowl.-Based Syst.* **2019**, *165*, 169–196. [CrossRef]
8. Cravo, G.L.; Amaral, A.R.S. A GRASP algorithm for solving large-scale single row facility layout problems. *Comput. Oper. Res.* **2019**, *106*, 49–61. [CrossRef]
9. Zhao, X.; Liang, J.; Dang, C. A stratified sampling based clustering algorithm for large-scale data. *Knowl.-Based Syst.* **2019**, *163*, 416–428. [CrossRef]
10. Yildiz, Y.E.; Topal, A.O. Large Scale Continuous Global Optimization based on micro Differential Evolution with Local Directional Search. *Inf. Sci.* **2019**, *477*, 533–544. [CrossRef]

11. Ge, Y.F.; Yu, W.J.; Lin, Y.; Gong, Y.J.; Zhan, Z.H.; Chen, W.N.; Zhang, J. Distributed differential evolution based on adaptive mergence and split for large-scale optimization. *IEEE Trans. Cybern.* **2017**. [CrossRef]

12. Wang, G.-G.; Deb, S.; Coelho, L.d.S. Elephant herding optimization. In Proceedings of the 2015 3rd International Symposium on Computational and Business Intelligence (ISCBI 2015), Bali, Indonesia, 7–9 December 2015; pp. 1–5.

13. Wang, G.-G.; Deb, S.; Gao, X.-Z.; Coelho, L.d.S. A new metaheuristic optimization algorithm motivated by elephant herding behavior. *Int. J. Bio-Inspired Comput.* **2016**, *8*, 394–409. [CrossRef]

14. Meena, N.K.; Parashar, S.; Swarnkar, A.; Gupta, N.; Niazi, K.R. Improved elephant herding optimization for multiobjective DER accommodation in distribution systems. *IEEE Trans. Ind. Inform.* **2018**, *14*, 1029–1039. [CrossRef]

15. Jayanth, J.; Shalini, V.S.; Ashok Kumar, T.; Koliwad, S. Land-Use/Land-Cover Classification Using Elephant Herding Algorithm. *J. Indian Soc. Remote Sens.* **2019**. [CrossRef]

16. Rashwan, Y.I.; Elhosseini, M.A.; El Sehiemy, R.A.; Gao, X.Z. On the performance improvement of elephant herding optimization algorithm. *Knowl.-Based Syst.* **2019**. [CrossRef]

17. Correia, S.D.; Beko, M.; da Silva Cruz, L.A.; Tomic, S. Elephant Herding Optimization for Energy-Based Localization. *Sensors* **2018**, *18*, 2849.

18. Jafari, M.; Salajegheh, E.; Salajegheh, J. An efficient hybrid of elephant herding optimization and cultural algorithm for optimal design of trusses. *Eng. Comput.-Ger.* **2018**. [CrossRef]

19. Hassanien, A.E.; Kilany, M.; Houssein, E.H.; AlQaheri, H. Intelligent human emotion recognition based on elephant herding optimization tuned support vector regression. *Biomed. Signal Process. Control* **2018**, *45*, 182–191. [CrossRef]

20. Wang, G.-G.; Deb, S.; Cui, Z. Monarch butterfly optimization. *Neural Comput. Appl.* **2015**. [CrossRef]

21. Yi, J.-H.; Lu, M.; Zhao, X.-J. Quantum inspired monarch butterfly optimization for UCAV path planning navigation problem. *Int. J. Bio-Inspired Comput.* **2017**. Available online: http://www.inderscience.com/info/ingeneral/forthcoming.php?jcode=ijbic (accessed on 30 March 2019).

22. Wang, G.-G.; Chu, H.E.; Mirjalili, S. Three-dimensional path planning for UCAV using an improved bat algorithm. *Aerosp. Sci. Technol.* **2016**, *49*, 231–238. [CrossRef]

23. Wang, G.; Guo, L.; Duan, H.; Liu, L.; Wang, H.; Shao, M. Path planning for uninhabited combat aerial vehicle using hybrid meta-heuristic DE/BBO algorithm. *Adv. Sci. Eng. Med.* **2012**, *4*, 550–564. [CrossRef]

24. Feng, Y.; Wang, G.-G.; Deb, S.; Lu, M.; Zhao, X. Solving 0-1 knapsack problem by a novel binary monarch butterfly optimization. *Neural Comput. Appl.* **2017**, *28*, 1619–1634. [CrossRef]

25. Feng, Y.; Yang, J.; Wu, C.; Lu, M.; Zhao, X.-J. Solving 0-1 knapsack problems by chaotic monarch butterfly optimization algorithm. *Memetic Comput.* **2018**, *10*, 135–150. [CrossRef]

26. Feng, Y.; Wang, G.-G.; Li, W.; Li, N. Multi-strategy monarch butterfly optimization algorithm for discounted {0-1} knapsack problem. *Neural Comput. Appl.* **2018**, *30*, 3019–3036. [CrossRef]

27. Feng, Y.; Yang, J.; He, Y.; Wang, G.-G. Monarch butterfly optimization algorithm with differential evolution for the discounted {0-1} knapsack problem. *Acta Electron. Sin.* **2018**, *46*, 1343–1350.

28. Feng, Y.; Wang, G.-G.; Dong, J.; Wang, L. Opposition-based learning monarch butterfly optimization with Gaussian perturbation for large-scale 0-1 knapsack problem. *Comput. Electr. Eng.* **2018**, *67*, 454–468. [CrossRef]

29. Wang, G.-G.; Zhao, X.; Deb, S. A novel monarch butterfly optimization with greedy strategy and self-adaptive crossover operator. In Proceedings of the 2015 2nd International Conference on Soft Computing & Machine Intelligence (ISCMI 2015), Hong Kong, 23–24 November 2015; pp. 45–50.

30. Wang, G.-G.; Deb, S.; Zhao, X.; Cui, Z. A new monarch butterfly optimization with an improved crossover operator. *Oper. Res. Int. J.* **2018**, *18*, 731–755. [CrossRef]

31. Wang, G.-G.; Hao, G.-S.; Cheng, S.; Qin, Q. A discrete monarch butterfly optimization for Chinese TSP problem. In Proceedings of the Advances in Swarm Intelligence: 7th International Conference, ICSI 2016, Part I, Bali, Indonesia, 25–30 June 2016; Tan, Y., Shi, Y., Niu, B., Eds.; Springer International Publishing: Cham, Switzerland, 2016; Volume 9712, pp. 165–173.

32. Wang, G.-G.; Hao, G.-S.; Cheng, S.; Cui, Z. An improved monarch butterfly optimization with equal partition and F/T mutation. In Proceedings of the Eight International Conference on Swarm Intelligence (ICSI 2017), Fukuoka, Japan, 27 July–1 August 2017; pp. 106–115.

33. Wang, G.-G. Moth search algorithm: A bio-inspired metaheuristic algorithm for global optimization problems. *Memetic Comput.* **2018**, *10*, 151–164. [CrossRef]

34. Feng, Y.; An, H.; Gao, X. The importance of transfer function in solving set-union knapsack problem based on discrete moth search algorithm. *Mathematics* **2019**, *7*, 17. [CrossRef]

35. Elaziz, M.A.; Xiong, S.; Jayasena, K.P.N.; Li, L. Task scheduling in cloud computing based on hybrid moth search algorithm and differential evolution. *Knowl.-Based Syst.* **2019**. [CrossRef]

36. Feng, Y.; Wang, G.-G. Binary moth search algorithm for discounted {0-1} knapsack problem. *IEEE Access* **2018**, *6*, 10708–10719. [CrossRef]

37. Gandomi, A.H.; Alavi, A.H. Krill herd: A new bio-inspired optimization algorithm. *Commun. Nonlinear Sci. Numer. Simulat.* **2012**, *17*, 4831–4845. [CrossRef]

38. Wang, G.-G.; Gandomi, A.H.; Alavi, A.H.; Hao, G.-S. Hybrid krill herd algorithm with differential evolution for global numerical optimization. *Neural Comput. Appl.* **2014**, *25*, 297–308. [CrossRef]

39. Wang, G.-G.; Gandomi, A.H.; Alavi, A.H. Stud krill herd algorithm. *Neurocomputing* **2014**, *128*, 363–370. [CrossRef]

40. Wang, G.-G.; Gandomi, A.H.; Alavi, A.H. An effective krill herd algorithm with migration operator in biogeography-based optimization. *Appl. Math. Model.* **2014**, *38*, 2454–2462. [CrossRef]

41. Guo, L.; Wang, G.-G.; Gandomi, A.H.; Alavi, A.H.; Duan, H. A new improved krill herd algorithm for global numerical optimization. *Neurocomputing* **2014**, *138*, 392–402. [CrossRef]

42. Wang, G.-G.; Deb, S.; Gandomi, A.H.; Alavi, A.H. Opposition-based krill herd algorithm with Cauchy mutation and position clamping. *Neurocomputing* **2016**, *177*, 147–157. [CrossRef]

43. Wang, G.-G.; Gandomi, A.H.; Alavi, A.H.; Deb, S. A hybrid method based on krill herd and quantum-behaved particle swarm optimization. *Neural Comput. Appl.* **2016**, *27*, 989–1006. [CrossRef]

44. Wang, G.-G.; Gandomi, A.H.; Yang, X.-S.; Alavi, A.H. A new hybrid method based on krill herd and cuckoo search for global optimization tasks. *Int. J. Bio-Inspired Comput.* **2016**, *8*, 286–299. [CrossRef]

45. Abdel-Basset, M.; Wang, G.-G.; Sangaiah, A.K.; Rushdy, E. Krill herd algorithm based on cuckoo search for solving engineering optimization problems. *Multimed. Tools Appl.* **2017**. [CrossRef]

46. Wang, G.-G.; Gandomi, A.H.; Alavi, A.H.; Deb, S. A multi-stage krill herd algorithm for global numerical optimization. *Int. J. Artif. Intell. Tools* **2016**, *25*, 1550030. [CrossRef]

47. Wang, G.-G.; Gandomi, A.H.; Alavi, A.H.; Gong, D. A comprehensive review of krill herd algorithm: Variants, hybrids and applications. *Artif. Intell. Rev.* **2019**, *51*, 119–148. [CrossRef]

48. Karaboga, D.; Basturk, B. A powerful and efficient algorithm for numerical function optimization: Artificial bee colony (ABC) algorithm. *J. Glob. Optim.* **2007**, *39*, 459–471. [CrossRef]

49. Wang, H.; Yi, J.-H. An improved optimization method based on krill herd and artificial bee colony with information exchange. *Memetic Comput.* **2018**, *10*, 177–198. [CrossRef]

50. Liu, F.; Sun, Y.; Wang, G.-G.; Wu, T. An artificial bee colony algorithm based on dynamic penalty and chaos search for constrained optimization problems. *Arab. J. Sci. Eng.* **2018**, *43*, 7189–7208. [CrossRef]

51. Kennedy, J.; Eberhart, R. Particle swarm optimization. In Proceedings of the IEEE International Conference on Neural Networks, Perth, Australia, 27 November–1 December 1995; pp. 1942–1948.

52. Helwig, S.; Branke, J.; Mostaghim, S. Experimental analysis of bound handling techniques in particle swarm optimization. *IEEE Trans. Evol. Comput.* **2012**, *17*, 259–271. [CrossRef]

53. Li, J.; Zhang, J.; Jiang, C.; Zhou, M. Composite particle swarm optimizer with historical memory for function optimization. *IEEE Trans. Cybern.* **2015**, *45*, 2350–2363. [CrossRef]

54. Gong, M.; Cai, Q.; Chen, X.; Ma, L. Complex network clustering by multiobjective discrete particle swarm optimization based on decomposition. *IEEE Trans. Evol. Comput.* **2014**, *18*, 82–97. [CrossRef]

55. Yuan, Y.; Ji, B.; Yuan, X.; Huang, Y. Lockage scheduling of Three Gorges-Gezhouba dams by hybrid of chaotic particle swarm optimization and heuristic-adjusted strategies. *Appl. Math. Comput.* **2015**, *270*, 74–89. [CrossRef]

56. Zhang, Y.; Gong, D.W.; Cheng, J. Multi-objective particle swarm optimization approach for cost-based feature selection in classification. *IEEE/ACM Trans. Comput. Biol. Bioinform.* **2017**, *14*, 64–75. [CrossRef]

57. Yang, X.-S.; Deb, S. Cuckoo search via Lévy flights. In Proceedings of the World Congress on Nature & Biologically Inspired Computing (NaBIC 2009), Coimbatore, India, 9–11 December 2009; pp. 210–214.

58. Wang, G.-G.; Gandomi, A.H.; Zhao, X.; Chu, H.E. Hybridizing harmony search algorithm with cuckoo search for global numerical optimization. *Soft Comput.* **2016**, *20*, 273–285. [CrossRef]

59. Wang, G.-G.; Deb, S.; Gandomi, A.H.; Zhang, Z.; Alavi, A.H. Chaotic cuckoo search. *Soft Comput.* **2016**, *20*, 3349–3362. [CrossRef]

60. Cui, Z.; Sun, B.; Wang, G.-G.; Xue, Y.; Chen, J. A novel oriented cuckoo search algorithm to improve DV-Hop performance for cyber-physical systems. *J. Parallel Distrib. Comput.* **2017**, *103*, 42–52. [CrossRef]

61. Li, J.; Li, Y.-X.; Tian, S.-S.; Zou, J. Dynamic cuckoo search algorithm based on Taguchi opposition-based search. *Int. J. Bio-Inspired Comput.* **2019**, *13*, 59–69. [CrossRef]

62. Baluja, S. *Population-Based Incremental Learning: A Method for Integrating Genetic Search Based Function Optimization and Competitive Learning*; CMU-CS-94-163; Carnegie Mellon University: Pittsburgh, PA, USA, 1994.

63. Storn, R.; Price, K. Differential evolution-a simple and efficient heuristic for global optimization over continuous spaces. *J. Glob. Optim.* **1997**, *11*, 341–359. [CrossRef]

64. Das, S.; Suganthan, P.N. Differential evolution: A survey of the state-of-the-art. *IEEE Trans. Evol. Comput.* **2011**, *15*, 4–31. [CrossRef]

65. Li, Y.-L.; Zhan, Z.-H.; Gong, Y.-J.; Chen, W.-N.; Zhang, J.; Li, Y. Differential evolution with an evolution path: A deep evolutionary algorithm. *IEEE Trans. Cybern.* **2015**, *45*, 1798–1810. [CrossRef] [PubMed]

66. Teoh, B.E.; Ponnambalam, S.G.; Kanagaraj, G. Differential evolution algorithm with local search for capacitated vehicle routing problem. *Int. J. Bio-Inspired Comput.* **2015**, *7*, 321–342. [CrossRef]

67. Beyer, H.; Schwefel, H. *Natural Computing*; Kluwer Academic Publishers: Dordrecht, The Netherlands, 2002.

68. Reddy, S.S.; Panigrahi, B.; Debchoudhury, S.; Kundu, R.; Mukherjee, R. Short-term hydro-thermal scheduling using CMA-ES with directed target to best perturbation scheme. *Int. J. Bio-Inspired Comput.* **2015**, *7*, 195–208. [CrossRef]

69. Gandomi, A.H.; Yang, X.-S.; Alavi, A.H. Mixed variable structural optimization using firefly algorithm. *Comput. Struct.* **2011**, *89*, 2325–2336. [CrossRef]

70. Yang, X.S. Firefly algorithm, stochastic test functions and design optimisation. *Int. J. Bio-Inspired Comput.* **2010**, *2*, 78–84. [CrossRef]

71. Wang, G.-G.; Guo, L.; Duan, H.; Wang, H. A new improved firefly algorithm for global numerical optimization. *J. Comput. Theor. Nanosci.* **2014**, *11*, 477–485. [CrossRef]

72. Zhang, Y.; Song, X.-F.; Gong, D.-W. A return-cost-based binary firefly algorithm for feature selection. *Inf. Sci.* **2017**, *418–419*, 561–574. [CrossRef]

73. Wang, G.-G.; Deb, S.; Coelho, L.d.S. Earthworm optimization algorithm: A bio-inspired metaheuristic algorithm for global optimization problems. *Int. J. Bio-Inspired Comput.* **2018**, *12*, 1–23. [CrossRef]

74. Goldberg, D.E. *Genetic Algorithms in Search, Optimization and Machine Learning*; Addison-Wesley: New York, NY, USA, 1998.

75. Sun, X.; Gong, D.; Jin, Y.; Chen, S. A new surrogate-assisted interactive genetic algorithm with weighted semisupervised learning. *IEEE Trans. Cybern.* **2013**, *43*, 685–698. [PubMed]

76. Garg, H. A hybrid PSO-GA algorithm for constrained optimization problems. *Appl. Math. Comput.* **2016**, *274*, 292–305. [CrossRef]

77. Dorigo, M.; Stutzle, T. *Ant Colony Optimization*; MIT Press: Cambridge, MA, USA, 2004.

78. Ciornei, I.; Kyriakides, E. Hybrid ant colony-genetic algorithm (GAAPI) for global continuous optimization. *IEEE Trans. Syst. Man Cybern. Part B Cybern.* **2012**, *42*, 234–245. [CrossRef] [PubMed]

79. Sun, X.; Zhang, Y.; Ren, X.; Chen, K. Optimization deployment of wireless sensor networks based on culture-ant colony algorithm. *Appl. Math. Comput.* **2015**, *250*, 58–70. [CrossRef]

80. Wang, G.-G.; Guo, L.; Gandomi, A.H.; Hao, G.-S.; Wang, H. Chaotic Krill Herd algorithm. *Inf. Sci.* **2014**, *274*, 17–34. [CrossRef]

81. Li, J.; Tang, Y.; Hua, C.; Guan, X. An improved krill herd algorithm: Krill herd with linear decreasing step. *Appl. Math. Comput.* **2014**, *234*, 356–367. [CrossRef]

82. Mehrabian, A.R.; Lucas, C. A novel numerical optimization algorithm inspired from weed colonization. *Ecol. Inform.* **2006**, *1*, 355–366. [CrossRef]

83. Sang, H.-Y.; Duan, P.-Y.; Li, J.-Q. An effective invasive weed optimization algorithm for scheduling semiconductor final testing problem. *Swarm Evol. Comput.* **2018**, *38*, 42–53. [CrossRef]

84. Sang, H.-Y.; Pan, Q.-K.; Duan, P.-Y.; Li, J.-Q. An effective discrete invasive weed optimization algorithm for lot-streaming flowshop scheduling problems. *J. Intell. Manuf.* **2015**, *29*, 1337–1349. [CrossRef]

85. Khatib, W.; Fleming, P. The stud GA: A mini revolution? In Proceedings of the 5th International Conference on Parallel Problem Solving from Nature, New York, NY, USA, 27–30 September 1998; pp. 683–691.

86. Simon, D. Biogeography-based optimization. *IEEE Trans. Evol. Comput.* **2008**, *12*, 702–713. [CrossRef]

87. Simon, D.; Ergezer, M.; Du, D.; Rarick, R. Markov models for biogeography-based optimization. *IEEE Trans. Syst. Man Cybern. Part B Cybern.* **2011**, *41*, 299–306. [CrossRef]

88. Geem, Z.W.; Kim, J.H.; Loganathan, G.V. A new heuristic optimization algorithm: Harmony search. *Simulation* **2001**, *76*, 60–68. [CrossRef]

89. Bilbao, M.N.; Ser, J.D.; Salcedo-Sanz, S.; Casanova-Mateo, C. On the application of multi-objective harmony search heuristics to the predictive deployment of firefighting aircrafts: A realistic case study. *Int. J. Bio-Inspired Comput.* **2015**, *7*, 270–284. [CrossRef]

90. Amaya, I.; Correa, R. Finding resonant frequencies of microwave cavities through a modified harmony search algorithm. *Int. J. Bio-Inspired Comput.* **2015**, *7*, 285–295. [CrossRef]

91. Yang, X.S.; Gandomi, A.H. Bat algorithm: A novel approach for global engineering optimization. *Eng. Comput.* **2012**, *29*, 464–483. [CrossRef]

92. Xue, F.; Cai, Y.; Cao, Y.; Cui, Z.; Li, F. Optimal parameter settings for bat algorithm. *Int. J. Bio-Inspired Comput.* **2015**, *7*, 125–128. [CrossRef]

93. Wu, G.; Shen, X.; Li, H.; Chen, H.; Lin, A.; Suganthan, P.N. Ensemble of differential evolution variants. *Inf. Sci.* **2018**, *423*, 172–186. [CrossRef]

94. Wang, G.-G.; Lu, M.; Zhao, X.-J. An improved bat algorithm with variable neighborhood search for global optimization. In Proceedings of the 2016 IEEE Congress on Evolutionary Computation (IEEE CEC 2016), Vancouver, BC, Canada, 25–29 July 2016; pp. 1773–1778.

95. Duan, H.; Zhao, W.; Wang, G.; Feng, X. Test-sheet composition using analytic hierarchy process and hybrid metaheuristic algorithm TS/BBO. *Math. Probl. Eng.* **2012**, *2012*, 712752. [CrossRef]

96. Pan, Q.-K.; Gao, L.; Wang, L.; Liang, J.; Li, X.-Y. Effective heuristics and metaheuristics to minimize total flowtime for the distributed permutation flowshop problem. *Expert Syst. Appl.* **2019**. [CrossRef]

97. Peng, K.; Pan, Q.-K.; Gao, L.; Li, X.; Das, S.; Zhang, B. A multi-start variable neighbourhood descent algorithm for hybrid flowshop rescheduling. *Swarm Evol. Comput.* **2019**, *45*, 92–112. [CrossRef]

98. Zhang, Y.; Gong, D.; Hu, Y.; Zhang, W. Feature selection algorithm based on bare bones particle swarm optimization. *Neurocomputing* **2015**, *148*, 150–157. [CrossRef]

99. Zhang, Y.; Gong, D.-W.; Sun, J.-Y.; Qu, B.-Y. A decomposition-based archiving approach for multi-objective evolutionary optimization. *Inf. Sci.* **2018**, *430–431*, 397–413. [CrossRef]

100. Logesh, R.; Subramaniyaswamy, V.; Vijayakumar, V.; Gao, X.-Z.; Wang, G.-G. Hybrid bio-inspired user clustering for the generation of diversified recommendations. *Neural Comput. Appl.* **2019**. [CrossRef]

101. Wang, G.-G.; Cai, X.; Cui, Z.; Min, G.; Chen, J. High performance computing for cyber physical social systems by using evolutionary multi-objective optimization algorithm. *IEEE Trans. Emerg. Top. Comput.* **2017**. [CrossRef]

102. Zou, D.; Li, S.; Wang, G.-G.; Li, Z.; Ouyang, H. An improved differential evolution algorithm for the economic load dispatch problems with or without valve-point effects. *Appl. Energ.* **2016**, *181*, 375–390. [CrossRef]

103. Rizk-Allah, R.M.; El-Sehiemy, R.A.; Wang, G.-G. A novel parallel hurricane optimization algorithm for secure emission/economic load dispatch solution. *Appl. Soft Compt.* **2018**, *63*, 206–222. [CrossRef]

104. Yi, J.-H.; Wang, J.; Wang, G.-G. Improved probabilistic neural networks with self-adaptive strategies for transformer fault diagnosis problem. *Adv. Mech. Eng.* **2016**, *8*, 1687814015624832. [CrossRef]

105. Sang, H.-Y.; Pan, Q.-K.; Li, J.-Q.; Wang, P.; Han, Y.-Y.; Gao, K.-Z.; Duan, P. Effective invasive weed optimization algorithms for distributed assembly permutation flowshop problem with total flowtime criterion. *Swarm Evol. Comput.* **2019**, *44*, 64–73. [CrossRef]

106. Yi, J.-H.; Xing, L.-N.; Wang, G.-G.; Dong, J.; Vasilakos, A.V.; Alavi, A.H.; Wang, L. Behavior of crossover operators in NSGA-III for large-scale optimization problems. *Inf. Sci.* **2018**. [CrossRef]

107. Yi, J.-H.; Deb, S.; Dong, J.; Alavi, A.H.; Wang, G.-G. An improved NSGA-III Algorithm with adaptive mutation operator for big data optimization problems. *Future Gener. Comput. Syst.* **2018**, *88*, 571–585. [CrossRef]

108. Liu, K.; Gong, D.; Meng, F.; Chen, H.; Wang, G.-G. Gesture segmentation based on a two-phase estimation of distribution algorithm. *Inf. Sci.* **2017**, *394–395*, 88–105. [CrossRef]

109. Wang, G.-G.; Guo, L.; Duan, H.; Liu, L.; Wang, H. The model and algorithm for the target threat assessment based on Elman_AdaBoost strong predictor. *Acta Electron. Sin.* **2012**, *40*, 901–906.

110. Wang, G.; Guo, L.; Duan, H. Wavelet neural network using multiple wavelet functions in target threat assessment. *Sci. World J.* **2013**, *2013*, 632437. [CrossRef] [PubMed]

111. Nan, X.; Bao, L.; Zhao, X.; Zhao, X.; Sangaiah, A.K.; Wang, G.-G.; Ma, Z. EPuL: An enhanced positive-unlabeled learning algorithm for the prediction of pupylation sites. *Molecules* **2017**, *22*, 1463. [CrossRef] [PubMed]

112. Zou, D.-X.; Deb, S.; Wang, G.-G. Solving IIR system identification by a variant of particle swarm optimization. *Neural Comput. Appl.* **2018**, *30*, 685–698. [CrossRef]

113. Rizk-Allah, R.M.; El-Sehiemy, R.A.; Deb, S.; Wang, G.-G. A novel fruit fly framework for multi-objective shape design of tubular linear synchronous motor. *J. Supercomput.* **2017**, *73*, 1235–1256. [CrossRef]

114. Sun, J.; Gong, D.; Li, J.; Wang, G.-G.; Zeng, X.-J. Interval multi-objective optimization with memetic algorithms. *IEEE Trans. Cybern.* **2019**. [CrossRef] [PubMed]

115. Liu, Y.; Gong, D.; Sun, X.; Zhang, Y. Many-objective evolutionary optimization based on reference points. *Appl. Soft Compt.* **2017**, *50*, 344–355. [CrossRef]

116. Gong, D.; Liu, Y.; Yen, G.G. A Meta-Objective Approach for Many-Objective Evolutionary Optimization. *Evol. Comput.* **2018**. [CrossRef]

117. Gong, D.; Sun, J.; Miao, Z. A set-based genetic algorithm for interval many-objective optimization problems. *IEEE Trans. Evol. Comput.* **2018**, *22*, 47–60. [CrossRef]

118. Yao, X.; Liu, Y.; Lin, G. Evolutionary programming made faster. *IEEE Trans. Evol. Comput.* **1999**, *3*, 82–102.

119. Yang, X.-S.; Cui, Z.; Xiao, R.; Gandomi, A.H.; Karamanoglu, M. *Swarm Intelligence and Bio-Inspired Computation*; Elsevier: Waltham, MA, USA, 2013.

120. Wang, G.; Guo, L.; Wang, H.; Duan, H.; Liu, L.; Li, J. Incorporating mutation scheme into krill herd algorithm for global numerical optimization. *Neural Comput. Appl.* **2014**, *24*, 853–871. [CrossRef]

121. Wu, G.; Mallipeddi, R.; Suganthan, P.N. *Problem Definitions and Evaluation Criteria for the CEC 2017 Competition on Constrained Real-Parameter Optimization*; National University of Defense Technology: Changsha, China; Kyungpook National University: Daegu, Korea; Nanyang Technological University: Singapore, 2017.

mathematics

MDPI

Article

Improved Whale Algorithm for Solving the Flexible Job Shop Scheduling Problem

Fei Luan [1,2,*], Zongyan Cai [1], Shuqiang Wu [1], Tianhua Jiang [3], Fukang Li [1] and Jia Yang [1]

[1] School of Construction Machinery, Chang'an University, Xi'an 710064, China; czyan@chd.edu.cn (Z.C.); wushuqiangjob@163.com (S.W.); fukangli198@163.com (F.L.); yangjialearning@163.com (J.Y.)
[2] College of Mechanical and Electrical Engineering, Shaanxi University of Science & Technology, Xi'an 710021, China
[3] School of Transportation, Ludong University, Yantai 264025, China; jth1127@163.com
* Correspondence: luanfei@sust.edu.cn

Received: 6 March 2019; Accepted: 24 April 2019; Published: 28 April 2019

Abstract: In this paper, a novel improved whale optimization algorithm (IWOA), based on the integrated approach, is presented for solving the flexible job shop scheduling problem (FJSP) with the objective of minimizing makespan. First of all, to make the whale optimization algorithm (WOA) adaptive to the FJSP, the conversion method between the whale individual position vector and the scheduling solution is firstly proposed. Secondly, a resultful initialization scheme with certain quality is obtained using chaotic reverse learning (CRL) strategies. Thirdly, a nonlinear convergence factor (NFC) and an adaptive weight (AW) are introduced to balance the abilities of exploitation and exploration of the algorithm. Furthermore, a variable neighborhood search (VNS) operation is performed on the current optimal individual to enhance the accuracy and effectiveness of the local exploration. Experimental results on various benchmark instances show that the proposed IWOA can obtain competitive results compared to the existing algorithms in a short time.

Keywords: whale optimization algorithm; flexible job shop scheduling problem; nonlinear convergence factor; adaptive weight; variable neighborhood search

---

1. Introduction

In recent years, scheduling played a crucial role in almost all manufacturing systems, as global competition became more and more intense. The classical job shop scheduling problem (JSP) is one of the most important scheduling forms existing in real manufacturing. It became a hotspot in the academic circle and received a large amount of attention in the research literature with its wide applicability and inherent complexity [1–3]. In JSP, a group of jobs need to be processed on a set of machines, where each job consists of a set of operations with a fixed order. The processing of each operation of the jobs must be performed on a given machine. Each machine is continuously available at time zero and can process only one operation at a time without interruption. The decision concerns how to sequence the operations of all the jobs on the machines, so that a given performance indicator can be optimized. Makespan is the time in which all the jobs need to be completed and is a typical performance indicator for the JSP.

The flexible job shop scheduling problem (FJSP) is an extension of the classical JSP, where each operation can be processed by any machine in a given set rather than one specified machine. The FJSP is closer to a real manufacturing environment compared with classical JSP. According to its practical applicability, the FJSP became very crucial in both academic and application fields. However, it is more difficult than classical JSP because it contains an additional decision problem, assigning operations to the appropriate machine. Therefore, the FJSP is a problem of challenging complexity and was proven to be non-deterministic polynomial-time (NP)-hard [4].

In the initial study, Brucker and Schlie firstly proposed a polynomial algorithm for solving the FJSP with two jobs [5]. During the past two decades, the FJSP attracted the interest of many researchers. There were many approximation algorithms, mainly metaheuristics, presented for solving the FJSP. Dauzere-Peres and Paulli [6] proposed a tabu search (TS) algorithm which was based on a new neighborhood structure for the FJSP. Mastrolilli and Gambardella [7] designed two neighborhood functions and presented an improved TS algorithm based on the original one which was proposed in literature [6]. Mati et al. [8] proposed a genetic algorithm for solving the FJSP with blocking constraints. Regarding the FJSP, Mousakhani. [9] developed a mixed-integer linear programming model (MILP) and designed an iterated local search algorithm to minimize total tardiness. Yuan et al. [10] designed a novel hybrid harmony search (HHS) algorithm based on the integrated approach for solving the FJSP with the objective to minimize makespan. Tao and Hua [11] presented an improved bacterial foraging algorithm (BFOA) based on cloud computing to solve the multi-objective flexible job shop scheduling problem (MOFJSP). Gong et al. [12] proposed a double flexible job shop scheduling problem (DFJSP) with flexible machines and workers, and then a new hybrid genetic algorithm (NHGA) was designed to solve the proposed DFJSP. Wang et al. [13] presented a two-stage energy-saving optimization algorithm for the FJSP. In their methods, the problem was divided into two subproblems: the machine assignment problem and the operation sequencing problem. An improved genetic algorithm was designed to solve the machine assignment problem and a genetic particle swarm hybrid algorithm was developed for the operation sequencing problem. An improved particle swarm optimization (PSO) was developed by Marzouki et al. [14]. Yuan and Xu [15] designed memetic algorithms (MAs) for solving the MOFJSP with three objectives, makespan, total workload, and critical workload. Gao et al. [16] proposed a discrete harmony search (DHS) to solve the MOFJSP with two objectives of makespan, the mean of earliness and tardiness. Piroozfard et al. [17] devised a novel multi-objective genetic algorithm (MOGA) for solving the problem with two conflicting objectives, total carbon footprint and total late work. Jiang et al. [18] pronounced a gray wolf optimization algorithm with a double-searching mode (DMGWO) to solve the energy-efficient job shop scheduling problem (EJSP). Singh and Mahapatra [19] proposed an improved particle swarm optimization (PSO) for the FJSP, in which quantum behavior and a logistic map were introduced. Wu and Sun. [20] presented a green scheduling algorithm for solving the energy-saving flexible job shop scheduling problem (EFJSP).

According to their potential advantages, many metaheuristic algorithms were proposed and improved to solve various problems [21–24]. The whale optimization algorithm (WOA) is a new metaheuristic algorithm which imitates the hunting behavior of humpback whales in nature [25]. Because of its characteristics (simple principle, fewer parameter settings, and strong optimization performance), WOA was applied to deal with various optimization problems in different fields, i.e., neural networks [26], feature selection [27], image segmentation [28], photovoltaic cells [29], the energy-efficient job shop scheduling problem [30], and the permutation flow shop scheduling problem [31]. This motivates us to present an improved whale optimization algorithm (IWOA) that can minimize the makespan of the FJSP. In our proposed IWOA, in order to make the whale optimization algorithm (WOA) adaptive to the FJSP, the conversion between the whale individual position vector and the scheduling solution is implemented by utilizing the converting method proposed in the literature [10]. Then, a resultful initialization scheme with certain quality is obtained by combining chaotic opposition-based learning strategies. To converge quickly, a nonlinear convergence factor and an adaptive weight are introduced to balance the abilities of exploitation and exploration of the algorithm. Furthermore, a variable neighborhood search operation is performed on the current optimal individual to enhance the accuracy and effectiveness of the local exploration. Experimental results on various benchmark instances show that the proposed IWOA can obtain competitive results compared to the existing algorithms in short time.

The rest of this paper is organized as follows: Section 2 introduces the definition of the problem. Section 3 illustrates the original whale optimization algorithm. In Section 4, the proposed IWOA is described in detail. Section 5 shows the empirical results of IWOA. Conclusions and suggestions for future works are provided in Section 6.

2. Problem Description

The FJSP is defined in this section. There are a set of n jobs $J = \{J_1, J_2, \ldots, J_n\}$ and a set of q machines $M = \{M_1, M_2, \ldots, M_q\}$, where n_i is the number of operations of job J_i, m is the total number of all operations, and O_{ij} represents the jth operation of job J_i. Each operation O_{ij} can be processed on one machine among a set of alternative machines of the jth operation of job J_i. The FJSP can be decomposed into two subproblems: the routing subproblem of assigning each operation to a machine among alternative machines M_{ij}, which is a subset of M, and the scheduling subproblem of sequencing the assigned operations on all alternative machines to attain a feasible schedule for optimizing a certain objective function.

The FJSP can be classified into total FJSP (TFJSP) and partial FJSP (PFJSP). For the TFJSP, each operation can be processed on all machines of M. For the PFJSP, each operation can only be processed on partial machine of M.

Moreover, the following assumptions are put forward in our study: all jobs are processable at time 0; all machines available at time 0; each machine can process at most one operation at a time; each operation must be completed once it starts; the transfer time between operations and the set-up time of machines are negligible.

In this study, the makespan was selected as the objective to be minimized. The mathematical model can be described as follows:

$$\min C_{\max} = \min(\max(C_i)), \tag{1}$$

$$ST.S_{ijh} - C_{i(j-1)k} \geq 0, \ Y_{ijh} = Y_{i(j-1)k} = 1, \tag{2}$$

$$C_{ijk} - S_{ijk} = L_{ijk}, Y_{ijk} = 1, \tag{3}$$

$$C_{egk} - C_{ijk} \geq L_{egk}, \ R_{ijegk} = 1, \ Y_{ijk} = Y_{egk} = 1, \tag{4}$$

$$\sum_{k=1}^{m} Y_{ijk} = 1, \ i = 1, 2, \ldots n; j = 1, 2, \ldots n_i, \tag{5}$$

$$Y_{ijk} \in \{0, 1\}, i = 1, 2, \ldots n; j = 1, 2, \ldots n_i; k = 1, 2, \ldots q, \tag{6}$$

$$R_{ijegk} \in \{0, 1\}, i, e = 1, 2, \ldots n; j = 1, 2, \ldots n_i; g = 1, 2, \ldots n_e; k = 1, 2, \ldots q, \tag{7}$$

$$1 \leq i, e \leq n, \ 1 \leq j, g \leq m, \ 1 \leq k, h \leq q, S_{ijk}, C_{ijk} \geq 0, \tag{8}$$

where $C_{\max}$ is the maximal completion time of jobs, C_i is the continuous variable for the completion time of job J_i, L_{ijk} denotes the processing time of operation O_{ij} on machine M_k, M_k denotes the kth machine of M, C_{ijk} is the continuous variable for the completion time of operation O_{ij} processing on machine M_k, S_{ijk} is the continuous variable for the start time of operation O_{ij} processing on machine M_k, and Y_{ijk} is a 0–1 variable; if operation O_{ij} is processed on machine M_k, $Y_{ijk} = 1$; otherwise, $Y_{ijk} = 0$. R_{ijegk} is a 0–1 variable; if operation O_{ij} is processed on machine M_k prior to operation O_{eg} as they both can be processed on it, $R_{ijegk} = 1$; otherwise, $R_{ijegk} = 0$.

Equation (1) indicates the optimizing objective. Equation (2) ensures the operation precedence constraint. Equation (3) states that each operation must be completed once it starts. Equation (4) ensures that each machine can processes only one operation at each time. Equation (5) ensures the operation can be processed only once. Equations (6) and (7) show the relevant 0–1 variables. Equation (8) denotes the non-negative feature of relevant variables.

3. Whale Optimization Algorithm

The whale optimization algorithm (WOA) is a new intelligent optimization algorithm that mimics the foraging behavior of humpback whales. After discovering the prey, the humpback whales swim in a spiral way toward the prey to surround it, at the same time emitting a bubble net for foraging. There are three kinds of predation methods, namely "encircling prey", "bubble-net attacking", and "search for prey"; among them, "bubble-net attacking" includes two kinds of approaches, namely "shrinking encircling mechanism" and "spiral updating position". Thus, the humpback whale's foraging method can be described mathematically as shown below.

3.1. Encircling Prey

Since the position of the prey (best position) is unknown in the search space, the WOA assumes that the current optimal individual is the target prey or is the closest individual to the target prey. After the optimal individual is discovered, other individuals will update their positions toward the optimal individual, and this behavior can be represented as follows

$$\vec{X}(t+1) = \vec{X}^*(t) - \vec{A} \cdot \vec{D}, \tag{9}$$

$$\vec{D} = |\vec{C} \cdot \vec{X}^*(t) - \vec{X}(t)|, \tag{10}$$

$$\vec{A} = 2\vec{a} \cdot \vec{r} - \vec{a}, \tag{11}$$

$$\vec{C} = 2\vec{r}, \tag{12}$$

where t defines the current iteration, $\vec{A}$ and $\vec{C}$ denote coefficient vectors, $\vec{D}$ represents the distance between the current optimal individual $\vec{X}^*(t)$ and the current individual $\vec{X}(t)$ at t iteration, $\vec{X}^*(t)$ represents the position vector of the optimal individual attained so far, $\vec{X}(t)$ defines the position vector of an individual whale, $\|$ represents the absolute value, and $\cdot$ means an element-by-element multiplication. Furthermore, $\vec{r}$ indicates a random vector in [0,1], and a is an element that linearly decreases from 2 to 0 according to Equation (13) over the course of an iteration, where t_{max} defines the maximum of the iteration.

$$a = 2 - \frac{2t}{t_{max}}. \tag{13}$$

The position of an individual whale can be updated according to the position of the current optimal individual. Different places around the current optimal individual can be obtained with regard to the current position by adjusting the values of $\vec{A}$ and $\vec{C}$. It is possible to reach any position within a feasible solution domain by defining the random vector r. Therefore, Equation (9) allows any individual whale to update its position in the neighborhood of the current optimal solution and simulates encircling the prey.

3.2. Bubble-Net Attacking

In the exploitation phase, the humpback whales swim around the prey in a shrinking circle and along a spiral path simultaneously. To model these two mechanisms, it is assumed that there is a probability of 50% to choose between them to update the position of whales during the optimization process.

3.2.1. Shrinking Encircling Mechanism

This behavior is obtained by decreasing the fluctuation range of A in Equation (9). According to Equation (11), the fluctuation range of A can be decreased by a. Specifically, A is a random value in the interval $[-a, a]$. Setting random values for A in $[-1,1]$, the new position of an individual whale can be defined anywhere in between the original position of the individual and the position of the current optimal individual.

3.2.2. Spiral Updating Position

To model this mechanism, the distance between the whale and the prey (current optimal individual position) is firstly calculated, and then a spiral path is achieved between the position of whale and the prey to simulate the helix-shaped movement of the humpback whales, which can be defined as follows:

$$\vec{X}(t+1) = \vec{D}' \cdot e^{bl} \cdot \cos(2\pi l) + \vec{X}^*(t), \tag{14}$$

$$\vec{D}' = |\vec{X}^*(t) - \vec{X}(t)|, \tag{15}$$

where $\vec{D}'$ is the absolute value for the distance between the current optimal individual $\vec{X}^*(t)$ and the current individual whale $\vec{X}(t)$ at t iteration, b is a constant and denotes the shape of the logarithmic spiral, l is a random number in $[-1, 1]$, and $\cdot$ is an element-by-element multiplication.

Thus, the mathematical model of the bubble-net attacking behavior of humpback whales can be defined by Equation (16), where p is a random number inside $[0, 1]$.

$$\vec{X}(t+1) = \begin{cases} \vec{X}^*(t) - \vec{A} \cdot \vec{D} & p < 0.5 \\ \vec{D}' \cdot e^{bl} \cdot \cos(2\pi l) + \vec{X}^*(t) & p \geq 0.5 \end{cases}. \tag{16}$$

3.3. Search for Prey

In contrast to the exploitation phase, the humpback whales also search for prey randomly; the mechanism is implemented by the variation of the vector A. When $|A| < 1$, the exploitation is achieved by updating the positions toward the current optimal individual; when $|A| \geq 1$, the exploration is adopted by updating the positions toward a randomly chosen individual to search for the global optimum, which can be denoted as follows:

$$\vec{X}(t+1) = \vec{X}_{rand}(t) - \vec{A} \cdot \vec{D}, \tag{17}$$

$$\vec{D} = |\vec{C} \cdot \vec{X}_{rand}(t) - \vec{X}(t)|, \tag{18}$$

where $\vec{X}_{rand}(t)$ is the individual position vector randomly selected from the current population.

4. The Proposed IWOA

4.1. Scheduling Solution Denotation

As mentioned above, the FJSP contains two subproblems, i.e., machine assignment and operation sequence. For this feature, a two-segment string with the size of $2mn$ is used to represent the scheduling solution. The first segment aims to choose an appropriate machine for each operation, and the second segment represents the processing sequence of operations on each machine. Taking a 3×3 (three jobs, three machines) FJSP as an example, each job has two operations. The scheduling solution is shown in Figure 1. For the first segment, the element j means the operation chooses the jth machine in the alternative machine set, where all elements are stored in a fixed order. For the second segment, each element represents the job code, where the elements with the same value i mean different operations of the same job i, and O_{ik} presents the kth operation of the job i.

O_{11}	O_{12}	O_{21}	O_{22}	O_{31}	O_{32}	O_{21}	O_{31}	O_{11}	O_{12}	O_{32}	O_{22}
1	2	3	2	1	2	2	3	1	1	3	2

Machine assignment | Operation sequence

Figure 1. Scheduling solution denotation.

4.2. Individual Position Vector

In our proposed IWOA, the individual position is still denoted as a multi-dimensional real vector, which also consists of two segments string with the size of mn, i.e., $X = \{x(1), x(2), \ldots x(mn), x(mn + 1), \ldots x(2mn)\}$, where $x(j) \in [x_{\min}(j), x_{\max}(j)], j = 1, 2, \ldots 2mn$. The first segment $X_1 = \{x(1), x(2), \ldots x(mn)\}$ denotes the information of machine assignment, and the second segment $X_2 = \{x(mn + 1), x(mn + 2), \ldots x(2mn)\}$ presents the information of operation sequencing. For the above 3×2 FJSP, the individual position vector can be represented by Figure 2, where element values are listed in the given order. In addition, the intervals $[x_{\min}(j), x_{\max}(j)]$ are all set as $[-\delta, \delta]$, where δ presents the number of the jobs.

O_{11}	O_{12}	O_{21}	O_{22}	O_{31}	O_{32}	O_{21}	O_{31}	O_{11}	O_{12}	O_{32}	O_{22}
2.7	1.2	-1.7	1.9	2.3	-1.2	2.2	0.2	-1.5	-1.0	1.3	2.0

Machine assignment | Operation sequence

Figure 2. Individual position vector.

4.3. Conversion Mechanism

Since the original WOA was proposed to tackle continuous problems, but the FJSP belongs to a discrete combinatorial problem, some measures should be implemented to construct the mapping relationship between the individual position vector and the discrete scheduling solution. In a previous study, Yuan et al. [10] proposed a method to implement the conversion between the continuous individual position vector and the discrete scheduling solution for the FJSP. Therefore, the conversion method in the literature [10] will be used in this study.

4.3.1. Conversion from Scheduling Solution to Individual Position Vector

For the machine assignment segment, the conversion process can be represented by Equation (19). Here, $x(i)$ denotes the ith element of the individual position vector, $s(i)$ presents the number of alternative machine set for the operation corresponding to the ith element, and $n(i)$ means the serial number of the chosen machine in its alternative machine set; if $s(i) = 1$, then $x(i)$ can be achieved by choosing a random value in the interval $[-\delta, \delta]$.

$$x(i) = \left[2\delta/(s(i) - 1)\right](n(i) - 1) - \delta, s(i) \neq 1. \tag{19}$$

For the operation sequence segment, firstly, it is needed to randomly generate mn real numbers in the range $[-\delta, \delta]$ corresponding to the scheduling solution. According to the ranked-order-value (ROV) rule, a unique ROV value is assigned to each random number in an increasing order, so that each ROV value can correspond to an operation. Secondly, the ROV value is rearranged according to the coding order of the operations, and the random number corresponding to the rearranged ROV value is the value of the element of the individual position vector. The conversion process is shown in Figure 3.

Random number	-3.0	2.8	2.1	-1.2	-0.8	1.1

Random number	-3.0	2.8	2.1	-1.2	-0.8	1.1
ROV value	1	6	5	2	3	4
Operation sequence	3	1	2	2	1	3

Opration ID	1	1	2	2	3	3
ROV value	6	3	5	2	1	4

Position element	2.8	-0.8	2.1	-1.2	-3.0	1.1

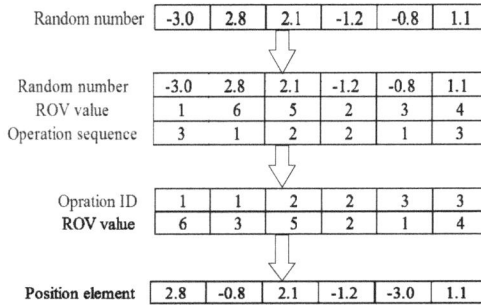

Figure 3. The conversion process from operation sequence to individual position vector.

4.3.2. Conversion from Individual Position Vector to Scheduling Solution

For the machine assignment segment, according to the reverse derivation of Equation (19), the conversion can be achieved, which can be denoted by Equation (20).

$$n(i) = round\left[\frac{(x(i) + \delta)(s(i) - 1)}{2\delta} + 1\right] \tag{20}$$

For the operation sequence segment, the ROV value is firstly increasingly assigned to each element of the individual position vector, and then used as the Fixed ID. Therefore, a new operation sequence can be obtained by corresponding the ROV value to the operations, which is shown in Figure 4.

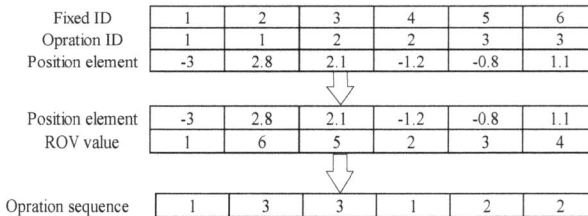

Fixed ID	1	2	3	4	5	6
Opration ID	1	1	2	2	3	3
Position element	-3	2.8	2.1	-1.2	-0.8	1.1

Position element	-3	2.8	2.1	-1.2	-0.8	1.1
ROV value	1	6	5	2	3	4

Opration sequence	1	3	3	1	2	2

Figure 4. The conversion from individual position vector to operation sequence.

4.4. Population Initialization

For a swarm intelligence optimization algorithm, the quality of the initial population is very crucial for the computational performance. In light of the characteristic of the FJSP, the population initialization process can be implemented in two phases. In the machine assignment phase, the better initial assignment schemes can be generated by utilizing a chaotic reverse learning method. In the operation sequence phase, some operation sequences are randomly generated. Combining each operation sequence with one of the initial assignment schemes, some scheduling solutions are generated and fitness function values of each scheduling solution are calculated. Then, the initial population can be achieved by choosing the scheduling solution with the best fitness value each time.

4.5. Nonlinear Convergence Factor

Like other swarm intelligence optimization algorithms, the coordination between the abilities of exploitation and exploration is important for the performance of the algorithm. In the original WOA, the abilities of exploitation and exploration mainly depend on the convergence factor *a*. The larger the value of *a* is, the stronger the ability of exploitation is, and then the WOA can exploit the optimal solution in a large space. The smaller the value of *a* is, the stronger the ability of exploration is, and then it can merely explore the optimal solution in a small space. Therefore, for improving the efficiency of exploitation, the value of *a* can be set to be larger in the early stage of iterations, which is beneficial to exploit the optimal solution in a larger space, and then it can be set to be smaller in the later stage of iterations, which is beneficial to concretely explore the better solution around the current optimal one. However, the value of *a* linearly decreases over the course of iterations by Equation (13), which cannot improve the efficiency of the nonlinear search of the algorithm. Therefore, a nonlinear improvement of *a* is adopted by Equation (21), where t_{max} and t denote the maximum iteration and current iteration, respectively.

$$a = \left(2 - \frac{2t}{t_{max}}\right)\left(1 - \frac{t^3}{t_{max}^3}\right) \tag{21}$$

4.6. Adaptive Weight

The improvement of *a* can improve the optimization ability of the algorithm to some extent, but it cannot achieve the purpose of effectively balancing the abilities of exploitation and exploration. Therefore, the adaptive weight and the nonlinear convergence factor *a* are cooperated to coordinate the abilities of exploitation and exploration of the algorithm. The adaptive weight proposed in the literature [32] is used to improve the optimization performance of the algorithm, with the formula shown by Equation (22), where t_{max} and t denote the maximum iteration and current iteration, respectively. The improved iterative formulas in the WOA can be defined by Equations (23) and (24).

$$\omega = \sin(\frac{\pi \cdot t}{2 \cdot t_{max}} + \pi) + 1. \tag{22}$$

$$\vec{X}(t+1) = \omega \cdot \vec{X}^*(t) - \vec{A} \cdot \vec{D}. \tag{23}$$

$$\vec{X}(t+1) = \vec{D}' \cdot e^{bl} \cdot \cos(2\pi l) + \omega \vec{X}^*(t). \tag{24}$$

4.7. Variable Neighborhood Search

In the local exploration phase, the whale individuals update their positions toward the current optimal individual X^* using Equation (16). Therefore, X^* determines the accuracy and effectiveness of the local exploration to some extent. Taking this into account, the variable neighborhood search strategy is used for improving the quality of the current optimal scheduling solution W^*, and then the quality of the current optimal individual X^* can be ameliorated as well. At the same time, an "iterative counter" is set for W^* and assigned 0 at the initial moment. If W^* does not change at each iteration, the "iterative counter" increases by 1; otherwise, it remains the same. When the "iterative counter" is equal to stability threshold *Ts* (15 in this paper), as the individuals reach the steady state, the variable neighborhood search strategy is performed on W^*, allowing it to escape from the local optimum. For implementing the strategy, three neighborhood structures were designed as outlined below.

For the neighborhood structure N_1, two random positions are chosen with different jobs in the second segment of the scheduling solution, exchanging the order of jobs from the second random position to the first random position.

For the neighborhood structure N_2, two random positions are chosen with different jobs in the second segment of the scheduling solution, inserting the job of the first random position in the position behind the second random position.

For the neighborhood structure N_3, a random position is chosen in the first segment of the scheduling solution, where the number of alternative machines is more than one, and then the current machine is replaced by another one of the alternative machines in the position.

The new scheduling solution is evaluated after each variable neighborhood search operation. If the new scheduling solution is better than the original one, then the new scheduling solution is set as the original one. The procedure of the variable neighborhood search operation can is illustrated in Algorithm 1.

Algorithm 1. The procedure of VNS.

Step 1: Set the current optimal scheduling solution W^* as the initial solution W, where $\lambda = 1$, $q = 1$, $q_{max} = 3$, and η_{max} represents the maximum iteration, at the initial moment.

Step 2: If $q = 1$, set $N_1(W)$ as W'; if $q = 2$, set $N_2(W)$ as W'; if $q = 3$, set $N_3(W)$ as W'; W' represents the new scheduling solution, and $N_i(W)$ represents employing the ith neighborhood structure operation on W, where I = 1, 2, or 3.

Step 3: Set W' as W, and then the local optimal scheduling solution W'' can be obtained by executing the local search operation.

Step 4: If W'' is better than W, then set W'' as W, and set $q = 1$; otherwise, set $q + 1$ as q.

Step 5: If $q > q_{max}$, then set $\eta + 1$ as η, and go to step 6; otherwise, go to step 3.

Step 6: If $\eta > \eta_{max}$, go to step 7; otherwise, go to step 2.

Step 7: End.

In this study, the threshold acceptance method is used for the local search operation, which is shown as Algorithm 2.

Algorithm 2. The procedure of the local search in VNS.

Step 1: Get the initial solution W', and set $\delta > 0$, $\gamma = 1$, $\rho = 1$, and maximum iteration γ_{max}.

Step 2: If $\rho = 1$, set $N_1(W') \cup N_3(W')$ as W''; if $\rho = 0$, set $N_2(W') \cup N_3(W')$ as W''.

Step 3: If $F_{max}(W'') - F_{max}(W') \leq \delta$, then set W'' as W'; otherwise, set $|\rho - 1|$ as ρ.

Step 4: Set $\gamma + 1$ as γ, if $\gamma > \gamma_{max}$, then set W'' as W', go to step 5; otherwise, go to step 2.

Step 5: End.

4.8. The Procedure of the Proposed IWOA

The detailed steps of the proposed IWOA can be described as Algorithm 3.

Algorithm 3. The procedure of IWOA.

Step 1: Set parameters and generate the initial population by utilizing the chaotic reverse learning strategy and search method.

Step 2 Calculate the fitness value of each scheduling solution in the population, and then find and retain the optimal scheduling solution W^*.

Step 3: Judge whether the termination conditions can be met. If not met, perform steps 4–7; otherwise, perform step 8.

Step 4: Judge whether the value of the "iterative counter" is equal to 15. If met, go to step 5; otherwise, go to step 6.

Step 5: Employ the variable neighborhood search operation on W^*, and update W^*.

Step 6: Execute the conversion from scheduling solution to individual position vector, and retain the optimal individual position vector X^* corresponding to W^*

Step 7: Update each individual position vector using Equations (17), (23) and (24), and execute the conversion from individual position vector to scheduling solution; set $t = t + 1$, and then go to step 2.

Step 8: The algorithm ends and outputs the optimal scheduling solution W^*.

---

5. Experimental Results

5.1. Experimental Settings

To evaluate the performance of the proposed IWOA for solving the FJSP, the algorithm was coded in MATLAB 2016a and run on a computer configured with an Intel Core i5-8250 central processing unit (CPU) with 1.80 GHz frequency, 8 GB random-access memory (RAM), and a Windows 10 Operating System. Fifteen famous benchmarks that included a set of 10 instances taken from Brandimarte (BRdata) [33] and five instances taken from Kacem et al (KAdata) [34] were chosen to test the proposed algorithm. These benchmark instances were used by many researchers to estimate their approaches. For each benchmark instance, experimental simulations were run 20 times using different algorithms. After several preliminary experiments, the parameters of the proposed IWOA were set as follows: a population size of 100, maximum iterations of 1000, spiral constant b of one, and η_{max} and γ_{max} both set to 10.

5.2. Effectiveness of the Improvement Strategies

In this paper, three strategies were employed to enhance the performance of the IWOA, i.e., CRL, NFC and AW, and VNS. In this subsection, the effectiveness of the three strategies is firstly evaluated. In Table 1, the first and second columns present the name and size of the problems, and computational data are listed in the following columns. "WOA" defines the original whale optimization algorithm. "IWOA-1" is the algorithm where the nonlinear convergence factor and adaptive weight are both applied to the WOA. "IWOA-2" is the whale optimization algorithm with the variable neighborhood search strategy introduced. "IWOA" is the presented algorithm in this study. In addition, "*Best*" represents the best result in the 20 runs. "*Avg*" means the average results value of the twenty runs. "*Time*" is the mean computational time (in seconds) in the 20 runs. "LB" denotes the optimal value of makespan found so far. Boldface denotes the best mean result in the 20 runs. To enhance the comparison, the same parameters were set for the compared algorithms; for instance, population size was 100 and maximum iterations were 1000.

From the experimental result in Table 1, the following conclusions can be obtained: (1) in comparisons of the "*Best*" value, the IWOA algorithm was better than the other three algorithms, which obtained seven optimal values, outperforming IWOA-1 in 12 out of 15 instances, IWOA-2 in nine out of 15 instances, and WOA in 13 out of 15 instances; (2) in comparisons of the "*Time*" value, WOA spent a shorter time than the other three algorithms. Compared with IWOA-1, the increase in computation time was mainly the result of the addition of the variable neighborhood search operation in WOA, which led to increased time complexity of the algorithm; (3) in comparisons of the "*Avg*" value, the IWOA algorithm obtained all optimal values, outperforming WOA and IWOA-1 in 15 out of 15 instances, and outperforming IWOA-2 in 13 out of 15 instances.

5.3. Effectiveness of the Proposed IWOA

To demonstrate the effectiveness of the proposed IWOA, the second experiment was executed on KAdata. In Table 2, the proposed algorithm is compared with the knowledge-based ant colony optimization (KBACO) [35], hybrid tabu search algorithm (TSPCB) [36], and the hybrid gray wolf optimization algorithm (HGWO) [37]. The first column presents the name of the problems. "*Best*" represents the best makespan. "*Avg*" means the average makespan. "*Time*" is the mean computational time of the instance. "LB" denotes the optimal value of makespan found so far. "*Avg-T*" is the mean computational time executed on KAdata. As can be seen, the proposed IWOA obtained three optimal values in solving KAdata, compared with five for ACO, five for HTS, and four for HGWO. However, the average computational time for the IWOA was very low, at only 4 s (on a Lenovo Thinkpad E480 with CPU i5-8250 @1.80GHz and 8 GB RAM) compared to 4978.8 s (in Matlab on a Dell Precision 650 workstation with a Pentium IV 2.4 GHz CPU and 1 GB RAM) for KBACO, 214.8 s (in C++ on a Pentium IV 1.6 GHz CPU and 512 MB RAM) for TSPCB, and 19 s (in Fortran on a Pentium CPU G2030@ 3.0

GHz and 2 GB) for HGWO. Because the computers applied for running the programs was different, the comparison among the running times of different algorithms was difficult. However, even if there exists some differences in the speed between the processors involved, IWOA was obviously faster than the other three algorithms.

Another experiment was implemented on BRdata. Table 3 compares our proposed IWOA with the following six algorithms: KBACO [35], TSPCB [36], HGWO [37], artificial immune algorithm (AIA) [38], particle swarm optimization combined with tabu search (PSO+TS) [39], and tabu search metaheuristic with a new neighborhood structure called "golf neighborhood" (TS3) [40]. The first column stands for the name of the problems, and the second column represents the optimal value found so far. "*Best*" represents the best makespan. "*Mean*" represents the average result of "*RPD*" in the 20 runs. Boldface denotes the best result of "*RPD*" in the 20 runs. "*RPD*" represents the relative percentage deviation to "*LB*" and is calculated as follows:

$$RPD = \frac{Best - LB}{LB} \times 100. \tag{25}$$

As can be seen from Table 3, the following conclusions can be easily obtained: (1) in comparisons of the "*Best*" value, the proposed IWOA showed competitive performance on BRdata, obtaining four optimal values, outperforming KBACO in seven out of 10 instances, TS3 and PSO+TS in nine out of 10 instances, and HGWO in eight out of 10 instances, while it was equal to both AIA and TSPCB in six out of 10 instances; (2) in comparisons of the "*RPD*" value, the proposed IWOA obtained five optimal values, outperforming KBACO in seven out of 10 instances, both TS3 and PSO+TS in nine out of 10 instances, and HGWO in eight out of 10 instances, while it was inferior to both AIA and TSPCB in three out of 10 instances; (3) in comparisons of the "*Mean*" value, the value for the proposed IWOA was very low at only 4.91, outperforming the 5.65 for KBACO, 10.12 for HGWO, 23.89 for PSO+TS, and 13.34 for TS3, while it was inferior to the 2.78 for TSPCB and 2.22 for AIA. However, by comparison, the IWOA obtained the best values in an acceptable time.

Table 1. Effectiveness analysis of improvement strategy. See Section 5.2 for column descriptions. WOA—whale optimization algorithm.

Instance	$n \times m$	LB	WOA			WOA-1			WOA-2			IWOA		
			Best	Avg	Time	Best	Avg	Time	Best	Avg	Time	Best	Avg	Time
Kacem01	4×5	11	11	11.7	0.2	11	11.3	0.2	11	11	1.8	11	11	1.8
Kacem02	8×8	14	22	26.3	0.4	16	17.4	0.4	14	15.4	2.9	14	14.8	2.9
Kacem03	10×7	11	17	19.5	0.5	14	16.1	0.6	13	14.1	3.1	13	13.6	3.3
Kacem04	10×10	7	13	15.8	0.9	7	7.5	1.0	7	7	3.8	7	7	4.1
Kacem05	15×10	11	24	28.7	1.4	19	21.5	1.6	14	14.5	7.6	14	14.2	7.9
MK01	10×6	39	40	42.3	1.4	40	41.9	1.4	40	40.5	7.4	40	40.2	8.2
MK02	10×6	26	34	36.6	1.2	34	35.2	1.2	26	29.7	7.9	26	28.1	8.8
MK03	15×8	204	235	2523	3.3	218	234.6	3.6	204	211.6	26.4	204	210.6	31.3
MK04	15×8	60	73	77.6	2.1	67	71.3	2.5	65	66.1	13.8	60	62.3	15.7
MK05	15×4	172	175	181.4	2.0	175	183.1	2.7	175	178.3	16.1	175	177.1	21.2
MK06	10×15	58	93	98.6	1.3	97	105.2	1.5	65	71.5	22.5	63	64.2	30.5
MK07	20×5	139	152	163.5	1.6	155	158.6	1.6	148	151.2	19.5	144	147.5	24.7
MK08	20×10	523	523	535.1	5.6	523	528.1	5.8	523	525.2	62.5	523	523	89.2
MK09	20×10	307	363	384.0	5.9	371	387.2	6.2	312	318.9	81.4	307	315.2	121.4
MK10	20×15	198	245	265.2	6.0	231	241.3	6.9	216	235.3	76.5	212	216.6	96.7

Table 2. Comparison between the proposed improved WOA (IWOA) and existing algorithms on the KAdata. See Section 5.3 for column descriptions.

Instance	LB	KBACO			TSPCB			HGWO			IWOA		
		Best	Avg	Time	Best	Avg	Time	Best	Avg	Time	Best	Avg	Time
Kacem01	11	11	11	900	11	11	2.5	11	11	5.6	11	11	1.8
Kacem02	14	14	14.3	3882	14	14.2	234	14	14.3	14.8	14	14.8	2.9
Kacem03	11	11	11	3966	11	11	260.5	11	11.6	16.3	13	13.6	3.3
Kacem04	7	7	7.4	6642	7	7	86	7	7.5	17.5	7	7	4.1
Kacem05	11	11	11.3	9504	11	11.7	491	13	14.1	40.7	14	14.2	7.9
Avg-T	-	-	-	4978.8	-	-	214.8	-	-	19.0	-	-	4.0

Table 3. Comparison between different algorithms on the BRdata.

Instancee	LB	KBACO		TSPCB		HGWO		AIA		PSO+TS		TS3		IWOA	
		Best	RPD	Best	RPD	Best	RPD	Best	RPD	Best	RPD	Best	RPD	Best	RPD
MK01	39	39	0	40	2.6	40	2.6	40	2.6	40	2.6	41	5.1	40	2.6
MK02	26	29	11.5	26	0	29	11.5	26	0	32	23.1	30	15.4	26	0
MK03	204	204	0	204	0	204	0	204	0	207	1.5	204	0	204	0
MK04	60	65	8.3	62	3.3	65	8.3	60	0	67	11.7	65	8.3	60	0
MK05	172	173	0.6	172	0	175	1.7	173	0.6	188	9.3	174	1.2	175	1.7
MK06	58	67	15.5	65	12.1	79	36.2	63	8.6	85	45.7	71	22.4	63	8.6
MK07	139	144	3.6	140	0.7	149	7.2	140	0.7	154	10.8	148	6.5	144	3.6
MK08	523	523	0	523	0	523	0	523	0	523	0	551	6.1	523	0
MK09	307	311	1.3	310	1.0	325	5.9	312	1.6	437	42.3	410	33.6	339	10.4
MK10	198	229	15.7	214	8.1	253	27.8	214	8.1	380	91.9	267	34.8	242	22.2
Mean	-	-	5.65	-	2.78	-	10.12	-	2.22	-	23.89	-	13.34	-	4.91

6. Conclusions

In this paper, a novel improved whale optimization algorithm (IWOA), based on the integrated approach, was presented for solving the flexible job shop scheduling problem (FJSP) with the objective of minimizing makespan. The conversion method between the whale individual position vector and the scheduling solution was firstly proposed. After that, three improvement strategies were employed in the algorithm, namely chaotic reverse learning (CRL), the nonlinear convergence factor (NFC) and adaptive weight (AW), and the variable neighborhood search (VNS). The CRL was employed to ensure the quality of the initial solutions. The NFC and AW were introduced to balance the abilities of exploitation and exploration. The VNS was adopted to enhance the accuracy and effectiveness of the local exploration.

Extensive experiments based on 15 benchmark instances were executed. The effectiveness of improvement strategies was firstly certified by a number of experiments. Then, the proposed IWOA was compared with six recently published algorithms. According to the comparison results, the proposed IWOA can obtain better results in an acceptable time.

In the future, we will concentrate on a more complex FJSP, such as the energy-efficient flexible job shop scheduling problem, the multi-objective flexible job shop scheduling problem, or the dynamic flexible job shop scheduling problem. Meanwhile, other effective improvement strategies in WOA will be studied to further improve the capacity of the algorithm for this FJSP.

Author Contributions: Conceptualization, methodology and writing—original manuscript, F.L. (Fei Luan); project management, supervision and writing—review, Z.C. and T.J.; experiments and result analysis, S.W. and F.L. (Fukang Li); investigation, formal analysis and editing, J.Y.

Funding: This work was supported by the National Natural Science Foundation of China under Grant 11072192, the Project of Shaanxi Province Soft Science Research Program under Grant 2018KRM090, the Project of Xi'an Science and Technology Innovation Guidance Program under Grant 201805023YD1CG7(1), and the Shandong Provincial Natural Science Foundation of China under Grant ZR2016GP02.

Conflicts of Interest: The authors declare no conflicts of interest.

References

1. Nowicki, E.; Smutnicki, C. A fast taboo search algorithm for the job shop problem. *Manag. Sci.* **1996**, *42*, 797–813. [CrossRef]

2. Gonc, J.F.; Magalhaes Mendes, J.J.; Resende, M.G.C. A hybrid genetic algorithm for the job shop scheduling problem. *Eur. J. Oper. Res.* **2005**, *167*, 77–95.

3. Lochtefeld, D.F.; Ciarallo, F.W. Helper-objective optimization strategies for the Job-Shop Scheduling Problem. *Appl. Soft Comput.* **2011**, *11*, 4161–4174. [CrossRef]

4. Garey, M.R.; Johnson, D.S.; Sethi, R. The complexity of flow shop and job shop scheduling. *Math. Oper. Res.* **1976**, *1*, 117–129. [CrossRef]

5. Brucker, P.; Schlie, R. Job-shop scheduling with multi-purpose machines. *Computing* **1990**, *45*, 369–375. [CrossRef]

6. Dauzere-Peres, S.; Paulli, J. An integrated approach for modeling and solving the general multi-processor job-shop scheduling problem using tabu search. *Ann. Oper. Res.* **1997**, *70*, 281–306. [CrossRef]

7. Mastrolilli, M.; Gambardella, L.M. Effective neighborhood functions for the flexible job shop problem. *J. Sched.* **2000**, *3*, 3–20. [CrossRef]

8. Mati, Y.; Lahlou, C.; Dauzere-Peres, S. Modelling and solving a practical flexible job shop scheduling problem with blocking constraints. *Int. J. Prod. Res.* **2011**, *49*, 2169–2182. [CrossRef]

9. Mousakhani, M. Sequence-dependent setup time flexible job shop scheduling problem to minimise total tardiness. *Int. J. Prod.* **2013**, *51*, 3476–3487. [CrossRef]

10. Yuan, Y.; Xu, H.; Yang, J. A hybrid harmony search algorithm for the flexible job shop scheduling problem. *Appl. Soft Comput.* **2013**, *13*, 3259–3272. [CrossRef]

11. Tao, N.; Hua, J. A cloud based improved method for multi-objective flexible job shop scheduling problem. *J. Intell. Fuzzy Syst.* **2018**, *35*, 823–829.

12. Gong, G.L.; Deng, Q.W.; Gong, X.R. A new double flexible job shop scheduling problem integrating processing time, green production, and human factor indicators. *J. Clean. Prod.* **2018**, *174*, 560–576. [CrossRef]

13. Wang, H.; Jiang, Z.G.; Wang, Y. A two-stage optimization method for energy-saving flexible job shop scheduling based on energy dynamic characterization. *J. Clean. Prod.* **2018**, *188*, 575–588. [CrossRef]

14. Marzouki, B.; Driss, O.B.; Ghédira, K. Multi Agent model based on Chemical Reaction Optimization with Greedy algorithm for Flexible Job shop Scheduling Problem. *Procedia Comput. Sci.* **2017**, *112*, 81–90. [CrossRef]

15. Yuan, Y.; Xu, H. Multiobjective flexible job shop scheduling using memetic algorithms. *IEEE Trans. Autom. Sci. Eng.* **2015**, *12*, 336–353. [CrossRef]

16. Gao, K.Z.; Suganthan, P.N.; Pan, Q.K.; Chua, T.J.; Cai, T.X.; Chong, C.S. Discrete Harmony Search Algorithm for Flexible Job Shop Scheduling Problem with Multiple Objectives. *J. Intell. Manuf.* **2016**, *27*, 363–374. [CrossRef]

17. Piroozfard, H.; Wong, K.Y.; Wong, W.P. Minimizing total carbon footprint and total late work criterion in flexible job shop scheduling by using an improved multi-objective genetic algorithm. *Resour. Conserv. Recycl.* **2018**, *128*, 267–283. [CrossRef]

18. Jiang, T.H.; Zhang, C.; Zhu, H.Q.; Deng, G.L. Energy-Efficient scheduling for a job shop using grey wolf optimization algorithm with double-searching mode. *Math. Probl. Eng.* **2018**, *2018*, 1–12. [CrossRef]

19. Singh, M.R.; Mahapatra, S. A quantum behaved particle swarm optimization for flexible job shop scheduling. *Comput. Ind. Eng.* **2016**, *93*, 36–44. [CrossRef]

20. Wu, X.L.; Sun, Y.J. A green scheduling algorithm for flexible job shop with energy-saving measures. *J. Clean. Prod.* **2018**, *172*, 3249–3264. [CrossRef]

21. Jiang, T.; Zhang, C. Application of grey wolf optimization for solving combinatorial problems: job shop and flexible job shop scheduling cases. *IEEE Access* **2018**, *6*, 26231–26240. [CrossRef]

22. Jiang, T.H.; Deng, G.L. Optimizing the low-carbon flexible job shop scheduling problem considering energy consumption. *IEEE Access* **2018**, *6*, 46346–46355. [CrossRef]

23. Jiang, T.H.; Zhang, C.; Sun, Q. Green job shop scheduling problem with discrete whale optimization algorithm. *IEEE Access* **2019**, *7*, 43153–43166. [CrossRef]

24. Li, X.Y.; Gao, L. An effective hybrid genetic algorithm and tabu search for flexible job shop scheduling problem. *Int. J. Prod. Econ.* **2016**, *174*, 93–110. [CrossRef]

25. Mirjalili, S.; Lewis, A. The whale optimization algorithm. *Adv. Eng. Soft.* **2016**, *95*, 51–67. [CrossRef]
26. Aljarah, I.; Faris, H.; Mirjalili, S. Optimizing connection weights in neural networks using the whale optimization algorithm. *Soft. Comput.* **2018**, *22*, 1–15. [CrossRef]
27. Mafarja, M.M.; Mirjalili, S. Hybrid whale optimization algorithm with simulated annealing for feature selection. *Neurocomputing* **2017**, *260*, 302–312. [CrossRef]
28. Aziz, M.A.E.; Ewees, A.A.; Hassanien, A.E. Whale optimization algorithm and moth-flame optimization for multilevel thresholding image segmentation. *Expert Syst. Appl.* **2017**, *83*, 242–256. [CrossRef]
29. Oliva, D.; Aziz, M.A.E.; Hassanien, A.E. Parameter estimation of photovoltaic cells using an improved chaotic whale optimization algorithm. *Appl. Energy* **2017**, *200*, 141–154. [CrossRef]
30. Jiang, T.H.; Zhang, C.; Zhu, H.Q.; Zhu, H.Q.; Gu, J.C.; Deng, G.L. Energy-efficient scheduling for a job shop using an improved whale optimization algorithm. *Mathematics* **2018**, *6*, 220. [CrossRef]
31. Abdel-Basset, M.; Manogaran, G.; El-Shahat, D.; Mirjalili, S.; Gunasekaran, M. A hybrid whale optimization algorithm based on local search strategy for the permutation flow shop scheduling problem. *Future Gener. Comput. Syst.* **2018**, *85*, 129–145. [CrossRef]
32. Guo, Z.Z.; Wang, P.; Ma, Y.F.; Wang, Q.; Gong, C.Q. Whale optimization algorithm based on adaptive weights and cauchy variation. *Micro Comput.* **2017**, *34*, 20–22. (In Chinese)
33. Brandimarte, P. Routing and scheduling in a flexible job shop by tabu search. *Ann. Oper.* **1993**, *41*, 157–183. [CrossRef]
34. Kacem, I.; Hammadi, S.; Borne, P. Correction to "Approach by localization and multiobjective evolutionary optimization for flexible job-shop scheduling problems". *IEEE Trans. Syst. Man Cybern. Part C* **2002**, *32*, 172. [CrossRef]
35. Xing, L.N.; Chen, Y.W.; Wang, P.; Zhao, Q.S.; Xiong, J. A knowledge-based ant colony optimiztion for flexible job shop scheduling problems. *Appl. Soft Comput.* **2010**, *10*, 888–896. [CrossRef]
36. Li, J.Q.; Pan, Q.K.; Suganthan, P.N.; Chua, T.J. A hybrid tabu search algorithm with an efficient neighborhood structure for the flexible job shop scheduling problem. *Int. J. Adv. Manuf. Technol.* **2011**, *52*, 683–697. [CrossRef]
37. Jiang, T.H. A hybrid grey wolf optimization algorithm for solving flexible job shop scheduling problem. *Control Decis.* **2018**, *33*, 503–508. (In Chinese)
38. Bagheri, A.; Zandieh, M.; Mahdavi, I.; Yazdani, M. An artificial immune algorithm for the flexible job-shop scheduling problem. *Future Gener. Comput. Syst.* **2010**, *26*, 533–541. [CrossRef]
39. Henchiri, A.; Ennigrou, M. *Particle Swarm Optimization Combined with Tabu Search in a Multi-Agent Model. for Flexible Job Shop Problem*; Springer Nature: Basingstoke, UK, 2013; Volume 7929, pp. 385–394.
40. Bozejko, W.; Uchronski, M.; Wodecki, M. *The New Golf Neighborhood for the Flexible Job Shop Problem*; ICCS, Elsevier Series; Elsevier: Amsterdam, The Netherlands, 2010; pp. 289–296.

MDPI

Article

Topology Structure Implied in β-Hilbert Space, *Heisenberg* Uncertainty Quantum Characteristics and Numerical Simulation of the *DE* Algorithm

Kaiguang Wang [1,2,†] **and Yuelin Gao** [2,*,†]

1   School of Mathematics and Information Science, North Minzu University, Yinchuan 750021, China; wkg13759842420@foxmail.com
2   Ningxia Province Key Laboratory of Intelligent Information and Data Processing, Yinchuan 750021, China
*   Correspondence: gaoyuelin@263.net; Tel.: +86-183-2997-0138 or +86-139-9510-0900
†   These authors contributed equally to this work.

Received: 9 March 2019; Accepted: 1 April 2019; Published: 4 April 2019

Abstract: The differential evolutionary (DE) algorithm is a global optimization algorithm. To explore the convergence implied in the *Hilbert* space with the parameter β of the DE algorithm and the quantum properties of the optimal point in the space, we establish a control convergent iterative form of a higher-order differential equation under the conditions of $P_{-\varepsilon}$ and analyze the control convergent properties of its iterative sequence; analyze the three topological structures implied in *Hilbert* space of the single-point topological structure, branch topological structure, and discrete topological structure; and establish and analyze the association between the *Heisenberg* uncertainty quantum characteristics depending on quantum physics and its topological structure implied in the β-Hilbert space of the DE algorithm as follows: The speed resolution Δ_v^2 of the iterative sequence convergent speed and the position resolution $\Delta_{x_\beta^\varepsilon}$ of the global optimal point with the swinging range are a pair of conjugate variables of the quantum states in β-Hilbert space about eigenvalues $\lambda_i \in \mathbb{R}$, corresponding to the uncertainty characteristics on quantum states, and they cannot simultaneously achieve bidirectional efficiency between convergent speed and the best point precision with any procedural improvements. Where $\lambda_i \in \mathbb{R}$ is a constant in the β-Hilbert space. Finally, the conclusion is verified by the quantum numerical simulation of high-dimensional data. We get the following important quantitative conclusions by numerical simulation: except for several dead points and invalid points, under the condition of spatial dimension, the number of the population, mutated operator, crossover operator, and selected operator are generally decreasing or increasing with a variance deviation rate $+0.50$ and the error of less than ± 0.5; correspondingly, speed changing rate of the individual iterative points and position changing rate of global optimal point β exhibit a inverse correlation in β-Hilbert space in the statistical perspectives, which illustrates the association between the *Heisenberg* uncertainty quantum characteristics and its topological structure implied in the β-Hilbert space of the DE algorithm.

Keywords: DE algorithm; β-Hilbert space; topology structure; quantum uncertainty property; numerical simulation

MSC: 81S10; 65L07; 46B28; 90C59; 54A05

1. Introduction

The differential evolutionary (DE) algorithm [1–3] is a global optimization algorithm with iterative search used to generate mutative individuals by differential operation, proposed by Storn and Price in 1995 to solve Chebyshev inequalities, which adopts floating-point vector coding to search in continuous space [4–6], is simple to operate and easy to achieve and offers better convergence, stronger robustness and other global optimization advantages [7–11]. In general, the minimization optimization problem of the *DE* algorithm is expressed as follows:

$$\min f(X_i^t + P_{-\varepsilon}), X_i^t = \{x_{ij}^t | i = 1, 2, \cdots, NP; j = 1, 2, \cdots, D\} \tag{1}$$

$$s.t. \, a_{ij} \le x_{ij}^t \le b_{ij}, i = 1, 2, \cdots, NP; j = 1, 2, \cdots, D \tag{2}$$

where the dimension (D) is the dimension of the decisional variable, number of population (NP) is the population size, $f(X_i + p_\varepsilon)$ is the fitness function, and $P_{-\varepsilon}$ is the individual perturbation variable with the relative error ε in the population, which is generally an infinitesimal and indicates the adjustable range of the optimal value when affected by some conditions. Conveniently, we assume that the perturbation variable $P_{-\varepsilon}$ of all individuals is the same when the external environment features perturbation. A larger perturbation variable $P_{-\varepsilon}$ indicates that the *DE* algorithm has a higher discrete degree for population individuals when generally approaching the optimal value.

A smaller perturbation variable $P_{-\varepsilon}$ indicates that the individual is less discrete when generally approaching the optimal value. Here, we assume that the infinitesimal has a fixed value, $X_i (i = 1, 2, \cdots, NP)$, is a D-dimensional vector, $x_{ij} (i = 1, 2, \cdots, NP; j = 1, 2, \cdots, D)$ is the jth components of the ith individual, and a_{ij}, b_{ij} are the upper bound and the lower bounds of the optimization range, respectively.

We are interested in the convergence of the *DE* algorithm in the optimization process and the spatial topological structure of the population in a closed ecological population [12,13], that is, the association between the population iterative sequence and population spatial topological structure. In this paper, the population is a closed ecological population, which generates an association of one-to-one correspondence between it and the population; thus, the population can be analyzed by the equivalent to the mathematical closed set. The assumptions are valid in theory. For the study of the dynamics of the *DE* algorithm, previous work [4] has analyzed the dynamics and behavior of the algorithm and provides a new direction for the dynamics of the algorithm. Numerical simulation of the route optimization and convergent problem of the *DE* algorithm has been performed [5], including studies of the convergence based on dynamics studies. Other researchers [6] have drawn comparisons regarding the convergence of various algorithm benchmark problems, and we can look at the corresponding relationship between convergence and the parameters. A parametric scheme for the algorithm dynamics research is provided for the *DE* algorithm to search and optimize the properties in the β-Hilbert space, and the study of the dynamic conditions of the *DE* algorithm is also performed.

In general, the spatial topology of a closed population is often associated with a composite operator topology on a defined function space [12,13]. One earlier study result is the isolated point theorem of the composite operator on H^2 given by Berkson [14], and MacCluer [15] and Shapiro J H [16] promote this conclusion. For the bounded analytic function space H^∞ on a unit circle or unit ball, previous work [16–18] studied the topology structure of $\mathcal{C}(H^\infty)$ and proved that the isolated composite operator of the operator topology on H^∞ is also isolated under the condition of essential norm topology. We now address the spatial topology implied in the limit point β of the convergent iterative sequence concerning the *DE* algorithm in the composite complete *Hilbert* space. Furthermore, the quantum characteristics of the *Heisenberg* uncertainty principle implied in *Hilbert* space or *Fock* [14,19–27] of the *DE* algorithm are one of the central issues studied in this paper. First, we solve the following problems:

1. The continuity of the closed population in the condition of $P_{-\varepsilon}$ and the control convergent properties of its iterative sequence;

2. The topological structure implied in the *Hilbert* space of the *DE* algorithm;

3. The *Heisenberg* uncertainty quantum characteristics implied in the β-Hilbert space of the *DE* algorithm;

4. High-dimensional numerical simulation of the quantum characteristics of the *DE* algorithm to determine the association between this algorithm and its topological structure.

2. Preparatory Knowledge

2.1. Basic Steps of the DE Algorithm

The basic operating principle of the *DE* algorithm is described as follows [1,4,7].

2.1.1. Initial Population

Let the population of the *DE* algorithm be $X(t)$; then, the population individuals can be expressed as

$$X_i^t = (x_{i1}^t, x_{i2}^t, \cdots, x_{iD}^t), i = 1, 2, \cdots, NP \tag{3}$$

where t is the evolutionary generation and NP is the population size.

Initial population: Determine the dimension D of the optimization problem. The maximum evolutionary generation is T, and the population size is NP. Let the initial value of the optimal vector be

$$X_i^0 = (x_{i1}^0, x_{i2}^0, \cdots, x_{iD}^0) \tag{4}$$

$$x_{ij}^0 = a_{ij} + rand(0,1) \cdot (b_{ij} - a_{ij}), i = 1, 2, \cdots, NP; j = 1, 2, \cdots, D \tag{5}$$

where, the range of individual variables is $a_{ij}, b_{ij} LeqR$, because of the randomness of iterative individuals in optimization process and real number coding for individuals.

2.1.2. Mutation Operation

The individual mutated component of the *DE* algorithm is the differential vector of the parental individuals, and the number of differential mutated individuals per time is derived from the two individual components $(x_{i_1}^t, x_{i_2}^t)$ in the tth generation parental population individuals, where $i_1, i_2 \in NP$. Then, the differential vector is defined as $D_{i_{1,2}} = (x_{i_1}^t - x_{i_2}^t)$. For any vector individual X_i^t, the mutation operation is defined as

$$V_i^{t+1} = X_{i_3}^t + F \cdot (X_{i_1}^t - X_{i_2}^t) \tag{6}$$

where $NP \geq 4$ is the population size, F is the contraction factor, and $i_1, i_2, i_3 \in \{1, 2, \cdots, NP\}$ and i_1, i_2, i_3 are mutually different so that we can obtain a mutated individual by differential operation by randomly selecting non-zero different vectors in the population, and the mutated individuals realize the possibility of adjusting the diversity of the population.

2.1.3. Crossover Operation

First, the test individual U_i^{t+1} is generated by crossing the target vector individual X_i^t and the mutated individual V_i^{t+1} in the population. To maintain population diversity, we can conduct crossover and selection operations for the mutated individual V_i^{t+1} and the target vector individual X_i^t by introducing the crossover probability CR and the random function $rand(0,1)$ to ensure that at least one of the test individuals U_i^{t+1} is contributed by the mutated individuals V_i^{t+1}. For other loci points, we can determine the contribution of certain sites of the test individual U_i^{t+1} that are determined by the mutation vector individuals V_i^{t+1} and target vector individual components (x_i^t) that are determined by the crossover probability. The experimental equation of the crossover operation is as follows:

$$(u_{ij}^{t+1}) = \begin{cases} (v_{ij}^{t+1}), if\ rand_j[0,1] \leq CR\ or\ j = j_{rand}, \\ (x_{ij}^t), otherwise. \\ i = 1, 2, \cdots, NP; j = 1, 2, \cdots, D \end{cases} \tag{7}$$

where $rand_j[0,1], CR \in (0,1)$ is the crossover probability above the formula (7). The larger the value of CR is, the greater the probability of generating new vector individuals by locating the crossover operation of different loci points for vector individuals in the population. When $CR = 0$, $U_i^{t+1} = X_i^t$, it indicates that no crossover occurred, which is beneficial to maintain the diversity of the population and the ability of global searching. When $CR = 1$, $U_i^{t+1} = V_i^{t+1}$, it indicates that crossover operations must occur at certain loci points, which helps maintain global searching and speed up convergence. $CR = 1$ *or* 1 represent the two extreme cases of crossover operation. $j = j_{rand}$ is a randomly selected loci point used to ensure that the test individuals U_i^t obtain at least one genetic locus of occurring mutation from the mutated individuals V_i^t and to ensure that the mutated individuals V_i^{t+1}, the target vector individuals X_i^t, and the test individuals U_i^{t+1} are different from each other, which indicates that this operation is an effective action in populations.

2.1.4. Selection Operation

The selection operation of the *DE* algorithm is a selected mechanism based on the greedy algorithm that the test individual U_i^{t+1} is generated by the mutation and selection operations, and the target vector individual X_i^t conducts competition and selection. If the fitness value of U_i^{t+1} is better than the fitness value of X_i^t, then U_i^{t+1} is inherited to the next generation as the best individual in the first iteration; otherwise, X_i^t remains in the next generation. The selection effect of the selection operator in the population is described by the following equation:

$$X_i^{t+1} = \begin{cases} U_i^{t+1}, if\ f(U_i^{t+1}) \le f(X_i^t) \\ X_i^t, otherwise \end{cases}, i = 1, 2, \cdots, NP \tag{8}$$

2.1.5. Compact Operator and *Fock* Space

Let H and L be the separable *Hilbert* space and $B(H, L)$ be the whole bounded linear operators from H to L; if the mapping $T(S)$ of the unit ball S of X in T satisfies relative compactness in Y, then $\forall T \in B(X, Y)$ is compact. In addition, the essential norm $\|T\|_e$ of operator $T \in B(X, Y)$ is the operator norm distance of all compact operators from T to $B(H, L)$. We also have $\|T\|_e \le T$ and

$$\|T\|_e = \sup_{f_\varepsilon^n \in \mathcal{U}} \left(\limsup_{k \to \infty} \|Tf_\varepsilon^n\|_l \right) \tag{9}$$

where $\mathcal{U}$ is all unit element sequences f_ε^n that are weakly convergent to 0.

Define the *Gaussian* measure dG on $\mathbb{C}^n$ as $d(G) = \frac{1}{\pi^n} e^{-|z|^2} dV(z), z \in \mathbb{C}^n$ where dV is the spatial measure on $\mathbb{C}^n$; then, *Fock* space $F^2 = F^2(\mathbb{C}^n)$ is the *Hilbert* space $L^2(G) \cap H(\mathbb{C}^n)$. Its inner product and norm are designated $\langle f, g \rangle = \int_{\mathbb{C}^n} f(z)\overline{g(z)}dG(z)$ and $\|f\|^2 = \int_{\mathbb{C}^n} |f(z)|^2 dG(z)$, respectively, where $f, g \in F^2$.

3. Continuity Structure of Closed Populations and Convergence of Iterative Sequences under P_ε

For any population existing in real space, the population individuals show discrete characteristics from the biological viewpoint but show continuous characteristics from a physical viewpoint in space. For the *DE* algorithm, the adaptive optimal individual in any population must be the limit value of the iterative sequence formed by all individuals in the population. Thus, an existing population perturbation $P_{-\varepsilon}$ is theoretically reasonable, which is described in the form of limitation as the following equation:

$$\min\ f(X_i^t + P_{-\varepsilon}), X_i^t = \{x_{ij}^t | i = 1, 2, \cdots, NP; j = 1, 2, \cdots, D\} \tag{10}$$

$$s.t.\ a_{ij} \le x_{ij}^t \le b_{ij}, i = 1, 2, \cdots, NP; j = 1, 2, \cdots, D \tag{11}$$

This formulation is equivalent to

$$\lim_{t \to +\infty} f((X_i^t)_\varepsilon) = \lim_{t \to +\infty} f_\varepsilon(X_i^t) = f_\varepsilon(X_i) \in (f(X_i) - \delta_\varepsilon, f(X_i) + \delta_\varepsilon) \tag{12}$$

$$s.t. \ a_{ij} \le x_{ij}^t \le b_{ij}, i = 1, 2, \cdots, NP; j = 1, 2, \cdots, D \tag{13}$$

where $f(X_i)$ is the optimal value of the *DE* algorithm as $t \to +\infty$ because the stability of the optimal value in space, $f_\varepsilon(X_i)$, must be between $(f_\varepsilon(X_i) - \delta_\varepsilon, f_\varepsilon(X_i) + \delta_\varepsilon)$, where δ_ε is the maximum range of the optimal value as being up and down. In the same population, there is only one optimal value, which inherits all the adaptive characteristics of population individuals in the space, and the fitness function $f_\varepsilon(X_i^t)$ corresponding to those individuals measures its adaptability in the population. We say that the former is an eigenvalue and that the latter is an eigenfunction. Then, we establish the continuity characteristic relationship and the uniform convergence of the iterative form of the population eigenvalue and eigenfunction.

3.1. Continuity Structure of the Closed Population Feature Quantity in Perturbation P_ε

Definition 1. *Assume that a population of size NP is the continuous real value of the complete real space $\mathbb{R}^+$, the population eigenvalue is $\lambda_k = X_i$, the population eigenfunction is $f((X_i^t)_\varepsilon)$, and $|\varepsilon| < \frac{1}{r}, r \in \mathbb{R}^+$, which is a convergent form that can converge in the perturbation variable P_ε with iteration numbers increasing. If*

$$\lim_{t \to +\infty} f((X_i^t)_\varepsilon) = \lim_{t \to +\infty} f_\varepsilon(X_i^t) = f_\varepsilon(X_i) \in (f(X_i) - \delta_\varepsilon, f(X_i) + \delta_\varepsilon) \tag{14}$$

$$s.t. \ a_{ij} \le x_{ij}^t \le b_{ij}, i = 1, 2, \cdots, NP; j = 1, 2, \cdots, D \tag{15}$$

then we find that $f((X_i^t)_\varepsilon)$ is continuous at the eigenvalue $\lambda_k = X_i$.

Property 1. *If $f((X_i^t)_\varepsilon)$ is continuous at the eigenvalue $\lambda_k = X_i$, then $f_\varepsilon(X_i) \in (f(X_i) - \delta_\varepsilon, f(X_i) + \delta_\varepsilon)$, that is, $f((X_i^t)_\varepsilon)$ is locally bounded.*

3.2. Uniform Convergence of the Differential Equation in Perturbation P_ε

In general, population individuals show discrete characteristics in space and continuous characteristics in time concerning the optimal process. Under the condition of the perturbation variable $P_{-\varepsilon}$, the convergent limit value is a bounded range, which is not a definite real value. To ensure that individuals can converge to a precise real value in the late iteration, the convergence of the differential equation must be uniformly converged under the condition of being the perturbation variable $P_{-\varepsilon}$ for all population individuals. First, we construct a continuous iterative form of error variable ε under the condition of perturbation P_ε:

$$\begin{cases} \varepsilon f_\varepsilon^{(n+1)''} p_1(x) f_\varepsilon^{(n+1)'} - q_\varepsilon(x, f_\varepsilon^{(n)}) = 0 \\ (0 < x < 1, q_\varepsilon(x, f_\varepsilon) = p_2(x) f_\varepsilon) \\ f_\varepsilon^{(n+1)}(0) = A, f_\varepsilon^{(n+1)}(1 + \varepsilon) = B, (A, B \in \mathbb{R}^+, n = 1, 2, \cdots) \\ f_\varepsilon^{(0)} \in \mathbf{V} = \{v \in C^2[0,1] / v(0) = A, v(1 + \varepsilon) = B\} \end{cases} \tag{16}$$

Second, we construct an approximate format (17) of $Il'InAM$ [28] of the perturbation error variable ε:

$$\begin{cases} r f_\varepsilon^{(n+1)l} + a(x) f_\varepsilon^{(n+1)l} = q_{\varepsilon(n+1)}(x, f_\varepsilon^{(n)}) \\ f_{\varepsilon 1}^{(n+1)l} = A, f_{\varepsilon(n+1)l}'' = B \\ r = \frac{a(x)h}{2} cth \frac{a(x)h}{2\varepsilon} \end{cases} \tag{17}$$

where $a(x)$ is a real-valued function.

Lemma 1. [29]. *For differential equations, we have the following:*

$$
\begin{cases}
\varepsilon f_\varepsilon'' + \alpha(x, f_\varepsilon, \varepsilon) f_\varepsilon' - \beta(x, f_\varepsilon, \varepsilon) = \gamma(x, f_\varepsilon, \varepsilon) \\
f_\varepsilon(a) = A(\varepsilon), f_\varepsilon(b) = B(\varepsilon), (a < x < b, a, b \in \mathbb{R}^+)
\end{cases}
\tag{18}
$$

Let $f_\varepsilon(x)$ be its solution; then, the following conditions are satisfied:
(i) $\alpha(x, f_\varepsilon, \varepsilon)$ is only a symbolic expression;
(ii) If $|\alpha(x, f_\varepsilon| + \beta(x, f_\varepsilon \geq a \geq 0$, then $\|f_\varepsilon\|_\infty \leq \max(|A(\varepsilon)| + |B(\varepsilon)|) + \frac{1}{a}[(b-a) \times (b-a+1)]\|\gamma(x, f_\varepsilon, \varepsilon)\|_\infty$.

Lemma 2. [28]. *Assume that there exists a constant $C > 0$ that satisfies $\|a(x)\|_\infty \leq C, \|q_{\varepsilon(n+1)}(x, f_\varepsilon^{(n)})\|_\infty \leq C, \max\{|A|, |B|\} \leq C$; then, there exists a constant $M > 0$ related to only C that satisfies $\|f_\varepsilon^{(n)} - f_\varepsilon^{(n)h}\|_\infty \leq Mh$, where h is the divided grid spacing, $f_\varepsilon^{(n)}$ is the solution of (16), and $f_\varepsilon^{(n)l}$ is the solution of (17).*

Theorem 1. (Theorem of Uniform Convergence). *For (16), if the Lipschitz condition and Lemmas 1 and 2 are satisfied, then*

$$
\|f_\varepsilon^{(n+1)l} - f_\varepsilon\|_\infty \leq \rho^{n+1}\|f_\varepsilon^{(0)l} - f_\varepsilon\|_\infty + \frac{M}{1-\rho}l
\tag{19}
$$

$$
\|f_\varepsilon^{(n+1)l} - f_\varepsilon\|_\infty \leq \frac{1}{1-\rho}\|f_\varepsilon^{(n)l} - f_\varepsilon^{(n+1)l}\|_\infty + \frac{M}{1-\rho}l
\tag{20}
$$

where $\rho = \frac{3L}{a} < 1$, L is the Lipschitz constant.

Proof. Let $\tilde{f}_\varepsilon^{(n+1)}$ be an iterative solution obtained by formulating $f_\varepsilon^{(n)l}$ as in (16). From Lemmas 1 and 2, we obtain

$$
\begin{aligned}
\|f_\varepsilon^{(n+1)l} - f_\varepsilon\|_\infty \quad & \leq \|f_\varepsilon^{(n+1)l} - \tilde{f}_\varepsilon^{(n+1)}\|_\infty + \|\tilde{f}_\varepsilon^{(n+1)} - f_\varepsilon\|_\infty \\
& \leq Ml + \rho\|f_\varepsilon^{(n)l} - f_\varepsilon\|_\varepsilon \\
& \leq \sum_{k=0}^{n} Ml\rho^k + \rho^{n+1}\|f_\varepsilon^{(0)} - f_\varepsilon\|_\infty \\
& \leq \frac{Ml}{1-\rho} + \rho^{n+1}\|f_\varepsilon^{(0)} - f_\varepsilon\|_\infty
\end{aligned}
\tag{21}
$$

and

$$
\lim_{\rho \to 0} \frac{Ml}{1-\rho} + \rho^{n+1}\|f_\varepsilon^{(0)} - f_\varepsilon\|_\infty = Ml
\tag{22}
$$

$$
\|f_\varepsilon^{(n+1)l} - f_\varepsilon\|_\infty \leq Ml
\tag{23}
$$

Thus, (19) is true. In addition,

$$
\begin{aligned}
\|f_\varepsilon^{(n)l} - f_\varepsilon\|_\infty \quad & \leq \|f_\varepsilon^{(n)l} - f_\varepsilon^{(n+1)l}\|_\infty + \|f_\varepsilon^{(n+1)l} - f_\varepsilon\|_\infty \\
& \leq \|f_\varepsilon^{(n)l} - f_\varepsilon^{(n+1)l}\|_\infty + Ml + \rho\|f_\varepsilon^{(n)l} - f_\varepsilon\|_\infty
\end{aligned}
\tag{24}
$$

In addition to the above,

$$
\lim_{\rho \to 0} \frac{1}{1-\rho}\|f_\varepsilon^{(n)l} - f_\varepsilon^{(n+1)l}\|_\infty + \frac{M}{1-\rho}l = Ml
\tag{25}
$$

$$
\|f_\varepsilon^{(n)l} - f_\varepsilon\|_\infty \leq Ml
\tag{26}
$$

$\square$

4. Topological Structure Implied in *Hilbert* Space of the *DE* Algorithm

4.1. Single-Point Topological Structure of Closed Populations in Hilbert Space

In the former part, we establish the nonlinear differential equation and its continuous iterative format according to the evolution process of the population, and we analyze the uniform convergence of the solution that illustrates the dynamical principle of population optimization in some way. In a closed ecological population of NP, which is necessarily bounded, we should further verify a situation logically if there exists an optimal solution under the condition of the perturbation variable $P_{-\varepsilon}$ after the population individuals are infinitely iterating. This is the single-point theorem that we introduce below. Since the closed population is a complete closed set under topological mapping, to analyze the topological properties conveniently, we introduce the inner product in the closed population so that the closed population is a *Hilbert* space. First, we introduce several lemmas.

Lemma 3. [30]. *The bounded set is a column compact set, and the arbitrary bounded closed set is a self-column compact set in $\mathbb{R}^n$.*

Lemma 4. [30]. *The arbitrary subset is a column compact set, and the arbitrary closed subspace is a self-column compact set in the column compact space.*

Lemma 5. [30]. *The column compact space must be a complete space.*

Lemma 6. (Brower Fixed-Point Theorem) [31]. *Let B be a closed unit ball, $T : B \to B$ be a continuous mapping, and $T(C)$ be column compact. Then, T must exist at a fixed point $x \in B$.*

Theorem 2. (Single-Point Theorem). *Let C be the closed population in $\mathbf{R}$; the mapping $T : C \to C$ is continuous. Then, there exists a single point in the closed population on C by the mapping T.*

Proof. To prove the theorem, we prove only that $T(C)$ is column-compact, as described in Lemma 6. □

Step 1 Because $T : C \to C$ is continuous and C is a compact set, we infer that T is uniformly continuous, that is, $\forall \varepsilon > 0, \exists \delta > 0$; then, $\|Tx - Tx'\| < \varepsilon, \forall x, x' \in C, \|x - x'\| < \delta$. If not, the above indicates that $\exists \varepsilon_0 > 0, \forall n \in \mathbb{N}, \exists x_n, x_n' \in C$ so that $\|x_n - x_n'\| < \frac{1}{n}$, but $\|Tx_n - Tx_n'\| \geq \varepsilon_0$. Because of C being a compact set, there exists a subsequence n_k so that $x_{n_k} \to x_0 \in C$. Since $\|x_{n_k} - x_{n_k}'\| < \frac{1}{n_k} \to 0$, then $x_{n_k'} \to x_0 \in C$. Since T is continuous, $Tx_{n_k} \to Tx_0, Tx_{n_k}' \to Tx_0, (k \to \infty)$, which implies that $\|Tx_{n_k} - Tx_{n_k}'\| \to 0, (k \to \infty)$, which contradicts $\|Tx_n - Tx_n'\| \geq \varepsilon_0$.

Step 2 To prove that $T(C)$ is column-compact, we prove only that there is a limited ϵ net on $T(C)$ $\forall \varepsilon > 0$. First, from the step 1 proof, we have $\forall \varepsilon > 0, \exists \delta > 0$ so that $\|Tx - Tx'\| < \varepsilon, \forall x, x' \in C, \|x - x'\| < \delta$. Second, due to C being a compact set, there is a limited ϵ net: $x_1, \cdots, x_n$ for $\delta > 0$. Third, we show that $\{Tx_1, \cdots, Tx_n\}$ is the limited ϵ net on $T(C)$. Actually, $\forall y \in TC, \exists x \in C$ so that $y = TC$. Let $\|x_i - x\| \leq \delta(1 \leq i \leq n)$ to obtain $\|Tx_i - Tx\| < \varepsilon$. In other words, the closed population has a single point on C by mapping T.

It is known that the complete space implied in the closed population includes only one single point that is considered the closed population optimal characteristic value according to the single-point theorem; then, the convergent iterative sequence generated by the algorithm itself can converge to a single point in the closed population. The theorem illustrates the inevitability of an existing optimal characteristic value in the complete closed population theoretically.

4.2. Branch Topological Structure of Closed Populations in Hilbert Space

There has been no definite research field focused on the route optimization branch theory of the closed population up until now. The single-point theorem indicates that there may be countless pieces of optimization routes, and it is not known how to associate the optimization routes with each other. However, it is certain that the different optimization routes are branched in *Hilbert* space implied in the closed population to generate the branch topology structure in *Hilbert* space, so that we can obtain the geometric structure of the closed population. First, we provide a fundamental theorem of *Fock* space F_m^2 [31,32] derived from *Hilbert* space; then, we can obtain the branch topological structure theorem of the *Hilbert* space implied in the closed population.

Theorem 3. *Let $\varphi : \mathbb{C} \to \mathbb{C}$ be an analytic mapping; for an arbitrary non-negative integer m, there exists the following:*

(a) C_φ is a bounded operator on F_m^2 if and only if $\varphi(z) = Az + B$. Here, $A \in \mathbb{M}_n, \|A\| \leq 1, B \in \mathbb{C}^n$, and when $\zeta \in \mathbb{C}^n$ and $|A\zeta| = |\zeta|, A\zeta \cdot \bar{B} = 0$.

(b) C_φ is a compact operator on F_m^2 if and only if $\varphi(z) = Az + B$. Here, $A \in \mathbb{M}_n, \|A\| < 1, B \in \mathbb{C}^n$.

We assume that for each positive integer k, we have $\mathbb{M}_k$ as the $k \times k$ complex matrix of the whole, which is equivalent to $A \in \mathbb{M}_k$ by a linear transformation $A : \mathbb{C}^k \to \mathbb{C}^k$.

Lemma 7. [31,32]*. Assume that $\varphi(z) = Az + B$ and $\psi(z) = A_1z + B_1$ cause the composite operators C_φ and C_ψ to be bounded on F_m^2 if there exists $\zeta \in \mathbb{C}^n$ that satisfies $|A\zeta| = |\zeta|$, but $|A\zeta| \neq |A_1\zeta|$. Then, there exists a positive constant $C_e \in \mathbf{R}$ that satisfies $\|C_\varphi - C_\psi\| \geq C_e$.*

Lemma 8. [31,32]*. Assume that $A \in \mathbb{M}_n, B \in \mathbb{C}^n$ causes C_{Az+B} to be bounded; then, C_{Az} and C_{Az+B} exist in the same path-connected branch of $\mathcal{C}_e(F_m^2)$.*

Theorem 4. (Theorem of a Branch Topological Structure). *Let C be the closed population in $\mathbf{R}$; the mapping $T : C \to C$ is continuous, and $\varphi(z) = Az + B$ and $\psi(z) = A_1z + B_1$ cause the composite operators C_φ and C_ψ to be bounded on F_m^2. Then, the necessary and sufficient condition of C_φ and C_ψ belonging to the same path-connected branch in Hilbert space is that for all $\zeta \in \mathbb{C}^n$ satisfied by $|A\zeta| = |\zeta|$ or $|A\zeta| \neq |A_1\zeta|$, there generally exists $|A\zeta| = |A_1\zeta|$.*

Proof. If we have C_φ and C_ψ in the same path-connected branch of $\mathcal{C}_e(F_m^2)$, then there exists a limited quantity of composite operators $C_{\varphi_i}{}_{i=1}^{k+1}$ that satisfy $C_{\varphi_{k+1}} = C_\varphi, C_{\varphi_1} = C_\psi$, and $\|C_\varphi - C_\psi\|_e < \frac{C_e}{2}, C_e \in \mathbf{R}, \forall i = 1, 2, \cdots, k$. Let $\varphi_i(z) = A_iz + B_i, i = 1, 2, \cdots, k+1, A_{k+1} = A, B_{k+1} = A$; then, for all $\zeta \in \mathbb{C}^n$ satisfied by $|A\zeta| = |\zeta|$ and $|A\zeta| \neq |A_1\zeta|$, there generally exists $|A_{i+1}\zeta| = |A_i\zeta|$. Thus, the necessary of the theorem is satisfied. Otherwise, we need only consider the case of $\|A\| = \|A_1\| = 1$. For all $\zeta \in \mathbb{C}^n$ satisfied by $|A\zeta| = |\zeta|$ and $|A\zeta| \neq |A_1\zeta|$, let there generally exist $|A\zeta| = |A_1\zeta|$. According to Lemma 8, we can prove the conclusion as follows: if the norm $\|D\| < 1, 1 \leq k \leq n-1$ of the matrix $D \in \mathbb{M}_{n-k}$ and $P = \begin{pmatrix} E_K & O \\ O & O \end{pmatrix}, P_1 = \begin{pmatrix} E_K & O \\ O & D \end{pmatrix}$, then C_{P_z} and C_{P_1z} exist in the same path-connected branch of $\mathcal{C}_e(F_m^2)$. From singular value decomposition (SVD) of matrix D, we need to prove only that C_{Q_z} and C_{Q_1z} exist in the same path-connected branch of $\mathcal{C}_e(F_m^2)$, where $Q = \begin{pmatrix} E_K & O \\ O & O \end{pmatrix}, Q_1 = \begin{pmatrix} E_K & O \\ O & \Lambda \end{pmatrix}$, where Λ is a diagonal matrix and the ith diagonal element is the ith singular value $\sigma_{k+i}, 0 \leq \sigma_{k+i} < 1, 1 \leq i \leq n-k$ of D. For $z \in \mathbb{C}^n$ where $z = (z'_k, z'_{n-k}), z'_k = (z_1, \cdots, z_1), z'_{n-k} = (z_{k+1}, \cdots, z_n)$. Let $\varphi_t(z) = tQ_1z + (1-t)Qz, y \in [0, 1]$; then, $\varphi_t(z) = (z'_k, t\Lambda z'_{n-k})$. To prove that the route $t \mapsto C_{\varphi_t}$ is continuous under the essential norm, note that $(C_{\varphi_t} - C_{\varphi_s})f_\varepsilon(z) = f_\varepsilon(z'_k, t\Lambda z'_{n-k}) - f_\varepsilon(z'_k, s\Lambda z'_{n-k}) = \Sigma_l a_l(t_{\lfloor l \rfloor} - s_{\lfloor l \rfloor})\sigma_{k+1}^{l_{k+1}} \cdots \sigma_n^{l_n} z_1^{l_1} \cdots z_n^{l_n}$, where $f_\varepsilon(z) = \Sigma_l a_l z_1^{l_1} \cdots z_n^{l_n}, \lfloor l \rfloor = k_{k+1} + \cdots + l_n$, then $|t^{\lfloor l \rfloor} - s^{\lfloor l \rfloor}| \leq \lfloor l \rfloor |t - s|, \forall t, s \in [0, 1]$. Because of $\forall a \in [0, 1)$, the function xa_x is bounded in $(0, \infty)$. Thus, there exists a positive constant M that satisfies $\lfloor l \rfloor \sigma_{k+1}^{l_{k+1}} \cdots \sigma_n^{l_n} \leq l_{k+1}\sigma_{k+1}^{l_{k+1}} + \cdots + l_n\sigma_n^{l_n} \leq M$. Consequently, $\|(C_{\varphi_t} - C_{\varphi_s})f_\varepsilon\|_m^2 \leq M^2|t - s|^2\|f_\varepsilon\|_m^2$. Combined with (9), we obtain $\|C_{\varphi_t} - C_{\varphi_s}\|_e \leq M|t - s|$, and the theorem is proven. $\square$

4.3. Discrete Topological Structure of Closed Populations in Hilbert Space

Theorem 5. C_φ *is discrete in Hilbert space implied in* $\mathcal{C}_e(F_m^2)$ *if and only if* $\varphi(z) = Uz$, $\varphi(z) = Uz$, *where* U *is the* U-matrix.

Proof. The adequacy of this theorem is obtained from Theorem 4; therefore, we prove only the necessity component of the theorem. If $A \in \mathbb{M}_n$ is a non-U-matrix and $\|A\| \leq 1$, from Lemma 8, we obtain that $\exists B \in \mathbb{C}^n \Rightarrow C_{Az+B}$ is bounded in F_m^2. Actually, we can consider the case of $\|A\| = 1$. Let the singular value decomposition (SVD) of A be $U\Lambda V$; then, Λ is a non-U-matrix. Furthermore, $\exists B' = 0 \in \mathbb{C}^n \Rightarrow C_{\Lambda z+B'}$ is bounded in F_m^2. Thus, C_φ is discrete in the *Hilbert* space as implied in $\mathcal{C}_e(F_m^2)$. □

Theorem 6. (Theorem of Discrete Topological Structure). *Let C be the closed population in* **R**, *the mapping $f_\varepsilon : C \to C$ be continuous, and β be a single point as described in Theorem 2, which is the optimal feature value of the convergent iterative sequence on the closed population C. Then, the single point must be a discrete point. Now, we can transform the original closed population C into a Hilbert space with the discrete parameter β by topological mapping; specifically, it is the β-Hilbert space.*

5. Quantum Characteristics of the *Heisenberg* Uncertainty Principle in β-Hilbert Space

The *Heisenberg* uncertainty principle is a fundamental principle of quantum mechanics that fundamentally illustrates that the position and momentum of a particle cannot be measured simultaneously in a quantum system; its basic form is $\Delta x \Delta p \geq \frac{h}{2}$, where h is the reduced *Planck* constant. When the *DE* algorithm pushes a closed population individual optimal process in β-Hilbert space, it measures the population individual in β-Hilbert space by the mutation, crossover, and selection of basic operational operators, which can screen the optimal characteristic value x^*. If we regard the entire β-Hilbert space as a complete space with the best signal source β, where each individual exhibits the characteristics of a better signal, then the signal source screened by the *DE* algorithm is the best of all of better signals, that is, it is the best signal source. Then, the quantity of information carried by each individual is related to the frequency of the best source and the information quantities of the best source retained by individuals that are convergent in probability F in the optimal time. With the optimization time gradually lengthening, the quantity of high-quality information carried by each individual in the convergent iterative sequence is continuously accumulated and gradually approaches that of the best signal source. There are two situations. One is that when slower the convergent speed of the iterative convergent sequence is slower, the speed of the high-quality information quantities carried by individuals accelerates is also slower, but the positional accuracy ε between the best source and population individuals is generally shrinking. Another situation is that when the convergent speed of the iterative convergent sequence is faster, and the high-quality information quantities carried by individuals in the population is reduced due to the spatial probability distributing unevenly, such that the positional accuracy ε between the best source and population individuals generally increases. Now, we provide a concrete representation of the quantum characteristics of the *Heisenberg* uncertainty principle of the *DE* algorithm in β-Hilbert space.

Definition 2. [33]. *For a $2n \times 2n$ matrix in symplectic groups,* $Q = \begin{pmatrix} A & B \\ C & D \end{pmatrix}$, *the linear canonical transformation of* $f(q') \in L^2(\mathbb{R}^n)$ *is defined as* $\hat{f}(q) = [C(M)f](q) = \int_{\mathbb{R}^n} C(M)(q,q')f(q')dq'$, *where* $C(M)(q,q') = \frac{e^{(-\frac{in\pi}{4})}}{(\sqrt{2\pi})^n \sqrt{\det(B)}} \cdot e^{i(\frac{q^\top DB^{-1}q}{2} - q^\top(B^\top)^{-1}q' + \frac{q'^\top B^{-1}Aq'}{2})}$. *Its inverse transform is* $f(q') = [C(M_{-1})\hat{f}](q') = \int_{\mathbb{R}^n} C(M_{-1})^*(Q,Q')\hat{f}(q)dq$.

Definition 3. (One-Dimensional Uncertainty Principle) [34]. *If f is a continuous function in Hilbert space, then its speed resolution Δ_v^2 and position resolution Δ_x^2 in Hilbert space are defined as*

$$\Delta_v^2 = \int_{\mathbb{R}_n}(v-v_0)^2|f(v)|^2 dv, \Delta_x^2 = \int_{\mathbb{R}_n}(x-x_0)^2|\hat{f}(x)|^2 dx$$

where $v_0 = \int_{\mathbb{R}_n}v|f(v)|^2 dv, x_0 = \int_{\mathbb{R}_n}x|\hat{f}(x)|^2 dx$; then, the Heisenberg uncertainty principle of the one-dimensional β-Hilbert space is $\Delta_v^2 \cdot \Delta_x^2 \geq \frac{b^2}{4}$.

Definition 4. *Let f_ε be a continuous-differential function defined in β-Hilbert space; then, its speed resolution Δ_v^2 and position resolution $\Delta_{x_\beta^\varepsilon}^2$ space are defined as*

$$\Delta_v^2 = \int_{\mathbb{R}_n}(v-v_0)^\top(v-v_0)|f_\varepsilon(v)|^2 dv, \Delta_{x_\beta^\varepsilon}^2 = \int_{\mathbb{R}_n}(x-x_0)^\top|\hat{f}_\varepsilon(x)|^2 dx \tag{27}$$

where $v^\top = (v_1, v_2, \cdots, v_n)^\top, v_0^\top = (\int_{\mathbb{R}_n}v_1|f(v)|^2 dv, \cdots, \int_{\mathbb{R}_n}v_n|f(v)|^2 dv)^\top$,
$x^\top = (x_1, x_2, \cdots, x_n)^\top, x_0^\top = (\int_{\mathbb{R}_n}x_1|f(x)|^2 dx, \cdots, \int_{\mathbb{R}_n}x_n|f(x)|^2 dx)^\top$.

Theorem 7. *Let f_ε be a continuous-differential function defined in β-Hilbert space and $f_\varepsilon(v_1, \cdots, v_n) \in L_2(\mathbb{R}_n), M \in Sp(2n, \mathbb{R})$. When $\det(B) \neq 0$, then we have the following equation:*

$$\begin{aligned}\Delta_v^2 \cdot \Delta_{x_\beta^\varepsilon}^2 &= \frac{\int_{\mathbb{R}_n}(v-v_0)^\top(v-v_0)|f_\varepsilon(v)|^2 dv}{\int_{\mathbb{R}_n}|f_\varepsilon(v)|^2 dv} \cdot \frac{\int_{\mathbb{R}_n}(x-x_0)^\top|\hat{f}_\varepsilon(x)|^2 dx}{\int_{\mathbb{R}_n}|\hat{f}_\varepsilon(x)|^2 dx} \\ &\geq (\frac{\sqrt{\lambda_1}}{2} + \cdots + \frac{\sqrt{\lambda_n}}{2})\end{aligned} \tag{28}$$

where $v_0^\top = (\int_{\mathbb{R}_n}v_1|f(v)|^2 dv, \cdots, \int_{\mathbb{R}_n}v_n|f(v)|^2 dv)^\top, x_0^\top = (\int_{\mathbb{R}_n}x_1|f(x)|^2 dx, \cdots, \int_{\mathbb{R}_n}x_n|f(x)|^2 dx)^\top$, and λ_i is an eigenvalue of $B^\top B$.

Proof. Under the conditions of $\det(B) \neq 0$, assume that $v_0 = 0, x_0 = 0$. Then, we can obtain by using a linear canonical transform [35] that

$$\begin{aligned}\Delta_{x_\beta^\varepsilon}^2 &= \int_{\mathbb{R}_n}x^\top x|\hat{f}_\varepsilon(u)|^2 dx \\ &= \int_{\mathbb{R}_n}x^\top x|\int_{\mathbb{R}_n}f_\varepsilon(u)\frac{e^{-\frac{in\pi}{4}}}{(\sqrt{2\pi})^n\sqrt{\det(B)}}e^{-ix^\top(B^\top)^{-1}u+i\frac{u^\top B^{-1}Au}{2}}du|^2 dx\end{aligned} \tag{29}$$

Let $t = B^{-1}x$; then, we can obtain from an integral transform that

$$\Delta_{x_\beta^\varepsilon}^2 = \int_{\mathbb{R}_n}t^\top B^\top Bt\frac{1}{(2\pi)^n}|\int_{\mathbb{R}_n}f_\varepsilon(u)e^{-it^\top u+i\frac{u^\top B^{-1}Au}{2}}du|^2 dt \tag{30}$$

Now, we set $\widetilde{f_\varepsilon(u)} = f_\varepsilon(u)e^{i\frac{u^\top B^{-1}Au}{2}}$; then, there exists

$$\Delta_{x_\beta^\varepsilon}^2 = \int_{\mathbb{R}_n}t^\top B^\top Bt\frac{1}{(2\pi)^n}|\int_{\mathbb{R}_n}\widetilde{f_\varepsilon(u)}e^{-it^\top u}du|^2 dt \tag{31}$$

and

$$\Delta_v^2 = \int_{\mathbb{R}_n}u^\top u|f_\varepsilon(u)|^2 du = \int_{\mathbb{R}_n}u^\top u\widetilde{f_\varepsilon(u)}|^2 du \tag{32}$$

Because $\det(B) \neq 0$ and because $B^\top B$ is a symmetric positive definite matrix, using matrix spectral decomposition, we find that the existing orthogonal matrix P satisfies

$$B^\top B = P^\top \Lambda P \tag{33}$$

where Λ is a diagonal matrix and where elements distributed on the diagonal are the eigenvalues of $B^\top B$; then, there exists

$$\Delta_{x_\beta^\varepsilon}^2 = \int_{\mathbb{R}_n} t^\top P^\top \Lambda P t \frac{1}{(2\pi)^n} | \int_{\mathbb{R}_n} \widetilde{f_\varepsilon(u)} e^{-it^\top u} du|^2 dt \tag{34}$$

Let $\omega = Pt$, conduct an integral transformation for (34), and let $u = P^\top y$; then, there exists

$$\begin{aligned}
\Delta_{x_\beta^\varepsilon}^2 &= \int_{\mathbb{R}_n} \omega^\top \Lambda \omega \frac{1}{(2\pi)^n} | \int_{\mathbb{R}_n} \widetilde{f_\varepsilon(u)} e^{-i\omega^\top Pu} du|^2 dt \\
&= \int_{\mathbb{R}_n} \omega^\top \Lambda \omega \frac{1}{(2\pi)^n} | \int_{\mathbb{R}_n} f_\varepsilon \widetilde{(P^\top y)} e^{-i\omega^\top y} dy|^2 dt
\end{aligned} \tag{35}$$

and

$$\Delta_v^2 = \int_{\mathbb{R}_n} u^\top u |\widetilde{f_\varepsilon(u)}|^2 du = \int_{\mathbb{R}_n} y^\top y |f_\varepsilon \widetilde{(P^\top y)}|^2 dy \tag{36}$$

Let $f_\varepsilon \widetilde{(P^\top y)} = h(y)$; then, there exists

$$\Delta_{x_\beta^\varepsilon}^2 = \int_{\mathbb{R}_n} \omega^\top \Lambda \omega \frac{1}{(2\pi)^n} | \int_{\mathbb{R}_n} h(y) e^{-i\omega^\top y} dy|^2 dt, \Delta_v^2 = \int_{\mathbb{R}_n} y^\top y |h(y)|^2 dy \tag{37}$$

Furthermore, there exists

$$\begin{aligned}
\Delta_v^2 \cdot \Delta_{x_\beta^\varepsilon}^2 &= \int_{\mathbb{R}_n} y^\top y |h(y)|^2 dy \cdot \int_{\mathbb{R}_n} \omega^\top \Lambda \omega \frac{1}{(2\pi)^n} | \int_{\mathbb{R}_n} h(y) e^{-i\omega^\top y} dy|^2 dt \\
&= \int_{\mathbb{R}_n} (y_1^2 + \cdots + y_n^2) |h(y)|^2 dy \\
&\quad \times \int_{\mathbb{R}_n} (\lambda_1 \omega_1^2 + \cdots + \lambda_n \omega_n^2) \frac{1}{(2\pi)^n} | \int_{\mathbb{R}_n} h(y) e^{-i\omega^\top y} dy|^2 |d\omega \\
&= \int_{\mathbb{R}_n} (y_1^2 + \cdots + y_n^2) |h(y)|^2 dy \times \int_{\mathbb{R}_n} (\lambda_1 |\omega_1 \\
&\quad \int_{\mathbb{R}_n} h(y) e^{2\pi i \omega^\top y} dy|^2 + \cdots + \lambda_n |\omega_n \int_{\mathbb{R}_n} h(y) e^{2\pi i \omega^\top y} dy|^2) d\omega
\end{aligned} \tag{38}$$

where λ_i is the *i*th eigenvalue of $B^\top B$. Let $h_i = \frac{\partial h(y)}{\partial y_i}$; then, from the *Fourier* transformation property [35], there exists

$$\begin{aligned}
\Delta_v^2 \cdot \Delta_{x_\beta^\varepsilon}^2 &= \int_{\mathbb{R}_n} (y_1^2 + \cdots + y_n^2) |h(y)|^2 dy \\
&\quad \times \int_{\mathbb{R}_n} (\lambda_1 | \int_{\mathbb{R}_n} h_1(y) e^{2\pi i \omega^\top y} dy|^2 + \cdots + \lambda_n | \int_{\mathbb{R}_n} h_n(y) e^{2\pi i \omega^\top y} dy|^2) d\omega
\end{aligned} \tag{39}$$

From the *Cauchy* inequality, we know that

$$\begin{aligned}
\Delta_v^2 \cdot \Delta_{x_\beta^\varepsilon}^2 &\geq ((\int_{\mathbb{R}_n} y_1^2 |h(y)|^2 dy \cdot \int_{\mathbb{R}_n} \lambda_1 | \int_{\mathbb{R}_n} h_1(y) e^{2\pi i \omega^\top y} dy|^2 dy)^{\frac{1}{2}} \\
&\quad + \cdots + (\int_{\mathbb{R}_n} y_n^2 |h(y)|^2 dy \cdot \int_{\mathbb{R}_n} \lambda_1 | \int_{\mathbb{R}_n} h_n(y) e^{2\pi i \omega^\top y} dy|^2 dy)^{\frac{1}{2}})^2
\end{aligned} \tag{40}$$

Then, using the *Cauchy* inequality of integral form, we know that

$$\begin{aligned}
\Delta_v^2 \cdot \Delta_{x_\beta^\varepsilon}^2 &\geq (\int_{\mathbb{R}_n} (|y_1 h(y) \sqrt{\lambda_1} h_1^*(y)| + \cdots + |y_n h(y) \sqrt{\lambda_n} h_n^*(y)|) dy)^2 \\
&= (\sqrt{\lambda_1} \int_{\mathbb{R}_n} (|y_1 h(y) h_1^*(y)| dy + \cdots + \sqrt{\lambda_n} \int_{\mathbb{R}_n} |y_n h(y) h_n^*(y)| dy)^2
\end{aligned} \tag{41}$$

From the one-dimensional uncertainty principle, we obtain

$$\int_{\mathbb{R}_n} |y_1 h(y) \sqrt{\lambda_1} h_1^*(y)| dy \geq \frac{1}{2} \tag{42}$$

To summarize, we obtain

$$\Delta_v^2 \cdot \Delta_{x_\beta^\varepsilon}^2 \geq (\frac{\sqrt{\lambda_1}}{2} + \cdots + \frac{\sqrt{\lambda_n}}{2})^2 \tag{43}$$

$\square$

6. Numerical Simulation

The above theorem fundamentally illustrates the geometric association between the convergent speed of the iterative sequence concerning the DE algorithm and the global optimal point precision. Specifically, Δ_v and $\Delta_{x_\beta^\varepsilon}$ are a pair of conjugate variables with quantum states, where the convergent speed of the iterative sequence caused by any improvement of the algorithm and the numerical accuracy of the global optimal point cannot be satisfied simultaneously. The above is a notably important conclusion for the DE algorithm. We use the SFEM (segmentation finite element method) to conduct a simple segmentation operation for β-Hilbert space and form a *Riemannian* manifold (Regarding *Riemannian* manifolds [36,37] in the β-Hilbert space, here, we mainly apply the space cluster caused by the wide-area property of *Riemannian* manifolds in *Hilbert* space. Then, we can improve the efficiency of algorithm optimization due to using the space cluster. In addition, because the *Riemannian* manifolds are more beneficial to the spatial segmentation operation by preventing the generation of singular points in space so that some points are omitted in the optimal process, we also consider the space quantum properties of *Riemannian* manifolds in the β-Hilbert space. Applying the *Riemannian* manifolds is a purely scientific method of mathematical physics in *Hilbert* space and is not intended to involve theoretical analysis of *Riemannian* manifolds) in the β-Hilbert region. The three topological structures implied in the β-Hilbert space of the DE algorithm are conducted by the operation of high-dimensional numerical simulation of quantum states to obtain the data of Tables 1–5 (In the Tables 1–5, $* \cdots *$ is the strength of the variable; a larger number implies a greater strength of the variable. Speed resolution is labeled as SR, position resolution as PR, relevancy of the finite unit element [38,39] as finite relevancy (FR), space dimension as (Dim), number population as (NP), mutational operation as (F), crossed operation as (CR), and elected operation as (X). The single-point topological structure, branch topological structure and discrete topological structure are labeled as $(SPTS)$, (BTS), and (DTS), respectively), (variance deviation rate (VDR) = (sample variance (SV)-population variance (PV))/PV; relevancy coefficient of the finite unit element as FR' = SR's VDR + PR's VDR. If $FR' \in (0,1)$, then FR is true value 1; If $FR' = 0$, then FR is partial truth value $1 - -$; If $FR' = 1$, then FR is absolute truth value $1 + +$; EB is the error bounds about iterative points in population), such that one can determine the association between quantum characteristics of the *Heisenberg* uncertainty principle implied in β-Hilbert space of the DE algorithm and its topological structure.

Table 1. Quantum simulation of high-dimensional data tables describing the three topological structures of the DE algorithm implied in β-Hilbert space about *Dim*.

(*Dim*)	(*SPTS*)			(*BTS*)			(*DTS*)		
	SR(%)	PR(%)	FR	SR(%)	PR(%)	FR	SR(%)	PR(%)	FR
10^2	9.500	0.400	1	6.660	0.558	1	8.214	0.325	1
10^3	9.230	0.422	1	7.242	0.500	1	8.225	0.214	1
10^4	8.471	0.500	1	8.011	0.500	1	9.535	0.110	1
10^5	7.620	0.558	1	9.763	0.500	1	9.774	0.012	1
10^6	6.101	0.660	1	9.763	0.500	1	9.896	0.011	1
10^7	5.310	0.793	1	9.880	0.500	1	9.977	0.010	1
SV	2.89009	0.02248	/	2.05701	0.00056	/	0.68463	0.01727	/
PV	2.40841	0.01874	/	1.71418	0.00047	/	0.57052	0.01439	/
DVR	+0.20	+0.32	1	+0.34	+0.19	1	+0.11	+0.16	1
EB	±0.5	±0.5	/	±0.5	±0.5	/	±0.5	±0.5	/

Table 2. Quantum simulation of high-dimensional data tables describing the three topological structures of the *DE* algorithm implied in β-Hilbert space about *NP*.

(NP)	(SPTS)			(BTS)			(DTS)		
	SR(%)	PR(%)	FR	SR(%)	PR(%)	FR	SR(%)	PR(%)	FR
10×10	8.687	0.793	1	5.243	0.021	1	8.688	0.029	1
10×20	8.756	0.660	1	7.242	0.021	1	8.744	0.025	1
10×30	8.863	0.558	1	9.011	0.020	1	8.880	0.013	1
10×40	8.880	0.500	1	9.763	0.015	1	8.863	0.009	1
10×50	9.000	0.500	1	9.841	0.010	1	9.010	0.005	1
10×60	9.010	0.500	1	9.865	0.010	1	9.101	0.004	1
SV	0.18447	0.01426	/	3.54164	0.00003	/	0.02428	0.00011	/
PV	0.15373	0.01188	/	2.95137	0.00002	/	0.02023	0.00009	/
DVR	+0.20	+0.20	1	+0.19	+0.50	1	+0.20	+0.22	1
EB	±0.5	±0.5	/	±0.5	±0.5	/	±0.5	±0.5	/

Table 3. Quantum simulation of high-dimensional data tables describing the three topological structures of the *DE* algorithm implied in β-Hilbert space about *F*.

(F)	(SPTS)			(BTS)			(DTS)		
	SR(%)	PR(%)	FR	SR(%)	PR(%)	FR	SR(%)	PR(%)	FR
0.2	6.786	0.660 ± 0.0001	1	2.652	2.558	1	11.749	0.005	1
0.3	6.020	0.877 ± 0.0001	1	2.641	2.560	1	11.744	0.009	1
0.4	6.020	0.966 ± 0.0001	1	2.633	2.559	1	11.126	0.010	1
0.5	5.852	1.101 ± 0.0001	1	2.620	2.660	1	9.535	0.010	1
0.6	5.633	1.210 ± 0.0001	1	2.619	2.676	1	9.535	0.015	1
0.7	5.330	1.220 ± 0.0001	1	2.618	2.881	1	9.535	0.055	1
SV	0.24052	0.04688	/	0.0002	0.0158	/	0.02428	0.00035	/
PV	0.20043	0.03907	/	0.00016	0.01316	/	0.02023	0.00029	/
DVR	+0.20	+0.19	1	+0.25	+0.20	1	+0.20	+0.20	1
EB	±0.5	±0.5	/	±0.5	±0.5	/	±0.5	±0.5	/

Table 4. Quantum simulation of high-dimensional data tables describing the three topological structures of the *DE* algorithm implied in β-Hilbert space about *CR*.

(CR)	(SPTS)			(BTS)			(DTS)		
	SR(%)	PR(%)	FR	SR(%)	PR(%)	FR	SR(%)	PR(%)	FR
0 *	0.430	0.991	$1 + +$	7.002	0.990	$1 + +$	/	0.000	$1 - -$
0 **	0.520	0.853	$1 + +$	7.242	0.960	$1 + +$	/	0.000	$1 - -$
0 ***	0.522	0.793	$1 + +$	7.620	0.960	$1 + +$	/	0.000	$1 - -$
SV	0.00276	0.01031	/	0.09707	0.0003	/	/	0	/
PV	0.00184	0.00687	/	0.06471	0.0002	/	/	0	/
DVR	+0.50	+0.50	$1 + +$	+0.50	+0.50	$1 + +$	/	0	$1 - -$
EB	±0.5	±0.5	/	±0.5	±0.5	/	/	±0.5	/
1 *	1.233	0.099	1	9.855	0.010 ± 0.0001	1	11.144	0.397	1
1 **	1.122	0.124	1	9.676	0.500 ± 0.0001	1	11.126	0.542	1
1 ***	1.010	0.500	1	9.110	0.500 ± 0.0001	1	10.250	0.633	1
SV	0.01243	0.05047	/	0.15124	0.08003	/	0.26116	0.01417	/
PV	0.00829	0.03364	/	0.10082	0.05336	/	0.1741	0.00944	/
DVR	+0.49	+0.49	1	+0.50	+0.03	1	+0.50	+0.20	1
EB	±0.5	±0.5	/	±0.5	±0.5	/	±0.5	±0.5	/

Table 5. Quantum simulation of high-dimensional data tables describing the three topological structures of the DE algorithm implied in β-Hilbert space about X.

(X)	(SPTS)			(BTS)			(DTS)		
	SR(%)	PR(%)	FR	SR(%)	PR(%)	FR	SR(%)	PR(%)	FR
0 *	0.523	0.796	1 + +	7.002	0.081	1 + +	/	0.000	1 − −
0 **	0.550	0.788	1 + +	7.242	0.055	1 + +	/	0.000	1 − −
0 ***	0.599	0.721	1 + +	7.620	0.001	1 + +	/	0.000	1 − −
SV	0.00035	0.0017	/	0.09707	0.00167	/	/	0	/
PV	0.00023	0.00113	/	0.06471	0.00111	/	/	0	/
VDR	+0.52	+0.50	1 + +	+0.50	+0.50	1 + +	/	0	1 − −
EB	±0.5	±0.5	/	±0.5	±0.5	/	/	±0.5	/
1 *	0.997	0.499	1	9.855	0.723	1 + +	11.144	0.551	1 + +
1 **	0.947	0.500	1	9.676	0.956	1 + +	11.126	0.640	1 + +
1 ***	0.930	0.110	1	9.110	0.990	1 + +	10.250	0.688	1 + +
SV	0.00121	0.05057	/	0.15124	0.02112	/	0.26116	0.00483	/
PV	0.00081	0.03371	/	0.10082	0.01408	/	0.1741	0.00322	/
VDR	+0.49	+0.50	1	+0.50	+0.50	1 + +	+0.50	+0.50	1 + +
EB	±0.5	±0.5	/	±0.5	±0.5	/	±0.5	±0.5	/

Because of the uncertainty of random algorithm, except for several dead points and invalid points, we conduct a quantitative analysis of Tables 1–5 to ensure the regularity of data analysis.

For **Dim**, we set the dimensions to increase with common ratio of 10. Firstly, we analyze the relationship between SR and PR about the SPTS: when the dimension increases, the SR decreases gradually with a variance deviation rate +0.20, then the SR of iterative individuals decreases gradually; and the PR increases gradually with a variance deviation rate +0.32, then the PR of iterative individual increases gradually. The above case shows that FR is true value 1, then SR and PR are completely inverse correlation.

Then, we analyze the relationship between SR and PR about the BTS: when the dimension increases, the SR increases gradually with a variance deviation rate +0.34, then the SR of iterative individuals increases gradually; and the PR decreases gradually with a variance deviation rate +0.19, then the PR of iterative individuals decreases gradually. The above case shows that FR is true value 1, then SR and PR are completely inverse correlation.

Finally, we analyze the relationship between SR and PR about the DTS: when the dimension increases, the SR increases gradually with a variance deviation rate +0.11, then the SR of iterative individuals increases gradually; and the PR decreases gradually with a variance deviation rate +0.16, then the PR of iterative individuals decreases gradually. The above case shows that FR is true value 1, then SR and PR are completely inverse correlation.

Similarly, we quantitatively analyze the **NP**. We set the NP to increase with tolerance of 100. Firstly, we analyze the relationship between SR and PR about the SPTS: when the NP increases, the SR increases gradually with a variance deviation rate +0.20, then the SR of iterative individuals increases gradually; and the PR decreases gradually with a variance deviation rate +0.20, then the PR of iterative individuals decreases gradually. The above case shows that FR is true value 1, then SR and PR are completely inverse correlation.

Then, we analyze the relationship between SR and PR about the BTS: when the NP increases, the SR increases gradually with a variance deviation rate +0.19, then the SR of iterative individuals increases gradually; and the PR decreases gradually with a variance deviation rate +0.50, then the PR of iterative individuals decreases gradually. The above case shows that FR is true value 1, then SR and PR are completely inverse correlation.

Finally, we analyze the relationship between SR and PR about the DTS: when the NP increases, the SR increases gradually with a variance deviation rate +0.20, then the SR of iterative individuals increases gradually; and the PR decreases gradually with a variance deviation rate +0.22, then the PR of iterative individuals decreases gradually. The above case shows that FR is true value 1, then SR and PR are completely inverse correlation.

Similarly, we quantitatively analyze the **F**. We set the F to increase with tolerance of 0.1. Firstly, we analyze the relationship between SR and PR about the SPTS: when the F increases, the SR decreases gradually with a variance deviation rate +0.20, then the SR of iterative individuals decreases gradually; and the PR increases gradually with a variance deviation rate +0.19, then the PR of iterative individual increases gradually. The above case shows that FR is true value 1, then SR and PR are completely inverse correlation.

Then, we analyze the relationship between SR and PR about the BTS: when the F increases, the SR decreases gradually with a variance deviation rate +0.25, then the SR of iterative individuals decreases gradually; and the PR increases gradually with a variance deviation rate +0.20, then the PR of iterative individual increases gradually. The above case shows that FR is true value 1, then SR and PR are completely inverse correlation.

Finally, we analyze the relationship between SR and PR about the DTS: when the F increases, the SR decreases gradually with a variance deviation rate +0.20, then the SR of iterative individuals decreases gradually; and the PR increases gradually with a variance deviation rate +0.20, then the PR of iterative individual increases gradually. The above case shows that FR is true value 1, then SR and PR are completely inverse correlation.

Similarly, we quantitatively analyze the **CR**. We divide the CR into two cases that the one is absolute crossover and the other is non-crossover, which are represented by '1' and '0' respectively. Firstly, we analyze the relationship between SR and PR about the SPTS: under the condition of the latter,when the intensity of CR increases gradually, the SR increases gradually with a variance deviation rate +0.50, then the SR of iterative individuals increases gradually; and the PR decreases gradually with a variance deviation rate +0.50, then the PR of iterative individuals decreases gradually. The above case shows that FR is absolute true value 1 + +, then SR and PR are absolutely and completely inverse correlation. Under the condition of the former, when the intensity of CR increases gradually, the SR decreases gradually with a variance deviation rate +0.49, then the SR of iterative individuals decreases gradually; and the PR increases gradually with a variance deviation rate +0.49, then the PR of iterative individual increases gradually. The above case shows that FR is true value 1, then SR and PR are completely inverse correlation.

Then, we analyze the relationship between SR and PR about the BTS: under the condition of the latter,when the intensity of CR increases gradually, the SR increases gradually with a variance deviation rate +0.50, then the SR of iterative individuals increases gradually; and the PR decreases gradually with a variance deviation rate +0.50, then the PR of iterative individuals decreases gradually. The above case shows that FR is absolute true value 1 + +, then SR and PR are absolutely and completely inverse correlation. Under the condition of the former, when the intensity of CR increases gradually, the SR decreases gradually with a variance deviation rate +0.50, then the SR of iterative individuals decreases gradually; and the PR increases gradually with a variance deviation rate +0.03, then the PR of iterative individual increases gradually. The above case shows that FR is true value 1, then SR and PR are completely inverse correlation.

Finally, we analyze the relationship between SR and PR about the DTS: under the condition of the latter, when the intensity of CR increases gradually, the variance deviation rate of SR does not exist, and the variance deviation rate of PR is 0, which shows that there are been no change in the individual diversity of the original population, and SR and PR are no change. The above case shows that FR is partial truth value 1 − −, then SR and PR are relatively inverse correlation. Under the condition of the former, when the intensity of CR increases gradually, the SR decreases gradually with a variance deviation rate +0.50, then the SR of iterative individuals decreases gradually; and the PR increases gradually with a variance deviation rate +0.20, then the PR of iterative individual increases gradually. The above case shows that FR is true value 1, then SR and PR are completely inverse correlation.

Similarly, we quantitatively analyze the **X**. We divide the CR into two cases that the one is absolute choice and the other is non-choice, which are represented by '1' and '0' respectively. Firstly, we analyze the relationship between SR and PR about the SPTS: under the condition of the latter, when the intensity of X increases gradually, the SR increases gradually with a variance deviation rate +0.52, then the SR of iterative individuals increases gradually; and the PR decreases gradually with a variance deviation rate +0.50, then the PR of iterative individuals decreases gradually. The above case shows that FR is absolute true value $1 + +$, then SR and PR are absolutely and completely inverse correlation. Under the condition of the former, when the intensity of X increases gradually, the SR decreases gradually with a variance deviation rate +0.49, then the SR of iterative individuals decreases gradually; and the PR increases gradually with a variance deviation rate +0.50, then the PR of iterative individual increases gradually. The above case shows that FR is true value 1, then SR and PR are completely inverse correlation.

Then, we analyze the relationship between SR and PR about the BTS: under the condition of the latter, when the intensity of X increases gradually, the SR increases gradually with a variance deviation rate +0.50, then the SR of iterative individuals increases gradually; and the PR decreases gradually with a variance deviation rate +0.50, then the PR of iterative individuals decreases gradually. The above case shows that FR is absolute true value $1 + +$, then SR and PR are absolutely and completely inverse correlation. Under the condition of the former, when the intensity of X increases gradually, the SR decreases gradually with a variance deviation rate +0.50, then the SR of iterative individuals decreases gradually; and the PR increases gradually with a variance deviation rate +0.50, then the PR of iterative individual increases gradually. The above case shows that FR is absolute true value $1 + +$, then SR and PR are absolutely and completely inverse correlation.

Finally, we analyze the relationship between SR and PR about the DTS: under the condition of the latter, when the intensity of X increases gradually, the variance deviation rate of SR does not exist, and the variance deviation rate of PR is 0, which shows that there are been no change in the individual diversity of the original population, and SR and PR are no change. The above case shows that FR is partial truth value $1 - -$, then SR and PR are relatively inverse correlation. Under the condition of the former, when the intensity of X increases gradually, the SR decreases gradually with a variance deviation rate +0.50, then the SR of iterative individuals decreases gradually; and the PR increases gradually with a variance deviation rate +0.50, then the PR of iterative individual increases gradually. The above case shows that FR is absolute true value $1 + +$, then SR and PR are absolutely and completely inverse correlation.

We conduct a qualitative analysis of Tables 1–5 as well. Except for several dead points and invalid points, under the condition of spatial dimension, the number of the population, mutated operator, crossover operator, and selected operator are generally decreasing or increasing; correspondingly, the speed changing rate of individual iterative points and the position changing rate of global optimal point β exhibit a inverse correlation in β-Hilbert space, which illustrates the association between the *Heisenberg* uncertainty quantum characteristics and its topological structure implied in the β-Hilbert space of the *DE* algorithm. Specifically, the association of the convergent iterative sequence and the global optimal point precision is a pair of conjugate variables on the quantum states in β-Hilbert space with the uncertainty characteristics on quantum states. It is fundamentally explained that any improvement in the algorithm cannot pursue the bidirectional efficiency between the convergent speed and the optimal point precision.

7. Conclusions

This paper mainly discusses the continuity structure of closed populations and the control convergent properties of the iterative sequences of the *DE* algorithm under the condition of $P_{-\varepsilon}$, establishes and analyzes the single-point topological structure, branch topological structure, and discrete topological structure implied in β-Hilbert space of the *DE* algorithm, verifies the association between the *Heisenberg* uncertainty quantum characteristics and its topological structure implied in the β-Hilbert space of the *DE* algorithm, and obtains the specific directions of the quantum uncertainty characters of the *DE* algorithm in β-Hilbert space by quantum simulation of high-dimensional data. The findings are that the speed resolution Δ_v^2 of the iterative sequence convergent speed and the position resolution $\Delta_{x_\beta^\varepsilon}$ of the global optimal point with the swinging range are a pair of conjugate variables of the quantum states in β-Hilbert space, corresponding to uncertainty characteristics of quantum states; they cannot simultaneously achieve bidirectional efficiency between the convergent speed and the best point precision with any procedural improvements. Because they are geometric features of *Riemannian* manifolds in the view of operator optimization in *Hilbert* space theoretically, however, which is only a theoretical guess, the quantum characters of the pair of conjugate variables in the *Riemannian* space require further exploration.

We all know that the most important theoretical research of meta-heuristic algorithm is how to balance the convergence speed and accuracy of the iterative points better to ensure that the iterative process is more efficient, when the iterative points approaches the global optimal point. We get the quantum uncertainty properties of the *DE*algorithm in the *beta-Hilbert*space by theoretical analysis. In the future, we will discuss the quantum estimation form and its asymptotic estimation form between convergent speed and convergent accuracy of iterative points by numerical simulation, which will lay a solid mathematical foundation for the convergent mechanism of meta-heuristic algorithm. Our second work in the future is to study the computational structure and physical structure of differential evolution algorithm, including computational complexity, spatial complexity, time tensor expansion, convergent analysis, quantum transformation state structure, *Heisenberg* uncertainty quantum state, dynamic torque analysis and so on, which will become the physical basis of the convergent theory of meta-heuristic algorithm.

Author Contributions: The first author has solved the following problems: 1. The continuity of the closed population in the condition of $P_{-\varepsilon}$ and the control convergent properties of its iterative sequence; 2. The topological structure implied in the *Hilbert* space of the *DE* algorithm; 3. The *Heisenberg* uncertainty quantum characteristics implied in the β-Hilbert space of the *DE* algorithm; And the same time,the first author implements numerical simulation of quantum inequalities for differential evolutionary algorithm. The correspondent Author reviewed the logical structure and data specification of the paper, and gave academic guidance to the first and second parts.

Funding: This work was supported by the Major Scientific Research Projects of North Minzu University (No: MSRPNMU2019003), the National Natural Science Foundation of China under Grant (No: 61561001), First-Class Disciplines Foundation of Ningxia (Grant No. NXYLXK2017B09) and Postgraduate Innovation Project of North Minzu University (No: YCX1932).

Acknowledgments: Acknowledgment for the financial support of the National Natural Science Foundation Project, the University-level Project of Northern University for Nationalities and the District-level Project of Ningxia. And the reviewers and instructors for the paper.

Conflicts of Interest: The authors declare no conflict of interest.

References

1. Storn, R.; Price, K. Differential Evolution—A Simple and Effcient Heuristic for global Optimization over Continuous Spaces. *J. Glob. Optim.* **1997**, *11*, 341–359. [CrossRef]
2. Price, K.; Price, K. *Differential Evolution—A Simple and Effcient Heuristic for Global Optimization over Continuous Spaces*; Kluwer Academic Publishers: Dordrecht, The Netherlands, 1997.
3. Reddy, S.S.; Bijwe, P.R. Differential evolution-based efficient multi-objective optimal power flow. *Neural Comput. Appl.* **2017**, 1–14. [CrossRef]
4. Liu, J.; Lampinen, J. A Fuzzy Adaptive Differential Evolution Algorithm. *Soft Comput.* **2005**, *9*, 448–462. [CrossRef]
5. Reddy, S.S.; Bijwe, P.R.; Abhyanka, A.R. Faster evolutionary algorithm based optimal power flow using incremental variables. *Int. J. Electr. Power Energy Syst.* **2014**, *54*, 198–210. [CrossRef]
6. Reddy, S.S. Optimal power flow using hybrid differential evolution and harmony search algorithm. *Int. J. Mach. Learn. Cybern.* **2018**, 1–15. [CrossRef]
7. Reddy, S.S.; Abhyankar, A.R.; Bijwe, P.R. Reactive power price clearing using multi-objective optimization. *Energy* **2011**, *36*, 3579–3589. [CrossRef]

8. Ilonen, J.; Kamarainen, J.K.; Lampinen, J. Differential Evolution Training Algorithm for Feed-Forward Neural Networks. *Neural Process. Lett.* **2003**, *17*, 93–105. [CrossRef]
9. Storn, R. System design by constraint adaptation and differential evolution. *IEEE Trans. Evol. Comput.* **1999**, *3*, 22–34. [CrossRef]
10. Reddy, S.S.; Panigrahi, B.K. Optimal power flow using clustered adaptive teaching learning-based optimisation. *Int. J. Bio-Inspir. Comput.* **2017**, *9*, 226. [CrossRef]
11. Yang, M.; Li, C.; Cai, Z.; Guan, J. Differential Evolution With Auto-Enhanced Population Diversity. *IEEE Trans. Cybern.* **2017**, *45*, 302–315. [CrossRef]
12. Eschenauer, H.A.; Olhoff, N. Topology optimization of continuum structures: A review. *Appl. Mech. Rev.* **2001**, *54*, 1453–1457. [CrossRef]
13. Albert, R.; Barabasi, A.L. Topology of evolving networks: Local events and universality. *Phys. Rev. Lett.* **2000**, *85*, 5234. [CrossRef] [PubMed]
14. Berkson, E. Composition operators isolated in the uniform operator topology. *Proc. Am. Math. Soc.* **1981**, *81*, 230–232. [CrossRef]
15. Maccluer, B.D. Components in the space of composition operators. *Integr. Equ. Oper. Theory* **1989**, *12*, 725–738. [CrossRef]
16. Shapiro, J.H.; Sundberg, C. Isolation amongst the composition operators. *Pac. J. Math.* **1990**, *145*, 179–185. [CrossRef]
17. Maccler, B.; Ohno, S.; Zhao, R. Topological structure of the space of composition operators on Hilbert. *Integr. Equ. Oper. Theory* **2001**, *40*, 481–494. [CrossRef]
18. Hosokawa, T.; Izuchi, K.; zheng, D. Isolated points and essential components of composition operators on Hilbert. *Integr. Equ. Oper. Theory* **2001**, *130*, 1765–1773.
19. Zhu, K. Uncertainty principles for the Fock space. *Sci. China-Math.* **2015**, *45*, 1847–1854.
20. Accardi, L.; Marek, B. Interacting Fock Spaces and Gaussianization of Probability Measures. *Infin. Dimens. Anal. Quantum Probab. Relat. Top.* **2014**, *1*, 9800036. [CrossRef]
21. Cho, H.R.; Zhu, K. Fock-Sobolev spaces and their Carleson measures. *J. Funct. Anal.* **2012**, *263*, 2483–2506. [CrossRef]
22. Popescu, G. Multi-analytic operators on Fock spaces. *Math. Ann.* **1995**, *303*, 31–46. [CrossRef]
23. Wang, X.; Cao, G.; Zhu, K. Boundedness and compactness of operators on the Fock space. *Integr. Equ. Oper. Theory* **2013**, *77*, 355–370. [CrossRef]
24. Gooch, J.W. *Heisenberg Uncertainty Principle*; Springer: New York, NY, USA, 2011.
25. Folland, G.B.; Sitaram, A. The uncertainty principle: A mathematical survey. *J. Fourier Anal. Appl.* **1997**, *3*, 207–238. [CrossRef]
26. Faris, W.G. Inequalities and uncertainty principles. *J. Math. Phys.* **1978**, *19*, 461–466. [CrossRef]
27. Donoho, D.L.; Stark, P.B. Uncertainty principles and signal recovery. *Siam J. Appl. Math.* **1989**, *49*, 906–931. [CrossRef]
28. Il'In, A.M. Differencing scheme for a differential equation with a small parameter affecting the highest derivative. *Math. Notes Acad. Sci. Ussr* **1969**, *6*, 596–602. [CrossRef]
29. Dorr, F.W.; Parter, S.V.; Shampine, L.F. Applications of the Maximum Principle to Singular Perturbation Problems. *Siam Rev.* **1973**, *15*, 43–88. [CrossRef]
30. Conway, J.B. A course in functional analysis. *Math. Gazette* **1990**, *75*, 698–698.
31. Dai, J. Topological Components of the Space of Composition Operators on Fock Spaces. *Complex Anal. Oper. Theory* **2015**, *9*, 201–212. [CrossRef]
32. Hong, R.C.; Choe, B.R.; Koo, H. Linear Combinations of Composition Operators on the Fock-Sobolev Spaces. *Potential Anal.* **2014**, *41*, 1223–1246.
33. Moshinsky, M.; Quesne, C. Linear Canonical Transformations and Their Unitary Representations. *J. Math. Phys.* **1971**, *12*, 1772–1780. [CrossRef]
34. Zhao, J.; Tao, R.; Wang, Y. On signal moments and uncertainty relations associated with linear canonical transform. *Signal Process.* **2010**, *90*, 2686–2689. [CrossRef]
35. Cattermole, K.W. *The Fourier Transform and Its Applications*; Osborne McGraw-Hill: London, UK, 2000.
36. Simons, J. Minimal Varieties in Riemannian Manifolds. *Ann. Math.* **1968**, *88*, 62–105. [CrossRef]
37. Cheng, S.Y.; Yau, S.T. Differential equations on riemannian manifolds and their geometric applications. *Commun. Pure Appl. Math.* **2010**, *28*, 333–354. [CrossRef]

38. Jin, J. The Finite Element Method in Electromagnetics. *J. Jpn. Soc. Appl. Electromagn.* **2002**, *1*, 39–40.
39. Satorra, A.; Bentler, P.M. A scaled difference chi-square test statistic for moment structure analysis. *Psychometrika* **2001**, *66*, 507–514. [CrossRef]

mathematics

MDPI

Article

An Improved Artificial Bee Colony Algorithm Based on Elite Strategy and Dimension Learning

Songyi Xiao [1,2], Wenjun Wang [3], Hui Wang [1,2,*], Dekun Tan [1,2], Yun Wang [1,2], Xiang Yu [1,2] and Runxiu Wu [1,2]

[1]   School of Information Engineering, Nanchang Institute of Technology, Nanchang 330099, China; speaknow@126.com (S.X.); dktan@nit.edu.cn (D.T.); wangyun@nit.edu.cn (Y.W.); xyuac@ust.hk (X.Y.); wurunxiu@tom.com (R.W.)
[2]   Jiangxi Province Key Laboratory of Water Information Cooperative Sensing and Intelligent Processing, Nanchang Institute of Technology, Nanchang 330099, China
[3]   School of Business Administration, Nanchang Institute of Technology, Nanchang 330099, China; wangwenjun881@126.com
*   Correspondence: huiwang@whu.edu.cn; Tel.: +86-0791-82086956

Received: 19 February 2019; Accepted: 13 March 2019; Published: 21 March 2019

Abstract: Artificial bee colony is a powerful optimization method, which has strong search abilities to solve many optimization problems. However, some studies proved that ABC has poor exploitation abilities in complex optimization problems. To overcome this issue, an improved ABC variant based on elite strategy and dimension learning (called ABC-ESDL) is proposed in this paper. The elite strategy selects better solutions to accelerate the search of ABC. The dimension learning uses the differences between two random dimensions to generate a large jump. In the experiments, a classical benchmark set and the 2013 IEEE Congress on Evolutionary (CEC 2013) benchmark set are tested. Computational results show the proposed ABC-ESDL achieves more accurate solutions than ABC and five other improved ABC variants.

Keywords: Artificial bee colony; swarm intelligence; elite strategy; dimension learning; global optimization

---

1. Introduction

In many real-world applications various optimization problems exist, which aim to select optimal parameters (variables) to maximize (minimize) performance indicators. In general, a minimization optimization problem can be defined by:

$$\min f(X), \tag{1}$$

where X is the vector of the decision variables.

To effectively solve optimization problems, intelligent optimization methods have been presented. Some representative algorithms are particle swarm optimization [1–5], artificial bee colony (ABC) [6,7], differential evolution [8,9], firefly algorithm [10–13], earthworm optimization algorithm [14], cuckoo search [15,16], moth search [17], pigeon inspired optimization [18], bat algorithm [19–23], krill herd algorithm [24–27], and social network optimization [28]. Among these algorithms, ABC has few control parameters and strong exploration abilities [29,30].

ABC simulates the foraging behaviors of bees in nature [6]. The processes of bees finding food sources are analogous to the processes of searching candidate solutions for a given problem. Although ABC is effective in many problems, it suffers from poor exploitation and slow convergence rates [31,32]. The possible reasons can be summarized in two ways: (1) offspring are in the neighborhood of their corresponding parent solutions and they are near to each other, and (2) offspring and their corresponding parent solutions are similar because of one-dimension perturbation.

In this work, a new ABC variant based on elite strategy and dimension learning (ESDL), called ABC-ESDL, is presented to enhance the performance of ABC. For the elite strategy, better solutions are chosen to guide the search. Moreover, the differences between different dimensions are used to generate candidate solutions with large dissimilarities. In the experiments, a classical benchmark set (with dimensions 30 and 100) and the 2013 IEEE Congress on Evolutionary (CEC 2013) benchmark set are tested. Results of ABC-ESDL are compared with ABC and five other modified ABCs.

The remainder of this work is organized as follows. In Section 2, the concept and definitions of ABC are introduced. Some recent work on ABC is given in Section 3. The proposed ABC-ESDL is described in Section 4. Test problems, results, and discussions are presented in Section 5. Finally, this work is summarized in Section 6.

2. Artificial Bee Colony

Like other bio-inspired algorithms, ABC is also a population-based stochastic method. Bees in the population try to find new food sources (candidate solutions). According to the species of bees, ABC consists of three types of bees: employed bees, onlooker bees, and scouts. The employed bees search the neighborhood of solutions in the current population, and they share their search experiences with the onlooker bees. Then, the onlooker bees choose better solutions and re-search their neighborhoods to find new candidate solutions. When solutions cannot be improved during the search, the scouts randomly initialize them [33].

Let $X_i = (x_{i1}, x_{i2}, \dots, x_{iD})$ be the i-th solution in the population at the t-th iteration. An employed bee randomly selects a different solution X_k from the current population and chooses a random dimension index j. Then, a new solution V_i is obtained by [33]:

$$v_{ij} = x_{ij} + \phi_{ij}(x_{ij} - x_{kj}),\qquad(2)$$

where $i = 1, 2, \dots, N$, and φ_{ij} is randomly chosen from $[-1.0, 1.0]$. As seen, the new solution V_i is similar to its parent solution X_i, and their differences are only on the j-th dimension. If V_i is better than X_i, X_i is updated by V_i. This means that bees find better solutions during the current search. However, this search process is slow, because the similarities between X_i and V_i are very large.

When employed bees complete the search around the neighborhood for solutions, all solutions will be updated by comparing each pair of $\{X_i, V_i\}$. Then, the selection probability p_i for each X_i is defined as follows [33]:

$$p_i = \frac{fit_i}{\sum_{j=1}^{N} fit_j},\qquad(3)$$

where fit_i is the fitness value of X_i and fit_i is calculated by:

$$fit_i = \begin{cases} 1/(1 + f_i), & \text{if } f_i \geq 0 \\ 1 + abs(f_i), & \text{otherwise} \end{cases},\qquad(4)$$

where f_i is the function value of X_i. It is obvious that a better solution will have a larger selection probability. So, the onlooker bees focus on searching the neighborhoods for better solutions. This may accelerate the convergence.

For a specific solution X, if employed bees and onlooker bees cannot find any new solutions in its neighborhood to replace it, the solutions maybe trapped into local minima. Then, a scout re-initializes it as follows [33]:

$$x_j = L_j + rand_j(U_j - L_j),\qquad(5)$$

where $j = 1, 2, \dots, D$, $[L_j, U_j]$ is the search range of the j-th dimension, and $rand_j$ is randomly chosen from $[0, 1.0]$ for the j-th dimension.

3. Related Work

Since the introduction of ABC, many different ABC variants and applications have been proposed. Some recent work on ABC is presented as follows.

Zhu and Kwong [31] modified the search model by introducing the global best solution (*Gbest*). Experiments confirmed that the modifications could improve the search efficiency. Karaboga and Gorkemli [34] presented a quick ABC (qABC) by employing a new solution search equation for the onlooker bees. Moreover, the neighborhood of *Gbest* was used to help the search. Gao and Liu [35] used the mutation operator in differential evolution (DE) to modify the solution search equation of ABC. Wang et al. [32] integrated multiple solution search strategies into ABC. It was expected that the multi-strategy mechanism could balance exploration and exploitation abilities. Cui et al. [36] proposed a new ABC with depth-first search framework and elite-guided search equation (DFSABC-elite), which assigned more computational resources to the better solutions. In addition, elite solutions were incorporated to modify the solution search equation. Li et al. [37] embedded a crossover operator into ABC to obtain a good performance. Yao et al. [38] used a multi-population technique in ABC. The entire population consisted of three subgroups, and each one used different evolutionary operators to play different roles in the search. Kumar and Mishra [39] introduced covariance matrices into ABC. Experiments on comparing continuous optimiser (COCO) benchmarks showed the approach was robust and effective. Yang et al. [40] designed an adaptive encoding learning based on covariance matrix learning. Furthermore, the selection was also adaptive according to the successful rate of candidate solutions. Chen et al. [41] firstly employed multiple different solution search models in ABC. Then, an adaptive method was designed to determine the chosen rate of each model.

In [42], a binary ABC was used to solve the spanning tree construction problem. Compared to the traditional Kruskal algorithm, the binary ABC could find sub-optimal spanning trees. In [43], a hybrid ABC was employed to tackle the effects of over-fitting in high dimensional datasets. In [44], chaos and quantum theory were used to improve the performance of ABC. Dokeroglu et al. [45] used a parallel ABC variant to optimize the quadratic assignment problem. Kishor et al. [46] presented a multi-objective ABC based on non-dominated sorting. A new method was used for employed bees to achieve convergence and diversity. The onlooker bees use similar operations with the standard ABC. Research on wireless sensor networks (WSNs) has attracted much attention [47–49]. Hashim et al. [50] proposed a new energy efficient optimal deployment strategy based on ABC in WSNs, in which ABC was used to optimize the network parameters.

4. Proposed Approach

In this section, a new ABC variant based on elite strategy and dimension learning (ABC-ESDL) is proposed. The proposed strategies and algorithm framework are described in the following subsections.

4.1. Elite Strategy

Many scholars have noticed that the original ABC was not good at exploitation during the search. To tackle this issue, several elite strategies were proposed. It is expected that elite solutions could help the search and save computational resources.

Zhu and Kwong used *Gbest* to modify the solution search model as below [31]:

$$v_{ij} = x_{ij} + \phi_{ij}(x_{ij} - x_{kj}) + \varphi_{ij}(Gbest_j - x_{ij}), \tag{6}$$

where φ_{ij} and φ_{ij} are two random values between -1.0 and 1.0.

Motivated by the mutation strategy of DE, new search equations were designed as follows [32,35]:

$$v_{ij} = Gbest_j + \phi_{ij}(x_{rj} - x_{kj}), \tag{7}$$

$$v_{ij} = Gbest_j + \phi_{ij}(Gbest_j - x_{kj}), \tag{8}$$

where X_r and X_k are two different solutions.

In our previous work [51], an external archive was constructed to store *Gbests* during the iterations. Then, these *Gbests* are used to guide the search:

$$v_{ij} = \widetilde{A}_j + \phi_{ij}(x_{rj} - x_{kj}), \tag{9}$$

where $\widetilde{A}$ is randomly chosen from the external archive.

Similar to [51], Cui et al. [36] designed an elite set E, which stores the best $\rho*N$ solutions in the current population, where $\rho \in (0,1)$. Based on the elite set, two modified search equations are defined as below:

$$v_{ij} = E_{lj} + \phi_{ij}(E_{lj} - x_{kj}), \tag{10}$$

$$v_{ij} = \frac{1}{2}(E_{lj} + Gbest_j) + \phi_{ij}(Gbest_j - x_{kj}), \tag{11}$$

where E_l is randomly chosen from the set E.

Inspired by the above work, a new search model for the employed bees is designed:

$$v_{ij} = \frac{1}{2}(E_{lj} + Gbest_j) + \phi_{ij}(x_{ij} - E_{lj}) + \varphi_{ij}(x_{ij} - Gbest_j), \tag{12}$$

where E_l is randomly chosen from the elite set E, φ_{ij} is a random value between -0.5 and 0.5, and φ_{ij} is a random value between 0 and 1.0.

As mentioned before, the onlooker bees re-search the neighborhoods of good solutions to find potentially better solutions. Therefore, further searching by the onlooker bees can be regarded as the exploitation phase. How to improve the effectiveness of the onlooker bees is important to the quality of exploitation. Thus, a different method is designed for the onlooker bees:

$$v_{ij} = \frac{1}{2}(E_{mj} + Gbest_j) + \phi_{ij}(x_{ij} - E_{lj}) + \varphi_{ij}(x_{ij} - Gbest_j), \tag{13}$$

where $m = 1, 2, ..., M$; M is the elite set size; and E_l is randomly chosen from the set E. If a solution X_i is selected based on the probability p_i, an onlooker bee generates M candidate solutions according to Equation (13). Each candidate solution is compared with X_i, and the better one is used as the new X_i. The size of the elite set should be small, because a large M will result in high computational time complexity.

To maintain the size of the elite set E, a simple replacement method is used. Initially, the best M solutions in the population are selected into E. During the search, if the offspring V_i is better than the worst solution E_w in the elite set E, we replace E_w with V_i. Then, the size of E will be M in the whole search.

4.2. Dimensional Learning

In ABC, a random dimension j is selected for conducting the solution search equation. Under this dimension, if their component values are similar, the difference $(x_{ij} - x_{kj})$ will be very small. This means that the step size $(x_{ij} - x_{kj})$ cannot help X_i jump to a far position. If the solution is trapped into local minima, it hardly escapes from the minima. In [52], a concept of dimension learning was proposed. The difference $(x_{ij} - x_{kh})$ between two different dimensions is used as the step size, where j and h are two randomly selected dimension indices and $j \neq h$. In general, the difference between two different dimensions is large. A large step size may help trapped solutions jump to better positions.

Based on the above analysis, dimension learning is embedded into Equations (12) and (13). Then, the new search models are rewritten as below:

$$v_{ij} = \frac{1}{2}(E_{lh} + Gbest_j) + \phi_{ij}(x_{ih} - E_{lj}) + \varphi_{ij}(x_{ih} - Gbest_j), \tag{14}$$

$$v_{ij} = \frac{1}{2}(E_{mj} + Gbest_h) + \phi_{ij}(x_{ij} - E_{lh}) + \varphi_{ij}(x_{ij} - Gbest_h), \tag{15}$$

where h is a random dimension and $j \neq h$.

4.3. Framework of Artificial Bee Colony-Elite Strategy and Dimension Learning

Our approach, ABC-ESDL, consists of four main operations: an elite set updating, an employed bee phase, an onlooker bee phase, and a scout bee phase. The first operation exists in the employed and onlooker bee phases. So, we only present the latter three operations.

In the employed bee phase, for each X_i, a new candidate solution V_i is created by Equation (12). The better one between V_i and X_i is chosen as X_i. If V_i is better than E_w in the elite set E, E_w is replaced by V_i. The procedure of the employed bee phase is presented in Algorithm 1, where *FEs* is the number of function evaluations.

Algorithm 1: Framework of the Employed bee phase

Begin
  for i = 1 to N **do**
    Generate V_i by Equation (14);
    Compute $f(V_i)$ and *FEs* = *FEs* + 1;
    if *f(V$_i$)* < *f(X$_i$)* **then**
        Update X_i by V_i, and set *trial$_i$*= 0;
        Update E_w, if possible;
    else
        trial$_i$ = *trial$_i$* + 1;
    end if
  end for
End

The onlooker bee phase is described in Algorithm 2, where *rand*(0,1) is a random value in the range [0, 1]. Compared to the employed bees, a different search model is employed for the onlooker bees. In Algorithm 1, an elite solution E_l is chosen from E randomly, and it is used for generating a new V_i. In Algorithm 2, all elite solutions in E are used to generate M new solutions V_i because there are M elite solutions. All M new solutions are compared with the original X_i, and the best one is used as the new X_i.

Algorithm 2: Framework of the Onlooker bee phase

Begin
  Calculate the probability p_i by Equation (3);
  I = 1, t = 1;
  while $t \leq N$ **do**
    if *rand(0,1)* < p_i**then**
      for h = 1 to M **do**
        Generate V_i by Equation (15);
        Compute $f(V_i)$ and *FEs* = *FEs* + 1;
        if *f(V$_i$)* < *f(X$_i$)* **then**

        Update X_i by V_i, and set $trial_i = 0$;
      Update E_w, if possible;
    else
       $trial_i = trial_i + 1$;
     end if
    end for
    t++;
   end if
   $i = (I + 1)\%N + 1$;
  end while
End

When $trial_i$ is set to 0, it means that the solution X_i has been improved. If the value of $trial_i$ exceeds a predefined value *limit*, it means that the solution X_i may fall into local minima. Thus, the current X_i should be reinitialized. The main steps of the scout bee phase are given in Algorithm 3.

Algorithm 3: Framework of the Scout bee phase

Begin
  if $trial_i \geq limit$ **then**
    Initialize X_i by Equation (5);
    Compute $f(X_i)$ and $FEs = FEs + 1$;
  end if
  Update the global best solution;
End

The framework of our approach, ABC-ESDL, is presented in Algorithm 4, where N represents the population size, M is the elite set size, and *MaxFEs* is the maximum value of *FEs*. To clearly illustrate the proposed ABC-ESDL, Figure 1 gives its flowchart.

Algorithm 4: Framework of ABC-ESDL

Begin
  Initialize N solution in the population;
  Initialize the elite set E;
  Set $trial_i = 0$, $I = 1,2, ..., N$;
  while $FEs \leq MaxFEs$ **do**
    Execute Algorithm 1;
    Execute Algorithm 2;
    Execute Algorithm 3;
    Update the global best solution;
  end while
End

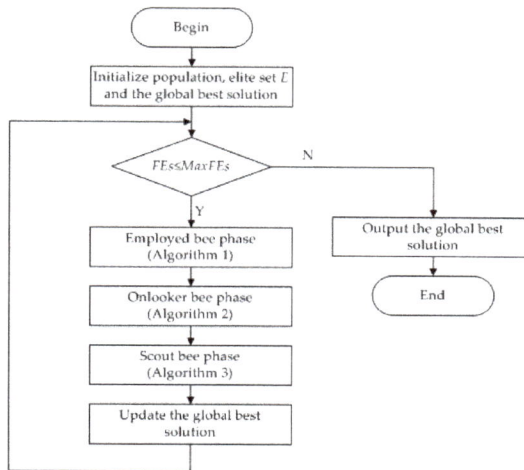

Figure 1. The flowchart of the proposed artificial bee colony-elite strategy and dimension learning (ABC-ESDL) algorithm.

5. Experimental Study

5.1. Test Problems

To verify the performance of ABC-ESDL, 12 benchmark functions with dimensions 30 and 100 were utilized in the following experiments. These functions were employed to test the optimization [53–58]. Table 1 presents the descriptions of the benchmark set where D is the dimension size, and the global optimum is listed in the last column.

Table 1. Benchmark problems.

Name	Function	Global Optimum
Sphere	$f_1(X) = \sum_{i=1}^{D} x_i^2$	0
Schwefel 2.22	$f_2(X) = \sum_{i=1}^{D} |x_i| + \prod_{i=1}^{D} |x_i|$	0
Schwefel 1.2	$f_3(X) = \sum_{i=1}^{D} \left(\sum_{j=1}^{i} x_j\right)^2$	0
Schwefel 2.21	$f_4(X) = \max\{|x_i|, 1 <= i <= D\}$	0
Rosenbrock	$f_5(X) = \sum_{i=1}^{D} [100(x_{i+1} - x_i^2)^2 + (1 - x_i^2)^2]$	0
Step	$f_6(X) = \sum_{i=1}^{D} \lfloor x_i + 0.5 \rfloor$	0
Quartic	$f_7(X) = \sum_{i=1}^{D} i \cdot x_i^4 + rand[0,1)$	0
Schwefel 2.26	$f_8(X) = \sum_{i=1}^{D} -x_i \sin(\sqrt{|x_i|})$	$-418.98*D$
Rastrigin	$f_9(X) = \sum_{i=1}^{D} [x_i^2 - 10 \cos 2\pi x_i + 10]$	0
Ackley	$f_{10}(X) = -20\exp(-0.2\sqrt{\frac{1}{D}\sum_{i=1}^{D} x_i^2}) - \exp(\frac{1}{D}\sum_{i=1}^{D} \cos(2\pi x_i)) + 20 + e$	0
Griewank	$f_{11}(X) = \frac{1}{4000}\sum_{i=1}^{D} (x_i)^2 - \prod_{i=1}^{D} \cos(\frac{x_i}{\sqrt{i}}) + 1$	0
Penalized	$f_{12}(X) = \frac{\pi}{D}\left\{\sum_{i=1}^{D} (y_i - 1)^2 [1 + \sin(\pi y_i + 1)] + (y_D - 1)^2 + (10\sin^2(\pi y_1))\right\}$ $+\sum_{i=1}^{D} u(x_i, 10, 100, 4),$ $y_i = 1 + \frac{x_i+1}{4}$ $u(x_i, a, k, m) = \begin{cases} u(x_i, a, k, m), x_i > a \\ 0, -a < x_i < a \\ k(-x_i - a)^m, x_i < -a \end{cases}$	0

167

5.2. Parameter Settings

In the experiments, ABC-ESDL was tested on the benchmark set with $D = 30$ and 100, respectively. Results of ABC-ESDL were compared with several other ABCs. The involved ABCs are listed as follows:

- ABC;
- *Gbest* guided ABC (GABC) [31];
- Improved ABC (IABC) [51];
- Modified ABC (MABC) [35];
- ABC with variable search strategy (ABCVSS) [59];
- ABC with depth-first search framework and elite-guided search equation (DFSABC-elite) [36];
- Our approach, ABC-ESDL.

To attain a fair comparison, the same parameter settings were used. For both $D = 30$ and 100, N and *limit* were equal to 100. For $D = 30$, *MaxFEs* was set to 1.5×10^5. For $D = 100$, *MaxFEs* was set to 5.0×10^5. The constant value $C = 1.5$ was used in GABC [31]. In MABC, the parameter $p = 0.7$ was used [35]. The archive size m was set to 5 in IABC [51]. The number of solution search equations used in ABCVSS was 5 [59]. In DFSABC-elite, p and r were set to 0.1 and 10, respectively [36]. In ABC-ESDL, the size (M) of the elite set was set to 5. All algorithms ran 100 times for each problem. The computing platform was with CPU Intel (R) Core (TM) i5-5200U 2.2 GHz, RAM 4 GB, and Microsoft Visual Studio 2010.

5.3. Comparison between ABC-ESDL and Other ABC Variants

Table 2 shows the results of ABC-ESDL and six other ABCs for $D = 30$, where "Mean" indicates the mean function value and "Std Dev" represents the standard deviation. The term "$w/t/l$" represents a summary for the comparison between ABC-ESDL and the six competitors. The symbol w represents that ABC-ESDL outperformed the compared algorithms on w functions. The symbol l means that ABC-ESDL was worse than its competitor on l functions. For the symbol t, ABC-ESDL and its compared algorithm obtained the same result on t functions. As shown, ABC-ESDL was better than ABC on all functions except for f_6. For this problem, all ABCs converged to the global minima. Compared to GABC, our approach ABC-ESDL performed better on nine functions. Both of them attained similar results on three functions. For ABC-ESDL, IABC, and ABCVSS, the same performances were achieved on four functions. ABC-ESDL found more accurate solutions than IABC and ABCVSS for the rest of the eight functions. DFSABC-elite outperformed ABC-ESDL on only one function, f_4, while ABC-ESDL was better than DFSABC-elite on seven functions.

Table 3 lists the results of ABC-ESDL and six other ABCs for $D = 100$. From the results, ABC-ESDL surpassed ABC on all problems. ABC-ESDL, ABC, and IABC retained the same results on f_6 and f_8. ABC-ESDL obtained better solutions for the rest of the ten functions. Compared to MABC and ABCVSS, ABC-ESDL was better on seven functions. Three algorithms had the same performance on five functions. DFSABC-elite outperformed ABC-ESDL on two functions, but ABC-ESDL was better than DFSABC-elite on five functions. Both of them obtained similar performances on five functions.

Table 2. Results of ABC-ESDL and six other ABC algorithms for $D = 30$.

Functions	ABC Mean	ABC Std Dev	GABC Mean	GABC Std Dev	IABC Mean	IABC Std Dev	MABC Mean	MABC Std Dev	ABCVSS Mean	ABCVSS Std Dev	DFSABC-Elite Mean	DFSABC-Elite Std Dev	ABC-ESDL Mean	ABC-ESDL Std Dev
f_1	1.14×10^{-15}	3.58×10^{-16}	4.52×10^{-16}	2.79×10^{-16}	1.67×10^{-35}	6.29×10^{-36}	9.63×10^{-42}	6.67×10^{-41}	1.10×10^{-36}	3.92×10^{-36}	4.72×10^{-75}	3.17×10^{-74}	2.30×10^{-82}	1.13×10^{-80}
f_2	1.49×10^{-10}	2.34×10^{-10}	1.43×10^{-15}	3.56×10^{-15}	3.09×10^{-19}	3.84×10^{-19}	1.5×10^{-21}	6.64×10^{-22}	8.39×10^{-20}	1.6×10^{-19}	6.01×10^{-38}	2.25×10^{-38}	3.13×10^{-41}	6.81×10^{-40}
f_3	1.05×10^4	3.37×10^3	4.26×10^3	2.17×10^3	5.54×10^3	2.71×10^3	1.48×10^4	1.44×10^4	9.92×10^3	9.36×10^3	4.90×10^3	9.80×10^3	3.61×10^3	1.28×10^3
f_4	4.07×10^1	1.72×10^1	1.16×10^1	6.32×10^0	1.06×10^1	4.26×10^1	5.54×10^{-1}	4.50×10^{-1}	4.36×10^{-1}	3.72×10^{-1}	2.60×10^{-2}	2.99×10^{-2}	2.11×10^{-1}	7.20×10^{-1}
f_5	1.28×10^0	1.05×10^0	2.30×10^{-1}	3.72×10^{-1}	2.36×10^{-1}	3.94×10^{-1}	1.10×10^0	3.45×10^0	1.20×10^0	1.03×10^0	1.58×10^1	1.00×10^2	1.16×10^{-3}	2.08×10^{-2}
f_6	0	0	0	0	0	0	0	0	0	0	0	0	0	0
f_7	1.54×10^{-1}	2.93×10^{-1}	5.63×10^{-2}	3.66×10^{-2}	4.23×10^{-2}	3.02×10^{-2}	2.77×10^{-2}	6.36×10^{-3}	3.25×10^{-2}	4.72×10^{-2}	1.64×10^{-2}	2.42×10^{-2}	1.46×10^{-2}	2.64×10^{-2}
f_8	$-12,490.5$	$5.87 \times 10^{+1}$	$-12,569.5$	3.25×10^{-10}	$-12,569.5$	1.31×10^{-10}	$-12,569.5$	1.97×10^{-13}	$-12,569.5$	1.94×10^{-11}	$-12,569.5$	1.97×10^{-11}	$-12,569.5$	4.65×10^{-11}
f_9	7.11×10^{-15}	2.28×10^{-15}	0	0	0	0	0	0	0	0	0	0	0	0
f_{10}	1.60×10^{-9}	4.32×10^{-9}	3.97×10^{-14}	2.83×10^{-14}	3.61×10^{-14}	1.76×10^{-14}	7.07×10^{-14}	2.36×10^{-14}	3.02×10^{-14}	2.04×10^{-14}	2.87×10^{-14}	1.46×10^{-14}	2.82×10^{-14}	2.00×10^{-14}
f_{11}	1.04×10^{-13}	3.56×10^{-13}	1.12×10^{-16}	2.53×10^{-16}	0	0	0	0	1.85×10^{-17}	3.87×10^{-16}	2.05×10^{-11}	6.04×10^{-10}	0	0
f_{12}	5.46×10^{-16}	3.46×10^{-16}	4.03×10^{-16}	2.39×10^{-16}	3.02×10^{-17}	0	1.57×10^{-32}	4.50×10^{-47}	1.57×10^{-32}	4.50×10^{-47}	1.57×10^{-32}	4.50×10^{-47}	1.57×10^{-32}	5.81×10^{-47}
$w/t/l$	11/1/0		9/3/0		8/4/0		7/5/0		8/4/0		7/4/1			

* The best result for each function is shown in boldface.

Table 3. Results of ABC-ESDL and six other ABC algorithms for $D = 100$.

Functions	ABC Mean	ABC Std Dev	GABC Mean	GABC Std Dev	IABC Mean	IABC Std Dev	MABC Mean	MABC Std Dev	ABCVSS Mean	ABCVSS Std Dev	DFSABC-Elite Mean	DFSABC-Elite Std Dev	ABC-ESDL Mean	ABC-ESDL Std Dev
f_1	7.42×10^{-15}	5.89×10^{-15}	3.37×10^{-15}	7.52×10^{-15}	3.23×10^{-33}	1.45×10^{-34}	7.98×10^{-38}	2.17×10^{-37}	6.18×10^{-35}	1.84×10^{-34}	1.04×10^{-73}	1.09×10^{-72}	6.82×10^{-85}	3.06×10^{-83}
f_2	1.09×10^{-9}	4.56×10^{-9}	6.54×10^{-15}	2.86×10^{-15}	4.82×10^{-18}	3.53×10^{-18}	2.68×10^{-20}	3.49×10^{-20}	6.18×10^{-20}	1.47×10^{-18}	2.80×10^{-37}	5.35×10^{-37}	1.02×10^{-82}	1.76×10^{-51}
f_3	1.13×10^5	2.62×10^4	9.28×10^4	2.71×10^4	9.76×10^4	2.81×10^4	1.58×10^5	9.39×10^4	1.10×10^5	5.12×10^4	6.42×10^4	5.17×10^4	7.82×10^4	8.55×10^4
f_4	8.91×10^1	4.37×10^1	8.37×10^1	3.68×10^1	8.29×10^1	1.28×10^1	3.88×10^1	3.70×10^0	3.82×10^0	1.09×10^0	7.32×10^{-1}	1.01×10^0	2.66×10^1	1.32×10^1
f_5	3.46×10^0	4.29×10^0	2.08×10^0	3.46×10^0	2.97×10^0	2.72×10^0	2.31×10^0	2.62×10^0	1.29×10^0	1.23×10^0	2.07×10^0	8.46×10^0	1.92×10^{-3}	3.22×10^{-2}
f_6	1.58×10^0	1.68×10^0	0	0	0	0	0	0	0	0	0	0	0	0
f_7	1.96×10^0	2.57×10^0	9.70×10^{-1}	7.32×10^{-1}	7.45×10^{-1}	2.27×10^{-1}	1.75×10^{-1}	1.67×10^{-2}	1.44×10^{-1}	1.72×10^{-1}	1.44×10^{-1}	8.16×10^{-2}	8.34×10^{-2}	9.15×10^{-2}
f_8	$-40,947.5$	7.34×10^2	$-41,898.3$	5.68×10^{-10}	$-41,898.3$	3.21×10^{-10}	$-41,898.3$	2.91×10^{-12}	$-41,898.3$	1.60×10^{-10}	$-41,898.3$	7.02×10^{-11}	$-41,898.3$	1.63×10^{-10}
f_9	1.83×10^{-11}	2.27×10^{-11}	1.95×10^{-14}	3.53×10^{-14}	1.42×10^{-14}	2.63×10^{-14}	0	0	0	0	0	0	0	0
f_{10}	3.54×10^{-9}	7.28×10^{-10}	1.78×10^{-13}	5.39×10^{-13}	1.50×10^{-13}	4.87×10^{-13}	3.58×10^{-11}	2.91×10^{-12}	1.32×10^{-13}	3.64×10^{-14}	1.25×10^{-13}	5.36×10^{-14}	1.25×10^{-13}	5.61×10^{-14}
f_{11}	1.12×10^{-14}	9.52×10^{-15}	1.44×10^{-15}	3.42×10^{-15}	7.78×10^{-16}	5.24×10^{-16}	0	0	0	0	1.81×10^{-16}	3.42×10^{-15}	1.81×10^{-33}	0
f_{12}	4.96×10^{-15}	3.29×10^{-15}	2.99×10^{-15}	4.37×10^{-15}	9.05×10^{-18}	0	4.71×10^{-33}	0	4.71×10^{-33}	0	4.71×10^{-33}	0	4.71×10^{-33}	0
$w/t/l$	12/0/0		10/2/0		10/2/0		7/5/0		7/5/0		5/5/2			

* The best result for each function is shown in boldface.

Figure 2 presents the convergence processes of ABC-ESDL, DFSABC-elite, MABC, and ABC on selected problems with $D = 30$. As seen, ABC-ESDL was faster than DFSABC-elite, MABC, and ABC. For f_1, f_2, f_{10}, and f_{12}, DFSABC-elite converged faster than MABC and ABC. For f_5, DFSABC-elite was the slowest algorithm. ABC was faster than DFSABC-elite on f_7. For f10, ABC-ESDL was slower than DFSABC-elite at the beginning search stage, and it was faster at the last search stage.

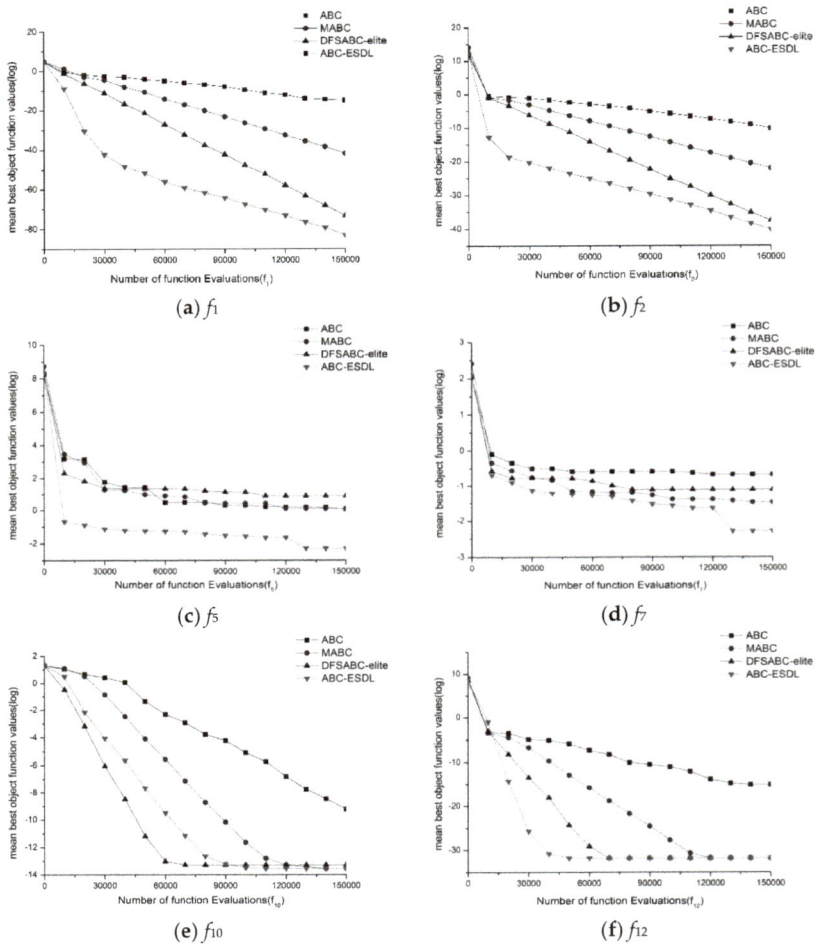

Figure 2. The convergence curves of ABC-ESDL, DFSABC-elite, MABC, and ABC on selected functions. (**a**) Sphere; (**b**) Schwefel 2.22; (**c**) Rosenbrock; (**d**) Quartic; (**e**) Ackley; and (**f**) Penalized.

By the suggestions of [53,56], a nonparametric statistical test was used to compare the overall performances of seven ABCs. In the following, the mean rank of each algorithm on the whole benchmark set was calculated by the Friedman test. Table 4 gives the mean rank values of seven ABCs for $D = 30$ and 100. The smallest rank value meant that the corresponding algorithm obtained the best performance. For $D = 30$ and 100, ABC-ESDL achieved the best performances, and DFSABC-elite was in second place. For $D = 30$, both MABC and ABCVSS had the same rank. When the dimension increased to 100, ABCVSS obtained a better rank than MABC.

Table 4. Mean ranks achieved by the Friedman test for $D = 30$ and 100.

Algorithms	Mean Rank	
	$D = 30$	$D = 100$
ABC	6.50	6.67
GABC	4.58	5.33
IABC	4.08	4.42
MABC	3.79	3.58
ABCVSS	3.79	3.29
DFSABC-elite	3.29	2.67
ABC-ESDL	**1.96**	**2.04**

* The best rank for each dimension is shown in boldface.

5.4. Effects of Different Strategies

There are two modifications in ABC-ESDL: elite strategy (ES) and dimension learning (DL). To investigate the effects of different strategies (ES and DL), we tested different combinations between ABC, ES, and DL on the benchmark set. The involved combinations are listed as below:

- ABC without ES or DL;
- ABC-ES: ABC with elite strategy;
- ABC-DL: ABC with dimension learning;
- ABC-ESDL: ABC with elite strategy and dimension learning.

For the above four ABC algorithms, the parameter settings were kept the same as in Section 5.3. The parameters *MaxFEs*, *N*, *limit*, and *M* were set to 5000*D, 100, 100, and 5, respectively. All algorithms ran 100 times for each problem for $D = 30$ and 100.

Table 5 presents the comparison of ABC-ESDL, ABC-ES, ABC-DL, and ABC for $D = 30$. The best result for each function is shown in boldface. From the results, all four algorithms obtained the same results on f_6. ABC was worse than ABC-ES on eight problems, but ABC-ES obtained worse results on three problems. ABC-DL outperformed ABC on ten problems, while ABC-DL was worse than ABC on only one problem. ABC-ESDL outperformed ABC-DL and ABC on 11 problems. Compared to ABC-ES, ABC-ESDL was better on ten problems, and both of them had the same performances on the rest of the two problems.

Table 5. Comparison of ABC with different strategies ($D = 30$).

Problems	ABC		ABC-ES		ABC-DL		ABC-ESDL	
	Mean	Std Dev	Mean	Std Dev	Mean	Std Dev	Mean	Std Dev
f_1	1.14×10^{-15}	3.58×10^{-16}	1.37×10^{-33}	2.51×10^{-34}	4.67×10^{-17}	4.78×10^{-17}	$\mathbf{2.30 \times 10^{-82}}$	1.13×10^{-80}
f_2	1.49×10^{-10}	2.34×10^{-10}	2.82×10^{-21}	3.23×10^{-21}	1.02×10^{-10}	3.46×10^{-11}	$\mathbf{3.13 \times 10^{-41}}$	6.81×10^{-40}
f_3	1.05×10^{4}	3.37×10^{3}	6.71×10^{3}	2.94×10^{3}	7.62×10^{3}	3.27×10^{3}	$\mathbf{3.61 \times 10^{3}}$	1.28×10^{3}
f_4	4.07×10^{1}	1.72×10^{1}	2.21×10^{0}	2.06×10^{0}	3.82×10^{1}	1.24×10^{1}	$\mathbf{2.11 \times 10^{-1}}$	7.20×10^{-1}
f_5	1.28×10^{0}	1.05×10^{0}	3.88×10^{1}	1.65×10^{1}	9.63×10^{-2}	1.09×10^{-2}	$\mathbf{1.16 \times 10^{-3}}$	2.08×10^{-2}
f_6	**0**	**0**	**0**	**0**	**0**	**0**	**0**	**0**
f_7	1.54×10^{-1}	2.93×10^{-1}	9.40×10^{-2}	$\mathbf{1.77 \times 10^{-2}}$	2.82×10^{-1}	2.51×10^{-2}	$\mathbf{1.46 \times 10^{-2}}$	2.64×10^{-2}
f_8	$-12,490.5$	$5.87 \times 10^{+1}$	-12557.8	1.62×101	$-12,533.1$	1.93×10^{2}	$\mathbf{-12,569.5}$	4.65×10^{-11}
f_9	7.11×10^{-15}	2.28×10^{-15}	7.94×10^{-14}	2.58×10^{-15}	2.43×10^{-15}	**0**	**0**	**0**
f_{10}	1.60×10^{-9}	4.32×10^{-9}	3.49×10^{-14}	$\mathbf{1.87 \times 10^{-14}}$	6.45×10^{-10}	1.99×10^{-14}	$\mathbf{2.82 \times 10^{-14}}$	2.00×10^{-14}
f_{11}	1.04×10^{-13}	3.56×10^{-13}	7.55×10^{-3}	6.38×10^{-3}	2.49×10^{-15}	1.52×10^{-15}	**0**	**0**
f_{12}	5.46×10^{-16}	3.46×10^{-16}	$\mathbf{1.57 \times 10^{-32}}$	**0**	1.56×10^{-19}	**0**	1.57×10^{-32}	5.81×10^{-47}
$w/t/l$	11/1/0		10/2/0		11/1/0		-	

* The best result for each function is shown in boldface.

Table 6 gives the results of ABC-ESDL, ABC-ES, ABC-DL, and ABC for $D = 100$. The best result for each function is shown in boldface. Similar to $D = 30$, we can get the same conclusion. ABC-ESDL performed better than ABC, ABC-ES, and ABC-DL. ABC-ES was better than ABC-DL on most test problems, and both of them outperformed the original ABC.

Table 6. Comparison of ABC with different strategies ($D = 100$).

Problems	ABC		ABC-ES		ABC-DL		ABC-ESDL	
	Mean	Std Dev	Mean	Std Dev	Mean	Std Dev	Mean	Std Dev
f_1	7.42×10^{-15}	5.89×10^{-15}	1.53×10^{-27}	2.87×10^{-26}	3.96×10^{-15}	1.91×10^{-14}	$\mathbf{6.82 \times 10^{-85}}$	$\mathbf{3.06 \times 10^{-83}}$
f_2	1.09×10^{-9}	4.56×10^{-9}	1.11×10^{-16}	1.18×10^{-15}	9.41×10^{-10}	1.86×10^{-9}	$\mathbf{1.02 \times 10^{-52}}$	$\mathbf{1.76 \times 10^{-51}}$
f_3	1.13×10^5	$\mathbf{2.62 \times 10^4}$	9.39×10^4	2.63×10^4	1.08×10^5	3.91×10^4	$\mathbf{7.82 \times 10^4}$	8.55×10^4
f_4	8.91×10^1	4.37×10^1	3.44×10^1	3.44×10^1	8.73×10^1	7.92×10^0	$\mathbf{2.66 \times 10^1}$	$\mathbf{1.32 \times 10^1}$
f_5	3.46×10^0	4.29×10^0	1.19×10^2	1.19×10^2	2.21×10^{-1}	1.83×10^0	$\mathbf{1.92 \times 10^{-3}}$	$\mathbf{3.22 \times 10^{-2}}$
f_6	1.58×10^0	1.68×10^0	$\mathbf{0}$	$\mathbf{0}$	3.13×10^0	6.59×10^0	$\mathbf{0}$	$\mathbf{0}$
f_7	1.96×10^0	2.57×10^0	1.87×10^{-1}	1.35×10^{-1}	1.43×10^0	1.04×10^0	$\mathbf{8.34 \times 10^{-2}}$	$\mathbf{9.15 \times 10^{-2}}$
f_8	$-40{,}947.5$	7.34×10^2	$-41{,}762.1$	5.95×10^2	$-41{,}240.7$	8.02×10^2	$\mathbf{-41{,}898.3}$	$\mathbf{1.63 \times 10^{-10}}$
f_9	1.83×10^{-11}	2.27×10^{-11}	1.29×10^{-9}	3.78×10^{-8}	2.07×10^{-6}	5.32×10^{-5}	$\mathbf{0}$	$\mathbf{0}$
f_{10}	3.54×10^{-9}	7.28×10^{-10}	1.57×10^{-13}	4.35×10^{-14}	2.17×10^{-9}	5.06×10^{-9}	$\mathbf{1.25 \times 10^{-13}}$	5.61×10^{-14}
f_{11}	1.12×10^{-14}	9.52×10^{-15}	9.13×10^{-4}	1.49×10^{-2}	1.89×10^{-15}	7.52×10^{-15}	$\mathbf{0}$	$\mathbf{0}$
f_{12}	4.96×10^{-15}	3.29×10^{-15}	4.29×10^{-28}	8.78×10^{-27}	3.21×10^{-18}	2.19×10^{-17}	$\mathbf{4.71 \times 10^{-33}}$	$\mathbf{7.50 \times 10^{-48}}$
$w/t/l$	12/0/0		11/1/0		12/0/0		-	

* The best result for each function is shown in boldface.

For the above analysis, ABC with a single strategy (ES or DL) achieved better results than the original ABC. By introducing ES and DL into ABC, the performance of ABC-ESDL was further enhanced, and it outperformed ABC and ABC with a single strategy. This demonstrated that both ES and DL were helpful in strengthening the performance of ABC.

5.5. Results of the CEC 2013 Benchmark Set

In Section 5.3, ABC-ESDL was tested on several classical benchmark functions. To verify the performance of ABC-ESDL on difficult functions, the 2013 IEEE Congress on Evolutionary (CEC 2013) benchmark set was utilized in this section [60].

In the experiments, ABC-ESDL was compared with ABC, GABC, MABC, ABCVSS, and DFSABC-elite on the CEC benchmark set with $D = 30$. By the suggestions of [60], *MaxFEs* was set to 10,000*D. For other parameters, the same settings were used as described in Section 5.3. For each test function, each algorithm was run 51 times. Throughout the experiments, the mean function error value ($f(X) - f(X^*)$) was reported, where X was the best solution found by the algorithm in a run, and X^* was the global optimum of the test function [60].

Table 7 presents the computational results of ABC-ESDL, DFSABC-elite, ABCVSS, MABC, GABC, and ABC on the 2013 IEEE Congress on Evolutionary (CEC 2013) benchmark set, where "Mean" indicates the mean function error values and "Std Dev" represents the standard deviation. The best result for each function is shown in boldface. From the results, ABC-ESDL outperformed ABC and GABC on 25 functions, but it was worse on the rest of the three functions. Compared to MABC, ABC-ESDL achieved better results on 20 functions, but MABC was better than ABC-ESDL on the rest of the eight functions. ABC-ESDL performed better than ABCVSS and DFSABC-elite on 21 and 22 functions, respectively. From the above analysis, even for difficult functions, ABC-ESDL still obtained better performances than the compared algorithms.

Table 7. Results on the CEC 2013 benchmark set.

Problems	ABC		GABC		MABC		ABCVSS		DFSABC-elite		ABC-ESDL	
	Mean	Std Dev	Mean	Std Dev	Mean	Std Dev	Mean	Std Dev	Mean	Std Dev	Mean	Std Dev
f_1	6.82×10^{-14}	2.18×10^{-13}	5.71×10^{-13}	9.44×10^{-14}	$\mathbf{4.55 \times 10^{-14}}$	1.33×10^{-12}	6.82×10^{-14}	1.88×10^{-12}	$\mathbf{4.55 \times 10^{-14}}$	1.19×10^{-12}	7.29×10^{-5}	1.66×10^{-3}
f_2	1.05×10^5	5.86×10^5	3.43×10^7	5.63×10^6	1.88×10^7	4.34×10^7	$\mathbf{2.54 \times 10^5}$	7.24×10^5	1.98×10^7	4.55×10^7	2.55×10^6	1.16×10^6
f_3	2.49×10^9	6.24×10^9	1.05×10^{10}	1.19×10^9	5.58×10^7	2.94×10^7	1.89×10^8	5.60×10^8	7.83×10^7	2.26×10^7	$\mathbf{1.95 \times 10^6}$	3.08×10^6
f_4	6.81×10^3	2.09×10^3	3.51×10^5	1.48×10^4	9.83×10^3	2.56×10^3	8.18×10^3	2.05×10^3	6.58×10^3	$\mathbf{1.79 \times 10^3}$	$\mathbf{5.94 \times 10^3}$	1.81×10^5
f_5	4.97×10^{-10}	8.53×10^{-9}	4.70×10^{-13}	$\mathbf{5.65 \times 10^{-14}}$	$\mathbf{5.68 \times 10^{-14}}$	1.52×10^{-12}	1.71×10^{-13}	6.98×10^{-12}	1.02×10^{-13}	2.50×10^{-13}	1.07×10^{-3}	3.00×10^{-3}
f_6	1.73×10^0	5.07×10^0	1.77×10^2	1.82×10^0	1.76×10^0	6.81×10^0	2.25×10^0	7.12×10^0	1.67×10^0	1.14×10^0	$\mathbf{8.58 \times 10^{-1}}$	$\mathbf{3.57 \times 10^{-1}}$
f_7	1.29×10^2	3.58×10^1	4.09×10^2	1.29×10^1	1.06×10^1	3.52×10^0	1.26×10^1	3.96×10^1	9.27×10^0	2.53×10^0	$\mathbf{7.16 \times 10^0}$	$\mathbf{2.15 \times 10^0}$
f_8	2.10×10^0	5.96×10^0	2.13×10^1	3.51×10^{-2}	2.09×10^0	5.97×10^0	2.11×10^0	5.97×10^0	2.10×10^0	5.96×10^0	$\mathbf{2.08 \times 10^0}$	$\mathbf{5.95 \times 10^0}$
f_9	3.02×10^1	8.69×10^0	1.40×10^2	2.43×10^1	$\mathbf{2.79 \times 10^1}$	8.69×10^0	3.05×10^1	8.81×10^0	3.04×10^1	8.47×10^0	2.97×10^1	$\mathbf{8.08 \times 10^0}$
f_{10}	3.40×10^{-1}	8.22×10^{-1}	1.43×10^0	8.48×10^{-1}	1.62×10^{-1}	4.60×10^{-1}	3.05×10^{-1}	1.32×10^{-1}	2.46×10^{-1}	5.59×10^{-1}	$\mathbf{2.51 \times 10^{-2}}$	$\mathbf{6.91 \times 10^{-2}}$
f_{11}	3.30×10^{-13}	4.73×10^{-13}	1.54×10^{-13}	2.86×10^{-14}	1.14×10^{-14}	3.24×10^{-14}	1.71×10^{-14}	4.30×10^{-14}	$\mathbf{5.68 \times 10^{-15}}$	$\mathbf{2.50 \times 10^{-15}}$	6.81×10^{-4}	1.91×10^{-4}
f_{12}	3.14×10^1	8.42×10^1	1.60×10^3	5.64×10^1	1.57×10^1	5.52×10^0	2.46×10^1	6.01×10^1	2.20×10^1	5.61×10^0	$\mathbf{1.51 \times 10^1}$	4.91×10^0
f_{13}	3.14×10^1	9.36×10^1	1.81×10^1	5.60×10^1	2.63×10^1	7.29×10^0	2.27×10^1	7.64×10^1	$\mathbf{2.13 \times 10^1}$	$\mathbf{6.65 \times 10^0}$	2.70×10^1	7.51×10^0
f_{14}	1.09×10^0	4.05×10^0	2.85×10^0	1.28×10^0	2.48×10^{-1}	5.23×10^{-1}	$\mathbf{6.25 \times 10^{-3}}$	$\mathbf{1.14 \times 10^{-2}}$	2.10×10^{-1}	1.80×10^{-1}	7.47×10^{-1}	3.05×10^{-1}
f_{15}	3.49×10^3	1.22×10^2	1.57×10^4	6.11×10^2	3.21×10^3	1.06×10^2	$\mathbf{2.65 \times 10^3}$	1.28×10^2	5.09×10^3	1.40×10^2	2.62×10^3	1.10×10^2
f_{16}	1.65×10^0	5.11×10^2	2.07×10^1	2.57×10^{-1}	1.57×10^0	3.98×10^0	2.15×10^0	5.66×10^0	2.49×10^0	5.97×10^0	$\mathbf{8.49 \times 10^{-1}}$	$\mathbf{3.75 \times 10^{-1}}$
f_{17}	3.11×10^0	8.81×10^1	1.07×10^2	1.06×10^1	$\mathbf{3.04 \times 10^0}$	8.67×10^0	3.04×10^0	8.66×10^0	3.27×10^0	8.66×10^0	3.09×10^0	8.76×10^0
f_{18}	3.88×10^2	1.01×10^2	1.76×10^3	5.02×10^2	1.90×10^2	6.59×10^1	3.45×10^2	9.27×10^1	2.84×10^2	7.53×10^1	$\mathbf{1.44 \times 10^2}$	4.99×10^1
f_{19}	1.07×10^{-1}	3.67×10^0	2.25×10^1	2.94×10^{-1}	6.81×10^{-2}	2.26×10^{-1}	1.54×10^{-1}	6.58×10^{-1}	4.56×10^{-2}	2.29×10^{-2}	$\mathbf{4.49 \times 10^{-2}}$	9.31×10^{-4}
f_{20}	1.54×10^1	4.17×10^0	5.00×10^1	6.93×10^0	1.46×10^1	4.11×10^0	1.48×10^1	4.11×10^0	1.46×10^1	4.02×10^0	$\mathbf{1.41 \times 10^1}$	3.82×10^0
f_{21}	2.01×10^2	5.59×10^1	3.67×10^2	9.04×10^1	2.06×10^2	5.77×10^1	2.18×10^2	6.27×10^1	2.00×10^2	9.20×10^1	$\mathbf{1.02 \times 10^2}$	5.31×10^1
f_{22}	1.16×10^2	3.71×10^1	7.47×10^1	2.64×10^1	1.05×10^2	3.06×10^1	1.15×10^2	4.13×10^1	1.19×10^2	3.20×10^1	$\mathbf{1.47 \times 10^0}$	2.04×10^0
f_{23}	5.53×10^3	1.52×10^2	2.16×10^4	8.78×10^3	4.11×10^3	1.35×10^2	5.57×10^3	1.63×10^2	5.87×10^3	1.74×10^2	$\mathbf{3.18 \times 10^3}$	1.26×10^2
f_{24}	3.02×10^2	8.33×10^1	6.00×10^2	7.56×10^1	2.88×10^2	8.13×10^1	2.86×10^2	8.27×10^1	2.86×10^2	8.09×10^1	$\mathbf{2.80 \times 10^2}$	8.07×10^1
f_{25}	3.15×10^2	8.88×10^1	7.16×10^2	9.05×10^0	$\mathbf{2.96 \times 10^2}$	$\mathbf{8.53 \times 10^1}$	3.02×10^2	8.67×10^1	3.00×10^2	8.53×10^1	3.00×10^2	8.56×10^1
f_{26}	2.01×10^2	5.73×10^1	2.07×10^2	6.27×10^2	2.01×10^2	5.72×10^1	2.01×10^2	5.73×10^1	2.01×10^2	5.72×10^1	$\mathbf{2.00 \times 10^2}$	5.71×10^1
f_{27}	4.02×10^2	1.34×10^1	3.81×10^3	5.67×10^2	1.11×10^3	3.00×10^2	4.02×10^2	1.90×10^1	4.02×10^2	$\mathbf{1.14 \times 10^1}$	4.00×10^2	1.26×10^1
f_{28}	1.64×10^2	7.10×10^1	4.27×10^3	—	3.00×10^2	8.54×10^1	3.13×10^2	9.29×10^1	3.00×10^2	8.70×10^1	$\mathbf{1.04 \times 10^2}$	8.45×10^1
w/t/l	25/0/3		25/0/3		20/0/8		21/0/7		22/1/5			

* The best result for each function is shown in boldface.

6. Conclusions

To balance exploration and exploitation, an improved version of ABC, called ABC-ESDL, is proposed in this paper. In ABC-ESDL, there are two modifications: elite strategy (ES) and dimension learning (DL). The elite strategy is used to guide the search. Good solutions are selected into the elite set. These elite solutions are used to modify the search model. To maintain the size of the elite set, a simple replacement method is employed. In dimension learning, the difference between different dimensions can achieve a large jump to help trapped solutions escape from local minima. The performance of our approach ABC-ESDL is verified on twelve classical benchmark functions (with dimensions 30 and 100) and the 2013 IEEE Congress on Evolutionary (CEC 2013) benchmark set.

Computational results of ABC-ESDL are compared with ABC, GABC, IABC, MABC, ABCVSS, and DFSABC-elite. For D = 30 and 100, ABC-ESDL is not worse than ABCVSS, MABC, IABC, GABC, and ABC. DFSABC-elite is better than ABC-ESDL on only one problem for D = 30 and two problems for D = 100. For the rest of problems, ABC-ESDL outperforms DFSABC-elite. For the 2013 IEEE Congress on Evolutionary (CEC 2013) benchmark set, ABC-ESDL still achieves better performances than the compared algorithms.

Another experiment investigates the effectiveness of ES and DL. Results show that ES or DL can achieve improvements. ABC with two strategies (both ES and DL) surpasses ABC and ABC with a single strategy (ES or DL). It confirms the effectiveness of our proposed strategies.

For the onlooker bees, offspring is generated for each elite solution in the elite set. So, an onlooker bee generates M new solutions when a parent solution X_i is selected. This complexity will increase the computational time. To reduce the effects of such computational effort, a small parameter M is used. In the future work, other strategies will be considered to replace the current method. In addition, more test functions [61] will be considered to further verify the performance of our approach.

Author Contributions: Writing—original draft preparation, S.X. and H.W.; writing—review and editing, W.W.; visualization, S.X.; supervision, H.W., D.T., Y.W., X.Y., and R.W.

Funding: This work was supported by the National Natural Science Foundation of China (Nos. 61663028, 61703199), the Distinguished Young Talents Plan of Jiang-xi Province (No. 20171BCB23075), the Natural Science Foundation of Jiangxi Province (No. 20171BAB202035), the Science and Technology Plan Project of Jiangxi Provincial Education Department (Nos. GJJ170994, GJJ180940), and the Open Research Fund of Jiangxi Province Key Laboratory of Water Information Cooperative Sensing and Intelligent Processing (No. 2016WICSIP015).

Conflicts of Interest: The authors declare no conflict of interest.

References

1. Kennedy, J. Particle Swarm Optimization. In Proceedings of the 1995 International Conference on Neural Networks, Perth, WA, Australia, 27 November–1 December 1995; Volume 4, pp. 1942–1948.
2. Wang, F.; Zhang, H.; Li, K.S.; Lin, Z.Y.; Yang, J.; Shen, X.L. A hybrid particle swarm optimization algorithm using adaptive learning strategy. *Inf. Sci.* **2018**, *436–437*, 162–177. [CrossRef]
3. Souza, T.A.; Vieira, V.J.D.; Souza, M.A.; Correia, S.E.N.; Costa, S.L.N.C.; Costa, W.C.A. Feature selection based on binary particle swarm optimisation and neural networks for pathological voice detection. *Int. J. Bio-Inspir. Comput.* **2018**, *11*, 91–101. [CrossRef]
4. Sun, C.L.; Jin, Y.C.; Chen, R.; Ding, J.L.; Zeng, J.C. Surrogate-assisted cooperative swarm optimization of high-dimensional expensive problems. *IEEE Trans. Evol. Comput.* **2017**, *21*, 644–660. [CrossRef]
5. Wang, H.; Wu, Z.J.; Rahnamayan, S.; Liu, Y.; Ventresca, M. Enhancing particle swarm optimization using generalized opposition-based learning. *Inf. Sci.* **2011**, *181*, 4699–4714.
6. Karaboga, D. *An Idea Based on Honey Bee Swarm for Numerical Optimization*; Technical Report-tr06; Engineering Faculty, Computer Engineering Department, Erciyes University: Kayseri, Turkey, 2005.
7. Amiri, E.; Dehkordi, M.N. Dynamic data clustering by combining improved discrete artificial bee colony algorithm with fuzzy logic. *Int. J. Bio-Inspir. Comput.* **2018**, *12*, 164–172. [CrossRef]

8. Meang, Z.; Pan, J.S. HARD-DE: Hierarchical archive based mutation strategy with depth information of evolution for the enhancement of differential evolution on numerical optimization. *IEEE Access* **2019**, *7*, 12832–12854. [CrossRef]

9. Meang, Z.; Pan, J.S.; Kong, L.P. Parameters with adaptive learning mechanism (PALM) for the enhancement of differential evolution. *Knowl.-Based Syst.* **2018**, *141*, 92–112. [CrossRef]

10. Yang, X.S. *Engineering Optimization: An Introduction with Metaheuristic Applications*; John Wiley & Sons: Etobicoke, ON, Canada, 2010.

11. Wang, H.; Wang, W.; Sun, H.; Rahnamayan, S. Firefly algorithm with random attraction. *Int. J. Bio-Inspir. Comput.* **2016**, *8*, 33–41. [CrossRef]

12. Wang, H.; Wang, W.J.; Cui, Z.H.; Zhou, X.Y.; Zhao, J.; Li, Y. A new dynamic firefly algorithm for demand estimation of water resources. *Inf. Sci.* **2018**, *438*, 95–106. [CrossRef]

13. Wang, H.; Wang, W.J.; Cui, L.Z.; Sun, H.; Zhao, J.; Wang, Y.; Xue, Y. A hybrid multi-objective firefly algorithm for big data optimization. *Appl. Soft Comput.* **2018**, *69*, 806–815. [CrossRef]

14. Wang, G.G.; Deb, S.; Coelho, L.S. Earthworm optimisation algorithm: A bio-inspired metaheuristic algorithm for global optimisation problems. *Int. J. Bio-Inspir. Comput.* **2018**, *12*, 1–22. [CrossRef]

15. Yang, X.S.; Deb, S. Cuckoo Search via Levy Flights. *Mathematics* **2010**, *1*, 210–214.

16. Zhang, M.; Wang, H.; Cui, Z.; Chen, J. Hybrid multi-objective cuckoo search with dynamical local search. *Memet. Comput.* **2018**, *10*, 199–208. [CrossRef]

17. Wang, G.G. Moth search algorithm: A bio-inspired metaheuristic algorithm for global optimization problems. *Memet. Comput.* **2016**, *10*, 1–14. [CrossRef]

18. Cui, Z.; Wang, Y.; Cai, X. A pigeon-inspired optimization algorithm for many-objective optimization problems. *Sci. China Inf. Sci.* **2019**, *62*, 070212. [CrossRef]

19. Yang, X.S. A new metaheuristic bat-inspired algorithm. *Comput. Knowl. Technol.* **2010**, *284*, 65–74.

20. Wang, Y.; Wang, P.; Zhang, J.; Cui, Z.; Cai, X.; Zhang, W.; Chen, J. A novel bat algorithm with multiple strategies coupling for numerical optimization. *Mathematics* **2019**, *7*, 135. [CrossRef]

21. Cai, X.J.; Gao, X.Z.; Xue, Y. Improved bat algorithm with optimal forage strategy and random disturbance strategy. *Int. J. Bio-Inspir. Comput.* **2016**, *8*, 205–214. [CrossRef]

22. Cui, Z.H.; Xue, F.; Cai, X.J.; Gao, Y.; Wang, G.G.; Chen, J.J. Detection of malicious code variants based on deep learning. *IEEE Trans. Ind. Inf.* **2018**, *14*, 3187–3196. [CrossRef]

23. Cai, X.; Wang, H.; Cui, Z.; Cai, J.; Xue, Y.; Wang, L. Bat algorithm with triangle-flipping strategy for numerical optimization. *Int. J. Mach. Learn. Cybern.* **2018**, *9*, 199–215. [CrossRef]

24. Wang, G.G.; Guo, L.H.; Gandomi, A.H.; Hao, G.S.; Wang, H.Q. Chaotic krill herd algorithm. *Inf. Sci.* **2014**, *274*, 17–34. [CrossRef]

25. Wang, G.G.; Gandomi, A.H.; Alavi, A.H. An effective krill herd algorithm with migration operator in biogeography-based optimization. *Appl. Math. Model.* **2014**, *38*, 2454–2462. [CrossRef]

26. Wang, G.G.; Gandomi, A.H.; Alavi, A.H. Stud krill herd algorithm. *Neurocomputing* **2014**, *128*, 363–370. [CrossRef]

27. Wang, G.G.; Guo, L.H.; Wang, H.Q.; Duan, H.; Luo, L.; Li, J. Incorporating mutation scheme into krill herd algorithm for global numerical optimization. *Neural Comput. Appl.* **2014**, *24*, 853–871. [CrossRef]

28. Grimaccia, F.; Gruosso, G.; Mussetta, M.; Niccolai, A.; Zich, R.E. Design of tubular permanent magnet generators for vehicle energy harvesting by means of social network optimization. *IEEE Trans. Ind. Electron.* **2018**, *65*, 1884–1892. [CrossRef]

29. Karaboga, D.; Akay, B. A survey: Algorithms simulating bee swarm intelligence. *Artif. Intell. Rev.* **2009**, *31*, 61–85. [CrossRef]

30. Kumar, A.; Kumar, D.; Jarial, S.K. A review on artificial bee colony algorithms and their applications to data clustering. *Cybern. Inf. Technol.* **2017**, *17*, 3–28. [CrossRef]

31. Zhu, G.; Kwong, S. Gbest-guided artificial bee colony algorithm for numerical function optimization. *Appl. Math. Comput.* **2010**, *217*, 3166–3173. [CrossRef]

32. Wang, H.; Wu, Z.J.; Rahnamayan, S.; Sun, H.; Liu, Y.; Pan, J.S. Multi-strategy ensemble artificial bee colony algorithm. *Inf. Sci.* **2014**, *279*, 587–603. [CrossRef]

33. Karaboga, D.; Akay, B. A comparative study of artificial bee colony algorithm. *Appl. Math. Comput.* **2009**, *214*, 108–132. [CrossRef]

34. Karaboga, D.; Gorkemli, B. A quick artificial bee colony (qABC) algorithm and its performance on optimization problems. *Appl. Soft Comput.* **2014**, *23*, 227–238. [CrossRef]

35. Gao, W.; Liu, S. A modified artificial bee colony algorithm. *Comput. Oper. Res.* **2012**, *39*, 687–697. [CrossRef]

36. Cui, L.Z.; Li, G.H.; Lin, Q.Z.; Du, Z.H.; Gao, W.F.; Chen, J.Y.; Lu, N. A novel artificial bee colony algorithm with depth-first search framework and elite-guided search equation. *Inf. Sci.* **2016**, *367–368*, 1012–1044. [CrossRef]

37. Li, G.H.; Cui, L.Z.; Fu, X.H.; Wen, Z.K.; Lu, N.; Lu, J. Artificial bee colony algorithm with gene recombination for numerical function optimization. *Appl. Soft Comput.* **2017**, *52*, 146–159. [CrossRef]

38. Yao, X.; Chan, F.T.S.; Lin, Y.; Jin, H.; Gao, L.; Wang, X.; Zhou, J. An individual dependent multi-colony artificial bee colony algorithm. *Inf. Sci.* **2019**. [CrossRef]

39. Kumar, D.; Mishra, K.K. Co-variance guided artificial bee colony. *Appl. Soft Comput.* **2018**, *70*, 86–107. [CrossRef]

40. Yang, J.; Jiang, Q.; Wang, L.; Liu, S.; Zhang, Y.; Li, W.; Wang, B. An adaptive encoding learning for artificial bee colony algorithms. *J. Comput. Sci.* **2019**, *30*, 11–27. [CrossRef]

41. Chen, X.; Tianfield, H.; Li, K. Self-adaptive differential artificial bee colony algorithm for global optimization problems. *Swarm Evol. Comput.* **2019**, *45*, 70–91. [CrossRef]

42. Zhang, X.; Zhang, X. A binary artificial bee colony algorithm for constructing spanning trees in vehicular ad hoc networks. *Ad Hoc Netw.* **2017**, *58*, 198–204. [CrossRef]

43. Zorarpacı, E.; Özel, S.A. A hybrid approach of differential evolution and artificial bee colony for feature selection. *Expert Syst. Appl.* **2016**, *62*, 91–103. [CrossRef]

44. Yuan, X.; Wang, P.; Yuan, Y.; Huang, Y.; Zhang, X. A new quantum inspired chaotic artificial bee colony algorithm for optimal power flow problem. *Energy Convers. Manag.* **2015**, *100*, 1–9. [CrossRef]

45. Dokeroglu, T.; Sevinc, E.; Cosar, A. Artificial bee colony optimization for the quadratic assignment problem. *Appl. Soft Comput.* **2019**, *76*, 595–606. [CrossRef]

46. Kishor, A.; Singh, P.K.; Prakash, J. NSABC: Non-dominated sorting based multi-objective artificial bee colony algorithm and its application in data clustering. *Neurocomputing* **2016**, *216*, 514–533. [CrossRef]

47. Wang, P.; Xue, F.; Li, H.; Cui, Z.; Xie, L.; Chen, J. A multi-objective DV-Hop localization algorithm based on NSGA-II in internet of things. *Mathematics* **2019**, *7*, 184. [CrossRef]

48. Pan, J.S.; Kong, L.P.; Sung, T.W.; Tsai, P.W.; Snasel, V. α-Fraction first strategy for hierarchical wireless sensor networks. *J. Internet Technol.* **2018**, *19*, 1717–1726.

49. Xue, X.S.; Pan, J.S. A compact co-evolutionary algorithm for sensor ontology meta-matching. *Knowl. Inf. Syst.* **2018**, *56*, 335–353. [CrossRef]

50. Hashim, H.A.; Ayinde, B.O.; Abido, M.A. Optimal placement of relay nodes in wireless sensor network using artificial bee colony algorithm. *J. Netw. Comput. Appl.* **2016**, *64*, 239–248. [CrossRef]

51. Wang, H.; Wu, Z.J.; Zhou, X.Y.; Rahnamayan, S. Accelerating artificial bee colony algorithm by using an external archive. In Proceedings of the IEEE Congress on Evolutionary Computation, Cancun, Mexico, 20–23 June 2013; pp. 517–521.

52. Li, B.; Sun, H.; Zhao, J.; Wang, H.; Wu, R.X. Artificial bee colony algorithm with different dimensional learning. *Appl. Res. Comput.* **2016**, *33*, 1028–1033.

53. Wang, H.; Wang, W.; Zhou, X.; Sun, H.; Zhao, J.; Yu, X.; Cui, Z. Firefly algorithm with neighborhood attraction. *Inf. Sci.* **2017**, *382*, 374–387. [CrossRef]

54. Wang, H.; Sun, H.; Li, C.H.; Rahnamayan, S.; Pan, J.S. Diversity enhanced particle swarm optimization with neighborhood search. *Inf. Sci.* **2013**, *223*, 119–135. [CrossRef]

55. Wang, G.G.; Tan, Y. Improving metaheuristic algorithms with information feedback models. *IEEE Trans. Cybern.* **2019**, *49*, 542–555. [CrossRef]

56. Wang, H.; Rahnamayan, S.; Sun, H.; Omran, M.G.H. Gaussian bare-bones differential evolution. *IEEE Trans. Cybern.* **2013**, *43*, 634–647. [CrossRef]

57. Wang, H.; Cui, Z.H.; Sun, H.; Rahnamayan, S.; Yang, X.S. Randomly attracted firefly algorithm with neighborhood search and dynamic parameter adjustment mechanism. *Soft Comput.* **2017**, *21*, 5325–5339. [CrossRef]

58. Sun, C.L.; Zeng, J.C.; Pan, J.S.; Xue, S.D.; Jin, Y.C. A new fitness estimation strategy for particle swarm optimization. *Inf. Sci.* **2013**, *221*, 355–370. [CrossRef]

59. Kiran, M.S.; Hakli, H.; Guanduz, M.; Uguz, H. Artificial bee colony algorithm with variable search strategy for continuous optimization. *Inf. Sci.* **2015**, *300*, 140–157. [CrossRef]

60. Liang, J.J.; Qu, B.Y.; Suganthan, P.N.; Hernández-Díaz, A.G. *Problem Definitions and Evaluation Criteria for the CEC 2013 Special Session and Competition on Real-Parameter Optimization*; Tech. Rep. 201212; Computational Intelligence Laboratory, Zhengzhou University, Zhengzhou China and Nanyang Technological University: Singapore, 2013.

61. Serani, A.; Leotardi, C.; Iemma, U.; Campana, E.F.; Fasano, G.; Diez, M. Parameter selection in synchronous and asynchronous deterministic particle swarm optimization for ship hydrodynamics problems. *Appl. Soft Comput.* **2016**, *49*, 313–334. [CrossRef]

![mathematics logo] *mathematics*

MDPI

Article

SRIFA: Stochastic Ranking with Improved-Firefly-Algorithm for Constrained Optimization Engineering Design Problems

Umesh Balande * and Deepti Shrimankar

Department of CSE, Visvesvaraya National Institute of Technology, Nagpur 440010, India;
dshrimankar@cse.vnit.ac.in
* Correspondence: umeshbalande30@gmail.com

Received: 2 February 2019; Accepted: 5 March 2019; Published: 11 March 2019

Abstract: Firefly-Algorithm (FA) is an eminent nature-inspired swarm-based technique for solving numerous real world global optimization problems. This paper presents an overview of the constraint handling techniques. It also includes a hybrid algorithm, namely the Stochastic Ranking with Improved Firefly Algorithm (SRIFA) for solving constrained real-world engineering optimization problems. The stochastic ranking approach is broadly used to maintain balance between penalty and fitness functions. FA is extensively used due to its faster convergence than other metaheuristic algorithms. The basic FA is modified by incorporating opposite-based learning and random-scale factor to improve the diversity and performance. Furthermore, SRIFA uses feasibility based rules to maintain balance between penalty and objective functions. SRIFA is experimented to optimize 24 CEC 2006 standard functions and five well-known engineering constrained-optimization design problems from the literature to evaluate and analyze the effectiveness of SRIFA. It can be seen that the overall computational results of SRIFA are better than those of the basic FA. Statistical outcomes of the SRIFA are significantly superior compared to the other evolutionary algorithms and engineering design problems in its performance, quality and efficiency.

Keywords: constrained optimization problems (COPs); evolutionary algorithms (EAs); firefly algorithm (FA); stochastic ranking (SR)

1. Introduction

Nature-Inspired Algorithms (NIAs) are very popular in solving real-life optimization problems. Hence, designing an efficient NIA is rapidly developing as an interesting research area. The combination of evolutionary algorithms (EAs) and swarm intelligence (SI) algorithms are commonly known as NIAs. The use of NIAs is popular and efficient in solving optimization problems in the research field [1]. EAs are inspired by Darwinian theory. The most popular EAs are genetic algorithm [2], evolutionary programming [3], evolutionary strategies [4], and genetic programming [5]. The term SI was coined by Gerardo Beni [6], as it mimics behavior of biological agents such as birds, fish, bees, and so on. Most popular SI algorithms are particle swarm optimization [7], firefly algorithm [8], ant colony optimization [9], cuckoo search [10] and bat algorithm [11]. Recently, many new population-based algorithms have been developed to solve various complex optimization problem such as killer whale algorithm [12], water evaporation algorithm [13], crow search algorithm [14] and so on. The No-Free-Lunch (NFL) theorem described that there is not a single appropriate NIA to solve all optimization problems. Consequently, choosing a relevant NIAs for a particular optimization problem involves a lot of trial and error. Hence, many NIAs are studied and modified to make them more powerful with regard to efficiency and convergence rate for some optimization problems. The primary factor of NIAs are intensification (exploitation) and diversification (exploration) [15].

Exploitation refers to finding a good solution in local search regions, whereas exploration refers to exploring global search space to generate diverse solutions [16].

Optimization algorithms can be classified in different ways. NIAs can be simply divided into two types: stochastic and deterministic [17]. Stochastic (in particular, metaheuristic) algorithms always have some randomness. For example, the firefly algorithm has "α" as a randomness parameter. This approach provides a probabilistic guarantee for a faster convergence of global optimization problem, usually to find a global minimum or maximum at an infinite time. In the deterministic approach, it ensures that, after a finite time, the global optimal solution will be found. This approach follows a detailed procedure and the path and values of both dimensions of problem and function are reputable. Hill-climbing is a good example of deterministic algorithm, and it follows same path (starting point and ending point) whenever the program is executed [18].

Real-world engineering optimization problems contain a number of equality and inequality constraints, which alter the search space. These problems are termed as Constrained-Optimization Problems (COPs). The minimization COPs defined as:

$$\text{Minimize: } f(\vec{z}) = (z_1, z_2, \ldots, z_n) \quad \vec{z} \in S, \tag{1}$$

$$g_j(\vec{z}) \leq 0 \quad j = 1, 2, 3, \ldots, m; \tag{2}$$

$$h_j(\vec{z}) = 0 \quad j = m+1, \ldots, q; \tag{3}$$

$$l_x \leq k \leq u_x \quad x = 1, 2, \ldots, n, \tag{4}$$

where $f(\vec{z})$ is the objective-function given in Equation (1), $(\vec{z}) = (z_1, z_2, z_3 \ldots, z_n)$ n-dimensional design variables, l_x and u_x are the lower and upper bounds, $g_j(\vec{z})$ inequality with m constraints and $h_j(\vec{z})$ equality with $q - 1$ constraints.

The feasible search space $F \subseteq S$ is represented as the equality (q) and inequality (m). Some point in the $z \in F$ contains feasible or infeasible solutions. The active constraint $(\vec{z}^*)$ is defined as inequality constraints that are satisfied when $g_j(z) \leq 0$ $(j = \{1,2,3,\ldots,m\})$ at given point $(\vec{z}^*) \in F$. In feasible regions, all constraints (i.e., equality constraints) were acknowledged as active constraints at all points.

In NIA problems, most of the constraint-handling techniques deal with inequality constraints. Hence, we have transformed equality constrained into equality using some tolerance value (ε):

$$|h_j(\vec{z})| - \varepsilon \leq 0, \tag{5}$$

where $j \in \{m+1, \ldots, q\}$ and 'ε' is tolerance allowed. Apply the value of tolerance ε for equality constraints for a given optimization problem. Then, the constraint-violation $CV_j(\vec{z})$ of an individual from the jth constraint can be calculated by

$$CV_j(\vec{z}) = \begin{cases} max\{g_j(\vec{z}), 0\} & 1 \leq j \leq m, \\ \max\{|h_j(\vec{z})| - \varepsilon, 0\} & m+1 \leq j \leq q. \end{cases} \tag{6}$$

The maximum constraint-violation of $\vec{z}$ of every constraint in the all individual or population is given as:

$$CV_j(\vec{z}) = \sum_{j=1}^{q} CV_j(\vec{z}). \tag{7}$$

With this background, the rest of paper is ordered as follows: Section 2 explains the classification of constrained-Handling Techniques (CHT); Section 3 deals with an overview of Constrained FA; Section 4 gives the outline of SR and OBL approaches. Section 5 described the proposed SRIFA with OBL; the experimental setup and computational outcomes of the SRIFA with 24 CEC 2006 benchmark test functions are illustrated in Section 6. The comparison of SRIFA with existing metaheuristic algorithms is also discussed with respect to its performance and effectiveness. The computational results of the SRIFA are examined with an engineering design problem in Section 7. Finally, in Section 8, conclusions of the paper are given.

2. Constrained-Handling Techniques (CHT)

Classification of CHT

In this section, we provide a literature survey of various CHT approaches that are adapted into NIAs to solve COPs. The classification of constrained handling approaches is shown in Figure 1. In the past few decades, various CHTs have been developed, particularly for EAs. Mezura-Montes and Coello conducted a comprehensive survey of NIA [19].

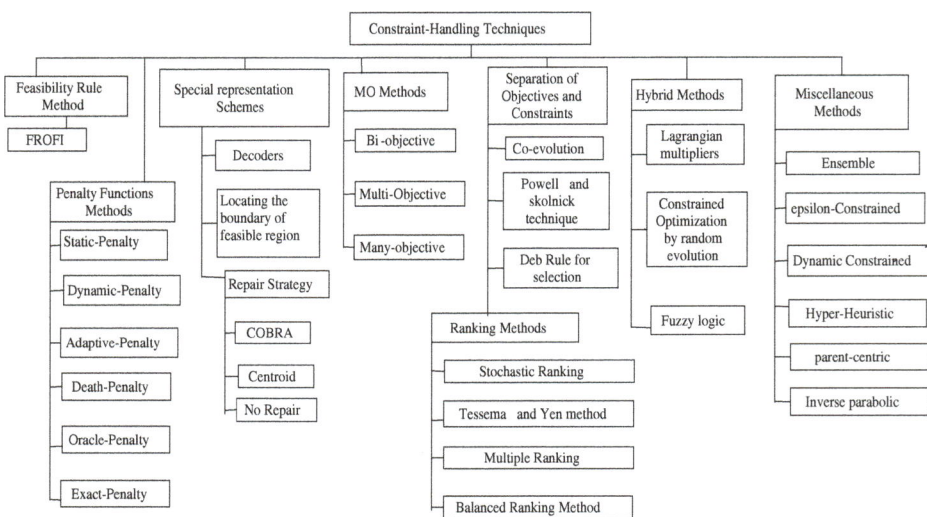

Figure 1. The classification of Constrained-Handling Techniques.

1. *Feasibility Rules Approach*: The most effective CHT was proposed by Deb [20]. Between any two solutions A_i and A_j compared, A_i is better than A_j, under the following conditions:

    (a) If A_i is a feasible solution, then A_j solution is not.
    (b) Between two A_i and A_j feasible solutions, if A_i has better objective value over A_j, then A_i is preferred.
    (c) Between two A_i and A_j infeasible solutions, if A_i has the lowest sum of constraint-violation over A_j, then A_i is preferred.

    Wang and Li [21] integrated a Feasibility-Rule integrated with Objective Function Information (FROFI), where Differential Evolution (DE) is used as a search algorithm along with feasibility rule.

2. *Penalty Function Method*: COPs can be transformed into unconstrained problems using penalty function. This penalty method includes various techniques such as static-penalty, dynamic-penalty [22], adaptive-penalty [23], death-penalty [24], oracle-penalty and exact-penalty methods.

3. *Special representation scheme*: This method includes decoders, locating the boundary of a feasible solution [25] and repair method [26]. The new repair methods classified into three types: Constrained Optimization by Radical basis Function Approximation (COBRA) [27], the centroid and No-pair method.

4. *Multi-objective Methods (MO) or Vector optimization or Pareto-optimization*: It is an optimization problems that has two or more objectives [28]. There are roughly two types of MO methods: bi-objective and many-objective.

5. *Split-up objective and constraints*: There are many techniques to handle split-up objective and constraints. These techniques are co-evolution, Powell and Skolnick technique, Deb-rule and ranking method. There are different types of ranking methods such as stochastic ranking, Tessema and Yen method, multiple ranking and the balanced ranking method.

6. *Hybrid Method*: The NIAs combined with a classical constrained method or heuristic method are called as hybrid methods. The hybrid method includes Lagrangian multipliers, constrained Optimization by random evolution and fuzzy logic [25].

7. *Miscellaneous Method*: These methods include ensemble [29], ϵ-constrained [30], dynamic constrained, hyper-heuristic, parent-centric and inverse parabolic [31].

3. Overview of Constrained FA

3.1. Basic FA

FA is a swarm-based NIAs proposed by Xin-she Yang [8]. Fister et al. [32] carried out in detail comprehensive review of FA. The basic FA pseudo-code is indicated in Algorithm 1. The mathematical formulation of the basic FA is as follows (Figure 2):

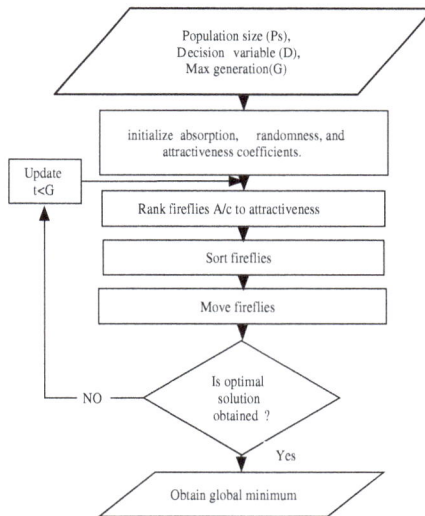

Figure 2. Basic Firefly Algorithm (BFA).

Let us consider that attractiveness of FA is assumed as brightness (i.e., fitness function). The distance between brightness of two fireflies (assume u and v) is given as:

$$I = I_0 e^{-\gamma r^2_{uv}} (z_u - z_{v,}) \tag{8}$$

where I is an intensity of light-source parameter, γ is an absorption coefficient, and (z_u) is distance between two fireflies u and v. I_0 is the intensity of light source parameter when r = 0. The attractiveness for two fireflies u and v (u is more attractive than v) is defined as:

$$\beta = \beta_0 e^{-\gamma r^2_{uv}} (z_u - z_v). \tag{9}$$

β_0 is attractiveness parameter when r = 0.

Movement of fireflies are basically based on the attractiveness, when a firefly u is less attractive than firefly v; then, firefly u moves towards firefly v and it is determined by Equation (10):

$$z_v = z_v + \beta_0 e^{-\gamma r^2_{uv}} (z_u - z_v) + \alpha \left(rand - \frac{1}{2} \right), \tag{10}$$

where the second term is an attractive parameter, the third term is a randomness parameter and *rand* is a vector of random-numbers generated uniform distribution between 0 and 1.

3.2. Constrained FA

The FA Combined with CHT has been widely used for solving COPs. Some typical constrained FA (CFA) has been briefly discussed below.

To solve engineering optimization problems, the adaptive-FA is designed has been discussed in [33]. Costa et al. [34] used penalty based techniques to evaluate different test functions for global optimization with FA. Brajevic et al. [35] developed feasibility-rule based with FA for COPs. Kulkarni et al. [36] proposed a modified feasibility-rule based for solving COPs using probability. The upgraded FA (UFA) is proposed to solve mechanical engineering optimization problem [37]. Chou and Ngo designed a multidimensional optimization structure with modified FA (MFA) [38].

Algorithm 1 Stochastic Ranking Approach (SRA)

1: Number of population (N), P_f balanced dominance of two solution of $f(\vec{z})$, CV_k $(\vec{z})$ is sum of constrained violation, m is individual who will be ranked
2: Rank the individual based on P_f and $f(\vec{z})$
3: Calculate $z_k = 1$,   */ $k \in 1, 2, 3, ..., \lambda$ and z_k is variable of f(z)
4: **for** i = 1 to n **do**
5:    **for** k = 1 to m-1 **do**
6:       Random R = U(0, 1) */random number generator
7:    **end for**
8:    **if** $(CV_k((z_k) = CV_k((z_{k+1}) = 0))$ or $R < P_f$ **then**
9:       **if** $(f(z_k) > f(z_{k+1}))$ **then**
10:          swap (z_k, z_{k+1})
11:       **end if**

12:    **else if** $(CV_k(z_k) > CV_k(z_{k+1}))$ **then**
13:          swap (z_k, z_{k+1})

          end if
14:    **end if**
15:    **if** no swapping **then** break;
16:    **end if**
17: **end for**

---

4. Stochastic Ranking and Opposite-Based Learning (OBL)

This section represents an overview of SR and OBL.

4.1. Stochastic Ranking Approach (SRA)

This approach, which was introduced by Runarsson and Yao [39], which balances fitness or (objective function) and dominance of a penalty approach. Based on this, the SRA uses a simple bubble sort technique to rank the individuals. To rank the individual in SRM, P_f is introduced, which is used to compare the fitness function in infeasible area of search space. Normally, when we take any two individuals for comparison, three possible solutions are formed.

(a) if both individuals are in a feasible region, then the smallest fitness function is given the highest priority; (b) For both individuals at an infeasible region, an individual having smallest constraint-violation (CV_k) is preferred to fitness function and is given the highest priority; and (c) if one individual is feasible and other is infeasible, then the feasible region individual is given highest priority. The pseudo code of SRM is given in Algorithm 1.

4.2. Opposition-Based Learning (OBL)

The OBL is suggested by Tizhoosh in the research industry, which is inspired by a relationship among the candidate and its opposite solution. The main aim of the OBL is to achieve an optimal solution for a fitness function and enhance the performance of the algorithm [40]. Let us assume that z ∈ $[x + y]$ is any real number, and the opposite solution of z is denoted as $ź$ and defined as

$$ź = x + y - z. \tag{11}$$

Let us assume that $Z = (z_1, z_2, z_3, \ldots, z_n)$ is an n-dimensional decision vector, in which $z_i \in [xi + yi]$ and $i = 1, 2, \ldots, n$. In the opposite vector, p is defined as $Ź = (ź_1, ź_2, ź_3, \ldots, ź_n)$, where $ź_i = ([x_i + y_i] - z_i)$.

5. The Proposed Algorithm

The most important factor in NIAs is to maintain diversity of population in search space to avoid premature convergence. From the intensification and diversification viewpoints, an expansion in diversity of population revealed that NIAs are in the phase of intensification, while a decreased population of diversity revealed that NIAs are in the phase of diversification. The adequate balance between exploration and exploitation is achieved by maintaining a diverse populations. To maintain balance between intensification and diversification, different approaches were proposed such as diversity maintenance, diversity learning, diversity control and direct approaches [16]. The diversity maintenance can be performed using a varying size population, duplication removal and selection of a randomness parameter.

On the other hand, when the basic FA algorithm is performed with insufficient diversification (exploration), it leads to a solution stuck in local optima or a suboptimal region. By considering these issues, a new hybridizing algorithm is proposed by improving basic FA.

5.1. Varying Size of Population

A very common and simple technique is to increase the population size in NIAs to maintain the diversity of population. However, due to an increase in population size, computation time required for the execution of NIAs is also increased. To overcome this problem, the OBL concept is applied to improve the efficiency and performance of basic FA at the initialization phase.

5.2. Improved FA with OBL

In the population-based algorithms, premature convergence in local optimum is a common problem. In the basic FA, every firefly moves randomly towards the brighter one. In that condition, population diversity is high. After some generation, the population diversity decreases due to a lack of selection pressure and this leads to a trap solution at local optima. The diversification of FA is reduced due to premature convergence. To overcome this problem, the OBL is applied to an initial phase of FA, in order to increase the diversity of firefly individuals.

In the proposed Improved Firefly Algorithm (IFA), we have to balance intensification and diversification for better performance and efficiency of the proposed FA. To perform exploration, a randomization parameter is used to overcome local optimum and to explore global search. To balance between intensification and diversification, the random-scale factor (R) was applied to generate randomly populations. Das et al. [41] used a similar approach in DE:

$$R_{u,v} = lb_v + 0.5(1 + rand(0, 1)) * (ub_v - lb_v), \tag{12}$$

where $R_{u,v}$ is a vth parameter of the uth firefly, ub_v is upper-bound, lb_v is a lower-bound of vth value and rand (0, 1) is randomly distributed of the random-number.

The movement of fireflies using Equation (10) will be modified as

$$z_v = z_v + \beta_0 e^{-\gamma r^2_{uv}} \left(z_u - z_v \right) + R_{u,v}. \tag{13}$$

5.3. Stochastic Ranking with an Improved Firefly Algorithm (SRIFA)

Many studies are published in literature for solving COPs using EAs and FA. However, it is quite challenging to apply this approach for constraints effectively handling optimization problems. FA produces admirable outcomes on COPs and it is well-known for having a quick convergence rate [42]. As a result of the quick convergence rate of FA and popularity of the stochastic-ranking for CHT, we proposed a hybridized technique for constrained optimization problems, known as Stochastic Ranking with an Improved Firefly Algorithm (SRIFA). The flowchart of SRIFA is shown in Figure 3.

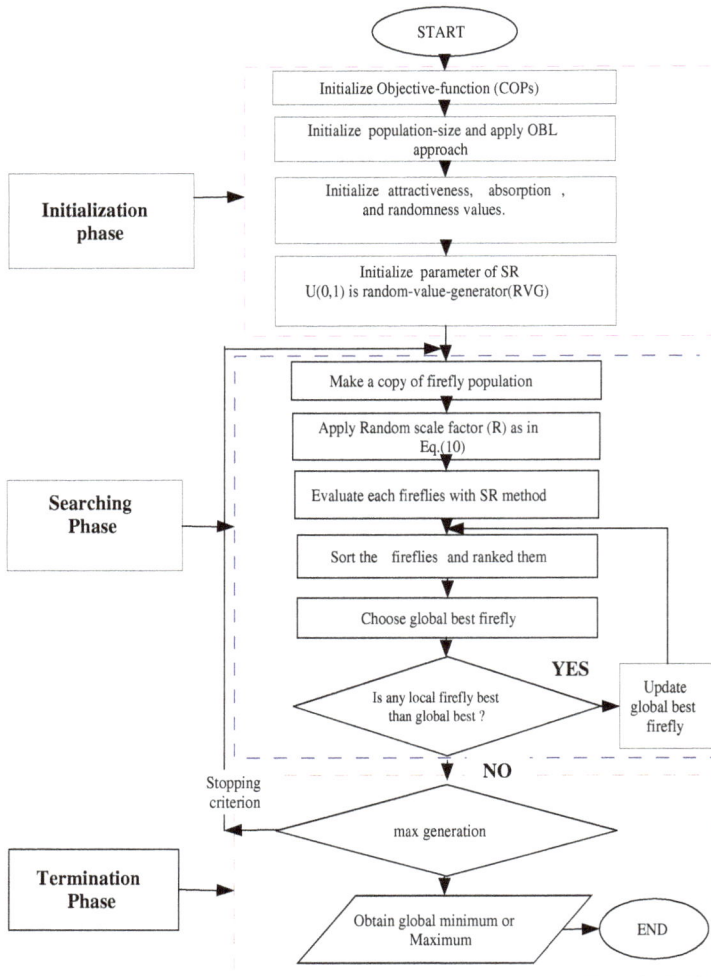

Figure 3. The flowchart of the SRIFA algorithm.

5.4. Duplicate Removal in SRIFA

The duplicate individuals in a population should be eliminated and new individuals should generated and inserted randomly into SRIFA. Figure 4 represents the duplication removal in SRIFA.

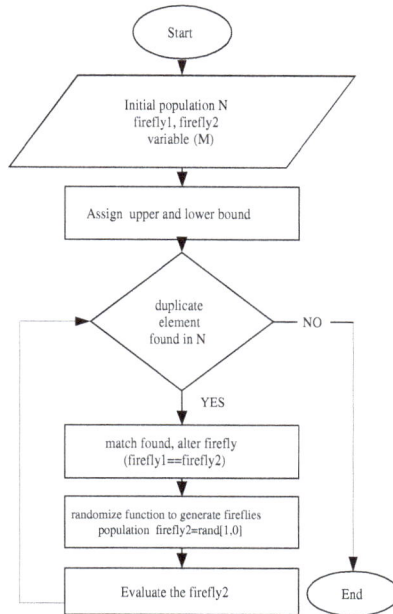

Figure 4. The flowchart of duplication removal in SRIFA.

6. Experimental Results and Discussions

To examine the performance of SRIFA with existing NIAs, the proposed algorithm is applied to 24 numerical benchmark test functions given in CEC 2006 [43]. This preferred benchmark functions have been thoroughly studied before by various authors.

In Table 1, the main characteristics of 24 test function are determined, where a fitness function (f(z)), number of variables or dimensions (D), $\rho = {}^{|Feas|}/_{|SeaR|}$ is expressed as a feasibility ratio between a feasible solution (Feas) with search region (SeaR), Linear-Inequality constraint (LI), Nonlinear Inequality constraint (NI), Linear-Equality constraint (LE), Nonlinear Equality constraint (NE), number active constraint represented as (a^*) and an optimal solution of fitness function denoted (OPT) are given. For convenience, all equality constraints, i.e., $h_j(z)$ are transformed into inequality constraint $h_j(z) - \varepsilon \leq 0$, where $\varepsilon = 10^{-4}$ is a tolerance value, and its goal is to achieve a feasible solution [43].

6.1. Experimental Design

To investigate the performance and effectiveness of the SRIFA, it is tested over 24 standard-functions and five well-known engineering design-problems. All experiments of COPs were experimented on an Intel core (TM) $i5 - 3570$ processor @3.40 GHz with 8 GB RAM memory, where an SRIFA algorithm was programmed with Matlab 8.4 (R2014b) under Win7 (x64). Table 2 shows the parameter used to conduct computational experiments of SRIFA algorithms. For all experiments, 30 independent runs were performed for each problem. To investigate efficiency and effectiveness of the SRIFA, various statistical parameters were used such as best, worst, mean, global optimum and standard-deviation (Std). Results in bold indicate best results obtained.

Table 1. Characteristic of 24 standard-functions.

Problems	Dimension	Types of Functions	ρ (%)	L-I	N-I	L-E	N-E	a^*	Opt
G01	13	Quadratic	0.0003	9	0	0	0	6	−15.0000
G02	20	Non-linear	99.9962	1	1	0	0	1	−0.8036
G03	10	Non-linear	0.0002	0	0	0	1	1	−1.0000
G04	5	Quadratic	26.9089	0	6	0	0	2	−30,655.5390
G05	4	nonlinear	0.0000	2	0	0	3	3	5126.4970
G06	2	Non-linear	0.0065	0	2	0	0	2	−6961.8140
G07	10	Quadratic	0.0010	3	5	0	0	6	24.3060
G08	2	Non-linear	0.8488	0	2	0	0	0	0.9583
G09	7	Non-linear	0.5319	0	4	0	0	2	680.6300
G10	8	Linear	0.0005	3	3	0	0	6	7049.2480
G11	2	Quadratic	0.0099	0	0	0	1	1	0.7499
G12	3	Quadratic	4.7452	0	9	0	0	0	−1.0000
G13	5	Non-linear	0.0000	0	0	1	2	3	0.0539
G14	10	Non-linear	0.0000	0	0	3	0	3	−47.7650
G15	3	Quadratic	0.0000	0	0	1	1	2	961.7150
G16	5	Non-linear	0.0204	4	34	0	0	4	−1.9050
G17	6	Non-linear	0.0000	0	0	4	4	4	8853.5397
G18	9	Quadratic	0.0000	0	13	0	0	6	−0.8660
G19	15	Non-linear	33.4761	0	5	0	0	0	32.6560
G20	24	Linear	0.0000	0	6	2	12	16	0.0205
G21	7	Linear	0.0000	0	1	0	5	6	193.7250
G22	22	Linear	0.0000	0	1	8	11	19	236.4310
G23	9	Linear	0.0000	0	2	3	1	6	−400.0050
G24	2	Linear	79.6556	0	2	0	0	2	−5.5080

Table 2. Experimental parameters for SRIFA.

Parameters	Value	Significances
Size of population (NP)	50	Gandomi [44] suggested that 50 fireflies are adequate to perform experiments for any application. If we increase the population size, the computational time of the proposed algorithm will be increased.
Initial randomization value (α_0)	0.5	In the literature, many authors suggested that a randomness parameter must used in range (0, 1). In our experiment, we have used a 0.5 value.
Initial attractiveness value (β_0)	0.2	The attractiveness parameter for our experiment is 0.2 value.
Absorption coefficient (γ)	4	The absorption value is crucial in our experiment. It determines convergence speed of algorithms. In most applications, the γ value in range (0.001, 100)
Number of iterations or generations (G)	4800	Total number of iterations.
Total number of function evaluation (NFEs)	240,000	The total number of objective function evaluations (50 × 4800 = 240,000 evaluations)
Constrained-handling values		Initial tolerance value: 0.5 (for equality)
Probability P_f	0.45	Final tolerance: 1×10^{-4} (for equality) It is used to rank objects. P_f is used to compare the fitness (objective) function in infeasible areas of the search space.
Varphi (ϕ)	1	Sum of constrained violation.

6.2. Calibration of SRIFA Parameters

In this section, we have to calibrate the parameter of the SRIFA. According to the strategy of the SRIFA, described in Figure 2, the SRIFA contains eight parameters: size of population (*NP*), initial randomization value (α_0), initial attractiveness value (β_0), absorption-coefficient (γ), max-generation (*G*), total number of function evaluations (*NFEs*), probability P_f and varphi (ϕ). To derive a suitable parameter, we have performed details of fine-tuning by varying parameters of SRIFA. The choice of each of these parameters as follows: (*NP*)∈ (5 to 100 with an interval of 5), (α_0) ∈ (0.10 to 1.00 with an interval of 0.10), (β_0) ∈ (0.10 to 1.00 with an interval 0.10), (γ) ∈ (0.01 to 100 with an interval of 0.01 until 1 further 5 to 100), (*G*) ∈ (1000 to 10,000 with an interval of 1000), *NFEs* ∈ (1000 to 240,000), P_f = ∈ (0.1 to 0.9 with an interval of 0.1) and (ϕ) ∈ (0.1 to 1.0 with an interval of 0.1). The best optimal solutions obtained by SRIFA parameter experiments from the various test functions. In Table 2, the best parameter value for experiments for the SRIFA are described.

6.3. Experimental Results of SRIFA Using a GKLS (GAVIANO, KVASOV, LERA and SERGEYEV) Generator

In this experiment, we have compared the proposed SRIFA with two novel approaches: Operational Characteristic and Aggregated Operational Zone. An operational characteristic approach is used for comparing deterministic algorithms, whereas an aggregated operational zone approach is used by extending the idea of operational characteristics to compare metaheuristic algorithms.

The proposed algorithm is compared with some widely used NIAs (such as DE, PSO and FA) and the well-known deterministic algorithms such as DIRECT, DIRECT-L (locally-biased version), and ADC (adaptive diagonal curves). The GKLS test classes generator is used in our experiments. The generator allows us to randomly generate 100 test instances having local minima and dimension. In this experiment, eight classes (small and hard) are used (with dimensions of n = 2, 3, 4 and 5) [45]. The control parameters of the GKLS-generator required for each class contain 100 functions and are defined by the following parameters: design variable or problem dimension (N), radius of the convergence region (ρ), distance from of the paraboloid vertex and global minimum (r) and tolerance (δ). The value of control parameters are given in Table 3.

Table 3. Control parameter of the GKLS generator.

N	Class	r	ρ	δ
2	Simple	0.9	0.2	10^4
2	Hard	0.9	0.1	10^4
3	Simple	0.66	0.2	10^5
3	Hard	0.9	0.2	10^5
4	Simple	0.66	0.2	10^6
4	Hard	0.9	0.2	10^6
5	Simple	0.66	0.3	10^7
5	Hard	0.9	0.2	10^7

From Table 4, we can see that the mean value of generations required for computation of 100 instances are calculated for each deterministic and metaheuristic algorithms using an GKLS generator. The values ">m(i)" indicate that the given algorithm did not solve a global optimization problem i times in 100 × 100 instances (i.e., 1000 runs for deterministic and 10,000 runs for metaheuristic algorithms). The maximum number of generations is set to be 10^6. The mean value of generation required for proposed algorithm is less than other algorithms, indicating that the performance SRIFA is better than the given deterministic and metaheuristic algorithms.

6.4. Experimental Results FA and SRIFA

In our computational experiment, the proposed SRIFA is compared with the basic FA. It differs from the basic FA in following few points. In the SRIFA, the OBL technique is used to enhance initial population of algorithm, while, in FA, fixed generation is used to search for optimal solutions. In the SRIFA, the chaotic map (or logistic map) is used to improve absorption coefficient γ, while, in the FA, fixed iteration is applied to explore the global solution. The random scale factor (R) was used to enhance performance in SRIFA. In addition, SRIFA uses Deb's rules in the form of the stochastic ranking method.

The experimental results of the SRIFA with basic FA are shown in Table 5. The comparison between SRIFA and FA are conducted using 24 CEC (Congress on Evolutionary-Computation) benchmark test functions [43]. The global optimum, CEC 2006 functions, best, worst, mean and standard deviation (Std) outcomes produced by SRIFA and FA in over 25 runs are described in Table 5.

In Table 5, it is clearly observed that SRIFA provides promising results compared to the basic FA for all benchmark test functions. The proposed algorithm found optimal or best solutions on all test functions over 25 runs. For two functions (G20 and G22), we were unable to find any optimal solution. It should be noted that 'N-F' refers to no feasible result found.

Table 4. Statistical results obtained by deterministic and metaheuristic algorithms using GKLS generator.

N	Class	Deterministic Algorithm (100 Runs for Each Algorithm and Class)			Metaheuristic Algorithms (10,000 Runs for Each Algorithm and Class)			
		DIRECT	**DIRECT-L**	**ADC**	**DE**	**PSO**	**FA**	**SRIFA**
2	Simple	198.9	292.8	176.3	>52,910.38 (511)	>110,102.74 (1046)	1190.3	1008
2	Hard	1063.8	1267.1	675.7	>357,467.49 (3556)	>247,232.35 (2282)	>4299.6 (3)	>3457.6 (3)
3	Simple	1117.7	1785.7	735.8	>165,125.02 (1515)	>170,320.10 (1489)	15,269.2	14,987
3	Hard	>42,322.7 (4)	4858.9	2006.8	>476,251.20 (4603)	>285,499.04 (2501)	>21,986.3 (1)	20,989
4	Simple	>47,282.9 (4)	18,983.6	5014.1	>462,401.52 (4546)	>303,436.36 (2785)	23,166.7	22,752.4
4	Hard	>95,708.3 (7)	68,754	16,473	>773,481.03 (7676)	>456,996.08 (4157)	40,380.7	38,123.2
5	Simple	>16,057.5 (1)	16,758.4	5129.9	>294,839.01 (2815)	>181,805.17 (1561)	>47,203.1 (16)	>45,892.8 (15)
5	Hard	>217,215.6 (16)	>269,064.4 (4)	30,471.8	>751,930.00 (7473)	>250,462.63 (2109)	>79,555.2 (38)	>76,564 (34)

Table 5. Statistical results obtained by SRIFA and FA on 24 benchmark functions over 25 runs.

Algo.	Functions	Global Opt	Best	Worst	Mean	Std
FA	G01	−15.000	−14.420072	−11.281250	−13.840104	1.16×10^{0}
SRIFA			**−15.000**	**−15.000**	**−15.000**	$\mathbf{7.86 \times 10^{-13}}$
FA	G02	−0.8036191	**−0.8036191**	−0.5205742	−0.7458475	6.49×10^{-2}
SRIFA			**−0.8036191**	**−0.800909**	**−0.80251**	$\mathbf{8.95 \times 10^{-4}}$
FA	G03	−1.000	**−1.0005**	**−1.0005**	**−1.0005**	9.80×10^{-7}
SRIFA			**−1.0005**	**−1.0005**	**−1.0005**	$\mathbf{6.54 \times 10^{-6}}$
FA	G04	−30,665.539	**−30,665.539**	**−30,665.539**	**−30,665.539**	2.37×10^{-9}
SRIFA			**−30,665.539**	**−30,665.54**	**−30,665.54**	$\mathbf{6.74 \times 10^{-11}}$
FA	G05	5126.49671	5126.49671	5144.3028	5233.2377	2.92×10^{1}
SRIFA			**5126.49671**	**5126.4967**	**5126.4967**	$\mathbf{1.94 \times 10^{-9}}$
FA	G06	−6961.8138	**−6961.81388**	**−6961.81388**	**−6961.81388**	1.76×10^{-7}
SRIFA			**−6961.8138**	**−6961.814**	**−6961.814**	$\mathbf{4.26 \times 10^{-8}}$
FA	G07	24.306	24.306283	24.310614	24.32652	3.80×10^{-3}
SRIFA			**24.306**	**24.306**	**24.306**	$\mathbf{2.65 \times 10^{-8}}$
FA	G08	−0.09582	**−0.09582504**	**−0.09582504**	**−0.09582504**	1.83×10^{-17}
SRIFA			**−0.09582**	**−0.09582**	**−0.09582**	$\mathbf{5.40 \times 10^{-20}}$
FA	G09	680.63	680.630058	680.630063	680.630082	7.11×10^{-6}
SRIFA			**680.6334**	**680.6334**	**680.6334**	$\mathbf{5.64 \times 10^{-7}}$
FA	G10	7049.248	7071.757586	7181.02714	7111.54937	3.00×10^{1}
SRIFA			**7049.2484**	**7049.2484**	**7049.2484**	$\mathbf{5.48 \times 10^{-4}}$
FA	G11	0.7499	**0.7499**	**0.7499**	**0.7499**	5.64×10^{-9}
SRIFA			**0.7499**	**0.7499**	**0.7499**	$\mathbf{8.76 \times 10^{-15}}$
FA	G12	−1.000	**−1.000**	**−1.000**	**−1.000**	5.00×10^{-2}
SRIFA			**−1.000**	**−1.000**	**−1.000**	$\mathbf{6.00 \times 10^{-3}}$
FA	G13	0.053942	**0.054**	0.439	0.131	$\mathbf{1.54 \times 10^{-1}}$
SRIFA			**0.053943**	**0.053943**	**0.053943**	0.00×10^{0}
FA	G14	−47.765	−47.764879	−47.764563	−47.762878	3.82×10^{-4}
SRIFA			**−47.7658**	**−47.7658**	**−47.7658**	$\mathbf{5.68 \times 10^{-6}}$
FA	G15	961.715	**961.715**	**961.715**	**961.715**	8.67×10^{-9}
SRIFA			**961.7155**	**961.7155**	**961.7155**	$\mathbf{6.34 \times 10^{-11}}$
FA	G16	−1.9050	−1.90515	−1.90386	−1.90239	8.76×10^{-5}
SRIFA			**−1.9050**	**−1.9050**	**−1.9050**	$\mathbf{2.55 \times 10^{-10}}$
FA	g17	8853.5397	8853.5339	8900.0831	9131.5849	5.52×10^{1}
SRIFA			**8853.5339**	**8853.5339**	**8853.5339**	$\mathbf{5.80 \times 10^{-3}}$
FA	G18	−0.8660	**−0.8660**	**−0.8660**	**−0.8660**	7.60×10^{-5}
SRIFA			**−0.8660**	**−0.8660**	**−0.8660**	$\mathbf{6.54 \times 10^{-10}}$
FA	G19	32.6560	32.7789	34.6224	38.3827	1.65×10^{0}
SRIFA			**32.6560**	**32.6560**	**32.6560**	$\mathbf{2.22 \times 10^{-6}}$
FA	G20	30.0967	'N-F'	'N-F'	'N-F'	'N-F'
SRIFA			'N-F'	'N-F'	'N-F'	'N-F'
FA	G21	193.7250	193.7245	683.1906	350.9696	5.41×10^{2}
SRIFA			**193.7240**	**193.7240**	**193.7240**	$\mathbf{4.26 \times 10^{-4}}$
FA	G22	236.4310	'N-F'	'N-F'	'N-F'	'N-F'
SRIFA			'N-F'	'N-F'	'N-F'	'N-F'
FA	G23	−400.0050	−347.917268	−347.9345669	−347.923470	7.54×10^{-3}
SRIFA			**−400.0050**	**−400.0052**	**−400.0050**	$\mathbf{5.65 \times 10^{-4}}$
FA	G24	−5.5080	**−5.5081**	**−5.5080**	**−5.5080**	1.11×10^{-5}
SRIFA			**−5.5081**	**−5.5080**	**−5.5080**	$\mathbf{1.21 \times 10^{-13}}$

6.5. Comparison of SRIFA with Other NIAs

To investigate the performance and effectiveness of the SRIFA, these results are compared with five metaheuristic algorithms. These algorithms are stochastic ranking with a particle-swarm-optimization (SRPSO) [46], self adaptive mix of particle-swarm-optimization (SAMO-PSO) [47], upgraded firefly algorithm (UFA) [37], an ensemble of constraint handling techniques for evolutionary-programming (ECHT-EP2) [48] and a novel differential-evolution algorithm (NDE) [49]. To evaluate proper comparisons of these algorithms, the same number of function evaluations (NFEs = 240,000) were chosen.

The statistical outcomes achieved by SRPSO, SAMO-PSO, UFA, ECHT-EP2 and NDE for 24 standard functions are listed in Table 6. The outcomes given in bold letter indicates best or optimal solution. N-A denotes "Not Available". The benchmark function G20 and G22 are discarded from the analysis, due to no feasible results were obtained.

On comparing SRIFA with SRPSO for 22 functions as described in Table 6, it is clearly seen that, for all test functions, statistical outcomes indicate better performance in most cases. The SRIFA obtained the best or the same optimal values among five metaheuristic algorithms. In terms of mean outcomes, SRIFA shows better outcomes to test functions G02, G14, G17, G21 and G23 for all four metaheuristic algorithms (i.e., SAMO-PSO, ECHT-EP2, UFA and NDE). SRIFA obtained worse mean outcomes to test function G19 than NDE. In the rest of all test functions, SRIFA was superior to all compared metaheuristic algorithms.

6.6. Statistical Analysis with Wilcoxon's and Friedman Test

Statistical analysis can be classified as parametric and non-parametric test (also known as distribution-free tests). In parametric tests, some assumptions are made about data parameters, while, in non-parametric tests, no assumptions are made for data parameters. We performed statistical analysis of data by non-parametric tests. It mainly consists of a Wilcoxon test (pair-wise comparison) and Friedman test (multiple comparisons) [50].

The outcomes of statistical analysis after conducting a Wilcoxon-test between SRIFA and the other five metaheuristic algorithms are shown in Table 7. The $R+$ value indicates that the first algorithm is significantly superior than the second algorithm, whereas $R-$ indicates that the second algorithm performs better than the first algorithm. In Table 7, it is observed that $R+$ values are higher than $R-$ values in all cases. Thus, we can conclude that SRIFA significantly outperforms compared to all metaheuristic algorithms.

The statistical analysis outcomes by applying Friedman test are shown in Table 8. We have ranked the given metaheuristic algorithms corresponding to their mean value. From Table 8, SRIFA obtained first ranking (i.e., the lowest value gets the first rank) compared to all metaheuristic algorithms over the 22 test functions. The average ranking of the SRIFA algorithm based on the Friedman test is described in Figure 5.

6.7. Computational Complexity of SRIFA

In order to reduce complexity of the given problem, constraints are normalized. Let n be population size and t is iteration. Generally in NIAs, at each iteration, a complexity is $O(n * FEs + Cof * FEs)$, where FEs is the maximum amount of function evaluations allowed and and Cof is the cost of objective function. At the initialization phase of SRIFA, the computational complexity of population generated randomly by the OBL technique is $O(nt)$. In a searching and termination phase, the computational complexity of two inner loops of FA and stochastic ranking using a bubble sort are $O(n^2t + n(log(n))) + O(nt)$. The total computational complexity of SRIFA is $O(n,t) = O(nt) + O(n^2t + nlogn) + O(nt) \approx O(n^2t)$.

Table 6. Statistical outcomes achieved by SRPSO, SAMO-PSO, ECHT-EP2, UFA, NDE AND SRIFA.

Fun	Features	SRPSO	SAMO-PSO	ECHT-EP2	UFA	NDE	SRIFA
G01	Best	−15.00	−15.00	−15.000	−15.000	−15.000	−15.000
	Mean	**−15.00**	**−15.00**	**−15.000**	**−15.000**	**−15.000**	**−15.000**
	Worst	−15.00	N-A	−15.000001	−15.000001	−15.000001	−15.000
	SD	5.27×10^{-12}	0.00×10^{0}	0.00×10^{0}	8.95×10^{-10}	0.00×10^{0}	7.86×10^{-13}
G02	Best	−0.80346805	0.8036191	−0.8036191	−0.8036191	−0.803480	−0.8036191
	Mean	−0.788615	−0.79606	−0.7998220	−0.7961871	−0.801809	−0.80251
	Worst	−0.7572932	N-A	−0.7851820	−0.7851820	−0.800495	**−0.800909**
	SD	1.31×10^{-3}	5.3420×10^{-3}	6.29×10^{-3}	7.48×10^{-3}	5.10×10^{-4}	8.95×10^{-4}
G03	Best	−0.9997	−1.0005	−1.0005	−1.0005	−1.0005001	−1.0005
	Mean	−0.9985	**−1.0005001**	**−1.0005**	**−1.0005**	**−1.0005001**	**−1.0005**
	Worst	−0.996532	N-A	−1.0005	−1.0005	−1.0005001	−1.0005
	SD	8.18×10^{-5}	0.02×10^{0}	0.02×10^{0}	1.75×10^{-6}	0.00×10^{0}	6.54×10^{-6}
G04	Best	−30,665.538	−30,665.539	−30,665.53867	−30,665.539	−30,665.539	−30,665.539
	Mean	−30,665.5386	**−30,665.539**	**−30,665.53867**	**−30,665.539**	**−30,665.539**	**−30,665.539**
	Worst	−30,665.536	N-A	−30,665.538	−30,665.539	−30,665.539	−30,665.539
	SD	4.05×10^{-5}	0.00×10^{0}	0.00×10^{0}	6.11×10^{-9}	0.00×10^{0}	6.74×10^{-11}
G05	Best	5126.4985	5126.4967	5126.4967	5126.4967	5126.4967	5126.4967
	Mean	5129.9010	**5126.496**	**5126.496**	**5126.496**	**5126.496**	**5126.496**
	Worst	5145.93	N-A	5126.496	5126.496	5126.496	5126.496
	SD	5.11	1.3169×10^{-10}	0.00×10^{0}	1.11×10^{-8}	0.00×10^{0}	1.94×10^{-9}
G06	Best	−6961.8139	−6961.8138	−6961.8138	−6961.8138	−6961.8138	−6961.8138
	Mean	−6916.1370	**−6961.8138**	**−6961.8138**	**−6961.8138**	**−6961.8138**	**−6961.8138**
	Worst	−6323.3140	N-A	−6961.8138	−6961.8138	−6961.8138	−6961.8138
	SD	138.331	0.00×10^{0}	0.00×10^{0}	3.87×10^{-8}	0.00×10^{0}	4.26×10^{-8}
G07	Best	24.312803	24.306209	24.3062	24.306209	24.306209	24.3062
	Mean	24.38	**24.306209**	**24.3063**	**24.306209**	**24.306209**	**24.306**
	Worst	24.885038	N-A	24.3063	24.306209	24.306209	24.306
	SD	1.13×10^{-2}	1.9289×10^{-8}	3.19×10^{-5}	1.97×10^{-9}	1.35×10^{-14}	2.65×10^{-8}
G08	Best	−0.09582	−0.095825	−0.09582504	−0.09582504	−0.095825	−0.09582
	Mean	−0.095823	**−0.095825**	**−0.095825**	**−0.095825**	**−0.095825**	**−0.09582**
	Worst	−0.095825	N-A	−0.09582504	−0.09582504	−0.095825	−0.09582
	SD	2.80×10^{-11}	0.00×10^{0}	0.00×10^{0}	1.70×10^{-17}	0.00×10^{0}	5.40×10^{-20}
G09	Best	680.63004	680.630057	680.630057	680.630057	680.630057	680.6334
	Mean	680.66052	**680.6300**	**680.6300**	**680.6300**	**680.6300**	**680.6300**
	Worst	680.766	N-A	680.6300	680.6300	680.6300	680.6300
	SD	3.33×10^{-3}	0.00×10^{0}	2.61×10^{-8}	5.84×10^{-10}	0.00×10^{0}	5.64×10^{-7}
G10	Best	7076.397	7049.24802	7049.2483	7049.24802	7049.24802	7049.2484
	Mean	7340.6964	**7049.2480**	**7049.249**	**7049.2480**	**7049.2480**	**7049.2480**
	Worst	8075.92	N-A	7049.2501	7049.24802	7049.24802	7049.2484
	SD	255.37	1.5064×10^{-5}	6.60×10^{-4}	2.26×10^{-7}	3.41×10^{-9}	5.48×10^{-4}
G11	Best	0.75	0.749999	0.749999	0.7499	0.749999	0.7499
	Mean	0.75	**0.749999**	**0.749999**	**0.7499**	**0.749999**	**0.7499**
	Worst	0.75	N-A	0.749999	0.7499	0.749999	0.7499
	SD	9.44×10^{-5}	0.00×10^{0}	3.40E-16	9.26E-16	0.00×10^{0}	8.76×10^{-15}
G12	Best	−1	−1.000	−1.000	−1.000	−1.000	−1.000
	Mean	−1	**−1.000**	**−1.000**	**−1.000**	**−1.000**	**−1.000**
	Worst	−1	N-A	−1.000	−1.000	−1.000	−1.000
	SD	2.62×10^{-11}	0.00×10^{0}	0.00×10^{0}	0.00×10^{0}	0.00×10^{0}	6.00×10^{-3}
G13	Best	N-A	0.053941	0.053941	0.053941	0.053941	0.053941
	Mean	N-A	**0.0539415**	**0.0539415**	**0.0539415**	**0.0539415**	**0.053943**
	Worst	N-A	N-A	0.0539415	0.0539415	0.0539415	0.053943
	SD	N-A	0.00×10^{0}	1.30×10^{-12}	1.43×10^{-12}	0.00×10^{0}	0.00×10^{0}
G14	Best	N-A	−47.7648	−47.7649	−47.76489	−47.7648	−47.7658
	Mean	N-A	−47.7648	−47.7648	−47.76489	−47.7648	**−47.7658**
	Worst	N-A	N-A	N-A	−47.76489	−47.7648	−47.7658
	SD	N-A	4.043×10^{-2}	N-A	2.34×10^{-6}	5.14×10^{-15}	5.68×10^{-6}

Table 6. *Cont.*

Fun	Features	SRPSO	SAMO-PSO	ECHT-EP2	UFA	NDE	SRIFA
G15	Best	961.7151	961.7150	961.7150	961.7150	961.7150	961.7150
	Mean	961.7207	**961.7150**	**961.7150**	**961.7150**	**961.7150**	**961.7150**
	Worst	961.7712	N-A	N-A	961.7150	961.7150	961.7155
	SD	1.12×10^{-2}	0.00×10^{0}	N-A	1.46×10^{-11}	0.00×10^{0}	6.34×10^{-11}
G16	Best	-1.9051	-1.9051	-1.9051	-1.9051	-1.9051	-1.9050
	Mean	$-\textbf{1.9050}$	-1.9051	$-\textbf{1.9050}$	$-\textbf{1.9050}$	$-\textbf{1.9050}$	$-\textbf{1.9050}$
	Worst	-1.9051	N-A	N-A	-1.9051	-1.9051	-1.9050
	SD	1.12×10^{-11}	1.15×10^{-5}	N-A	1.58×10^{-11}	0.00×10^{0}	2.55×10^{-10}
G17	Best	N-A	8853.5338	8853.5397	8853.5338	8853.5338	8853.5338
	Mean	N-A	8853.5338	8853.8871	8853.5338	8853.5338	**8853.5338**
	Worst	N-A	N-A	N-A	8853.5338	8853.5338	8853.5338
	SD	N-A	0.00×10^{0}	N-A	2.18×10^{-8}	0.00×10^{0}	5.80×10^{-3}
G18	Best	N-A	-0.8660	-0.8660	-0.8660	-0.8660	-0.8660
	Mean	N-A	$-\textbf{0.8660}$	$-\textbf{0.8660}$	$-\textbf{0.8660}$	$-\textbf{0.8660}$	$-\textbf{0.8660}$
	Worst	N-A	N-A	N-A	-0.8660	-0.8660	-0.8660
	SD	N-A	7.0436×10^{-7}	N-A	3.39×10^{-10}	0.00×10^{0}	6.54×10^{-10}
G19	Best	N-A	32.6555	32.6555	32.6555	32.6555	32.6560
	Mean	N-A	32.6556	36.4274	32.6555	**32.6556**	32.6560
	Worst	N-A	N-A	N-A	32.6555	32.6557	32.6560
	SD	N-A	6.145×10^{-2}	N-A	1.37×10^{-8}	3.73×10^{-5}	2.22×10^{-6}
G20	Best	N-A	N-A	N-A	N-A	N-A	N-A
	Mean	N-A	N-A	N-A	N-A	N-A	N-A
	Worst	N-A	N-A	N-A	N-A	N-A	N-A
	SD	N-A	N-A	N-A	N-A	N-A	N-A
G21	Best	N-A	193.7255	193.7251	266.5	193.72451	193.7250
	Mean	N-A	193.7251	246.0915	255.5590	193.7251	**193.7250**
	Worst	N-A	N-A	N-A	520.1656	193.724	193.7260
	SD	N-A	1.9643×10^{-2}	N-A	9.13×10^{1}	6.26×10^{-11}	4.26×10^{-4}
G22	Best	N-A	N-A	N-A	N-A	N-A	N-A
	Mean	N-A	N-A	N-A	N-A	N-A	N-A
	Worst	N-A	N-A	N-A	N-A	N-A	N-A
	SD	N-A	N-A	N-A	N-A	N-A	N-A
G23	Best	N-A	-400.0551	-355.661	-400.0551	-400.0551	-400.005
	Mean	N-A	-400.0551	-194.7603	-400.0551	-400.0551	$-\textbf{400.0050}$
	Worst	N-A	N-A	N-A	-400.0551	-400.0551	-400.0052
	SD	N-A	1.96×10^{1}	N-A	5.08×10^{-8}	3.45×10^{-9}	5.65×10^{-4}
G24	Best	-5.5080	-5.5080	-5.5080	-5.5080	-5.5080	-5.5080
	Mean	$-\textbf{5.5080}$	$-\textbf{5.5080}$	$-\textbf{5.5080}$	$-\textbf{5.5080}$	$-\textbf{5.5080}$	$-\textbf{5.5080}$
	Worst	-5.5080	N-A	N-A	-5.5080	-5.5080	-5.5080
	SD	2.69×10^{-11}	0.00×10^{0}	N-A	5.37×10^{-13}	0.00×10^{0}	1.21×10^{-13}

Table 7. Results obtained by a Wilcoxon-test for SRIFA against SRPSO, SAMO-PSO, ECHT-EP2, UFA and NDE.

Algorithms	$R+$	$R-$	*p*-value	Best	Equal	Worst	Decision
SRIFA versus SRPSO	176	3	0.465	17	3	2	+
SRIFA versus SAMO-PSO	167	38	0.363	14	7	1	+
SRIFA versus ECHT-EP2	142	17	0.002	15	3	4	+
SRIFA versus UFA	45	19	0.016	10	8	4	$\approx$
SRIFA versus NDE	67	25	0.691	14	4	6	$\approx$

Table 8. Results obtained Friedman test for all metaheuristic algorithms.

Functions	SRPSO	SAMO-PSO	ECHT-EP2	UFA	NDE	SRIFA
G01	4.5	4.5	4.5	4.5	4.5	4.5
G02	6	7	5	4.5	3	3
G03	4	4.5	6	6	4	4
G04	4.5	4.5	4.5	4.5	4.5	4.5
G05	4.5	4.5	4.5	4.5	4.5	3
G06	4.5	6	6	4.5	4.5	4.5
G07	3.5	4.5	3.5	3.5	3.5	3.5
G08	4.5	7	4.5	4.5	4.5	4.5
G09	6	4.5	6	6	4.5	4.5
G10	6	4.5	4.5	4.5	3	3
G11	4.5	7	3.5	4.5	4.5	3
G12	4.5	4.5	4.5	4.5	4.5	4.5
G13	4	4.5	4	4	4	4
G14	3.5	4	3.5	3.5	3.5	3.5
G15	3	7	4.5	4.5	4.5	4.5
G16	4.5	8	4.5	4.5	4.5	4.5
G17	4.5	4.5	4.5	2	2	2
G18	4.5	3.5	3	3.5	3	3
G19	5.5	8	4	5	5	4
G20	N-A	N-A	N-A	N-A	N-A	N-A
G21	6	6	4.5	2.5	2.5	2.5
G22	N-A	N-A	N-A	N-A	N-A	N-A
G23	8	7	6	3.5	1.5	1.5
G24	6	8	4.5	4.5	4.5	3.5
Avearge rank	4.8409091	5.613636364	4.5454545	4.25	3.8409091	**3.6136**

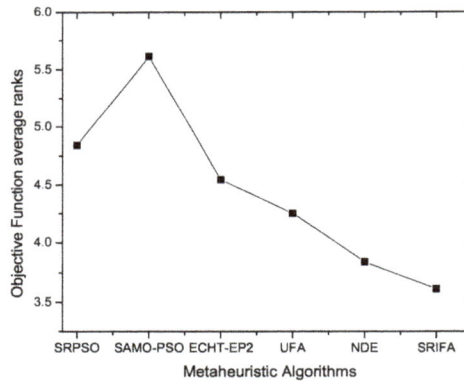

Figure 5. Average ranking of the proposed algorithm with various metaheuristic algorithms.

7. SRIFA for Constrained Engineering Design Problems

In this section, we evaluate the efficiency and performance of SRIFA by solving five widely used constrained engineering design problems. These problems are: (i) tension or compression spring design [51]; (ii) welded-beam problem [52]; (iii) pressure-vessel problem [53]; (iv) three-bar truss problem [51]; and (v) speed-reducer problem [53]. For every engineering design problem, statistical outcomes were calculated by executing 25 independent runs for each problem. The mathematical formulation of all five constrained engineering design problems are given in "Appendix A".

Every engineering problem has unique characteristics. The best value of constraints, parameter and objective values obtained by SRIFA for all five engineering problems are listed in Table 9. The statistical outcomes and number of function-evaluations (NFEs) of SRIFA for all five engineering design problems are listed in Table 10. These results were obtained by SRIFA over 25 independent runs.

Table 9. Best outcomes of parameter objective and constraints values for over engineering-problems.

	Tension/Compression	Welded-Beam	Pressure-Vessel	Three-Truss-Problem	Speed-Reducer
x1	0.0516776638592	0.205729638946844	0.8125	0.788675145296995	3.50000000002504
x2	0.3567324816961	3.47048866663245	0.4375	0.40824826019360	0.70000000000023
x3	11.2881015418157	9.03662391025916	-	-	17
x4	-	0.20572963979284	42.0984455958043	-	7.30000000000014
x5	-	-	176.63659584313	-	7.71531991152672
x6	-	-	-	-	3.35021466610421
x7	-	-	-	-	5.28665446498064
x8	-	-	-	-	-
x9	-	-	-	-	-
x10	-	-	-	-	-
F(x)	0.012665232805563	1.72485231254328	6059.714335	263.895843376515	2994.47106614799
G1(x)	0	−0.000063371885873	−0.0000000000000873	−0.070525402833398	−0.073915280394101
G2(x)	−1.216754326628263	−0.000002714066983	−0.00035880820872	−1.467936135628140	−0.197998527141053
G3(x)	−4.0521785529112	−0.000000000839532	−0.000000016701007	−0.602589267205258	−0.499172248101033
G4(x)	−0.727728835000534	−3.432983781912125	−0.633634041562312	-	−0.904643904554311
G5(x)	-	−0.080729638942761	-	-	−0.000000000000654
G6(x)	-	−0.235540322583421	-	-	−0.000000000000212
G7(x)	-	−0.000000209274321	-	-	−0.702499999999991
G8(x)	-	-	-	-	−0.000000000000209
G9(x)	-	-	-	-	−0.795833333333279
G10(x)	-	-	-	-	−0.051325753542591
G11(x)	-	-	-	-	−0.000000000001243

Table 10. Statistical outcomes achieved by SRIFA for all five engineering problems over 25 independent runs.

Problems	Best-Value	Mean-Value	Worst-Value	SD	NFEs
Tension/Compression	0.0126652328	0.0126652329	0.0126652333	6.54×10^{-10}	2000
Welded-beam	1.7248523087	1.7248523087	1.7248523089	8.940×10^{-12}	2000
pressure-vessel	6059.7143350561	6059.7143351	6059.7143352069	6.87×10^{-8}	2000
Three-truss	263.8958433765	263.8958433768	263.8958433770	6.21×10^{-11}	1500
Speed-reducer	2996.348165	2996.348165	2996.348165	8.95×10^{-12}	3000

7.1. Tension/Compression Spring Design

A tension/compression spring-design problem is formulated to minimize weight with respect to four constraints. These four constraints are shear stress, deflection, surge frequency and outside diameter. There are three design variables, namely: mean coil (D), wire-diameter d and the amount of active-coils N.

This proposed SRIFA approach is compared to SRPSO [46], MVDE [54], BA [44], MBA [55], JAYA [56], PVS [57], UABC [58], IPSO [59] and AFA [33]. The comparative results obtained by SRIFA for nine NIAs are given in Table 11. It is clearly observed that SRIFA provides the most optimum results over nine metaheuristic algorithms. The mean, worst and SD values obtained by SRIFA are superior to those for other algorithms. Hence, we can draw conclusions that SRIFA performs better in terms of statistical values. The comparison of the number of function evacuations (NFEs) with various NIAs is plotted in Figure 6.

Table 11. Statistical results of comparison between SRIFA and NIAs for tension/compression spring design problem.

Algorithm	Best-Value	Mean-Value	Worst-Value	SD	NFEs
SRPSO	**0.012668**	0.012678	0.012685	7.05×10^{-6}	20,000
MVDE	0.012665272	0.012667324	0.012719055	2.45×10^{-6}	10,000
BA	0.01266522	0.01350052	0.0168954	3.09×10^{-6}	24,000
MBA	**0.012665**	0.012713	0.0129	6.30×10^{-5}	7650
Jaya	**0.012665**	0.012666	0.012679	4.90×10^{-4}	10,000
PVS	0.01267	0.012838	0.013141	N-A	10,000
UABC	**0.012665**	0.012683	N-A	3.31×10^{-5}	15,000
IPSO	0.01266523	0.013676527	0.01782864	1.57×10^{-3}	4000
AFA	0.012665305	0.126770446	0.000128058	0.012711688	50,000
SRIFA	**0.0126652328**	**0.0126652329**	**0.0126652333**	**6.54×10^{-10}**	**2000**

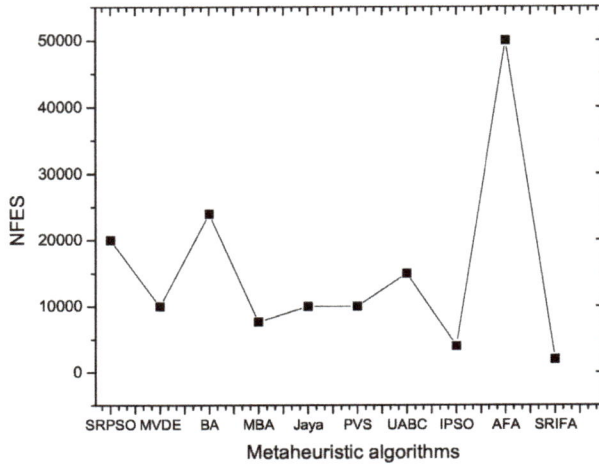

Figure 6. NIAs with NFEs for the tension/compression problem.

7.2. Welded-Beam Problem

The main objective of the welded-beam problem is to minimize fabrication costs with respect to seven constraints. These constraints are bending stress in the beam (σ), shear stress (τ), deflection of beam (δ), buckling load on the bar (P_c), side constraints, weld thickness and member thickness (L).

Attempts have been made by many researchers to solve the welded-beam-design problem. The SRIFA was compared with SRPSO, MVDE, BA, MBA, JAYA, MFA, FA, IPSO and AFA. The statistical results obtained by SRIFA on comparing with nine metaheuristic algorithms are described in Table 12. It can be seen that statistical results obtained from SRIFA performs better than all metaheuristic algorithms.

Table 12. Statistical results of comparison between SRIFA and NIAs for welded-beam problem.

Algorithm	Best-Value	Mean-Value	Worst-Value	SD	NFEs
SRPSO	1.72486658	1.72489934	1.72542212	1.12×10^{-6}	20,000
MVDE	1.7248527	1.7248621	1.7249215	7.88×10^{-6}	15,000
BA	1.7312065	1.878656	2.3455793	0.2677989	50,000
MBA	1.724853	1.724853	1.724853	$\mathbf{6.94 \times 10^{-19}}$	47,340
JAYA	**1.724852**	1.724852	1.724853	3.30×10^{-2}	10,000
MFA	1.7249	1.7277	1.7327	2.40×10^{-3}	50,000
FA	1.7312065	1.878665	2.3455793	2.68×10^{-1}	50,000
IPSO	1.7248624	1.7248528	1.7248523	2.02×10^{-6}	12,500
AFA	**1.724853**	**1.724853**	1.724853	0.00×10^{0}	50,000
SRIFA	**1.7248523087**	**1.7248523087**	**1.7248523089**	8.940×10^{-12}	**2000**

The results obtained by the best optimum value for SRIFA performs superior to almost all of the seven algorithms (i.e., SRPSO, MVDE, BA, MBA, MFA, FA, and IPSO) but almost the same optimum value for JAYA and AFA. In terms of mean results obtained by SRIFA, it performs better than all metaheuristic algorithms except AFA as it contains the same optimum mean value. The standard deviation (SD) obtained by SRIFA is slightly worse than the SD obtained by MBA. From Table 12, it can be seen that SRIFA is superior in terms of SD for all remaining algorithms. The smallest NFE result is obtained by SRIFA as compared to all of the metaheuristic algorithms. The comparisons of NFEs with all NIAs are shown in Figure 7.

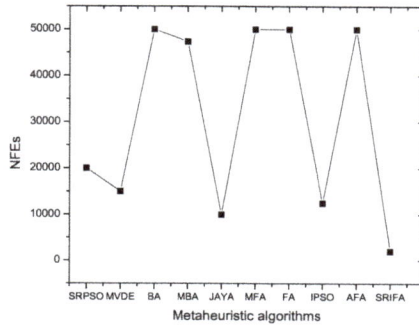

Figure 7. NIAs with NFEs for welded-beam problem.

7.3. Pressure-Vessel Problem

The main purpose of the pressure-vessel problem is to minimize the manufacturing cost of a cylindrical-vessel with respect to four constraints. These four constraints are thickness of head (T_h), thickness of pressure vessel (T_s), length of vessel without head (L) and inner radius of the vessel (R).

The SRIFA is optimized with SRPSO, MVDE, BA, EBA [60], FA [44], PVS, UABC, IPSO and AFA. The statistical results obtained by SRIFA for nine metaheuristic algorithms are listed in Table 13. It is clearly seen that SRIFA has the same best optimum value when compared to six algorithms (MVDE, BA, EBA, PVS, UABC and IPSO). The mean, worst, SD and NFE results obtained by SRIFA are superior to all NIAs. The comparisons of NFEs with all NIAs are shown in Figure 8.

Table 13. Statistical results of comparison between SRIFA and NIAs for the pressure-vessel problem.

Algorithm	Best-Value	Mean-Value	Worst-Value	SD	NFEs
SRPSO	6086.20	6042.84	6315.01	8.04×10^1	20,000
MVDE	**6059.714387**	6059.997236	6090.533528	2.91×10^0	15,000
BA	**6059.71**	6179.13	6318.95	1.37×10^2	15,000
EBA	**6059.71**	6173.67	6370.77	1.42×10^2	15,000
FA	5890.383	5937.3379	6258.96825	1.65×10^2	25,000
PVS	**6059.714**	6063.643	6090.526	N-A	42,100
UABC	**6059.714335**	6192.116211	N-A	2.04×10^2	15,000
IPSO	**6059.7143**	6068.7539	6090.5314	1.40×10^1	7500
AFA	6059.71427196	6090.52614259	6064.33605261	1.13×10^1	50,000
SRIFA	**6059.7143350561**	**6059.7143351**	**6059.7143352069**	$\mathbf{6.87 \times 10^{-8}}$	**2000**

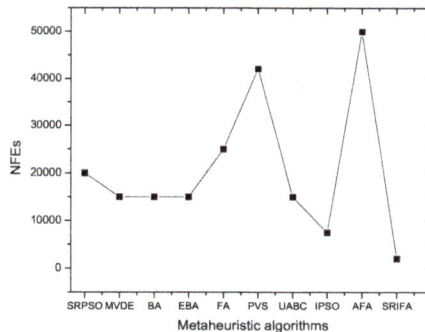

Figure 8. NIAs with NFEs for pressure-vessel problem.

7.4. Three-Bar-Truss Problem

The main purpose of the given three-bar-truss problem is to minimize the volume of a three-bar truss with respect to three stress constraints.

The SRIFA is compared with SRPSO, MVDE, NDE [49], MAL-FA [61], UABC, WCA [62] and UFA. The statistical results obtained by SRIFA in comparison with the seven NIAs are described in Table 14. It is clearly seen that SRIFA has almost the same best optimum value except with the UABC algorithm. In terms of mean and worst results obtained, SRIFA performed better compared to all metaheuristic algorithms except NDE and UFA, which contain the same optimum mean and worst value. The standard deviation (SD) obtained by SRIFA is superior to all metaheuristic algorithms. The smallest NFE value is obtained by SRIFA compared to all other metaheuristic algorithms. The comparisons of NFEs with all other NIAs are shown in Figure 9.

Table 14. Statistical results of comparison between SRIFA and NIAs for the three-bar truss problem.

Algorithm	Best-Value	Mean-Value	Worst-Value	SD	NFEs
SRPSO	**263.8958440**	263.8977800	263.9079550	3.02×10^{-5}	20,000
MVDE	**263.8958434**	263.8958434	263.8958548	2.55×10^{-6}	7000
NDE	**263.8958434**	**263.8958434**	**263.8958434**	0.00×10^{0}	4000
MAL-FA	**263.895843**	263.896101	263.895847	9.70×10^{-7}	4000
UABC	263.895843	263.895843	N-A	0.00×10^{0}	12,000
WCA	**263.895843**	263.896201	263.895903	8.71×10^{-5}	5250
UFA	**263.8958433765**	**263.8958433768**	**263.8958433770**	1.92×10^{-10}	4500
SRIFA	**263.8958433765**	**263.8958433768**	**263.8958433770**	$\mathbf{6.21 \times 10^{-11}}$	**1500**

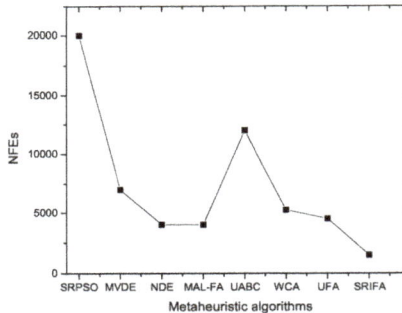

Figure 9. NIAs with NFEs for three-bar-truss problem.

7.5. Speed-Reducer Problem

The goal of the given problem is to minimize the speed-reducer of weight with respect to eleven constraints. This problem has seven design variables that are gear face, number of teeth in pinion, teeth module, length of first shaft between bearings. diameter of first shaft, length of second shaft between bearings, and diameter of second shaft.

The proposed SRIFA approach is compared with SRPSO, MVDE, NDE, MBA, JAYA, MBA, UABC, PVS, IPSO and AFA. The statistical results obtained by SRIFA for nine metaheuristic algorithms are listed in Table 15. It can be observed that the SRIFA provides the best optimum value among all eight metaheuristic algorithms except JAYA (they have the same optimum value). The statistical results (best, mean and worst) value obtained by SRIFA and JAYA algorithm is almost the same, while SRIFA requires less NFEs for executing the algorithm. Hence, we can conclude that SRIFA performed better in terms of statistical values. Comparisons of number of function evacuations (NFEs) with various metaheuristic algorithm are plotted in Figure 10.

Table 15. Statistical results of comparison between SRIFA and NIAs for the speed-reducer problem.

Algorithm	Best-Value	Mean-Value	Worst-Value	SD	NFEs
SRPSO	2514.97	2700.10	2860.13	8.73×10^1	20,000
MVDE	2994.471066	2994.471066	2994.471069	2.819316×10^{-7}	30,000
NDE	2994.471066	2994.471066	2994.471066	4.17×10^{-12}	18,000
MBA	2994.482453	2996.769019	2999.652444	1.56×10^0	6300
JAYA	**2996.348**	**2996.348**	**2996.348**	0.00×10^0	10,000
UABC	2994.471066	2994.471072	N-A	5.98×10^{-6}	15,000
PVS	2994.47326	2994.7253	2994.8327	N-A	6000
IPSO	2994.471066	2994.471066	2994.471066	2.65×10^{-9}	5000
AFA	2996.372698	2996.514874	2996.514874	9.00×10^{-2}	50,000
SRIFA	**2996.348165**	**2996.348165**	**2996.348165**	8.95×10^{-12}	**3000**

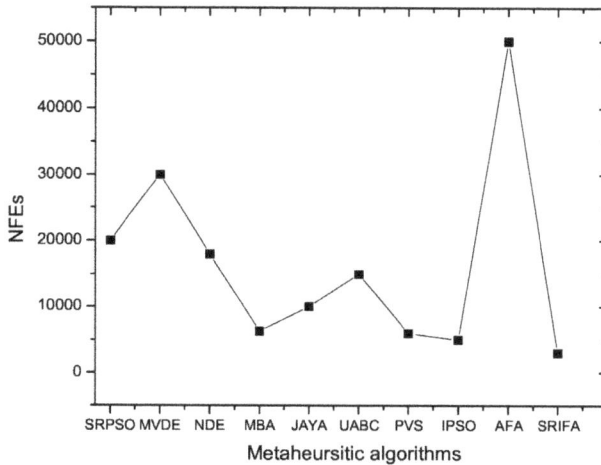

Figure 10. NIAs with NFEs for the speed-reducer problem.

8. Conclusions

This paper proposes a review of constrained handling techniques and a new hybrid algorithm known as a Stochastic Ranking with an Improved Firefly Algorithm (SRIFA) to solve a constrained optimization problem. In population-based problems, stagnation and premature convergence occurs due to imbalance between exploration and exploitation during the development process that traps the solution in the local optimal. To overcome this problem, the Opposite Based Learning (OBL) approach was applied to basic FA. This OBL technique was used at an initial population, which leads to increased diversity of the problem and improves the performance of the proposed algorithm.

The random scale factor was incorporated into basic FA, for balancing intensification and diversification. It helps to overcome the premature convergence and increase the performance of the proposed algorithm. The SRIFA was applied to 24 CEC benchmark test functions and five constrained engineering design problems. Various computational experiments were conducted to check the effectiveness and quality of the proposed algorithm. The statistical results obtained from SRIFA when compared to those of the FA clearly indicated that our SRIFA outperformed in terms of statistical values.

Furthermore, the computational experiments demonstrated that the performance of SRIFA was better compared to five NIAs. The performance and efficiency of the proposed algorithm were significantly superior to other metaheuristic algorithms presented from the literature. The statistical analysis of SRIFA was conducted using the Wilcoxon's and Friedman test. The results obtained proved that efficiency, quality and performance of SRIFA was statistically superior compared to NIAs.

Moreover, SRIFA was also applied to the five constrained engineering design problems efficiently. In the future, SRIFA can be modified and extended to explain multi-objective problems.

Author Contributions: Conceptualization, U.B.; methodology, U.B. and D.S.; software, U.B. and D.S.; validation, U.B. and D.S.; formal analysis, U.B. and D.S.; investigation, U.B.; resources, D.S.; data curation, D.S.; writing—original draft preparation, U.B.; writing—review and editing, U.B. and D.S.; supervision, D.S.

Acknowledgments: The authors would like to thank Dr. Yaroslav Sergeev for sharing GKLS generator. The authors would also like to thank Mr. Rohit for his valuable suggestions, which contribute a lot to technical improvement of the manuscript.

Conflicts of Interest: The authors have declared no conflict of interest.

Abbreviations

The following abbreviations are used in this manuscript:

COPs   Constrained Optimization Problems
CHT   Constrained Handling Techniques
EAs   Evolutionary Algorithms
FA   Firefly Algorithm
OBL   Opposite-Based Learning
NIAs   Nature Inspired Algorithms
NFEs   Number of Function Evaluations
SRA   Stochastic Ranking Approach
SRIFA   Stochastic Ranking with Improved Firefly Algorithm

Appendix A

Appendix A.1. Tension/Compression Spring Design Problem

$$f\left(\vec{Z}\right) = (z_3 + 2)\, z_2 z_1^2,$$

subject to:

$$G_1\left(\vec{z}\right) = 1 - \frac{z_2^3 z_3}{71{,}785 x_1^4},$$

$$G_2\left(\vec{z}\right) = 4 z_2^2 - z_1 z_2 / 12{,}566 \left(z_2 z_1^3 - z_1^4\right) +$$

$$1/5108 x_1^2 \ -1 \le 0,$$

$$G_3\left(\vec{z}\right) = 1 - 140.45 z_1 / z_2^2 z_3 \ \le 0,$$

$$G_4\left(\vec{z}\right) = z_2 + z_1 / 1.5 \ -1 \le 0,$$

$$0.05 \le z_1 \le 2,$$

$$0.25 \le z_2 \le 1.3,$$

$$2 \le z_3 \le 15.$$

Appendix A.2. Welded-Beam-Design Problem

$$f\left(\vec{Z}\right) = 1.10471z_1^2 z_2 + 0.04811z_3 z_4 \left(14 + z_2\right),$$

subject to:

$$G_1\left(\vec{z}\right) = \tau\left(\vec{z}\right) - \tau_{\max} \leq 0,$$

$$G_2\left(\vec{z}\right) = \sigma\left(\vec{z}\right) - \sigma_{\max} \leq 0,$$

$$G_3\left(\vec{z}\right) = z_1 - z_4 \leq 0,$$

$$G_4\left(\vec{z}\right) = 0.10471z_1^2 + 0.4811z_3 z_4 \left(14 + z_2\right) - 5 \leq 0,$$

$$G_5\left(\vec{z}\right) = 0.125 - z_1 \leq 0,$$

$$G_6\left(\vec{z}\right) = \delta\left(\vec{z}\right) - \delta_{\max} \leq 0,$$

$$G_7\left(\vec{z}\right) = P - P_c\left(\vec{z}\right) \leq 0,$$

$$0.1 \leq z_i \leq 2 \quad i = 1, 4,$$

$$0.1 \leq z_i \leq 10 \quad i = 2, 3,$$

where

$$\tau\left(\vec{z}\right) = \sqrt{(\tau^l) + 2\tau^l \tau^{ll} \frac{z_2}{2R} \tau^{ll}} \quad, \tau^l = \frac{P}{\sqrt{2}z_1 z_2}, \tau^{ll} = \frac{MR}{J},$$

$$M = P\left(L + \frac{z_2}{2}\right), R = \sqrt{\frac{x_2^2}{4} + \left(\frac{z_1 + z_2}{2}\right)^2},$$

$$J = 2\left\{\sqrt{2}z_1 z_2 \left[\sqrt{\frac{z_2^2}{4} + \left(\frac{z_1 + z_2}{2}\right)^2}\right]\right\}, \sigma\left(\vec{z}\right) = \frac{6PL}{z_4 z_3^2},$$

$$\delta\left(\vec{z}\right) = \frac{4PL^3}{Ex_3^3 z_4}, P_c\left(\vec{z}\right) = \frac{4.013E\sqrt{\frac{x_3^2 x_4^6}{36}}}{L^2} \left(1 - \frac{z_3}{2L}\sqrt{\frac{E}{4G}}\right),$$

$$P = 6000\,\text{lb}, \ L = 14\,\text{in}, \ E = 30 \times 10^6\,\text{psi},$$

$$G = 12 \times 10^6\,\text{psi},$$

$$\tau_{\max} = 13,600\,\text{psi}, \ \sigma_{\max} = 30,000\,\text{psi}, \ \delta_{\max} = 0.25\,\text{in}.$$

Appendix A.3. Pressure-Vessel Design Problem

$$f\left(\vec{Z}\right) = 0.6224z_1 z_3 z_4 + 1.7781z_2 z_3^2 + 3.1661z_1^2 z_4 +$$

$$19.84z_1^2 z_3,$$

subject to:

$$G_1\left(\vec{z}\right) = -z_1 + 0.0193z,$$

$$G_2\left(\vec{z}\right) = -z_2 + 0.00954z_3 \leq 0,$$

$$G_3\left(\vec{z}\right) = -\pi z_3^2 z_4 - \frac{4}{3}\pi z_3^2 + 12,96,000 \leq 0,$$

$$G_4\left(\vec{z}\right) = z_4 - 240 \leq 0,$$

$$0 \leq z_1 \leq 100 \quad i = 1, 2,$$

$$10 \leq z_1 \leq 200 \quad i = 3, 4.$$

Appendix A.4. Three-Bar-Truss Design Problem

$$f\left(\vec{Z}\right) = \left(2 + \sqrt{2}z_1z_2\right) \times l,$$

subject to:

$$G_1\left(\vec{z}\right) = \frac{\sqrt{2}z_1 + z_2}{\sqrt{2}z_1{}^2 + 2z_1z_2}p - \rho \le 0,$$

$$G_2\left(\vec{z}\right) = \frac{z_2}{\sqrt{2}z_1{}^2 + 2z_1z_2}p - \rho \le 0,$$

$$G_1\left(\vec{z}\right) = \frac{1}{\sqrt{2}z_1{}^2 + 2z_1z_2}p - \rho \le 0,$$

$$0 \le z_1 \le 1 \quad i = 1,2,$$

$$l = 100\,\text{cm}, P = 2\,\text{kN}/\text{cm}^2, \sigma = 2\,\text{kN}/\text{cm}^3.$$

Appendix A.5. Speed-Reducer-Design Problem

$$f\left(\vec{Z}\right) = 0.7854z_1z_2^2\left(3.3333z_3^2 + 14.933z_3 - 43.0934\right) -$$
$$1.508z_1\left(z_6^2 + z_7^2\right) + 7.4777\left(z_6^3 + z_7^3\right) + 0.7854\left(z_4x_6^2 + z_5z_7^2\right),$$

subject to

$$G_1\left(\vec{z}\right) = \frac{27}{z_1z_2^2z_3} - 1 \le 0,$$

$$G_2\left(\vec{z}\right) = \frac{397.5}{z_1z_2^2z_3^2} - 1 \le 0,$$

$$G_3\left(\vec{z}\right) = \frac{1.93z_4^3}{z_2z_6^4z_3} - 1 \le 0,$$

$$G_4\left(\vec{z}\right) = \frac{1.93z_5^3}{z_2z_7^4z_3} - 1 \le 0,$$

$$G_5\left(\vec{z}\right) = \frac{\left[(745z_4/z_2z_3)^2 + 16.9 \times 10^6\right]^{1/2}}{110x_6^3} - 1 \le 0,$$

$$G_6\left(\vec{z}\right) = \frac{\left[(745z_5/z_2z_3)^2 + 157.5 \times 10^6\right]^{1/2}}{85z_7^3} - 1 \le 0,$$

$$G_7\left(\vec{z}\right) = \frac{z_2z_3}{40} - 1 \le 0,$$

$$G_8\left(\vec{z}\right) = \frac{5z_2}{z_1} - 1 \le 0,$$

$$G_9\left(\vec{z}\right) = \frac{z_1}{12z_2} - 1 \le 0,$$

$$G_{10}\left(\vec{z}\right) = \frac{15z_6 + 1.9}{z_4} - 1 \le 0,$$

$$G_{11}\left(\vec{z}\right) = \frac{11z_7 + 1.9}{z_5} - 1 \le 0,$$

where

$$2.6 \le z_1 \le 3.6,\ 0.7 \le z_2 \le 0.8,\ 17 \le z_3 \le 28,$$
$$7.3 \le z_4, z_5 \le 8.3,\ 2.9 \le z_6 \le 3.9,\ 5 \le z_7 \le 5.5.$$

References

1. Slowik, A.; Kwasnicka, H. Nature Inspired Methods and Their Industry Applications—Swarm Intelligence Algorithms. *IEEE Trans. Ind. Inform.* **2018**, *14*, 1004–1015. [CrossRef]
2. Goldberg, D.E.; Holland, J.H. Genetic Algorithms and Machine Learning. *Mach. Learn.* **1988**, *3*, 95–99. [CrossRef]
3. Fogel, D.B. An introduction to simulated evolutionary optimization. *IEEE Trans. Neural Netw.* **1994**, *5*, 3–14. [CrossRef] [PubMed]
4. Beyer, H.G.; Schwefel, H.P. Evolution strategies—A comprehensive introduction. *Nat. Comput.* **2002**, *1*, 3–52. [CrossRef]
5. Koza, J.R. *Genetic Programming: On the Programming of Computers by Means of Natural Selection*; MIT Press: Cambridge, MA, USA, 1992.
6. Beni, G. From Swarm Intelligence to Swarm Robotics. In *Swarm Robotics*; Springer: Berlin/Heidelberg, Germany, 2005; pp. 1–9.
7. Kennedy, J.; Eberhart, R. Particle swarm optimization. In Proceedings of the ICNN'95—International Conference on Neural Networks, Perth, Australia, 27 November–1 December 1995; Volume 4, pp. 1942–1948.
8. Yang, X.S. Firefly algorithm, stochastic test functions and design optimisation. *Int. J. Bio-Inspired Comput.* **2010**, *2*, 78–84. [CrossRef]
9. Dorigo, M.; Birattari, M. Ant Colony Optimization. In *Encyclopedia of Machine Learning*; Springer: Boston, MA, USA, 2010; pp. 36–39.
10. Yang, X.; Deb, S. Cuckoo Search via Lévy flights. In Proceedings of the 2009 World Congress on Nature Biologically Inspired Computing (NaBIC), Coimbatore, India, 9–11 December 2009; pp. 210–214.
11. Yang, X.S. A New Metaheuristic Bat-Inspired Algorithm. In *Nature Inspired Cooperative Strategies for Optimization (NICSO 2010)*; Springer: Berlin/Heidelberg, Germany, 2010; pp. 65–74.
12. Biyanto, T.R.; Irawan, S.; Febrianto, H.Y.; Afdanny, N.; Rahman, A.H.; Gunawan, K.S.; Pratama, J.A.; Bethiana, T.N. Killer Whale Algorithm: An Algorithm Inspired by the Life of Killer Whale. *Procedia Comput. Sci.* **2017**, *124*, 151–157. [CrossRef]
13. Saha, A.; Das, P.; Chakraborty, A.K. Water evaporation algorithm: A new metaheuristic algorithm towards the solution of optimal power flow. *Eng. Sci. Technol. Int. J.* **2017**, *20*, 1540–1552. [CrossRef]
14. Abdelaziz, A.Y.; Fathy, A. A novel approach based on crow search algorithm for optimal selection of conductor size in radial distribution networks. *Eng. Sci. Technol. Int. J.* **2017**, *20*, 391–402. [CrossRef]
15. Blum, C.; Roli, A. Metaheuristics in Combinatorial Optimization: Overview and Conceptual Comparison. *ACM Comput. Surv.* **2003**, *35*, 268–308. [CrossRef]
16. Črepinšek, M.; Liu, S.H.; Mernik, M. Exploration and Exploitation in Evolutionary Algorithms: A Survey. *ACM Comput. Surv.* **2013**, *45*, 35:1–35:33. [CrossRef]
17. Sergeyev, Y.D.; Kvasov, D.E.; Mukhametzhanov, M.S. Emmental-Type GKLS-Based Multiextremal Smooth Test Problems with Non-linear Constraints. In *Learning and Intelligent Optimization*; Battiti, R., Kvasov, D.E., Sergeyev, Y.D., Eds.; Springer International Publishing: Cham, Switzerland, 2017; pp. 383–388.
18. Kvasov, D.E.; Mukhametzhanov, M.S. Metaheuristic vs. deterministic global optimization algorithms: The univariate case. *Appl. Math. Comput.* **2018**, *318*, 245–259. [CrossRef]
19. Mezura-Montes, E.; Coello, C.A.C. Constraint-handling in nature-inspired numerical optimization: Past, present and future. *Swarm Evol. Comput.* **2011**, *1*, 173–194. [CrossRef]
20. Deb, K. An efficient constraint handling method for genetic algorithms. *Comput. Methods Appl. Mech. Eng.* **2000**, *186*, 311–338. [CrossRef]
21. Wang, Y.; Wang, B.; Li, H.; Yen, G.G. Incorporating Objective Function Information Into the Feasibility Rule for Constrained Evolutionary Optimization. *IEEE Trans. Cybern.* **2016**, *46*, 2938–2952. [CrossRef] [PubMed]
22. Tasgetiren, M.F.; Suganthan, P.N. A Multi-Populated Differential Evolution Algorithm for Solving Constrained Optimization Problem. In Proceedings of the 2006 IEEE International Conference on Evolutionary Computation, Vancouver, BC, Canada, 16–21 July 2006; pp. 33–40.
23. Farmani, R.; Wright, J.A. Self-adaptive fitness formulation for constrained optimization. *IEEE Trans. Evol. Comput.* **2003**, *7*, 445–455. [CrossRef]
24. Kramer, O.; Schwefel, H.P. On three new approaches to handle constraints within evolution strategies. *Natural Comput.* **2006**, *5*, 363–385. [CrossRef]

25. Coello, C.A.C. Theoretical and numerical constraint-handling techniques used with evolutionary algorithms: A survey of the state of the art. *Comput. Methods Appl. Mech. Eng.* **2002**, *191*, 1245–1287. [CrossRef]

26. Chootinan, P.; Chen, A. Constraint handling in genetic algorithms using a gradient-based repair method. *Comput. Oper. Res.* **2006**, *33*, 2263–2281. [CrossRef]

27. Regis, R.G. Constrained optimization by radial basis function interpolation for high-dimensional expensive black-box problems with infeasible initial points. *Eng. Optim.* **2014**, *46*, 218–243. [CrossRef]

28. Mezura-Montes, E.; Reyes-Sierra, M.; Coello, C.A.C. Multi-objective Optimization Using Differential Evolution: A Survey of the State-of-the-Art. In *Advances in Differential Evolution*; Springer: Berlin/Heidelberg, Germany, 2008; pp. 173–196.

29. Mallipeddi, R.; Das, S.; Suganthan, P.N. Ensemble of Constraint Handling Techniques for Single Objective Constrained Optimization. In *Evolutionary Constrained Optimization*; Springer: New Delhi, India, 2015; pp. 231–248.

30. Takahama, T.; Sakai, S.; Iwane, N. Solving Nonlinear Constrained Optimization Problems by the e Constrained Differential Evolution. In Proceedings of the 2006 IEEE International Conference on Systems, Man and Cybernetics, Taipei, Taiwan, 8–11 October 2006; Volume 3, pp. 2322–2327.

31. Padhye, N.; Mittal, P.; Deb, K. Feasibility Preserving Constraint-handling Strategies for Real Parameter Evolutionary Optimization. *Comput. Optim. Appl.* **2015**, *62*, 851–890. [CrossRef]

32. Fister, I.; Yang, X.S.; Fister, D. Firefly Algorithm: A Brief Review of the Expanding Literature. In *Cuckoo Search and Firefly Algorithm: Theory and Applications*; Springer International Publishing: Cham, Switzerland, 2014; pp. 347–360.

33. Baykasoğlu, A.; Ozsoydan, F.B. Adaptive firefly algorithm with chaos for mechanical design optimization problems. *Appl. Soft Comput.* **2015**, *36*, 152–164. [CrossRef]

34. Costa, M.F.P.; Rocha, A.M.A.C.; Francisco, R.B.; Fernandes, E.M.G.P. Firefly penalty-based algorithm for bound constrained mixed-integer nonlinear programming. *Optimization* **2016**, *65*, 1085–1104. [CrossRef]

35. Brajevic, I.; Tuba, M.; Bacanin, N. Firefly Algorithm with a Feasibility-Based Rules for Constrained Optimization. In Proceedings of the 6th WSEAS European Computing Conference, Prague, Czech Republic, 24–26 September 2012; pp. 163–168.

36. Deshpande, A.M.; Phatnani, G.M.; Kulkarni, A.J. Constraint handling in Firefly Algorithm. In Proceedings of the 2013 IEEE International Conference on Cybernetics (CYBCO), Lausanne, Switzerland, 13–15 July 2013; pp. 186–190.

37. Brajević, I.; Ignjatović, J. An upgraded firefly algorithm with feasibility-based rules for constrained engineering optimization problems. *J. Intell. Manuf.* **2018**. [CrossRef]

38. Chou, J.S.; Ngo, N.T. Modified Firefly Algorithm for Multidimensional Optimization in Structural Design Problems. *Struct. Multidiscip. Optim.* **2017**, *55*, 2013–2028. [CrossRef]

39. Runarsson, T.P.; Yao, X. Stochastic ranking for constrained evolutionary optimization. *IEEE Trans. Evol. Comput.* **2000**, *4*, 284–294. [CrossRef]

40. Tizhoosh, H.R. Opposition-Based Learning: A New Scheme for Machine Intelligence. In Proceedings of the International Conference on Computational Intelligence for Modelling, Control and Automation and International Conference on Intelligent Agents, Web Technologies and Internet Commerce (CIMCA-IAWTIC'06), Vienna, Austria, 28–30 November 2005; Volume 1, pp. 695–701.

41. Das, S.; Konar, A.; Chakraborty, U.K. Two Improved Differential Evolution Schemes for Faster Global Search. In Proceedings of the 7th Annual Conference on Genetic and Evolutionary Computation, Washington, DC, USA, 25–29 June 2005; pp. 991–998.

42. Ismail, M.M.; Othman, M.A.; Sulaiman, H.A.; Misran, M.H.; Ramlee, R.H.; Abidin, A.F.Z.; Nordin, N.A.; Zakaria, M.I.; Ayob, M.N.; Yakop, F. Firefly algorithm for path optimization in PCB holes drilling process. In Proceedings of the 2012 International Conference on Green and Ubiquitous Technology, Jakarta, Indonesia, 30 June–1 July 2012; pp. 110–113.

43. Liang, J.; Runarsson, T.P.; Mezura-Montes, E.; Clerc, M.; Suganthan, P.N.; Coello, C.C.; Deb, K. Problem definitions and evaluation criteria for the CEC 2006 special session on constrained real-parameter optimization. *J. Appl. Mech.* **2006**, *41*, 8–31.

44. Gandomi, A.H.; Yang, X.S.; Alavi, A.H. Mixed variable structural optimization using Firefly Algorithm. *Comput. Struct.* **2011**, *89*, 2325–2336. [CrossRef]

45. Sergeyev, Y.D.; Kvasov, D.; Mukhametzhanov, M. On the efficiency of nature-inspired metaheuristics in expensive global optimization with limited budget. *Sci. Rep.* **2018**, *8*, 453. [CrossRef] [PubMed]

46. Ali, L.; Sabat, S.L.; Udgata, S.K. Particle Swarm Optimisation with Stochastic Ranking for Constrained Numerical and Engineering Benchmark Problems. *Int. J. Bio-Inspired Comput.* **2012**, *4*, 155–166. [CrossRef]

47. Elsayed, S.M.; Sarker, R.A.; Mezura-Montes, E. Self-adaptive mix of particle swarm methodologies for constrained optimization. *Inf. Sci.* **2014**, *277*, 216–233. [CrossRef]

48. Mallipeddi, R.; Suganthan, P.N. Ensemble of Constraint Handling Techniques. *IEEE Trans. Evol. Comput.* **2010**, *14*, 561–579. [CrossRef]

49. Mohamed, A.W. A novel differential evolution algorithm for solving constrained engineering optimization problems. *J. Intell. Manuf.* **2018**, *29*, 659–692. [CrossRef]

50. Derrac, J.; García, S.; Molina, D.; Herrera, F. A practical tutorial on the use of nonparametric statistical tests as a methodology for comparing evolutionary and swarm intelligence algorithms. *Swarm Evol. Comput.* **2011**, *1*, 3–18. [CrossRef]

51. Ray, T.; Liew, K.M. Society and civilization: An optimization algorithm based on the simulation of social behavior. *IEEE Trans. Evol. Comput.* **2003**, *7*, 386–396. [CrossRef]

52. zhuo Huang, F.; Wang, L.; He, Q. An effective co-evolutionary differential evolution for constrained optimization. *Appl. Math. Comput.* **2007**, *186*, 340–356. [CrossRef]

53. Lee, K.S.; Geem, Z.W. A new meta-heuristic algorithm for continuous engineering optimization: Harmony search theory and practice. *Comput. Methods Appl. Mech. Eng.* **2005**, *194*, 3902–3933. [CrossRef]

54. de Melo, V.V.; Carosio, G.L. Investigating Multi-View Differential Evolution for solving constrained engineering design problems. *Expert Syst. Appl.* **2013**, *40*, 3370–3377. [CrossRef]

55. Sadollah, A.; Bahreininejad, A.; Eskandar, H.; Hamdi, M. Mine blast algorithm: A new population based algorithm for solving constrained engineering optimization problems. *Appl. Soft Comput.* **2013**, *13*, 2592–2612. [CrossRef]

56. Rao, R.V.; Waghmare, G. A new optimization algorithm for solving complex constrained design optimization problems. *Eng. Optim.* **2017**, *49*, 60–83. [CrossRef]

57. Savsani, P.; Savsani, V. Passing vehicle search (PVS): A novel metaheuristic algorithm. *Appl. Math. Model.* **2016**, *40*, 3951–3978. [CrossRef]

58. Brajevic, I.; Tuba, M. An upgraded artificial bee colony (ABC) algorithm for constrained optimization problems. *J. Intell. Manuf.* **2013**, *24*, 729–740. [CrossRef]

59. Guedria, N.B. Improved accelerated PSO algorithm for mechanical engineering optimization problems. *Appl. Soft Comput.* **2016**, *40*, 455–467. [CrossRef]

60. Yılmaz, S.; Küçüksille, E.U. A new modification approach on bat algorithm for solving optimization problems. *Appl. Soft Comput.* **2015**, *28*, 259–275. [CrossRef]

61. Balande, U.; Shrimankar, D. An oracle penalty and modified augmented Lagrangian methods with firefly algorithm for constrained optimization problems. *Oper. Res.* **2017**. [CrossRef]

62. Eskandar, H.; Sadollah, A.; Bahreininejad, A.; Hamdi, M. Water cycle algorithm—A novel metaheuristic optimization method for solving constrained engineering optimization problems. *Comput. Struct.* **2012**, *110–111*, 151–166. [CrossRef]

mathematics

MDPI

Article

A Novel Hybrid Algorithm for Minimum Total Dominating Set Problem

Fuyu Yuan, Chenxi Li, Xin Gao, Minghao Yin * and Yiyuan Wang *

School of Computer Science and Information Technology, Northeast Normal University, Changchun 130000, China; yuanfuyu@aliyun.com (F.Y.); icx935@nenu.edu.cn (C.L.); gaolzzxin@gmail.com (X.G.)
* Correspondence: ymh@nenu.edu.cn (M.Y.); yiyuanwangjlu@126.com (Y.W.)

Received: 14 January 2019; Accepted: 24 February 2019; Published: 27 February 2019

Abstract: The minimum total dominating set (MTDS) problem is a variant of the classical dominating set problem. In this paper, we propose a hybrid evolutionary algorithm, which combines local search and genetic algorithm to solve MTDS. Firstly, a novel scoring heuristic is implemented to increase the searching effectiveness and thus get better solutions. Specially, a population including several initial solutions is created first to make the algorithm search more regions and then the local search phase further improves the initial solutions by swapping vertices effectively. Secondly, the repair-based crossover operation creates new solutions to make the algorithm search more feasible regions. Experiments on the classical benchmark DIMACS are carried out to test the performance of the proposed algorithm, and the experimental results show that our algorithm performs much better than its competitor on all instances.

Keywords: minimum total dominating set; evolutionary algorithm; genetic algorithm; local search

1. Introduction

Given an undirected graph $G = (V, E)$, a dominating set (DS) is a subset of vertices $S \in V$ that each vertex in $V \backslash S$ is adjacent to at least one vertex in S. For each vertex $v \in V$, vertex v must have a neigbor in S, and this dominating set is called a total dominating set (TDS). We can easily conclude that TDS is a typical variant of DS. The minimum total dominating set (MTDS) problem aims to identify the minimum size of TDS in a given graph. MTDS has many applications in various fields, such as sensor and ad hoc communications and networks as well as gateway placement problems [1–3].

MTDS is proven to be NP-hard (non-deterministic polynomial) [4], which means unless P = NP, there is no polynomial time to solve this problem. At present, Zhu proposed a novel one-stage analysis for greedy algorithms [5] with approximation ratio $ln(\delta - 0.5) + 1.5$ where δ is the maximum degree of the given graph. This algorithm also used a super-modular greedy potential function, which was a desirable property in mathematics. However, in real life and industrial production, the size of problems is always very large. When the size of problems is increased [6–9], the approximation algorithm will be invalid. Considering these circumstances, researchers often use heuristic algorithms [10–13] to deal with these problems. Although the heuristic algorithms cannot guarantee the optimality of the solution they obtain, they can find high-quality solutions effectively within a reasonable time. Thus, in this paper we propose a hybrid evolution method combining local search and genetic algorithm (HELG) to solve MTDS.

Evolutionary algorithms include genetic algorithm, genetic programming, evolution strategies and evolution programming, etc. Among them, genetic algorithm as a classical method is the most widely used. Genetic algorithm is a computational model to simulate the natural selection and genetic mechanism of Darwin's biological evolution theory. It is a method to search the optimal solution by simulating the natural evolution process. Genetic algorithm begins with a population representing the potential solution set of the problem, and a population consists of a certain number of individuals encoded by genes. After the first generation of the population, according to the principle of survival of the fittest, the evolution of generations produces more and better approximate solutions. In each generation, the selection is based on the fitness of the individual in the problem domain. Individuals, by means of genetic operators of natural genetics, perform crossovers and mutations to produce populations representing new solution sets.

Recently, evolutionary algorithms play an important role in solving optimization problems. It is common to adjust evolutionary algorithm to solve problems by adding a different problem-related mechanism. One possible improvement is the hybrid of evolutionary method and local search algorithm. Przewozniczek et al. investigated the pros and cons of hybridization on the baseof a hard practical and up-to-date problem and then proposed an effective optimization method for solving the routing and spectrum allocation of multicast flows problem in elastic optical networks [14]. Połap et al. proposed three proposition to increase the efficiency of classical meta-heuristic methods [15]. In this paper, the proposed algorithm takes advantage of local search framework as well as genetic algorithm.

Firstly, the algorithm creates a population including several individuals as initial solutions in our algorithm. Then, for each initial solution, we prove its solution via the local search. After local search, a repair-based crossover operation is proposed to improve the searchability of the algorithm. The algorithm randomly selects two solutions in the population and randomly exchanges several vertices of them. After crossover, if the obtained solutions are infeasible, the algorithm will repair them. This operation enables the algorithm to search larger areas, resulting in obtaining more feaisble solutions. In addition, we use a scoring function to help the method choose vertices more effective. In detail, each vertex is assigned to a cost value, and then we calculate the scoring value of every vertex by the cost value. The scoring function is used to measure the benefits of the state changing of a vertex. Whenever the algorithm swaps a pair of vertices, we should try to increase the benefits of the candidate solution and reduce the loss. This scoring value makes our algorithm efficient. When the original HELG fails to find improved solutions, this heuristic can make the algorithm escape from the local optimal.

Based on the above strategies, we design a hybrid evolutionary algorithm HELG for MTDS. Since we are the first to solve MTDS with a heuristic algorithm, in order to evaluate the efficiency of HELG, we carry out some experiments to compare HELG with a greedy algorithm and a classical metaheuristics algorithm, the ant colony optimization (ACO) algorithm [16,17] for MTDS. The experimental results show that on most instances our algorithm performs much better than the greedy algorithm.

The remaining sections are arranged as follows: in Section 2, we give some necessary notations and definitions. In Section 3, we introduce the novel scoring heuristic. The evolution algorithm HELG for MTDS is described in Section 4. Section 5 gives the experimental evaluations and the experimental results analysis. Finally, we summarize this paper and list future work.

2. Preliminaries

Given an undirected graph $G = (V, E)$ with vertex set V and edge set E, each edge is a 2-element subset of V. For an edge $e = (u, v)$, vertices u and v are the endpoints of e, and u is adjacent to v. The distance between u and v means the number of edges from the shortest path of u to v and is defined by $dist(u, v)$. Then the ith level neighborhood of a vertex v is defined by $N_i(v) = \{u | dist(u, v) = i\}$. Specifically, $N(v) = N_1(v)$. The *degree* of a vertex v is $deg(v) = |N(v)|$.

A set D of V is a dominating set if each vertex not in D is adjacent to at least one vertex in D. Total dominating set is a variant of dominating set. The definition of total dominating set is as follows.

Definition 1. *(total dominating set) Given an undirected graph $G = (V, E)$, a total dominating set (TDS) is a subset D of V that every vertex of G is adjacent to some vertices in D, whether it is in D or not.*

The minimum total dominating set (MTDS) problem aims at identifying the TDS with the minimum size in a given graph.

3. Scoring Heuristic

In our algorithm, we need to maintain a total dominating set as a candidate solution CS during the construction and local search phases. It is important to decide which vertex should be added into or removed from CS. In this section, we will introduce an effective scoring heuristic that can be used to decide how to choose the vertices to be added and removed.

Given an undirected graph $G = (V, E)$, a cost value denoted by $\omega(v)$ is applied to each vertex $v \in V$, which will be maintained during the local search. For each vertex $v \in V$, $\omega(v)$ is initialized as 1 in the construction phase. In the local search phase, if CS uncovers the vertex v, the value of $\omega(v)$ will be increased by 1 such that these uncovered vertices will have more opportunities to be selected.

Based on the cost value described above, the vertex scoring method will be defined. We denote the scoring function as s, the scoring method is divided into two cases as follows:

- $v \notin CS$. The value of $s(v)$ is the total cost value of vertices which will become dominated by CS after adding vertex v into CS.
- $v \in CS$. The value of $s(v)$ is the opposite number of the total cost value of vertices which will become not dominated by CS after removing vertex v from CS.

The vertex scoring method is used to measure the benefits of changing the state of v. Obviously, if $v \in CS$, $s(v) \leq 0$. Otherwise, $s(v) \geq 0$. This method can decide how to choose the vertices to be added and removed.

4. Evolution Algorithm HELG for MTDS

In this section, we propose a hybrid evolutionary algorithm combining local search and genetic algorithm. The algorithm includes three important phases: generation of a population including n initial solutions, local search, and a repair-based crossover operation.

4.1. Population Initialization

We first generate a population including n initial solutions. We use a preprocessing method and the restricted candidate list (RCL) to initialize a population. In the process of preprocessing, based on the definition of MTDS, if the degree of a vertex is 1, its neighborhood will be added into CS and forbidden to be removed from CS during the search. A function *probid* is used to implement this process. For a vertex v, if $probid(v) = 1$, v will be forbidden to be removed from CS in the subsequent search. RCL contains the vertices with good benefits. The algorithm chooses vertices from RCL randomly to construct an initial candidate solution.

The pseudo code of construction phase is shown in Algorithm 1.

At first, the index k and population Pop are initialized (line 1). Then the n individuals are created (lines 2–18). For each individual, the scoring function s of each vertex, the *probid* value of each vertex and the candidate solution CS are initialized (lines 3–5). Then, the algorithm starts the preprocessing process (lines 6–9). If the degree of a vertex corresponds to 1, the algorithm adds its neighborhood into CS. Then the *probid* value of them will be assigned to 1 which means that they will be forbidden to be removed from CS in the following phase. The algorithm then enters a loop to add vertices into CS until it becomes a TDS (lines 10–16). The maximum value s_{max} and the minimum value s_{min} of s are calculated in lines 11 and 12. The vertices whose scoring values are not less than $s_{min} + \mu(s_{max} - s_{min})$ comprise the RCL (line 13). Here, μ is a parameter that belongs to $[0, 1]$. Then we choose a vertex u from RCL randomly and add it into CS (lines 14–15). The scoring values of u and the 1st and 2nd level neighborhood are updated in line 16. Then the just created individual is added into the Pop (line 17). Then, k is updated in line 18. In the end, we return the population Pop (line 19).

Algorithm 1: create_population (*n*)

Input: the population size *n*

Output: the initial population *Pop*

1 $k \leftarrow 0, Pop \leftarrow \{\}$;

2 **while** $k < n$ **do**

3   Initialize the scoring function *s* of each vertex based on the cost value ω;

4   Initialize the *probid* value of each vertex as 0;

5    $CS \leftarrow \{\}$;

6   **for** *each* $v \in V$ **do**

7   **if** $deg(v) == 1$ **then**

8      $CS \leftarrow CS \cup N(v)$;

9      $probid(N(v)) \leftarrow 1$;

10   **while** *CS is not a TDS* **do**

11    $s_{max} \leftarrow \text{MAX}\{s(v) > 0, v \in V/CS\}$;

12    $s_{min} \leftarrow \text{MIN}\{s(v) > 0, v \in V/CS\}$;

13    $RCL \leftarrow \{v | s(v) \geq s_{min} + \mu(s_{max} - s_{min}), v \in V/CS\}$;

14    $u \leftarrow$ choose a vertex *u* from RCL randomly;

15    $CS \leftarrow CS \cup \{u\}$;

16   update $s(u), s(v_1),$ and $s(v_2)$ for each vertex $v_1 \in N(u)$ and $v_2 \in N_2(u)$;

17   $Pop \leftarrow Pop \cup CS$;

18   $k++$;

19 **return** *Pop*;

4.2. Local Search Phase

Several candidate solutions are built in *create_population* phase. The local search phase explores the neighborhood of the initial candidate solution to improve the solution quality and obtain a smaller one. If no better solution is found, the algorithm will return the current solution as a local optimum. Otherwise, the improved solution will be the new best candidate solution. This phase is executed iteratively.

The pseudo code of local search phase is shown in Algorithm 2.

At the beginning of the algorithm, the number of iterations *k* and the local optimal solution CS^* are initialized (line 1). Then the algorithm enters a loop until the maximum number of iterations *mstep* is reached (line 2). In the search process, once a better solution is found, the algorithm updates CS^* by *CS* and removes a vertex with the highest *s* from *CS* (lines 3–9). The *probid* value of the selected vertex is forbidden to be 1. The number of iterations *k* is set to 0 (line 4). The scoring values of the just removed vertex and its 1st and 2nd level neighborhood are updated in line 8. Otherwise, the algorithm will remove a vertex *v* with the highest *s* and $probid(v) \neq 1$ from *CS*, breaking ties randomly (line 9). The corresponding scoring values are updated in line 11. Subsequently, the algorithm selects a vertex with the highest score and adds it to *CS* (lines 12–13). After that, the cost value of each undominated vertex *v* and the scoring function of each $v \in V$ are updated (lines 14–15). The step of iteration is increased by 1 (line 16). In the end, the algorithm returns CS^* as a local optimal solution.

Algorithm 2: LocalSearch (*CS*, *mstep*)

Input: an initial candidate solution *CS*, the maximum number of iterations *mstep*
Output: an improved candidate solution *CS**

1   Initialize $k \leftarrow 0, CS^* \leftarrow CS$;
2   **while** $k < mstep$ **do**
3   **if** *CS is a TDS* **then**
4          $k = 0$;
5          $CS^* \leftarrow CS$;
6          $v \leftarrow$ select v from CS with the highest $s(v)$, and $probid(v) \neq 1$, breaking ties randomly;
7          $CS \leftarrow CS/\{v\}$;
8   update $s(v), s(u_1)$ and $s(u_2)$ for each $u_1 \in N(v)$ and $u_2 \in N_2(v)$;
9      $v \leftarrow$ select v from CS with the highest $s(v)$ and $probid(v) \neq 1$, breaking ties randomly;
10      $CS \leftarrow CS/\{v\}$;
11   update $s(v), s(u_1)$, and $s(u_2)$ for each $u_1 \in N(v)$ and $u_2 \in N_2(v)$;
12      $v_1 \leftarrow$ randomly select undominated vertex v' and select v_1 from $N(v')$ with the highest $s(v_1)$, breaking ties randomly;
13      $CS \leftarrow CS \cup \{v_1\}$;
14      $\omega(v)$++ for each undominated vertex v;
15   update $s(u)$ for each $u \in V$;
16   k++;
17   **return** CS^*;

4.3. Repair-Based Crossover Operation

Genetic algorithms often use crossover to increase the diversity of the algorithm. The central role of biological evolution in nature is the recombination of biological genes (plus mutations). Similarly, the central role of genetic algorithms is the crossover operator of genetic algorithm. The so-called crossover refers to the operation of replacing the partial structure of two parent individuals to generate a new individual. Through crossover, the searchability of genetic algorithm has been greatly improved.

In this paper, we propose a repair-based crossover operation. After local search, our algorithm obtains an improved population. We choose two solutions from the population randomly, and then exchange the vertices in the two solutions with probability 0.5.

Because of the particularity of MTDS, the obtained solutions after crossover may be infeasible. So we should repair the infeasible solutions and make them become total dominating sets. After crossover, we check if the obtained solutions are total dominating sets. If a solution is infeasible, we add some reasonable vertices through population initialization phase until it is feasible. Then we perform a redundancy remove operation. For the solution obtained by crossover, we remove a vertex and check whether the solution is feasible. If it is feasible, the vertex will be removed, and otherwise it will be added back. The redundancy remove operation performs iteratively until every vertex has been checked.

The solution obtained by the repair-based crossover operation will replace the two old solutions into the population.

4.4. The Framework of HELG

In this paper, we propose a hybrid evolutionary algorithm HELG that combines local search and genetic algorithm. The algorithm first generates a population including n initial solutions, and then applies local search to improve each solution. The obtained n solutions will perform crossover to produce new solutions. This algorithm will perform iteratively until time limit is satisfied.

The framework of HELG is shown in Algorithm 3 and described as below.

Algorithm 3: HELG(G)

    Input: an undirected graph G
    Output: the best solution CS^* found
1   initialize CS^*;
2   $Pop = \{S_1, ..., S_n\} \leftarrow create\_population(n)$;
3   **while** *time limit is not satisfied* **do**
4   **for** *each* $CS \in Pop$ **do**
5         $CS \leftarrow LocalSearch(CS, mstep)$;
6   **if** $|CS| < |CS^*|$ **then**
7           $CS^* \leftarrow CS$;

8      $Pop \leftarrow Crossover(Pop)$;
9   choose the best CS from Pop;
10   **if** $|CS| < |CS^*|$ **then**
11       $CS^* \leftarrow CS$;

12 **return** CS^*;

At first, the algorithm initializes the best solution CS^* and a population Pop (lines 1–2). Then the algorithm performs a loop (lines 3–11). For each solution CS of Pop, we use local search to improve the quality of solution and if CS is better than CS^*, CS^* is updated by CS (lines 4–7). For the obtained n solutions, we perform the repair-based crossover operation to generate new solutions (line 8). If the best solution CS among Pop is better than CS^*, CS^* is updated by CS (lines 9–11) When the time limit is satisfied, the best solution CS^* is returned (line 12).

5. Experiments

In this section, we carry out a series of experiments to evaluate the efficiency of our algorithm. The experiments are carried out on a classical benchmark DIMACS (the Center for Discrete Mathematics and Theoretical Computer Science) [18]. We select 61 instances in the DIMACS benchmark. The instances are from industry and generated by various models.

HELG is implemented in C++ and compiled by g++ with the -O3 option. We run the algorithm on a machine with Intel(R) Xeon(R) CPU E7-4830 @2.13Ghz and 4GB memory under Linux. For each instance, the HELG algorithm runs 30 times independently with different random seeds, until the time limit (100 s) is satisfied. HELG has three important parameters (i.e., n, μ and $mstep$). In the population creation phase, we set the RCL parameter $\mu = 0.1$ and $n = 10$. In the local search phase, we set the maximum number of iterations $mstep = 100$.

Since we are the first to solve MTDS with a heuristic algorithm, in order to evaluate the efficiency of HELG, a greedy algorithm is as our control method which uses the same scoring heuristic. At first, the candidate solution CS is empty. The greedy algorithm selects the vertex with the highest score value and adds it into CS every time. When CS becomes a TDS, the algorithm stops and returns CS as the optimal solution. Another comparison algorithm we use is a classical metaheuristics algorithm, the ant colony optimization (ACO) algorithm [16,17], which uses the same initialization procedure with our algorithm. We use this algorithm as a comparison algorithm to evaluate the effectiveness of our algorithm. For each instance, the ACO also runs 30 times independently with different random seeds, until the time limit (100 s) is satisfied.

For each instance, MIN is the minimum total dominating set found, AVG is the average size of 10 solutions, and $Time$ is the running time when the algorithm gets the minimum total dominating set. Because the greedy algorithm is only executed once for each instance, it has no average value and the time is less than 1. Better solutions and time are expressed in bold.

The experimental results are shown in Tables 1 and 2. Compared with the greedy algorithm, HELG can obtain better minimum solutions in 52 instances. In the remaining 9 instances, HELG gets the same minimum solutions with the greedy algorithm. Compared with ACO, HELG can obtain better minimum solutions in 40 instances. In the remaining 21 instances, HELG gets the same minimum size with ACO. Among them, HELG gets better average values than ACO in 16 instances, and gets the same average value with ACO but performs faster in 5 instances. The DIMACS benchmark is divided into 10 groups. We choose 1 instance from every group. Every instance is run 30 times independently. The visualized comparisons of ACO and HELG can be seen by Kolmogorov-Smirnov test in Figure 1, which shows the distribution of the total dominating set values. From these, we can observe that HELG performs much better than ACO and greedy algorithm.

Table 1. Experimental results of greedy algorithm, ACO, and HELG on DIMACS I. The better minimum or average solution values are in bold.

Instances	Vertices	Edges	Greedy		ACO			HELG		
			MIN	Time	MIN	AVG	Time	MIN	AVG	Time
brock200_2	200	10,024	7	<1	6	6	0.3	6	6	**0.21**
brock200_4	200	6811	10	<1	9	9	2.28	9	9	**1.93**
brock400_2	400	20,014	19	<1	16	16.3	6.28	**15**	**15**	5.39
brock400_4	400	20,035	19	<1	14	16.3	34.6	14	15.1	25.16
brock800_2	800	111,434	15	<1	13	14.2	15.2	**12**	**12.3**	10.5
brock800_4	800	111,957	18	<1	15	16.2	7.63	**13**	**13**	6.75
c-fat200-1.CLQ	200	1534	23	<1	22	22	48.62	**20**	**20.3**	43.93
c-fat200-2.CLQ	200	3235	12	<1	10	11	8.45	10	**10**	4.02
c-fat200-5.CLQ	200	8473	5	<1	5	5.3	13.51	**4**	**4**	2.46
c-fat500-1.CLQ	500	4459	53	<1	50	51.2	20.73	**47**	**48.1**	10.41
c-fat500-2.CLQ	500	9139	26	<1	24	25.9	79.62	**23**	**23.5**	74.99
c-fat500-5.CLQ	500	23,191	10	<1	9	9	11.12	9	9	**8.08**
C1000.9	1000	49,421	47	<1	45	45.6	68.45	**44**	**44.3**	64.97
C125.9	125	787	30	<1	21	22	45.2	**20**	**21.1**	38.16
C2000.5	2000	999,164	**9**	<1	10	10	46.7	**9**	**9.5**	43.37
C2000.9	2000	199,468	52	<1	51	52.4	17.89	**50**	**51.6**	11.64
C250.9	250	3141	33	<1	28	29	80.34	**27**	**28.2**	76.81
C4000.5	4000	3,997,732	**11**	<1	**11**	12.4	160.8	**11**	**11.6**	154.97
C500.9	500	12,418	39	<1	37	38.3	97.7	**34**	**36.6**	91.68
DSJC1000.5	1000	249,674	9	<1	8	9.1	11.6	8	**8.8**	7.38
DSJC500.5	500	62,126	9	<1	8	8	27.84	**7**	**7.4**	24.5
gen200_p0.9_44	200	1990	33	<1	26	27.2	77.1	**26**	**26.6**	76.25
gen200_p0.9_55	200	1990	31	<1	29	30.2	78.65	**25**	**26.2**	71.54
gen400_p0.9_55	400	7980	38	<1	35	35.5	53.2	**34**	**35**	51.65
gen400_p0.9_65	400	7980	38	<1	33	35	66.4	33	**34.3**	58.49
gen400_p0.9_75	400	7980	38	<1	35	36	27.5	**33**	**35.2**	20.4
hamming10-4	1024	89,600	23	<1	23	23	45.2	**21**	**22.6**	41.77
hamming6-2	64	192	19	<1	15	16.2	39.7	15	**15.6**	35.86
hamming6-4	64	1312	3	<1	3	3	0.01	3	3	**0.01**
hamming8-2	256	1024	69	<1	63	65	17.9	**62**	**64.4**	11.33
hamming8-4	256	11,776	9	<1	6	7.2	59.6	6	**6.5**	57.68

To further illustrate the efficiency contribution of our algorithm, we show the time-to-target plot [12,19] in Figure 2 to compare HELG with ACO on brock400_4 and its target 14. To obtain the plot, we performed 100 independent runs of each algorithm on brock400_4. The figure shows that the probabilities of finding a solution of the target value by HELG are approximately 30% and 70% in at most 13.78 and 33.43 s, respectively, whereas the probabilities of finding a solution of the target value by ACO are approximately 30% and 70% in at most 23.62 and 43.47 s respectively, considerably longer than HELG. From that, we can observe that the strategies we used in HELG are very effective.

The experimental results show that our algorithm performs much better than the comparison algorithms. This proves that the genetic algorithm and local search in our algorithm are both very effective.

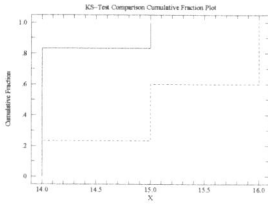

brock400_4

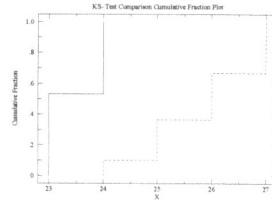

c-fat500-2.CLQ

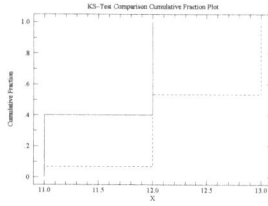

C4000.5

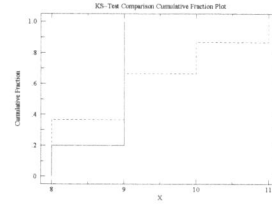

DSJC1000.5

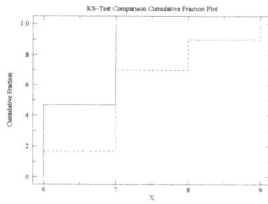

hamming8-4

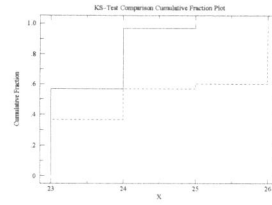

johnson32-2-4

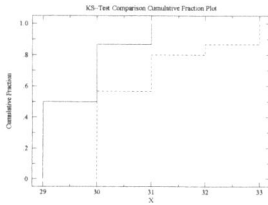

keller6

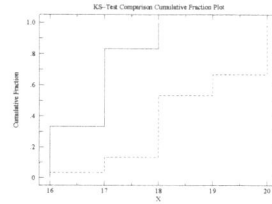

MANN_a9

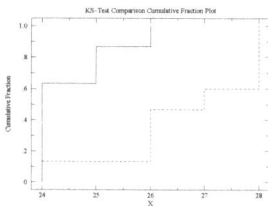

p_hat1500-1.CLQ

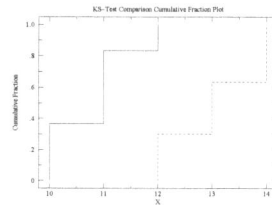

san400_0.7_3

Figure 1. The total dominating set values obtained by $\cdots$: ACO; $-$: HELG. Kolmogorov-Smirnov test can be applied to display the distribution of these values.

Table 2. Experimental results of greedy algorithm, ACO, and HELG on DIMACS II. The better minimum or average solution values are in bold.

Instances	Vertices	Edges	Greedy		ACO			HELG		
			MIN	Time	MIN	AVG	Time	MIN	AVG	Time
johnson16-2-4	120	1680	11	<1	**10**	11.2	32.6	**10**	10.4	30.72
johnson32-2-4	496	14,880	24	<1	**23**	24.5	34.8	**23**	23.5	33.38
johnson8-2-4	28	168	**5**	<1	**5**	5.8	0.01	**5**	**5**	0
johnson8-4-4	70	560	12	<1	8	8	16.43	**7**	7.5	9.81
keller4	171	5100	8	<1	8	8	10.45	**7**	7.4	4.52
keller5	776	74,710	16	<1	17	17	8.28	**15**	16.5	2.16
keller6	3361	1,026,582	30	<1	30	30.8	61.2	**29**	29.6	54.68
MANN_a27	378	702	113	<1	75	77.4	100.63	**67**	79.5	81.69
MANN_a45	1035	1980	**90**	<1	**90**	**90**	3.25	**90**	**90**	0.03
MANN_a81	3321	6480	**162**	<1	**162**	163	1.28	**162**	**162**	0.17
MANN_a9	45	72	20	<1	**16**	18.6	40.5	**16**	16.8	34.05
p_hat1500-1.CLQ	1500	839,327	26	<1	**24**	26.7	26.5	**24**	24.5	18.01
p_hat1500-2.CLQ	1500	555,290	11	<1	13	14.1	98.6	**8**	9.2	87.62
p_hat1500-3.CLQ	1500	277,006	5	<1	5	6.1	63.2	**4**	4.9	59.48
p_hat300-1.CLQ	300	33,917	16	<1	17	18.4	126.3	**14**	15.2	97.86
p_hat300-2.CLQ	300	22,922	**6**	<1	7	7	4.58	**6**	6.5	0.54
p_hat300-3.CLQ	300	11,460	4	<1	4	5	35.48	**3**	3.1	31.86
p_hat700-1.CLQ	700	183,651	21	<1	21	21	74.69	**17**	20	64.01
p_hat700-2.CLQ	700	122,922	9	<1	11	13	54.31	**8**	10.2	48.35
p_hat700-3.CLQ	700	61,640	**4**	<1	**4**	5	6.43	**4**	4.4	3.25
san1000	1000	249,000	**6**	<1	7	8	8.45	**6**	6.5	4.73
san200_0.7_1	200	5970	9	<1	**8**	9	6.7	**8**	8.8	1.57
san200_0.7_2	200	5970	10	<1	9	9.4	75.12	**7**	7.9	63.8
san200_0.9_1	200	1990	29	<1	24	24	30.5	**22**	23.9	25.07
san200_0.9_2	200	1990	30	<1	27	27.8	86.5	**24**	25.9	77.4
san200_0.9_3	200	1990	31	<1	27	27.6	45.65	**25**	26.6	34.13
san400_0.5_1	400	39,900	5	<1	6	6	8.41	**4**	5.4	2.12
san400_0.7_1	400	23,940	13	<1	13	13	27.4	**9**	10.5	19.49
san400_0.7_2	400	23,940	11	<1	**10**	11	100.6	9	10.7	92.22
san400_0.7_3	400	23,940	14	<1	12	13.1	87.45	**10**	10.8	69.77

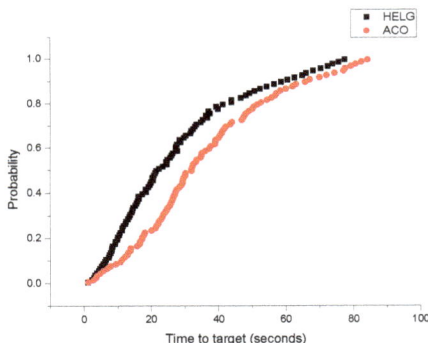

Figure 2. Time-to-target plot comparing HELG with ACO on instance brock400_4 and its target 14.

6. Summary and Future Work

This paper proposes a hybrid evolutionary algorithm combining local search and genetic algorithm to solve MTDS. A scoring heuristic is used to improve the efficiency of the algorithm. In the population initialization phase, we create a population including several initial solutions. In the local search phase, the algorithm improves the initial solutions by adding and removing operations. After that, we propose a repair-based crossover operation to increase the diversity of our algorithm. A series of experiments are carried out to evaluate the algorithm. The experimental results show that HELG performs well in solving MTDS.

In the future, we would like to design more efficient evolutionary algorithms to solve MTDS. We would like to relax this problem, such as the found MTDS is missing the adjacency of one vertex.

We will study evolutionary algorithms in more MDS-related problems, for example, the multi-objective MDS optimization problem.

Author Contributions: Software, X.G. and C.L.; Methodology, F.Y. and Y.W.; Writing–original draft preparation, F.Y. and C.L.; Writing—review and editing, C.L. and M.Y.

Funding: This research and the APC was supported by the Fundamental Research Funds for the Central Universities 2412018QD022 and Education Department of Jilin Province JJKH20190289KJ, NSFC (under grant nos. 61502464, 61503074, 61806050) and China National 973 Program 2014CB340301.

Conflicts of Interest: The authors declare no conflict of interest.

References

1. Subhadrabandhu, D.; Sarkar, S.; Anjum, F. Efficacy of misuse detection in ad hoc networks. In Proceedings of the 2004 First IEEE Communications Society Conference on Sensor and Ad Hoc Communications and Networks, IEEE SECON 2004, Santa Clara, CA, USA, 4–7 October 2004; pp. 97–107.

2. Aoun, B.; Boutaba, R.; Iraqi, Y.; Kenward, G. Gateway Placement Optimization in Wireless Mesh Networks With QoS Constraints. *IEEE J. Sel. Areas Commun.* **2006**, *24*, 2127–2136. [CrossRef]

3. Chen, Y.P.; Liestman, A.L. Approximating minimum size weakly-connected dominating sets for clustering mobile ad hoc networks. In Proceedings of the International Symposium on Mobile Ad Hoc Networking and Computing, Lausanne, Switzerland, 9–11 June 2002; pp. 165–172.

4. Garey, M.R.; Johnson, D.S. *Computers and Intractability: A Guide to the Theory of NP-Completeness*; W.H. Freeman and Company: New York, NY, USA, 1990.

5. Zhu, J. Approximation for minimum total dominating set. In Proceedings of the International Conference on Interaction Sciences: Information Technology, Culture and Human, Seoul, Korea, 24–26 November 2009; pp. 119–124.

6. Cai, S.; Su, K.; Luo, C.; Sattar, A. NuMVC: An efficient local search algorithm for minimum vertex cover. *J. Artif. Intell. Res.* **2013**, *46*, 687–716. [CrossRef]

7. Ping, H.; Yin, M.H. An upper (lower) bound for Max (Min) CSP. *Sci. China (Inf. Sci.)* **2014**, *57*, 1–9.

8. Cai, S.; Lin, J.; Luo, C. Finding A Small Vertex Cover in Massive Sparse Graphs: Construct, Local Search, and Preprocess. *J. Artif. Intell. Res.* **2017**, *59*, 463–494. [CrossRef]

9. Wang, Y.; Cai, S.; Yin, M. Two Efficient Local Search Algorithms for Maximum Weight Clique Problem. In Proceedings of the Thirtieth AAAI Conference on Artificial Intelligence, Phoenix, AZ, USA, 12–17 February 2016; pp. 805–811.

10. Wang, Y.; Cai, S.; Chen, J.; Yin, M. A Fast Local Search Algorithm for Minimum Weight Dominating Set Problem on Massive Graphs. In Proceedings of the Twenty-Seventh International Joint Conference on Artificial Intelligence (IJCAI-18), Stockholm, Sweden, 13–19 July 2018; pp. 1514–1522.

11. Wang, Y.; Cai, S.; Yin, M. Local search for minimum weight dominating set with two-level configuration checking and frequency based scoring function. *J. Artif. Intell. Res.* **2017**, *58*, 267–295. [CrossRef]

12. Wang, Y.; Li, C.; Sun, H; Yin, M. MLQCC: An improved local search algorithm for the set k covering problem. *Int. Trans. Oper. Res.* **2019**, *26*, 856–887. [CrossRef]

13. Wang, Y.; Cai, S.; Yin, M. New heuristic approaches for maximum balanced biclique problem. *Inf. Sci.* **2018**, *432*, 362–375. [CrossRef]

14. Przewozniczek, M.W.; Walkowiak, K.; Aibin, M. The evolutionary cost of baldwin effect in the routing and spectrum allocation problem in elastic optical networks. *Appl. Soft Comput.* **2017**, *52*, 843–862. [CrossRef]

15. Połap, D.; Kęsik, K.; Woźniak, M.; Damaševičius, R. Parallel Technique for the Metaheuristic Algorithms Using Devoted Local Search and Manipulating the Solutions Space. *Appl. Sci.* **2018**, *8*, 293. [CrossRef]

16. Romania, Q.S. Ant colony optimization applied to minimum weight dominating set problem. In Proceedings of the 12th WSEAS International Conference on Automatic Control, Modelling & Simulation, Catania, Italy, 29–31 May 2010.

17. Dorigo, M.; Stützle, T. Ant Colony Optimization. In Proceedings of the 1999 Congress on Evolutionary Computation-CEC99 (Cat. No. 99TH8406), Washington, DC, USA, 6–9 July 1999; pp. 1470–1477.

18. Johnson, D.S.; Trick, M.A. *Cliques, Coloring, and Satisfiability: Second DIMACS Implementation Challenge, October 11–13, 1993*; American Mathematical Soc.: Providence, RI, USA, 1996; Volume 26.

19. Aiex, R.M.; Resende, M.G.; Ribeiro, C.C. TTT plots: A perl program to create time-to-target plots. *Optim. Lett.* **2007**, *1*, 355–366. [CrossRef]

Article

First-Arrival Travel Times Picking through Sliding Windows and Fuzzy C-Means

Lei Gao [1],*, Zhen-yun Jiang [1] and Fan Min [1,2]

[1]   School of Computer Science, Southwest Petroleum University, Chengdu 610500, China;
       jiangzyswpu@163.com (Z.-y.J.); minfanphd@163.com (F.M.)
[2]   Institute for Artificial Intelligence, Southwest Petroleum University, Chengdu 610500, China
*   Correspondence: largeleier@163.com; Tel.: +86-137-0802-3716

Received: 27 January 2019; Accepted: 25 February 2019; Published:27 February 2019

Abstract: First-arrival picking is a critical step in seismic data processing. This paper proposes the first-arrival picking through sliding windows and fuzzy c-means (FPSF) algorithm with two stages. The first stage detects a range using sliding windows on vertical and horizontal directions. The second stage obtains the first-arrival travel times from the range using fuzzy c-means coupled with particle swarm optimization. Results on both noisy and preprocessed field data show that the FPSF algorithm is more accurate than classical methods.

Keywords: first-arrival picking; fuzzy c-means; particle swarm optimization; range detection

1. Introduction

Seismic refraction data analysis is one of the principal methods for near-surface modeling [1–4]. A critical step of the method is first-arrival picking for direct and head waves. It influences the effectiveness of many steps such as static correction [5,6] and velocity modeling [7]. Misidentifications of these arrival times may have significant effects on the hypocenters [8]. However, the raw seismic traces are always contaminated by strong background noise with complex near-surface conditions [7]. The main challenge is to accurately extract the first arrivals under the noise interference [9,10] and irregular topography [11]. According to Akram and Eaton [8], there is an urgent need for automatic picking methods as the scale of seismic data continues to grow.

There are three main types of first-arrival picking methods. The first is *Coppens'* method [12] and its variants. It uses energy ratios within two amplitude windows to process the data [12]. Al-Ghamdi and Saeed [13] improved this method by using adaptive thresholds. The multi-window algorithm [14] uses three moving windows instead. Moreover, it distinguishes signals from noise using the average of the absolute amplitudes in each window. Sabbione and Velis [15] used a modified form of *Coppens'* method along with entropy and fractal-dimension methods to pick first-arrival travel times. The second is the direct correlation method [16]. The direct correlation method was proposed by Molyneux and Schmitt [16]. It uses the maximum cross correlation value as a criterion [16], which fails in data sets with low signal-to-noise ratio (S/N). The third is the backpropagation neural networks method [17]. It applies backpropagation neural networks in first-arrival refraction event picking and seismic data trace editing [17].

Recently, a new algorithm based on fuzzy c-means is proposed to deal with low signal-to-noise ratio data [18]. It divides microseismic data into two clusters according to the different levels of similarity between the signals and noise. Thus, the initial time of the signal cluster is regarded as the first-arrival time. Others have reported many automatic picking schemes such as digital image segmentation [19], STA/LTA method [20,21], Akaike information criterion [22], fractal-based algorithm [23] and TDEE [11]. However, the detection accuracy of existing algorithms is still unsatisfactory.

In this paper, we propose the first-arrival picking through sliding windows and fuzzy c-means (FPSF) algorithm. Figure 1 illustrates the overall structure of our algorithm through an example. In the range detection stage, each trace is processed with a vertical sliding window to seek the first-arrival interval of each trace. Then, the horizontal window is employed to adjust the window for each trace. In this way, the first-arrival range is identified. In the first-arrival travel times picking stage, a particle swarm optimization (PSO) algorithm seeks original cluster centers. Finally, a fuzzy c-means (FCM) picks the first-arrival travel times.

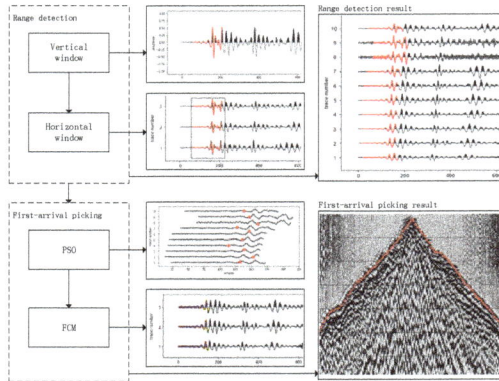

Figure 1. The framework of the FPSF algorithm (In the middle part, red indicates first-arrival intervals or initial clustering centers; In the right-top part, red indicates the first-arrival range; In the right-bottom part, red indicates first arrivals).

FPSF presents two new features to handle the challenge mentioned earlier. One is to introduce a range detection stage before first-arrival picking. We design a range detection technique using sliding windows on vertical and horizontal directions. On the one hand, the energy of single trace will abruptly shift in the first-arrival interval. Hence, a vertical sliding window is employed for keeping track of inner-trace change. The quality of the window is measured by the difference between its upper and lower parts, and its position in the trace. It finds the interval where the energy values suddenly shift the most. On the other hand, the first-arrival travel times of adjacent traces are approximate. Hence, a horizontal sliding window is employed for keeping track of inter-trace change. It adjusts the locations of vertical windows to ensure their similarities. All vertical windows of each trace consist of first-arrival range. With this technique, the data size is decreased dramatically, and the accuracy can be improved.

The other is to employ PSO and FCM for clustering seismic data. FCM is successful in image processing, so the application to our data is expected [18]. The data is restricted to the first-arrival range. First, an improved particle swarm optimization is used to determine original cluster centers. Second, an improved fuzzy c-means which has original cluster centers will pick up first-arrival travel times among the range.

Experiments are undertaken on two field data sets. We compare FPSF with some methods including modified Coppens' method (MCM) [15], the direct correlation method (DC) [16], and backpropagation neural networks method (BNN) [17]. Results show that FPSF is accurate. FPSF can be used in many domains such as image processing and seismic data processing.

The rest of the paper is organized as follows. In Section 2, we review some related works. In Section 3, we build a data model, define the first-arrival picking problem and introduce some concepts. In Section 4, we elaborate on the principle of the new method of this paper. In the sequel, two field data sets are experimented to verify the effectiveness of the method in Section 5. Finally, Section 6 summarizes this paper.

2. Related Works

The history of seismic refraction data analysis can be traced back to the 1920s [6]. Seismic refraction data analysis tasks include deconvolution [24], dynamic correction [25], static correction [26,27], speed analysis [28], and migration [29]. Picking first arrivals [19] is an important pre-processing stage for these tasks. For example, the effectiveness of static corrections depends on the precise of the first arrivals [6,15].

There are three strategies in the development of picking first arrivals. The manual strategy relies solely on the experts, therefore it is time-consuming and occasionally inaccurate [15,19,23,30]. To make the matter worse, this strategy can lead to biased and inconsistent picks because it relies on the subjectivity of the selection operator [15].

The man–machine interaction strategy provides experts with software for visual inspection [19,23,30]. The expert should identify a few first arrivals, and then the software will pick the others. In case of some difficult situations, the expert interfere with the process. Naturally, this strategy is more efficient. However, the whole procedure is still very time consuming and subjective.

The automatic strategy [15,16,23] aims to provide more efficient and intelligent solution. It requires the development of advanced machine learning and data mining algorithms. Note that this strategy does not prevent experts from intervening. Experts should check the result and correct it if necessary. Naturally, if the algorithm works well, manual intervention is rare.

Currently, there are many well-known seismic data processing systems, such as Promax [31], CGG (refering to wikipedia), Focus and Grisys [32]. They all contain the key step of picking first arrivals. Affected by data quality and parameter setting, the results of each software program are very different. Therefore, an accurate, efficient and stable algorithm for this problem is needed.

3. Preliminaries

In this section, we build a data model, define the first-arrival picking problem and introduce some basic concepts. Table 1 lists notations used throughout the paper.

3.1. Data Model

Single shot gather is a basic concept in the field of geophysical prospecting.

Definition 1. *A single shot gather is an $m \times n$ matrix $S = [s_{i,j}]$, where n is the number of traces, m is the number of samples for each trace, and $s_{i,j}$ is the energy value of the ith sample of the jth trace.*

Figure 2a illustrates an original single shot gather with 1000 samples and 800 traces. The horizontal coordinate is the number of traces. The vertical coordinate is the number of samples. The amplitude values of samples are the energy values. Black and white points correspond to positive and negative amplitudes, respectively. To cope with different geophones and different traces, we pre-process our data normalized to the range $[-1, 1]$.

3.2. First-Arrival Picking

We consider the following problem.

Problem 1. *The first-arrival picking problem.*
Input: A single shot gather S;
Output: $F = [f(1), f(2), \ldots, f(n)]$, where $\forall j \in [1..n]$, $f(j) \in [1..m]$ is the first-arrival time of the j-th trace.

Table 1. Notations.

Notation	Meaning
S	The single shot gather
F	The first-arrival times
l	The vertical window size
b	The horizontal window size
λ	The starting index of the current window
a	The energy ratio weight
k	The search step size
Λ	The starting index array of result windows
n	The number of traces
m	The number of samples for each trace
R	The first-arrival range matrix
θ	The parameters of fitness function
π	The parameters of fitness function
B	The boundary matrix
B_1	The position boundary matrix
B_2	The velocity range matrix
g	The dimension of the input problem
f	The fitness function
M	The number of particles
δ_1	The inertia weight of each particle's velocity
δ_2	The global influence weight
w	The inertia weight
T	The maximum iteration times
ε	The convergent error
x^*	The solution of the best particle
U	The first-arrival range matrix
e	The number of clustering centers
γ	The fuzzy indicator
J_{FCM}	The objective function
c_k	The center of the k-th cluster
$d_k(i,j)$	The distance between $s_{i,j}$ and c_k

That is, for any trace j, there is exactly one sample $f(j)$ corresponding to the first-arrival travel time. Figure 2b shows the output of first-arrival picking with the input shown in Figure 2a. Every trace has one first-arrival travel time. The red point is the first-arrival location of every trace.

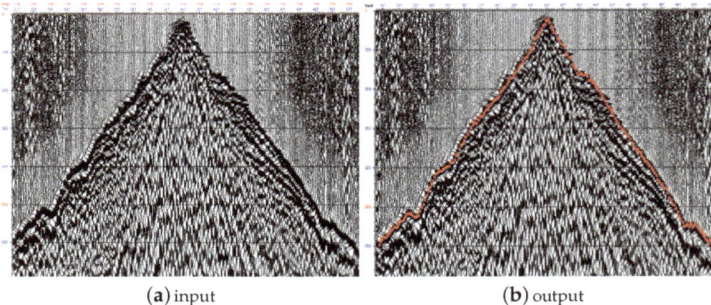

(**a**) input (**b**) output

Figure 2. Field data and first-arrival travel times.

3.3. Fuzzy c-Means

FCM algorithm was proposed by Dunn [33]. It was promoted as the general FCM clustering algorithm by Bezdek [34]. FCM has been used in many fields such as image segmentation [35–37].

The fuzzy set was conceived as a result of an attempt to come to grips with the problem of imprecisely defined categories [34]. K-means determines whether or not a group of objects form a cluster. Different from k-means, FCM determines the belonging of an object to a class with a matter of degree [34,38]. When objects X consists of k compact well separated clusters, k-means generates a limiting partition with membership functions which closely approximate the characteristic functions of the clusters. However, when X is not the union of k compact well separated clusters, the limiting partition is truly fuzzy in the sense that the values of its component membership functions differ substantially from 0 or 1 over certain regions of X. The fuzzy algorithm seems significantly less prone to the "cluster-splitting" tendency and may also be less easily diverted to uninteresting locally optimal partitions Dunn [33].

Let $O = \{o_1, o_2, \ldots, o_N\}$ be a set of objects. The standard fuzzy c-means objective function for partitioning O into e clusters is given by [37,39]:

$$\min J_\alpha = \sum_{i=1}^{e} \sum_{k=1}^{N} u_{ik}^{\alpha} d_{ik}^{\beta}, \tag{1}$$

where $u_{ik} \in [0, 1]$ is the degree of o_k belonging to the i-th cluster, α is the fuzzy indicator, $d_{ik} = \|o_k - c_i\|$ is the distance between o_k and c_i, and c_i is the center of the i-th cluster.

We also require the sum of the degrees of each object be 1, i.e., $\sum_{i=1}^{e} u_{ik} = 1 \ \forall k \in [1..N]$. In many applications, the Euclidean distance is employed to compute d_{ik}, and $\beta = 2$.

3.4. Particle Swarm Optimization

Particle swarm optimization was first proposed by [40]. It is a nature-inspired meta-heuristic approach applied to solve optimisation problems in different fields [41–45].

For an optimisation problem with M particles. Let $x_i'(t)$ denote the best solution that particle i has obtained until iteration t. Let $x^*(t)$ denote the best solution obtained among all the particles so far. To search for the optimal solution, each particle updates its velocity $v_i(t)$ and position $x_i(t)$ according to Equation (2) and (3) [46]:

$$v_i(t+1) = wv_j(t) + \delta_1 rand_1 \times (x_i'(t) - x_i(t)) + \delta_2 rand_2 \times (x^*(t) - x_i(t)), \tag{2}$$

$$x_i(t+1) = x_i(t) + v_i(t), \tag{3}$$

where t is the current iteration, $rand_1$ and $rand_2$ are the random variables in the range of $[0,1]$, δ_1 and δ_2 are the two positive acceleration constants to adjust relative velocity with respect to best global and local positions and the inertia weight w is used to balance the capabilities of the global exploration.

4. The Proposed Algorithm

The algorithm framework has been illustrated in Figure 1. The range detection stage is composed of vertical window sliding and horizontal window sliding. The first-arrival picking stage is composed of PSO and FCM. This section describes each stage in detail.

4.1. Range Detection

This subsection explains the range detection stage with a vertical sliding window and a horizontal sliding window.

We apply a vertical sliding window to capture the energy which is large, early and shift abrupt in each trace. Let the window size be l and the starting index of the current window of the j-th trace be λ_j. We design the following optimization objective function:

$$\min r(l, \lambda_j) = a \times \frac{(1 + \sum_{i=1}^{l/2} s_{i+\lambda_j,j})}{(1 + \sum_{i=l/2+1}^{l} s_{i+\lambda_j,j})} + (1 - a) \times l\lambda_j, \tag{4}$$

where a is the energy ratio weight. Here, the first part expresses the ratio between the upper and lower part of the window. The smaller the value, the larger the shift. The second part expresses the evaluation of the position of the window. The smaller the value, the earlier the travel time. The weight a is used to obtain a trade-off between these two values.

Algorithm 1 lists the vertical window sliding process. Line 1 initializes the minimal value of the object function value r^*. Lines 2 to 10 show the process of sliding a vertical window with a step size of k. Line 3 calculates the sum of the upper part of the window us. Line 4 calculates the sum of the lower part of the window ls. Line 5 computes object function value of the current window according to Equation (4). Lines 6 to 9 determine if the update condition has been reached. If so, update the minimal value of the object function value r^* and the starting index of the vertical window λ_j in Lines 7 and 8.

Algorithm 1: Vertical Window Sliding for One Trace

Input: The j-th trace $s_{.,j} = [s_{1,j}, s_{2,j}, \ldots, s_{m,j}]$, window size l, ratio weight a and search step size k.

Output: The starting index of the result window λ_j.

Method: verticalSliding.

1: $r^* = +\infty$; // Initialize

2: **for** $(i = 1$ step k to $m - l)$ **do**

3:     $us = \sum_i^{i+\frac{l}{2}-1} s_{i,j}$; // The sum of the upper part

4:     $ls = \sum_{i+\frac{l}{2}}^{i+l} s_{i,j}$; // The sum of the lower part

5:     $r = a \times (1 + us)/(1 + ls) + (1 - a) \times l \times i$; // Compute r

6:     **if** $(r < r^*)$ **then**

7:         $r^* = r$;

8:         $\lambda_j = i$; // Update the starting index

9:     **end if**

10: **end for**

11: **return** λ_j;

We apply a horizontal window to adjust the neighboring first-arrival intervals determined by the vertical windows. Median filtering is employed to smooth the first-arrival intervals in the window to ensure their similarity.

Definition 2. *First-arrival range matrix is an $l \times n$ matrix $R = [s_{i,j}]$, where l is the size of vertical window and n is the number of traces. It saves the first-arrival range including first arrivals. This matrix stores the range of the first arrivals. The size of the original data set S has been reduced from $m \times n$ to $l \times n$.*

Algorithm 2 lists the horizontal window sliding process. Lines 1 to 9 show the process of sliding a horizontal window. Line 2 moves the horizontal window in step size b. Line 3 obtains the median of the window m. Lines 4 to 7 determine whether the difference between each element in the window and the median is too large. If so, update this value with a large difference to the median in Line 6.

After the above steps, range detection stage has been completed and the first-arrival range expressed by the first-arrival range matrix (R) has been confirmed. This is one kind of dimensionality reduction techniques [47] and data size is directly reduced by 90%. Just as pre-processing can help increase the accuracy [48], range detection stage can be viewed as a pre-processing stage.

Algorithm 2: Horizontal Window Sliding

Input: The starting index array of result windows $\Lambda = [\lambda_1, \lambda_2, \ldots, \lambda_n]$, vertical window size l and horizontal window size b.

Output: Range starting index array Λ.

Method: horizontalSliding.

1: **for** $(i = 1$ to $\frac{n}{b})$ **do**

2:     $H = [\lambda_{(i-1) \times b+1}, \lambda_{(i-1) \times b+2}, \ldots, \lambda_{i \times b}]$; // Move the horizontal window

3:     $m = median(H)$; // Get the median of the window

4:     **for** $(j = 1$ to $b)$ **do**

5:        **if** $(abs(H[j] - m) \geqslant \frac{l}{2})$ **then**

6:           $\Lambda[j + (i-1) \times b] = m$; // Update the value with a large difference

7:        **end if**

8:     **end for**

9: **end for**

10: **return** Λ;

4.2. First-Arrival Picking from the Range

This subsection explains the first-arrival picking stage with PSO and FCM. The data field at this stage is the first-arrival range confirmed by range detection stage.

We employ PSO to find the original clustering centers of FCM according to the advantages of PSO including global optimization and fast convergence [49]. Specifically, we use the following fitness function:

$$f(x_i) = \frac{\theta}{\pi + J_{FCM}}, \tag{5}$$

where θ and π are the parameters of fitness function with the constraint $\theta \leq \pi$. The J_{FCM} is the objective function of the FCM clustering method we employed. The parameter θ was usually proposed as 2 [39].

The particle swarm velocity iterative update formula is Equation (2). The particle swarm position iterative update formula is Equation (3).

Definition 3. *Boundaries are represented by a $g \times 2$ matrix $B = [b_{i,j}]$, where $b_{i,1}$ is the lower bound, and $b_{i,2}$ is the respective upper bound.*

Here, we have two boundaries, the position boundary and the velocity boundary. Let B_1 be the position boundary, and let B_2 be the velocity boundary.

Algorithm 3 lists the process of particle swarm optimization. Lines 1 to 4 initialize each of the particle's position x_i and velocity v_i with random values. Line 5 initializes iteration times t. Lines 6 to 11 calculate the fitness function and record best solution of each particle according to Equation (5). Line 9 records best solution of each particle itself. Line 12 finds the optimal particle. Line 14 updates the global optimal particle. Lines 16 to 21 update velocity and position of each particle. Line 17 updates the velocity of each particle according to Equation (2). Lines 18 and 20 determine whether the velocity and position are out of boundaries. Line 19 updates the position of each particle according to Equation (3).

We employ FCM to pick first arrivals according to the similarity of the first-arrival energy values of adjacent traces. The fuzzy c-means algorithm iteratively calculates on the seismic data set to obtain the clustering center that minimize the objective function.

Algorithm 3: PSO

Input: The fitness function f, the matrices of position boundary B_1 and velocity boundary B_2, the number of particles M, the inertia weight of each particle's velocity δ_1, the global influence weight δ_2, the inertia weight w, the maximum iteration times T and the convergent error ε.
Output: Solution of the best particle x^*.
Method: particleSwarmOptimization.

1: **for** $(i = 1$ to $M)$ **do**
2:   $x_i = rand(B_1);$ // Initialize position x_i and velocity v_i
3:   $v_i = rand(B_2);$
4: **end for**
5: $t = 1;$ // Initialize iteration times t
6: **while** $(t \leq T$ && $!check\_convergence(\varepsilon))$ **do**
7:   **for** $(i = 1$ to $M)$ **do**
8:     **if** $(f(x_i) > p_i)$ **then**
9:       $[p_i, x_i'] = f(x_i);$ // Record optimal solution of each particle
10:     **end if**
11:   **end for**
12:   $[p_{cbest}, x^*] = find\_best([p_1, p_2, \ldots, p_M]);$ // Find optimal particle
13:   **if** $(p_{gbest} < p_{cbest})$ **then**
14:     $update\_pbest(p_{gbest}, x^*);$ // Update global optimal particle
15:   **end if**
16:   **for** $(i = 1$ to $M)$ **do**
17:     $v_i = update\_velocity(v_i, x_i', w, \delta_1, \delta_2);$ // Update particle velocity v_i
18:     $v_i = check\_and\_adjust(v_i, B_2);$ // Check and adjust
19:     $x_i = update\_position(v_i);$ // Update particle position x_i.
20:     $x_i = check\_and\_adjust(x_i, B_1);$ // Check and adjust
21:   **end for**
22:   $t = t + 1;$
23: **end while**
24: **return** x^*;

Definition 4. *First-arrival range matrix is an $l \times n \times e$ matrix $U = [u_{i,j,k}]$, where l is the height of the first-arrival range, n is the number of traces, e is the number of clustering centers, and $u_{i,j,k}$ is the membership degree of $s_{i,j}$ belonging to the k-th cluster.*

We use the following objective function:

$$J_{FCM} = \sum_{i,j} \sum_{k=1}^{e} u_{i,j,k}^{\gamma} d_k(i,j)^2, \tag{6}$$

where γ is the fuzzy indicator, $d_k(i,j) = \|s_{i,j} - c_k\|$ is the distance between $s_{i,j}$ and c_k, and c_k is the center of the k-th cluster.

The membership degree is updated according to

$$u_{i,j,k} = \frac{1}{\sum_{p=1}^{c}(d_k(i,j)/d_p(i,j))^{2/(\gamma-1)}}. \tag{7}$$

The clustering center is updated according to

$$c_k = \frac{\sum_{i,j}(u_{i,j,k})^{\gamma} s_{i,j}}{\sum_{i,j}(u_{i,j,k})^{\gamma}}. \tag{8}$$

Algorithm 4 lists the process of fuzzy c-means. Line 1 initializes membership matrix U. Line 2 computes clustering objective function value J according to Equation (6). Lines 3 to 6 iteratively update membership matrix U and clustering center array x^*. Line 4 updates membership matrix U according to Equation (7). Line 5 updates clustering center array x^* according to Equation (8).

Algorithm 4: FCM

Input: Original clustering center array x^*, the first-arrival range matrix R, the number of clusters e, the fuzzy indicator γ and the convergent error σ.
Output: Membership matrix U and clustering center array x^*.
Method: fuzzyClusterMethod.

1: $U = update\_matrix\_U(x^*, R, e, \gamma)$; // // Initialize membership matrix U
2: $J = J(U, x^*)$; // Compute function value J according to Equation (6)
3: **while** (*!check_convergence*(δ)) **do**
4:    $U = update\_matrix\_U(x^*, R, e, \gamma)$; // Update U according to Equation (7)
5:    $x^* = update\_centers\_X(U, R, \gamma)$; // Update x^* according to Equation (8)
6: **end while**// Check the convergence
7: **return** U, x^*;

After the FCM processing, e clustering centers can be fixed. The data is divided into 10 classes and one of the classes is the result of first-arrival picking. After the above steps, the first-arrival picking stage has been completed and the first-arrival travel times have been confirmed.

5. Experimental Results

This section shows the experimental results with two data sets.

Figure 3a shows the field microseismic data consists of 280 shots from *Xinjiang*, China. Every shot has about 400 traces and time sampling interval is 2 ms. Figure 3b shows the result of range detection. Figure 3c shows the result of FPSF.

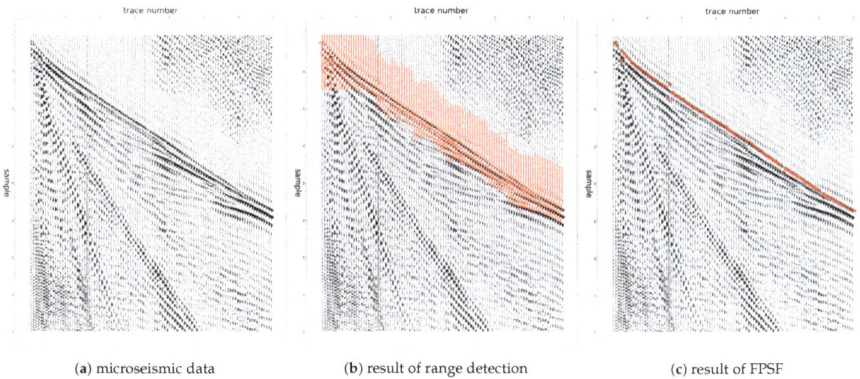

(a) microseismic data   (b) result of range detection   (c) result of FPSF

Figure 3. First arrivals picked by FPSF for field microseismic record. (**a**) field microseismic record; (**b**) the result of range detection; (**c**) the result of FPSF.

Figure 4a shows the field microseismic data consists of 150 shots from *Sichuan*, China. Every shot has about 500 traces and time sampling interval is 2 ms. Figure 4b shows the result of range detection. Figure 4c shows the result of FPSF.

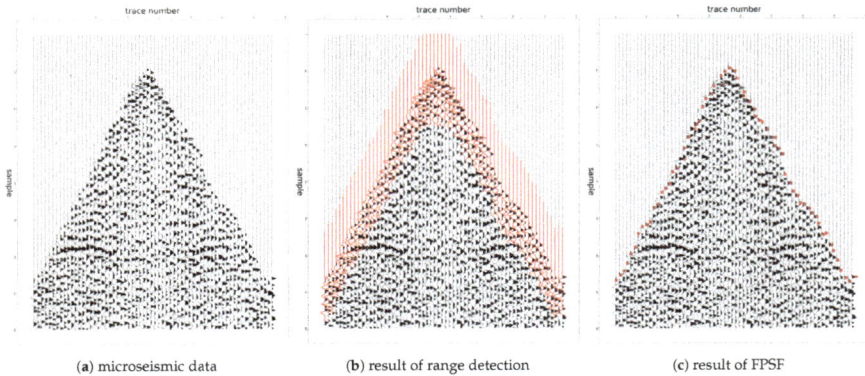

(**a**) microseismic data   (**b**) result of range detection   (**c**) result of FPSF

Figure 4. First arrivals picked by FPSF for field microseismic record. (**a**) field microseismic record; (**b**) the result of range detection; (**c**) the result of FPSF.

Figure 5a shows *Xinjiang* field microseismic data. Figure 5b shows the comparison of the difference among the values by the MCM method (purple spots), BNN method (blue spots), DC method (green spots) and FPSF (red spots).

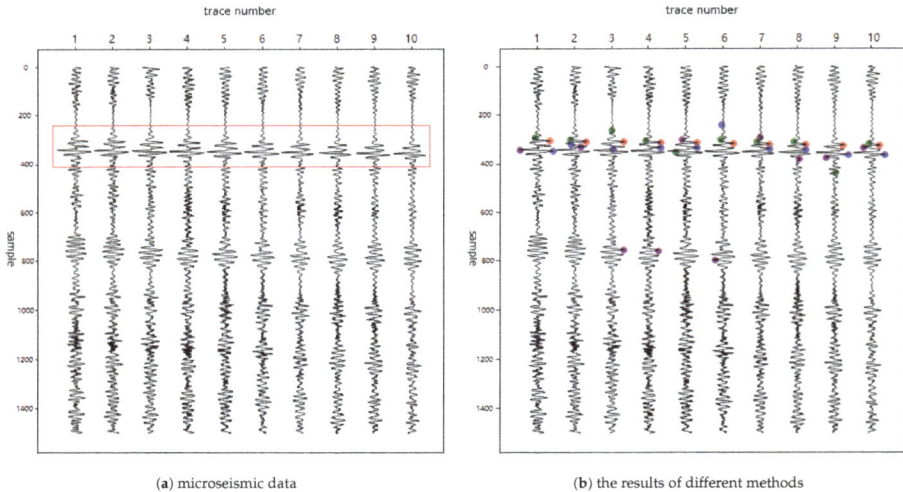

(**a**) microseismic data   (**b**) the results of different methods

Figure 5. First arrivals picked by FPSF for field microseismic record. (**a**) field microseismic record; (**b**) the result of different methods: MCM (purple), BNN (blue), DC (green), FPSF (red).

Table 2 shows the accuracy of each method for different data sets. We can find out that FPSF is more accurate than BNN, DC on the two data sets and MCM on one data set. In general, FPSF shows superiority over BNN, DC and MCM on the two data sets.

Table 2. Accuracy comparison.

Data Sources (Data Length)	BNN	DC	MCM	FPSF
Xinjiang (300200)	31.5%	81%	84%	96.5%
Sichuan (160064)	3.125%	4.69%	82.81%	81.25%

6. Conclusions

At the current stage of developments of artificial intelligence, we urgently need new theories applied to seismic exploration rather than sticking to the theory of seismic exploration itself. In this paper, we propose the first-arrival picking through sliding windows and fuzzy c-means (FPSF) algorithm. The biggest difference from the conventional methods that are currently available is that our method does not directly pick up the first-arrival travel times but determines a first-arrival range before picking. Combined with the characteristics of seismic data, we have improved some methods, such as particle swarm optimization and fuzzy c-means. Experiments with Xinjiang's data set with 280 shots and Sichun's data set with 150 shots show that the proposed method can significantly improve accuracy and stability. In the future, we will apply the idea of FPSF to other domains such as image processing.

Author Contributions: Conceptualization, L.G.; Data curation, L.G.; Formal analysis, L.G. and F.M.; Funding acquisition, L.G.; Investigation, L.G.; Methodology, L.G. and F.M.; Project administration, L.G. and F.M.; Resources, Z.-y.J.; Software, Z.-y.J.; Supervision, Z.-y.J. and F.M.; Validation, F.M.; Visualization, F.M.; Writing—original draft, L.G.; Writing—review and editing, L.G., Z.-y.J. and F.M.

Funding: This work is supported by the Natural Science Foundation of China under Grant No. 41604114.

Conflicts of Interest: The authors declare no conflict of interest.

References

1. Waheed, U.B.; Flagg, G.; Yarman, C.E. First-arrival traveltime tomography for anisotropic media using the adjoint-state method. *Geophysics* **2016**, *81*, R147–R155. [CrossRef]
2. Zelt, C.; Haines, S.; Powers, M.; Sheehan, J.; Rohdewald, S.; Link, C.; Hayashi, K.; Zhao, D.; Zhou, H.W.; Burton, B.; et al. Blind test of methods for obtaining 2-D near-surface seismic velocity models from first-arrival traveltimes. *J. Environ. Eng. Geophys.* **2013**, *18*, 183–194. [CrossRef]
3. Sun, M.Y.; Zhang, J.; Zhang, W. Alternating first-arrival traveltime tomography and waveform inversion for near-surface imaging. *Geophysics* **2017**, *82*, R245–R257. [CrossRef]
4. Zhu, X.H.; Valasek, P.; Roy, B.; Shaw, S.; Howell, J.; Whitney, S.; Whitmore, N.D.; Anno, P. Recent applications of turning-ray tomography. *Geophysics* **2008**, *73*, VE243–VE254. [CrossRef]
5. Kahrizi, A.; Hashemi, H. Neuron curve as a tool for performance evaluation of MLP and RBF architecture in first break picking of seismic data. *J. Appl. Geophys.* **2014**, *108*, 159–166. [CrossRef]
6. Yilmaz, Ö. *Seismic Data Analysis: Processing, Inversion, and Interpretation of Seismic Data*; Society of Exploration Geophysicists: Tulsa, OK, USA, 2001; Volume 1. [CrossRef]
7. An, S.; Hu, T.Y.; Peng, G.X. Three-Dimensional cumulant-based coherent integration method to enhance first-break seismic signals. *IEEE Trans. Geosci. Remote Sens.* **2017**, *55*, 2089–2096. [CrossRef]
8. Akram, J.; Eaton, D.W. A review and appraisal of arrival-time picking methods for downhole microseismic data. *Geophysics* **2016**, *81*, KS71–KS91. [CrossRef]
9. Li, Y.; Wang, Y.; Lin, H.B.; Zhong, T. First arrival time picking for microseismic data based on DWSW algorithm. *J. Seismol.* **2018**, *22*, 833–840. [CrossRef]
10. Hu, R.Q.; Wang, Y.C. A first arrival detection method for low SNR microseismic signal. *Acta Geophys.* **2018**, *66*, 945–957. [CrossRef]
11. Lan, H.Q.; Zhang, Z.J. A high-order fast-sweeping scheme for calculating first-arrival travel times with an trregular surface. *Bull. Seismol. Soc. Am.* **2013**, *103*, 2070–2082. [CrossRef]
12. Coppens, F. First arrival picking on common-offset trace collections for automatic estimation of static corrections. *Geophys. Prospect.* **1985**, *33*, 1212–1231. [CrossRef]
13. Al-Ghamdi.; Saeed, A. *Automatic First Arrival Picking Using Energy Ratios*; ProQuest: Ann Arbor, MI, USA, 2007.
14. Chen, M.; Li, Y.; Xie, J. A novel SVM-Based method for seismic first-arrival detecting. *Appl. Mech. Mater.* **2010**, *29–32*, 973–978. [CrossRef]
15. Sabbione, J.I.; Velis, D. Automatic first-breaks picking: New strategies and algorithms. *Geophysics* **2010**, *75*, V67–V76. [CrossRef]

16. Molyneux, J.B.; Schmitt, D.R. First-break timing: Arrival onset times by direct correlation. *Geophysics* **1999**, *64*, 1492–1501. [CrossRef]

17. McCormack, M.D.; Zaucha, D.E.; Dushek, D.W. First-break refraction event picking and seismic data trace editing using neural networks. *Geophysics* **1993**, *58*, 67–78. [CrossRef]

18. Zhu, D.; Li, Y.; Zhang, C. Automatic time picking for microseismic data based on a fuzzy c-means clustering algorithm. *IEEE Geosci. Remote Sens. Lett.* **2016**, *13*, 1900–1904. [CrossRef]

19. Mousa, W.A.; Al-Shuhail, A.A.; Al-Lehyani, A. A new technique for first-arrival picking of refracted seismic data based on digital image segmentation. *Geophysics* **2011**, *76*, V79–V89. [CrossRef]

20. Allen, R.V. Automatic earthquake recognition and timing from single traces. *Bull. Seismol. Soc. Am.* **1978**, *68*, 1521–1532.

21. Wong, J.; Han, L.J.; Bancroft, J.C.; Stewart, R.R. Automatic time-picking of first arrivals on noisy microseismic data. *Can. Soc. Explor. Eeophys. Conf. Abstr.* **2009**, *1*, 1–4.

22. Takanami, T.; Kitagawa, G. Estimation of the arrival times of seismic waves by multivariate time series model. *Ann. Inst. Stat. Math.* **1991**, *43*, 407–433. [CrossRef]

23. Boschetti, F.; Dentith, M.D.; List, R.D. A fractal-based algorithm for detecting first arrivals on seismic traces. *Geophysics* **1996**, *61*, 1095–1102. [CrossRef]

24. Tian, N.; Fan, T.G.; Hu, G.Y.; Zhang, R.W.; Zhou, J.N.; Le, J. The roles of the spatial regularization in seismic deconvolution. *Acta Geod. Geophys.* **2016**, *51*, 43–55. [CrossRef]

25. Bertrand, A.; MacBeth, C. Repeatability enhancement in deep-water permanent seismic installations: A dynamic correction for seawater velocity variations. *Geophys. Prospect.* **2005**, *53*, 229–242. [CrossRef]

26. Strong, S.; Hearn, S. Statics correction methods for 3D converted-wave (PS) seismic reflection. *Explor. Geophys.* **2017**, *48*, 237–245. [CrossRef]

27. Cox, M. *Static Corrections for Seismic Reflection Surveys*; Society of Exploration Geophysicists: Tulsa, OK, USA, 1999. [CrossRef]

28. Naus-Thijssen, F.M.J.; Goupee, A.J.; Vel, S.S.; Johnson, S.E. The influence of microstructure on seismic wave speed anisotropy in the crust: Computational analysis of quartz-muscovite rocks. *Geophys. J. Int.* **2011**, *185*, 609–621. [CrossRef]

29. Li, Q.H.; Jia, X.F. Generalized staining algorithm for seismic modeling and migration. *Geophysics* **2017**, *82*, T17–T26. [CrossRef]

30. Hatherly, P.J. A computer method for determining seismic first arrival times. *Geophysics* **1982**, *47*, 1431–1436. [CrossRef]

31. Fajaryanti, R.; Manik, H.M.; Purwanto, C. Application of multichannel seismic reflection method to measure temperature in Sulawesi Sea. *IOP Conf. Ser. Earth Environ. Sci.* **2018**, *176*, 012044. [CrossRef]

32. Wang, F.Y.; Zhao, C.B.; Feng, S.Y.; Ji, J.F.; Tian, X.F.; Wei, X.Q.; Li, Y.Q.; Li, J.C.; Hua, X.S. Seismogenic structure of the 2013 Lushan M (s) 7. 0 earthquake revealed by a deep seismic reflection profile. *Chin. J. Geophys. Chin. Ed.* **2015**, *58*, 3183–3192.

33. Dunn, J.C. A fuzzy relative of the ISODATA process and its use in detecting compact well-separated clusters. *J. Cybern.* **1973**, *3*, 32–57. [CrossRef]

34. Bezdek, J.C. Pattern recognition with fuzzy objective function algorithms. *Adv. Appl. Pattern Recognit.* **1981**, *22*, 203–239.

35. Jia, Z.X.; Xia, Y.; Chen, Q.; Sun, Q.S.; Xia, D.S.; Feng, D.D. Fuzzy c-means clustering with weighted image patch for image segmentation. *Appl. Soft Comput.* **2012**, *12*, 1659–1667. [CrossRef]

36. Shafei, B.; Steidl, G. Segmentation of images with separating layers by fuzzy c-means and convex optimization. *J. Vis. Commun. Image Represent.* **2012**, *23*, 611–621. [CrossRef]

37. Szilágyi, L.; Szilágyi, S.M.; Benyó, Z. A modified FCM algorithm for fast segmentation of brain MR images. In *Analysis and Design of Intelligent Systems using Soft Computing Techniques*; Springer: Berlin/Heidelberg, Germany, 2007; pp. 119–127. [CrossRef]

38. Ahmed, M.N.; Yamany, S.M.; Mohamed, N.; Farag, A.A.; Moriarty, T. A modified fuzzy c-means algorithm for bias field estimation and segmentation of MRI data. *IEEE Trans. Med. Imaging* **2002**, *21*, 193–199. [CrossRef] [PubMed]

39. Bezdek, J.C.; Pal, S.K. *Fuzzy Models for Pattern Recognition*; IEEE Press: New York, NY, USA, 1992; Volume 56.

40. Kennedy, J. Particle swarm optimization. In *Encyclopedia of Machine Learning*; Springer: Boston, MA, USA, 2011; pp. 760–766. [CrossRef]

41. Ganguly, S.; Sahoo, N.C.; Das, D. Multi-objective particle swarm optimization based on;fuzzy-pareto-dominance for possibilistic planning of electrical;distribution systems incorporating distributed generation. *Fuzzy Sets Syst.* **2013**, *213*, 47–73. [CrossRef]

42. Tsekouras, G.E.; Tsimikas, J. On training RBF neural networks using input-output fuzzy clustering and particle swarm optimization. *Fuzzy Sets Syst.* **2013**, *221*, 65–89. [CrossRef]

43. Wang, G.G.; Guo, L.H.; Gandomi, A.H.; Hao, G.S.; Wang, H.Q. Chaotic krill herd algorithm. *Inf. Sci.* **2014**, *274*, 17–34. [CrossRef]

44. Wang, G.G.; Tan, Y. Improving metaheuristic algorithms with information feedback models. *IEEE Trans. Cybern.* **2017**, *49*, 542–555. [CrossRef] [PubMed]

45. Fang, Y.; Liu, Z.H.; Min, F. A PSO algorithm for multi-objective cost-sensitive attribute reduction on numeric data with error ranges. *Soft Comput.* **2017**, *21*, 7173–7189. [CrossRef]

46. Ding, S.C.; Hang, J.; Wei, B.L.; Wang, Q.J. Modelling of supercapacitors based on SVM and PSO algorithms. *IET Electr. Power Appl.* **2018**, *12*, 502–507. [CrossRef]

47. Keogh, E.; Chakrabarti, K.; Pazzani, M.; Mehrotra, S. Dimensionality reduction for fast similarity search in large time series databases. *Knowl. Inf. Syst.* **2001**, *3*, 263–286. [CrossRef]

48. Mousas, C.; Anagnostopoulos, C.N. Learning motion features for example-based finger motion estimation for virtual characters. *3D Res.* **2017**, *8*, 25. [CrossRef]

49. Liu, H.; Xiao, G.F. Improved fuzzy clustering image segmentation algorithm based on particle swarm optimization. *Comput. Eng. Appl.* **2013**, *49*, 37–52. [CrossRef]

![Sigma mathematics logo] *mathematics*

MDPI

Article

A Multi-Objective DV-Hop Localization Algorithm Based on NSGA-II in Internet of Things

Penghong Wang [1], Fei Xue [2], Hangjuan Li [1], Zhihua Cui [1,*], Liping Xie [1,*] and Jinjun Chen [3]

[1] Complex System and Computational Intelligent Laboratory, Taiyuan University of Science and Technology, Taiyuan 030024, China; penghongwang@sina.cn (P.W.); L15536914519@163.com (H.L.)
[2] School of Information, Beijing Wuzi University, Beijing 101149, China; xuefei2004@126.com
[3] Department of Computer Science and Software Engineering, Swinburne University of Technology, Melbourne 3000, Australia; jinjun.chen@gmail.com
* Correspondence: cuizhihua@tyust.edu.cn (Z.C.); lipingxie1978@163.com (L.X.); Tel.: +86-138-3459-9274 (Z.C.); +86-136-4341-3592 (L.X.)

Received: 17 December 2018; Accepted: 7 February 2019; Published: 15 February 2019

Abstract: Locating node technology, as the most fundamental component of wireless sensor networks (WSNs) and internet of things (IoT), is a pivotal problem. Distance vector-hop technique (DV-Hop) is frequently used for location node estimation in WSN, but it has a poor estimation precision. In this paper, a multi-objective DV-Hop localization algorithm based on NSGA-II is designed, called NSGA-II-DV-Hop. In NSGA-II-DV-Hop, a new multi-objective model is constructed, and an enhanced constraint strategy is adopted based on all beacon nodes to enhance the DV-Hop positioning estimation precision, and test four new complex network topologies. Simulation results demonstrate that the precision performance of NSGA-II-DV-Hop significantly outperforms than other algorithms, such as CS-DV-Hop, OCS-LC-DV-Hop, and MODE-DV-Hop algorithms.

Keywords: wireless sensor networks (WSNs); DV-Hop algorithm; multi-objective DV-Hop localization algorithm; NSGA-II-DV-Hop

1. Introduction

As the hottest research topics currently, internet of things (IoT) contains many technologies such as cyber physical systems [1,2], embedded system technology, network information technology, and so on. And wireless sensor networks (WSNs) [3,4], as an important branch of cyber physical systems, have become an innovation and area of research under the spotlight worldwide. Moreover, WSNs technology is so popular that it has been applied in various fields, including the military and national defense, industry [5], disaster relief, medical treatment, environmental monitoring [6], and so on [7]. However, for most WSNs applications, sensor node location information plays a key role; generally, the information obtained from WSNs would be meaningless, if sensor node locations were unknown in applications such as smart grid, object tracking, and location-based routing. Hence, the sensor node localization technology is a critical issue in the rapid development of WSNs and even IoT technology.

Currently, the BeiDou navigation satellite system (BDS) [8,9] and global positioning system (GPS) [10] are generally considered to be the most capable systems for obtaining the exact location. However, it's worth mentioning that due to the expensive cost, it is almost impossible to complete the full coverage installation of BDS equipment in the whole WSNs. Besides, its positioning accuracy is invariably not satisfactory enough in some special contexts, including the indoor, mine tunnel, canyon, and other complex environments. As a result, it has begun to receive researchers' attention that the use of interactions and connectivity information between sensor nodes for positioning. Using this information, researchers have proposed a series of localization algorithms. These algorithms are generally classified as a range-based localization algorithm or range-free localization algorithm, depending on whether they are independent of the additional hardware devices. These hardware devices are necessary to obtain the requisite information for the range-based localization algorithm, such as point-to-point distances and angles between sensor nodes. The information between the sensor nodes ensures that the range-based algorithm can achieve accurate positioning, including RSSI [11], ToA [12], and AoA [13], but it requires extensive CPU time and a mass of energy. In contrast, the range-free localization algorithm only needs to ensure the connectivity between sensor nodes, including APIT [14], Centroid [15], Amorphous [16], and DV-Hop [17]. Due to cost constraints, it's widely used in a large and complex network.

DV-Hop localization algorithm, as a representative range-free positioning algorithm, has garnered extensive attention because of its simple positioning principle. Its main principle is that the beacon nodes (node location information is known) use the connectivity between nodes to send packets to other nodes in the network to obtain the minimum hop count between the beacon nodes and unknown nodes (node location information is unknown). And then, the average distances per hop of beacon nodes are calculated using the position and hop count information of the beacon nodes. Finally, the locations of the unknown nodes are estimated by calculating the distance between the unknown nodes to each beacon node. Compared to other range-free positioning algorithms, it is easier to bring into operation, but the low positioning accuracy has become a problem to be solved. For this reason, scholars propose various improved algorithms based on DV-Hop localization algorithm, including the deterministic algorithms [18–20] and bio-inspired optimization algorithms. In addition, Mobility-Assisted Localization in WSNs has also been widely studied by scholars, such as Rezazadeh [21], who proposed a path planning mechanism to improve the accuracy of mobile assisted localization. Alomari improved the path planning method and proposed a path planning strategy based on dynamic fuzzy-logic [22], and proposed an obstacle avoidance strategy based on swarm intelligence optimization [23].

In recent years, with the excellent performance of intelligent computing in various complex optimization problems, various bionic algorithms have been proposed, such as particle swarm optimization (PSO) [24], ant colony optimization (ACO) [25], bat algorithm (BA) [26–28], Differential Evolution (DE) [29], Firefly algorithm (FA) [30–32], and so on [33]. Compared with the mathematics optimization methods, these biological inspired algorithms show some unique advantages. First, they don't depend on the requirement of any gradient information in the variable space; in addition, they are insensitive to the initial value and insusceptible to local entrapment. These optimization algorithms play a very good role in practical applications, such as [34–40], however, with the increasing amount of data in the IoT era, many problems in the real world include multiple decision variables and evaluation indicators. Single-objective optimization has gradually revealed defects for solving such problems. For this reason, multi-objective optimization algorithms based on bionics have also been proposed and are used in various fields, including Multi-Objective Particle Swarm Optimizers (MOPSO) [41,42], multi-objective evolutionary algorithm based on decomposition (MOEA/D) [43], hybrid multi-objective cuckoo search (HMOCS) [44,45], and so on [46,47].

In this paper, we propose a multi-objective DV-Hop localization algorithm based on NSGA-II [48] to solve the sensor node localization problem in WSNs. The remainder of this paper is arrayed as follows. In Section 2, DV-Hop with optimization algorithms and problems are reviewed. In Section 3, standard DV-Hop and NSGA-II are presented. In Section 4, a multi-objective DV-Hop localization model is structured and NSGA-II-DV-Hop is proposed. Simulation results and performance analysis are summarized in Section 5. Lastly, the conclusion is summarized in Section 6.

2. Related Works

In the last few years, with the maturity of various stochastic optimization algorithms in theory, more attention has been paid to the practical application of the algorithm. In 1975, Holland [49] proposed the theory and method of genetic algorithm by studying the genetic evolution process in the natural environment. And after a series of research work, Goldberg [50] formally presented the genetic algorithm (GA) in 1989. In 2007, on the basis of solving the numerical optimization by genetic algorithm, Nan [51] proposed to apply the real-coded GA to WSNs. And in 2010, Gao [52] developed an improved GA to solve wireless sensor localization problem in WSNs. Moreover, Bo [53] also applied GA to solve the problem of WSNs location, and proposed a population constraint strategy based on three beacon nodes to solve the feasible domain of the population.

Furthermore, Yang [54] presented a cuckoo search (CS) algorithm based on Levy flights in 2009. In 2014, Sun [55] developed the CS algorithm and applied it to the DV-Hop positioning algorithm and achieved good positioning results. Based on this, Zhang [56] proposed a weight-oriented CS algorithm (WOCS), and combined it with DV-Hop to locate the unknown sensor nodes in WSNs. The paper improved the search ability of the CS algorithm for unknown nodes by limiting the hop count (which is the minimum hop count between the unknown nodes and each beacon node) in the DV-Hop algorithm. Furthermore, Cui [57] further developed the WOCS algorithm, and proposed an oriented CS algorithm based on the Lévy-Cauchy distribution (OCS-LC) in 2017. This improved strategy is applied to solve the positioning problem of sensor nodes in WSN, and compared with the CS algorithm, there is a large performance improvement when the number of sensor nodes is small. However, these studies were based on the study of the location performance of sensor nodes in a large area, but ignored the positioning performance of sensor nodes in complex terrain. In response to this phenomenon, Cui [58] studied the positioning performance of sensor nodes in C-shaped random and C-shaped grids in 2018. Nevertheless, in this research, the nodes in the network are required to obey Uniform distribution, which is unimaginable in practical production applications. Not only is this so, a common feature of these studies is that more effort is devoted to the improvement of algorithmic search strategies, while ignoring improvements to the original model.

In these studies, although the positioning accuracy has been improved, there are some defects. According to the calculation formula (Equation (7)) of the single-objective model, the population gradually converges to the estimated position as the number of iterations increases, as shown in Figure 4 of the part IV. The actual position of the unknown node is UN, but the population will converge to the UN^{*1} and UN^{*2} points, which will bring a large error.

To solve this problem, we propose three other complex terrains for research, including coal mine tunnels [59,60], lake terrain, and canyons terrain. In these specific cases, for the distribution of sensor nodes some new features emerge. For instance, in the coal mine tunnel, the nodes are distributed in narrow tunnels that are interlaced, and the nodes are densely distributed. This requires the algorithm to have a good positioning effect when the number of nodes and the number of beacon nodes are large. However, in the lake terrain, the nodes are distributed around the lake, which leads to communication difficulty when the communication radius is small. Therefore, the algorithm is required to have a strong positioning capability when the communication radius is small and the number of nodes is small. And in the canyons terrain, the nodes are distributed in the canyon among several mountains. In this case, the algorithm is required to have better stable positioning accuracy when the radius and the beacon nodes are small. So, in this paper, we propose a multi-objective DV-Hop localization algorithm based on NSGA-II. The biggest highlight of this paper is to abandon the idea that scholars blindly improve the algorithm search strategy, and change the objective function model in the algorithm to achieve more precise positioning of unknown nodes. A constraint strategy based on all beacon nodes is proposed based on the three beacon nodes constraint strategy.

3. DV-Hop Algorithm and NSGA-II Algorithm

3.1. DV-Hop Algorithm

In this subsection, we will detail the specific implementation process of the DV-Hop algorithm.

Phase 1: Communication detection and broadcasting phase.

At this stage, it is mainly to detect whether direct communication between any two nodes is possible, and also to record the minimum hops count that nodes can communicate with each other. The specific process is that each beacon node broadcasts a packet to the network (the packet includes its location and its own minimum hop count information to other nodes), and the initialization value of each node hop count information is 0. Each time the packet is forwarded, the number of hop count is increased by one. Among them, each node only records the minimum hop count information between it and other nodes.

Phase 2: Distance estimation phase.

Since the position information of the beacon node is known, the $Hopsize_i$ (the average distance per hop between any two beacon nodes) can be obtained by Equation (1).

$$HopSize_i = \frac{\sum\limits_{j \neq i} \sqrt{(x_i - x_j)^2 + (y_i - y_j)^2}}{\sum\limits_{j \neq i} h_{ij}} \tag{1}$$

where (x_i, y_i), (x_j, y_j) are the coordinates of beacon nodes i and j respectively, and h_{ij} is the minimum hop count between the beacon nodes which is calculated by **Phase 1**.

And then, the d_{ik} (the distance between beacon node i and unknown node k) is estimated by Equation (2).

$$d_{ik} = Hopsize_i \times h_{ik} \tag{2}$$

where h_{ik} is the minimum hop count between the beacon node i and unknown node k.

Phase 3: Unknown node coordinate estimation phase.

For the unknown node k, if more than three distances have been estimated by Equation (2), the position of the unknown node k can be calculated mathematically, such as the trilateral measuring method. The computational equation is

$$\begin{cases} (x_1 - x)^2 + (y_1 - y)^2 = d_1^2 \\ \cdots \\ (x_n - x)^2 + (y_n - y)^2 = d_n^2 \end{cases} \tag{3}$$

where (x, y) represents the unknown nodes' coordinates, (x_n, y_n) denotes the coordinates of beacon node n, and d_n denotes the distance estimated by Equation (2).

Convert Equation (3) to a matrix form $AX = b$, where A, b, and X are described as the following Equations (4) and (5), respectively.

$$A = \begin{pmatrix} 2(x_1 - x_n) & 2(y_1 - y_n) \\ \cdots & \cdots \\ 2(x_{n-1} - x_n) & 2(y_{n-1} - y_n) \end{pmatrix}, X = \begin{pmatrix} x \\ y \end{pmatrix} \tag{4}$$

$$b = \begin{pmatrix} x_1^2 - x_n^2 + y_1^2 - y_n^2 + d_n^2 - d_1^2 \\ \cdots \\ x_{n-1}^2 - x_n^2 + y_{n-1}^2 - y_n^2 + d_n^2 - d_{n-1}^2 \end{pmatrix} \tag{5}$$

Based on Equations (4) and (5), the location of the unknown node can be obtained by the least square method. The calculation equation can be expressed as Equation (6).

$$\hat{X} = (A^T A)^{-1} A^T b \tag{5}$$

The flowchart of DV-Hop algorithm is introduced in Figure 1.

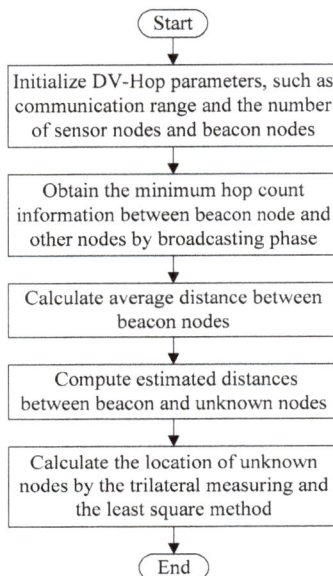

Figure 1. The distance vector-hop technique (DV-Hop) flowchart.

3.2. NSGA-II Algorithm

A non-dominated sorting genetic algorithm II (NSGA-II) was first proposed in [48] as a biological heuristics algorithm which usually used to solve complex industrial optimization problems. The algorithm has been widely concerned by scholars since its invention due to its faster convergence speed, stronger robustness, and better draw near the true Pareto-optimal front. In NSGA-II algorithm, its core operation contains two parts. One part includes the three traditional operation processes in GA, such as crossover, selection, and mutation; the other part refers to the unique non-dominated sorting operation in the multi-objective optimization algorithm. Therein, the selection operation will retain some of the better individuals with their fitness values (which refer to the non-dominated sorted value). The mutation operation is designed according to the genetic mutation in the biology, in order to ensure that the algorithm has strong global convergence ability. Conversely, the crossover operation is designed based on the principle that homologous chromosomes cross to generate new species to improve the algorithm search ability.

The pseudo-code of NSGA-II algorithm is introduced in Algorithm 1.

Algorithm 1: The pseudo-code of NSGA-II

Begin

    Input: Population: *NP*; Dimension: *D*; Maximum Generation: *Gmax*; Cross
    probability: *Pc*; mutation probability: *Pm*.

  Initialization: compute objective values, fast non-dominated sort, selection,
    crossover and mutation.

  Generation = 1;

    While *Generation < Gmax* do

        Combine parent and offspring population, compute objective values
        and fast non-dominated sort.

        Selection operation.

        If rand() < *Pc*

        Crossover operation;

        End

        If rand() < *Pm*

          Mutation operation;

        End

        Generation = Generation+1;

    End

    Output: The best individuals

End

---

4. The Proposed Multi-Objective Algorithm

In this paper, we propose a multi-objective DV-Hop localization algorithm based on NSGA-II, which achieves the purpose of improving the positioning accuracy by adopting multi-objective improvement on the original objective.

4.1. The Multi-Objective Model

In the traditional DV-Hop algorithm based on optimization algorithm, Equation (7) is recognized as the most typical objective function.

$$fitness_1 = \min(\sum_{i=1}^{m} |\sqrt{(x_i - x)^2 + (y_i - y)^2} - d_i|) \tag{7}$$

where d_i denotes the estimated distance in the simulation experiment between beacon node i and an unknown node, (x_i, y_i) represents the location of the beacon node i, (x, y) denotes the location of the unknown node, $fitness_1$ denotes the objective (which refers to one of the objective functions in this paper).

However, this objective function is determined by Equation (8), and Equation (8) is the core theory of the combination of the optimization algorithm and the DV-Hop algorithm. For unknown node j, assume (x, y) is the actual location, and the estimated distances are $d_1, d_2, \ldots, d_n$ in the simulation experiment for all beacon nodes, the corresponding errors are $\delta_1, \delta_2, \ldots, \delta_n$. Then, the relationship among them can be expressed as follows: under the premise that the value of $\sqrt{(x_n - x)^2 + (y_n - y)^2}$ is constant, the smaller $\delta_1, \delta_2, \ldots, \delta_n$, the more accurate the positioning accuracy. Therefore, convert Equation (8) to a function form $y = ax$, the objective function is expressed as Equation (7).

$$\begin{cases} \sqrt{(x_1 - x)^2 + (y_1 - y)^2} = d_1 + \delta_1 \\ \sqrt{(x_2 - x)^2 + (y_2 - y)^2} = d_2 + \delta_2 \\ \qquad \cdots \\ \sqrt{(x_n - x)^2 + (y_n - y)^2} = d_n + \delta_n \end{cases} \tag{8}$$

Nevertheless, (x, y) is the unknown node estimated position rather than the actual position, and $d_1, d_2, \ldots, d_n$ are obtained in the second phase of the DV-Hop algorithm, and are constant. This means that the position obtained by Equation (7) (the objective function) is closer to the position under the estimated distance, rather than the true exact position. Based on this phenomenon, we present to add an objective function to strengthen the search constraint on the exact position.

Suppose there are some sensor nodes in the detected area, which contain the beacon nodes and unknown nodes, such as Figure 2. In Figure 2, BN denotes the beacon node; UN_1, UN_2, UN_3 denote the unknown nodes, respectively; R denotes the communication radius; and Dis_i is the actual distance between UN_i and BN; the circular area is the communication area of the BN. When the number of unknown nodes is enough to fill the entire the circular area, the average distance between UN_i and BN is calculated as Equation (9).

$$avg\_dis = \frac{\int_0^R 2\pi r^2 dr}{\int_0^R 2\pi r dr} = \frac{2}{3}R \tag{9}$$

avg_dis denotes the average distance between UN_i and BN, and also represents the average distance per hop between sensor nodes. Particularly, different from $HopSize_i$ is that the calculation result of avg_dis is the theoretical value of the average distance from the unknown node to the beacon node in per hop. Therefore, the theoretical distance dis_{ik} from the unknown node k to each beacon node i is calculated as Equation (10).

$$dis_{ik} = avg\_dis \times h_{ik} \tag{10}$$

where h_{ik} is the minimum hop count between the beacon node i and unknown node k.

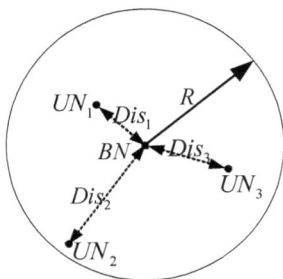

Figure 2. Distance relationship of sensor nodes.

Similarly, we define the second objective function as follows:

$$fitness_2 = \min\left(\sum_{i=1}^{m} \left| \sqrt{(x_i - x)^2 + (y_i - y)^2} - dis_i \right| \right) \tag{11}$$

For the sake of clarity, we will elaborate on the difference between our proposed multi-objective and traditional single-objective (We use three beacon nodes BN_1, BN_2, BN_3 and one unknown node UN for analysis). Figure 3 shows the constraint principle of the objective function in ideal conditions. At the moment, the unknown node i in the population finally converges the location of the UN, and this location is the exact position.

However, the estimated distance is usually accompanied by errors. Therefore, the single-objective function constraint principle in the estimated distance is shown in Figure 4. Where, UN is defined as the actual location of the unknown node, BN_1, BN_2 indicate the beacon nodes, d_1, d_2 are calculated by Equation (2), UN^{*1}, UN^{*2} represent the estimated location of the unknown node which calculated with Equation (7) in ideal circumstances. It is not difficult to see that the error between the potential optimal solution set found by the single-objective function model and the real position is still large.

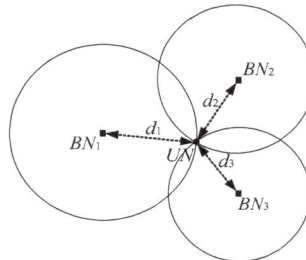

Figure 3. Constraint principle in ideal conditions.

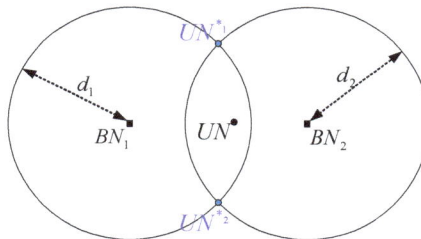

Figure 4. Single-objective constraint principle.

For the defects of single-objective optimization, we propose to use the multi-objective optimization method to reduce the error, such as Figure 5. Figure 5 is composed of two parts, one part is the decision space on the left side and the other part is the objective space on the right side. Where, f_1, f_2 respectively represent two contradictory objective function models that we proposed, dis_1, dis_2 are calculated by Equation (10). As can be seen from the Figure 5, a solution in the objective space corresponds to multiple potential optimal solutions in the decision space. That means that multi-objective models can find more potential optimal solutions in the decision space than the single objective model. Meanwhile, it contains the potential optimal solution that the single-objective model can find. According to this theory, the error of the estimated position obtained by using the multi-objective model must be less than or equal to the single-objective model.

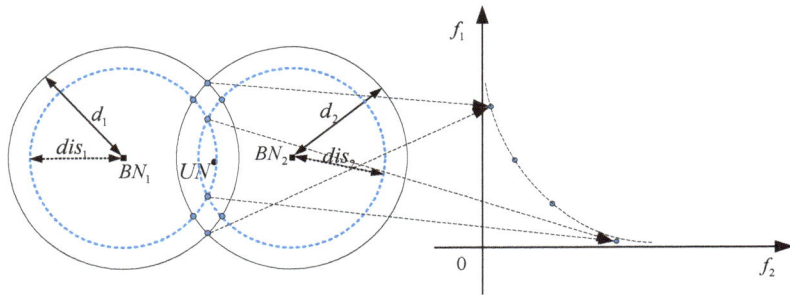

Figure 5. Multi-objective constraint principle.

4.2. Population Constraint Strategy

In addition to improvements to the model, this paper also improves the algorithm's search strategy. In reference [34], the author proposed a population constraint strategy based on three beacon nodes to solve the feasible domain of the population, such as Figure 6a (where BN_1, BN_2, BN_3 denote the beacon nodes, UN is the unknown node, H_1, H_2, H_3 represent the minimum hop count, R represents the radius and the shadow area is the feasible domain). The expression is as follows:

$$\begin{cases} \max\limits_{i=1,2,3} (x_i - RH_i) \le x_{UN} \le \min\limits_{i=1,2,3} (x_i + RH_i) \\ \max\limits_{i=1,2,3} (y_i - RH_i) \le y_{UN} \le \min\limits_{i=1,2,3} (y_i + RH_i) \end{cases} \tag{12}$$

However, when the distance among the three beacon nodes is relatively close and they are located on the same side of the unknown node, the feasible domain of the population is still larger. In this situation, the robustness of the positioning accuracy deteriorates. In this paper, we propose a population constraint strategy based on all beacon nodes, such as Figure 6b. The expression is as follows:

$$\begin{cases} \max\limits_{i=1,2,\dots,n} (x_i - RH_i) \le x'_{UN} \le \min\limits_{i=1,2,\dots,n} (x_i + RH_i) \\ \max\limits_{i=1,2,\dots,n} (y_i - RH_i) \le y'_{UN} \le \min\limits_{i=1,2,\dots,n} (y_i + RH_i) \end{cases} \tag{13}$$

As the number of beacon nodes increases, the probability of the beacon nodes being on the same side of the unknown node decreases correspondingly. This means that the constraint enhancement from the beacon nodes has different directions, and thus the population feasible region decreases. As shown in Figure 6b, the feasible domain of population is significantly reduced compared to Figure 6a. By reducing the feasible region of the population, the convergence speed of the algorithm can be accelerated and the positioning accuracy improved.

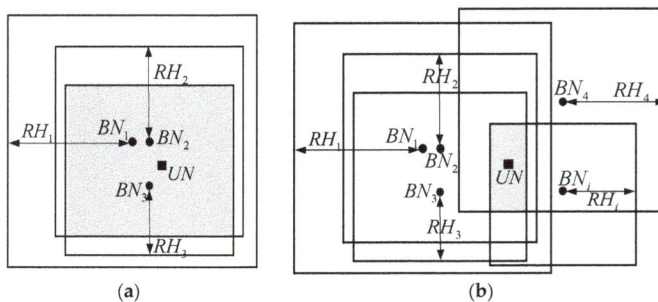

Figure 6. Population constraint strategy. (**a**) Population constraint based on three beacon nodes; (**b**) Population constraint based on all beacon nodes.

4.3. NSGA-II-DV-Hop Algorithm

The construction process of the multi-objective model was introduced before. In this section, the solution process of the model will be introduced. NSGA-II is considered by this paper to be a feasible and reliable algorithm for solving multi-objective models. The pseudo-code of NSGA-II-DV-Hop is introduced in Algorithm 2.

Algorithm 2: The pseudo-code of NSGA-II-DV-Hop

Begin

Input: Communication radius, number of nodes, beacon nodes, and the location of beacon nodes; Population: NP; Dimension: D; Maximum Generation: $Gmax$; Cross probability: Pc; mutation probability: Pm.

DV-Hop algorithm with Figure 1.

Initialization: Compute objective values with Equation (7) and Equation (11), fast non-dominated sort, selection, crossover and mutation.

Population constraint strategy with Equation (13).

Generation = 1;

While *Generation < Gmax* do

Combine parent and offspring population; compute objective values with Equation (7), Equation (11), and fast non-dominated sort.

Selection operation.

    If rand() < *Pc*

Perform cross-operations on the positions of different individuals in the population;

    End

    If rand() < *Pm*

Randomly generate a position that satisfies the boundary condition;

    End

    If (the position is contradictory with the boundary condition)

Randomly generate a position that satisfies the boundary condition.

    end

    Generation = Generation+1;

End

Calculate average localization error with Equation (14).

Output: The best location and average localization error.

End

---

5. Experimental Results and Analysis

5.1. Experimental Environment and Evaluation Criteria

To verify the effectiveness of NSGA-II-DV-Hop, extensive experiments were conducted in MATLAB 2016a. Experimental results will be compared with other three algorithms, including the DV-Hop, CS-DV-Hop, OCS-LC-DV-Hop, and MODE-DV-Hop. Experiment content tests the four different complex networks, including the Random, C-shaped random, O-shaped random, and X-shaped random, as shown in Figure 7. These different network topologies represent different application backgrounds, including plain terrain, canyons terrain, lake terrain, and coal mine tunnels (where all nodes are randomly employed). In addition, the detected area is a 100 × 100 m square region, and other parameters are listed in Table 1.

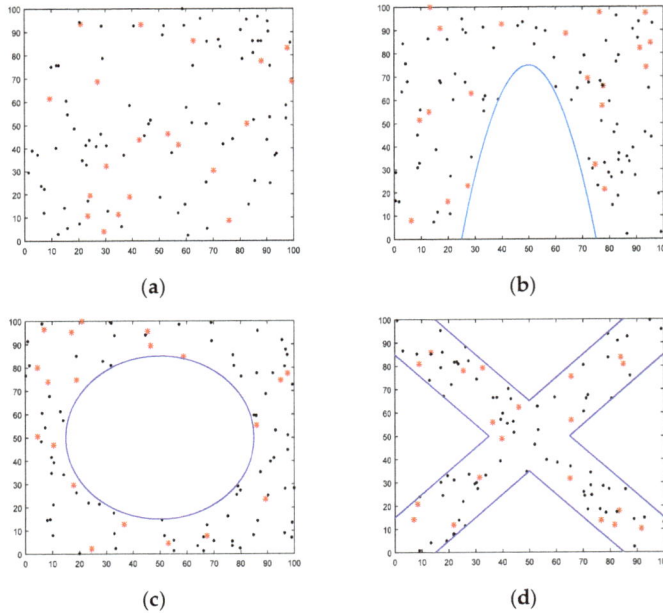

Figure 7. Four different complex networks topologies. (**a**) The random topology; (**b**) The C-shaped random topology; (**c**) The O-shaped random topology; (**d**) The X-shaped random topology.

Table 1. Parameter settings.

Parameter	Value
Pc	1
Pm	1/c (c refers to the variable dimension)
Population	20
Largest iterations	500
R(m)	25
Nodes	100
Beacon nodes	20

In order to compare the positioning performance of different algorithms more fairly, the average localization error (ALE) of unknown nodes is employed as the evaluation criterion. The specific calculation formula is as follows:

$$ALE = \frac{100}{M \times R} \sum_{i=1}^{M} \sqrt{(x_i' - x_i)^2 + (y_i' - y_i)^2} \tag{14}$$

where M and R note the number of unknown nodes and communication radius respectively; (x_i', y_i') represents the estimated location and (x_i, y_i) denotes the exact location.

5.2. Two Objective Function Relationships

In order to verify whether the multi-objective DV-Hop localization algorithm based on NSGA-II proposed in this paper is feasible, we performed the relationship between two objective functions in different network topologies. The results are shown in Figure 8. In Figure 8a–d respectively show the relationship between the two objective functions in four network topologies, and these relationships are contradictory. The experimental results also demonstrate that the method we proposed is feasible.

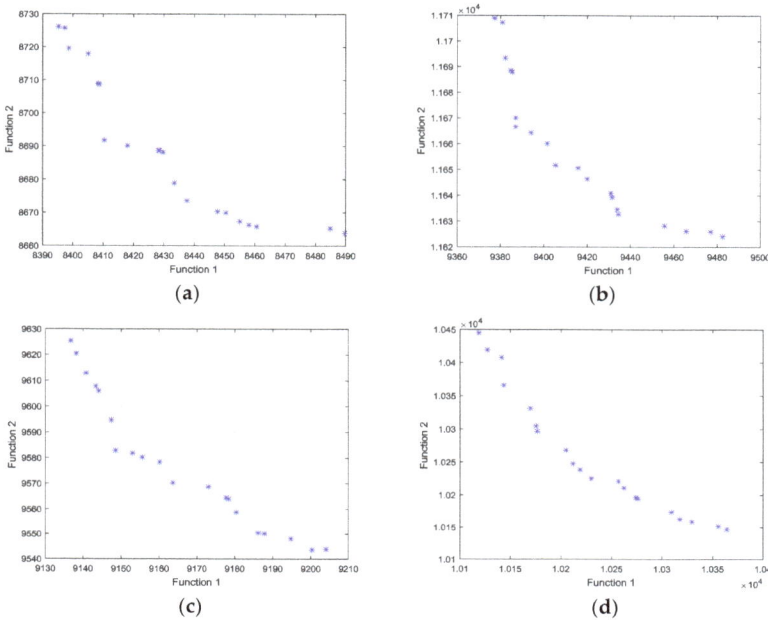

Figure 8. Two objective function relationships in four different network topologies. (**a**) The random topology; (**b**) The C-shaped random topology; (**c**) The O-shaped random topology; (**d**) The X-shaped random topology.

In addition, in multi-objective optimization, the solutions obtained after the optimization completed are the Pareto-optimal solutions. These equivalent solutions can be selected according to the actual situation. In this paper, to make the operation simpler, the minimum value of the sum of the two objective values in the solution set is identified as the optimal solution for comparison.

5.3. Influence of Communication Radius

In this experimental phase, the influences of different communication radius on the localization performance are performed. And the communication radius will change from 15 to 40, when the number of nodes and the beacon nodes remain unchanged. The simulation results are shown in Table 2 and Figure 9a–d.

Figure 9a shows the ALE of four algorithms in random topology, and in this topology, NSGA-II-DV-Hop is slightly inferior to the CS-DV-Hop and OCS-LC-DV-Hop algorithm, but significantly better than the DV-Hop algorithm. However, in the other three network topologies (Figure 9b–d, the ALE of NSGA-II-DV-Hop always has the lowest localization error no matter what kind of communication radius.

From Table 2, compared with DV-Hop, NSGA-II-DV-Hop can reduce a maximum of 21.91%, 114.77%, 69.71%, and 39.29% on localization errors, respectively. In particular, in the C-shaped random network topology, compared with CS-DV-Hop and OCS-LC-DV-Hop, the positioning accuracy of the NSGA-II-DV-Hop algorithm is improved by 26.74% and 24.42%, respectively. In addition, the performance of MODE-DV-Hop is similar to NSGA-II-DV-Hop.

Table 2. Average localization error (ALE) of different algorithms in different network topologies and communication radius.

Communication Radius		15	20	25	30	35	40
random topology	DV-Hop	65.24	46.14	33.25	28.92	27.59	26.54
	CS-DV-Hop	48.17	26.52	23.58	22.15	21.44	18.54
	OCS-LC-DV-Hop	**38.52**	24.58	21.83	20.84	**19.01**	**17.65**
	MODE-DV-Hop	52.71	24.84	21.30	20.32	19.93	18.13
	NSGAII-DV-Hop	52.57	**24.23**	22.09	21.46	20.19	18.06
C-shaped random topology	DV-Hop	172.33	112.53	63.73	49.78	44.81	41.62
	CS-DV-Hop	84.30	62.38	38.17	31.25	31.42	29.93
	OCS-LC-DV-Hop	81.98	58.59	37.35	30.46	32.09	29.36
	MODE-DV-Hop	66.80	51.23	34.20	30.44	**27.74**	28.72
	NSGAII-DV-Hop	**57.56**	**49.54**	**32.89**	**28.89**	28.87	**28.37**
O-shaped random topology	DV-Hop	117.88	56.50	44.77	39.39	29.24	31.28
	CS-DV-Hop	48.27	30.51	31.83	26.72	20.44	21.38
	OCS-LC-DV-Hop	49.32	31.05	23.77	26.86	20.85	21.98
	MODE-DV-Hop	**47.81**	27.44	23.67	23.24	18.48	19.97
	NSGAII-DV-Hop	48.17	**25.78**	**22.59**	**22.96**	**17.80**	**19.06**
X-shaped random topology	DV-Hop	80.18	54.22	43.49	39.39	37.15	36.29
	CS-DV-Hop	42.84	32.54	34.51	30.46	30.55	26.28
	OCS-LC-DV-Hop	45.68	33.60	35.84	32.43	30.41	26.60
	MODE-DV-Hop	43.04	**31.37**	29.65	**27.88**	**24.93**	26.38
	NSGAII-DV-Hop	**40.89**	32.49	**29.18**	29.39	27.30	**25.93**

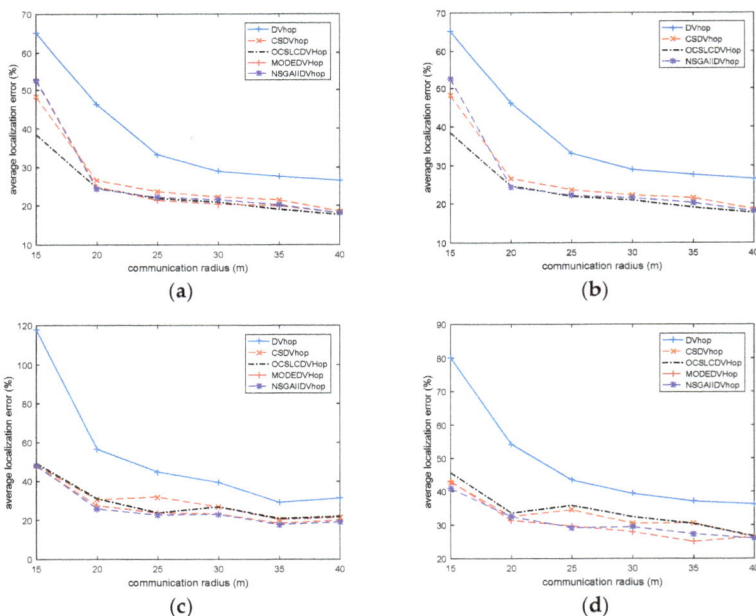

Figure 9. The ALE of four network topologies in different communication radius. (**a**) The random topology; (**b**) The C-shaped random topology; (**c**) The O-shaped random topology; (**d**) The X-shaped random topology.

5.4. Influence of Nodes

The number of nodes incrementally increases from 50 to 100 in this simulation phase, and the number of beacon nodes and communication radius stay the same. The experiment results are given in Table 3 and Figure 10.

Table 3. ALE of different algorithms in different network topologies and number of nodes.

Number of Nodes		50	60	70	80	90	100
random topology	DV-Hop	51.70	43.60	30.56	32.57	33.13	33.25
	CS-DV-Hop	26.98	25.65	24.94	24.78	24.99	23.58
	OCS-LC-DV-Hop	**24.35**	**24.17**	**23.57**	23.39	22.43	21.83
	MODE-DV-Hop	27.41	27.83	26.98	23.29	**21.89**	**21.30**
	NSGAII-DV-Hop	27.95	25.82	26.31	**22.84**	22.64	22.09
C-shaped random topology	DV-Hop	76.27	75.39	70.34	66.42	65.12	63.73
	CS-DV-Hop	46.12	45.19	41.73	41.18	39.21	38.17
	OCS-LC-DV-Hop	43.98	43.07	40.63	39.64	38.68	37.35
	MODE-DV-Hop	39.05	42.74	36.04	36.01	36.18	34.20
	NSGAII-DV-Hop	**34.01**	**37.24**	**34.56**	**34.92**	**33.52**	**32.89**
O-shaped random topology	DV-Hop	33.92	40.59	40.82	41.80	42.46	44.77
	CS-DV-Hop	22.54	21.16	22.20	22.66	22.06	31.83
	OCS-LC-DV-Hop	21.63	23.48	23.12	23.31	22.84	23.77
	MODE-DV-Hop	20.18	**20.47**	22.60	22.75	23.56	23.67
	NSGAII-DV-Hop	**18.79**	21.78	**22.03**	**21.70**	**22.16**	**22.59**
X-shaped random topology	DV-Hop	34.16	36.47	38.00	40.31	40.30	43.49
	CS-DV-Hop	33.98	31.64	32.58	33.74	33.68	34.51
	OCS-LC-DV-Hop	35.34	34.21	35.27	35.86	35.13	35.84
	MODE-DV-Hop	**29.03**	27.90	29.21	**28.20**	**27.52**	29.65
	NSGAII-DV-Hop	30.07	**27.27**	**28.55**	28.25	27.54	**29.18**

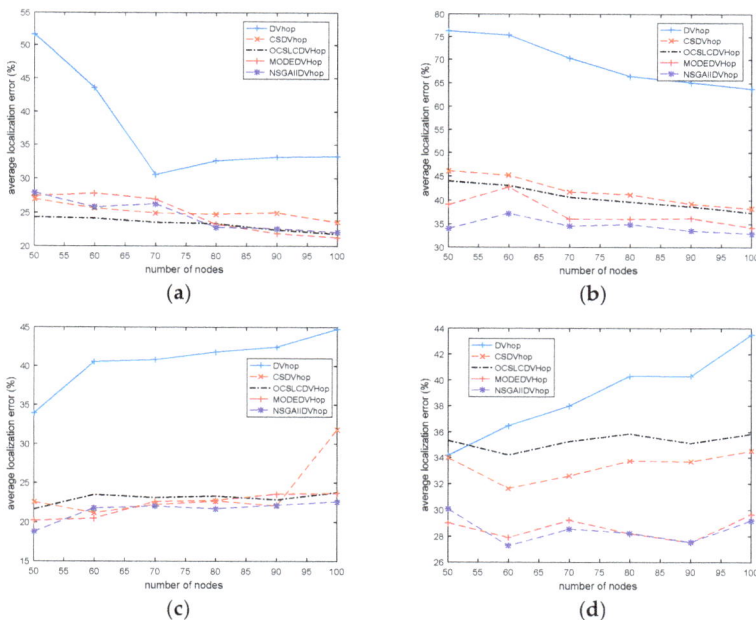

Figure 10. The ALE of four network topologies in different number of nodes. (**a**) The random topology; (**b**) The C-shaped random topology; (**c**) The O-shaped random topology; (**d**) The X-shaped random topology.

From Figure 10, we can see that in C-shaped and X-shaped random network topologies, the localization accuracy of NSGA-II-DV-Hop and MODE-DV-Hop algorithms are significantly superior to CS-DV-Hop, OCS-LC-DV-Hop, and DV-Hop algorithms. And in the Random or O-shaped network topologies, the performance of NSGA-II-DV-Hop is slightly better than the CS-DV-Hop and OCS-LC-DV-Hop, but always superior to the DV-Hop algorithm.

As depicted in Table 3, NSGA-II-DV-Hop has excellent positioning performance. Compared with the DV-Hop localization algorithm, the ALEs of NSGA-II-DV-Hop are less than 4.25–23.75%, 30.84–42.26%, 15.13–22.18%, and 4.09–14.31% respectively. The most conspicuous improvement occurs in X-shaped and C-Shaped topologies, and the ALEs are reduced by 7.61% and 9.97% more than OCS-LC-DV-Hop algorithm, respectively. Compared with the MODE-DV-Hop, the precision of the NSGA-II-DV-Hop is slightly better.

5.5. Influence of Beacon Nodes

In this simulation phase, the number of beacon nodes incrementally increases from 5 to 20, and the number of nodes and communication radius remain the same. The experiment results are given in Table 4 and Figure 11.

Table 4. ALE of different algorithms in different network topologies and number of beacon nodes.

Number of Baecon Nodes		5	10	15	20	25	30
random topology	DV-Hop	49.21	38.21	38.77	33.25	28.31	32.48
	CS-DV-Hop	38.76	29.67	28.59	23.58	22.88	20.94
	OCS-LC-DV-Hop	36.98	28.72	26.80	21.83	21.01	19.22
	MODE-DV-Hop	35.99	24.41	23.62	**21.30**	**20.11**	**17.49**
	NSGAII-DV-Hop	**34.74**	**23.25**	**21.90**	22.09	20.81	19.43
C-shaped random topology	DV-Hop	88.45	67.42	69.45	63.73	64.88	69.80
	CS-DV-Hop	101.44	48.14	42.49	38.17	49.41	53.24
	OCS-LC-DV-Hop	102.36	49.62	41.73	37.35	51.77	52.90
	MODE-DV-Hop	74.48	37.55	40.08	34.20	37.43	36.11
	NSGAII-DV-Hop	**67.25**	**34.78**	**36.83**	**32.89**	**35.34**	**34.63**
O-shaped random topology	DV-Hop	98.08	79.95	38.47	44.77	38.28	40.49
	CS-DV-Hop	42.65	36.22	30.35	31.83	34.84	37.10
	OCS-LC-DV-Hop	45.15	36.60	33.17	23.77	34.99	35.72
	MODE-DV-Hop	42.59	35.76	23.97	23.67	23.86	21.88
	NSGAII-DV-Hop	**41.14**	**30.23**	**23.46**	**22.59**	23.38	**21.47**
X-shaped random topology	DV-Hop	58.46	59.14	47.89	43.49	46.66	48.57
	CS-DV-Hop	51.90	40.74	41.54	34.51	47.54	44.36
	OCS-LC-DV-Hop	48.83	39.74	46.47	35.84	45.32	45.87
	MODE-DV-Hop	45.76	**34.70**	32.19	29.65	**28.96**	**25.75**
	NSGAII-DV-Hop	**42.74**	35.03	**30.81**	**29.18**	29.29	27.25

As shown in Figure 11, we can see that the positioning accuracy of NSGA-II-DV-Hop always has an advantage over the other three localization algorithms no matter which topologies. Furthermore, as the number of beacon nodes increases, the ALEs of NSGA-II-DV-Hop present a declining trend, but the ALE of the other three algorithms fluctuate upwards and downwards. The reason causing this kind of phenomenon is that in the complex network topology, the unknown nodes at the edge of the detected area increases, and the feasible domain of the unknown node satisfies the probability increase of Figure 6a, so that the positioning performance deteriorates. Inversely, the NSGA-II-DV-Hop algorithm proposed in this paper adopts the principle of Figure 6b, which reduces the feasible domain of the unknown node, so that the algorithm has more reliable positioning performance.

As shown in Table 4, the original DV-Hop always has the worst localization performance; and NSGA-II-DV-Hop algorithm has the greatest degree of enhancement no matter which network topologies. Especially, compared with the OCS-LC-DV-Hop, NSGA-II-DV-Hop positioning accuracy increased by up to 35.11% and 18.62% respectively in C-shaped and X-shaped network topologies. And the minimum ALEs always are in NSGA-II-DV-Hop and MODE-DV-Hop.

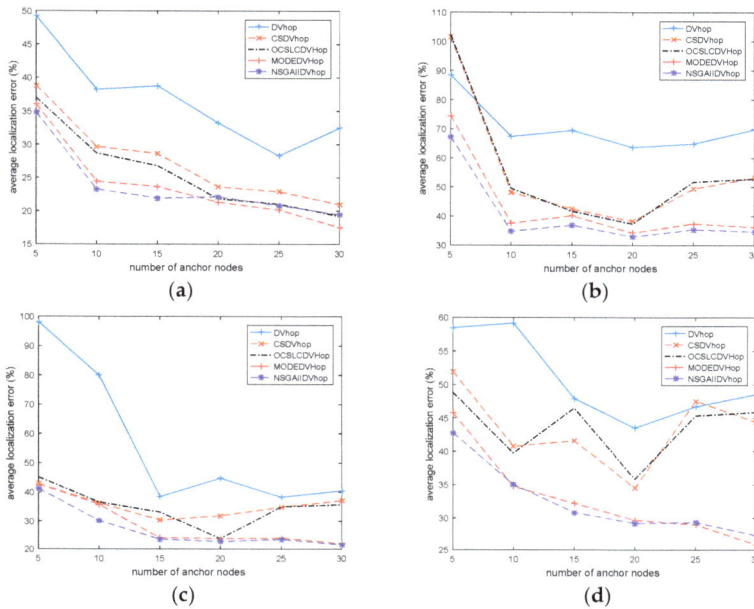

Figure 11. The ALE of four network topologies in different number of beacon nodes. (**a**) The random topology; (**b**) The C-shaped random topology; (**c**) The O-shaped random topology; (**d**) The X-shaped random topology.

5.6. The Standard Deviation and the Confidence Intervals

As can be seen from Table 5, the standard deviations of the NSGA-II-DV-Hop and MODE-DV-Hop are larger than the CS-DV-Hop and OCS-LC-DV-Hop, which is because the multi-objective model has more potential optimal solutions, such as Figure 5. However, it is worth paying attention that the confidence intervals of NSGA-II-DV-Hop and MODE-DV-Hop are less than the CS-DV-Hop and OCS-LC-DV-Hop in most cases, which means that the performance of the multi-objective model is reliable.

Table 5. The standard deviation and confidence intervals of different algorithms in four network topologies.

		Random Topology	C-Shaped Random Topology	O-Shaped Random Topology	X-Shaped Random Topology
the standard deviation and the confidence intervals (probably at 95%)	CS-DV-Hop	0.5636 [0.46, 0.67] 23.5816 [23.12, 24.03]	0.5241 [0.41, 0.70] 38.1680 [37.97, 38.36]	0.1390 [0.11, 0.19] 31.8336 [31.78, 31.89]	0.2150 [0.17, 0.29] 34.5050 [34.42, 34.59]
	OCS-LC-DV-Hop	0.9243 [0.67, 1.31] 21.8342 [21.04, 22.21]	0.4277 [0.34, 0.58] 37.3458 [37.19, 37.51]	0.6448 [0.51, 0.87] 23.7727 [23.53, 24.01]	0.1736 [0.13, 0.23] 35.8445 [35.77, 35.91]
	MODE-DV-Hop	1.2770 [1.02, 1.71] 21.3018 [20.82, 21.77]	0.7446 [0.59, 1.00] 34.2048 [33.92, 34.48]	0.6658 [0.53, 0.89] 23.6688 [23.42, 23.91]	1.1133 [0.88, 1.49] 29.6472 [29.23, 30.06]
	NSGA-II-DV-Hop	0.7005 [0.55, 0.94] 22.0850 [21.82, 22.35]	0.4887 [0.38, 0.66] 32.8934 [32.71, 33.08]	0.4911 [0.39, 0.66] 22.5942 [22.41, 22.77]	0.8246 [0.65, 1.11] 29.1820 [28.87, 29.48]

6. Conclusions

This paper proposes a multi-objective DV-Hop localization algorithm based on NSGA-II called NSGA-II-DV-Hop. To further reduce the positioning error, the traditional DV-Hop localization algorithm based on single-objective optimization algorithm is transformed into a multi-objective DV-Hop localization algorithm. We use the multi-objective constraint approach to reduce the convergence domain of unknown nodes and achieve the purpose of improving positioning accuracy. In addition, we also improve the search strategy of the algorithm, changing the population constraint strategy based on three beacon nodes to the population constraint strategy based on all beacon nodes. The simulation results demonstrate that this improved strategy can effectively reduce the sensitivity of the algorithm positioning performance to the number of beacon nodes. Furthermore, this paper also tests four complex network topologies in different backgrounds, and the experimental results show that NSGA-II-DV-Hop significantly outperforms original DV-Hop, CS-DV-Hop, OCS-LC-DV-Hop, and MODE-DV-Hop in all topologies, which also validates the practicability and reliability of this multi-objective model.

And in the future, we will continue to study the error distribution characteristics of the estimated distance in different network topologies and the construction of multi-objective models when there are obstacles in the network.

Author Contributions: Conceptualization, P.W.; Data curation, P.W. and F.X.; Formal analysis, P.W., H.L., Z.C. and J.C.; Funding acquisition, Z.C.; Methodology, P.W.; Project administration, Z.C.; Writing—original draft, P.W. and H.L.; Writing—review & editing, P.W., H.L. and Z.C.

Acknowledgments: This work is supported by the National Natural Science Foundation of China under Grant No.61806138, No.U1636220, No.61663028 and No.61403271, Natural Science Foundation of Shanxi Province under Grant No.201801D121127, Scientific and Technological innovation Team of Shanxi Province under Grant No.201805D131007, PhD Research Startup Foundation of Taiyuan University of Science and Technology under Grant No.20182002.

Conflicts of Interest: The authors declare no conflict of interest.

References

1. Khaitan, S.K.; Mccalley, J.D. Design Techniques and Applications of Cyber physical Systems: A Survey. *IEEE Syst. J.* **2015**, *9*, 350–365. [CrossRef]
2. Lee, E.A. Cyber Physical Systems: Design Challenges. In Proceedings of the IEEE International Symposium on Object Oriented Real-Time Distributed Computing, Orlando, FL, USA, 5–7 May 2008; pp. 363–369.
3. Akyildiz, I.F.; Su, W.; Sankarasubramaniam, Y.; Cayirci, E. A survey on sensor networks. *IEEE Commun. Mag.* **2002**, *40*, 102–114. [CrossRef]
4. Guo, P.; Wang, J.; Li, B.; Lee, S.Y. A variable threshold-value authentication architecture for wireless mesh networks. *J. Internet Technol.* **2014**, *15*, 929–936.
5. Wang, Z.; Wang, X.; Liu, L.; Huang, M.; Zhang, Y. Decentralized feedback control for wireless sensor and actuator networks with multiple controllers. *Int. J. Mach. Learn. Cybern.* **2017**, *8*, 1471–1483. [CrossRef]
6. Chandanapalli, S.B.; Reddy, E.S.; Lakshmi, D.R. DFTDT: Distributed functional tangent decision tree for aqua status prediction in wireless sensor networks. *Int. J. Mach. Learn. Cybern.* **2017**, *9*, 1419–1434. [CrossRef]
7. Suo, H.; Wan, J.; Huang, L.; Zou, C. Issues and Challenges of Wireless Sensor Networks Localization in Emerging Applications. In Proceedings of the International Conference on Computer Science and Electronics Engineering, Hangzhou, China, 23–25 March 2012; IEEE Computer Society: Washington, DC, USA, 2012; pp. 447–451.
8. Yang, Y.X.; Li, J.L.; Xu, J.Y.; Tang, J.; Guo, H.; He, H. Contribution of the Compass satellite navigation system to global PNT users. *Sci. Bull.* **2011**, *56*, 2813. [CrossRef]
9. Montenbruck, O.; Steigenberger, P.; Hugentobler, U.; Teunissen, P.; Nakamura, S. Initial assessment of the COMPASS/BeiDou-2 regional navigation satellite system. *GPS Solut.* **2013**, *17*, 211–222. [CrossRef]
10. Kaplan, E.D. Understanding GPS: Principles and Application. *J. Atmos. Sol.-Terr. Phys.* **1996**, *59*, 598–599.

11. Girod, L.; Bychkovskiy, V.; Elson, J.; Estrin, D. Locating tiny sensors in time and space: A case study. In Proceedings of the IEEE International Conference on Computer Design: VLSI in Computers and Processors, Freiberg, Germany, 18 September 2002; IEEE Computer Society: Washington, DC, USA, 2002; pp. 214–219.

12. Harter, A.; Hopper, A.; Steggles, P.; Ward, A.; Webster, P. The anatomy of a context-aware application. *Wirel. Netw.* **2002**, *8*, 187–197. [CrossRef]

13. Niculescu, D.; Nath, B. Ad hoc positioning system (APS) using AOA. In Proceedings of the Joint Conference of the IEEE Computer and Communications, San Francisco, CA, USA, 30 March–3 April 2003; pp. 1734–1743.

14. He, T.; Huang, C.; Blum, B.M.; Stankovic, J.A.; Abdelzaher, T. Range-free localization schemes in large scale sensor networks. In Proceedings of the IEEE Mobicom, San Diego, CA, USA, 14–19 September 2003; pp. 81–95.

15. Capkun, S.; Hamdi, M.; Hubaux, J.P. GPS-free positioning in mobile ad-hoc networks. In Proceedings of the Hawaii International Conference on System Sciences, Maui, HI, USA, 6 January 2002; p. 10.

16. Nagpal, R. Organizing a Global Coordinate System from Local Information on an Amorphous Computer. Available online: https://dspace.mit.edu/handle/1721.1/5926 (accessed on 16 December 2018).

17. Niculescu, D.; Nath, B. DV Based Positioning in Ad Hoc Networks. *Telecommun. Syst.* **2003**, *22*, 267–280. [CrossRef]

18. Zhao, J.; Jia, H. A hybrid localization algorithm based on DV-Distance and the twice-weighted centroid for WSN. In Proceedings of the IEEE International Conference on Computer Science and Information Technology, Chengdu, China, 9–11 July 2010; pp. 590–594.

19. Hou, S.; Zhou, X.; Liu, X. A novel DV-Hop localization algorithm for asymmetry distributed wireless sensor networks. In Proceedings of the IEEE International Conference on Computer Science and Information Technology, Chengdu, China, 9–11 July 2010; pp. 243–248.

20. Qian, Q.; Shen, X.; Chen, H. An Improved Node Localization Algorithm Based on DV-Hop for Wireless Sensor Networks. *Comput. Sci. Inf. Syst.* **2011**, *8*, 953–972. [CrossRef]

21. Rezazadeh, J.; Moradi, M.; Ismail, A.S.; Dutkiewicz, E. Superior Path Planning Mechanism for Mobile Beacon-Assisted Localization in Wireless Sensor Networks. *IEEE Sens. J.* **2014**, *14*, 3052–3064. [CrossRef]

22. Alomari, A.; Phillips, W.; Aslam, N.; Comeau, F. Dynamic Fuzzy-Logic Based Path Planning for Mobility-Assisted Localization in Wireless Sensor Networks. *Sensors* **2017**, *17*, 1904. [CrossRef] [PubMed]

23. Alomari, A.; Phillips, W.; Aslam, N.; Comeau, F. Swarm Intelligence Optimization Techniques for Obstacle-Avoidance Mobility-Assisted Localization in Wireless Sensor Networks. *IEEE Access* **2017**, 2169–3536. [CrossRef]

24. Kennedy, J.; Eberhart, R. Particle swarm optimization. In Proceedings of the IEEE International Conference on Neural Networks, Perth, WA, Australia, 27 November–1 December 1995; Volume 4, pp. 1942–1948.

25. Dorigo, M.; Birattari, M.; Stutzle, T. Ant colony optimization-artificial ants as a computational intelligence technique. *IEEE Comput. Intell. Mag.* **2006**, *1*, 28–39. [CrossRef]

26. Cai, X.; Wang, H.; Cui, Z.; Cai, J.; Xue, Y.; Wang, L. Bat algorithm with triangle-flipping strategy for numerical optimization. *Int. J. Mach. Learn. Cybern.* **2018**, *9*, 199–215. [CrossRef]

27. Cai, X.; Gao, X.; Xue, Y. Improved bat algorithm with optimal forage strategy and random disturbance strategy. *Int. J. Bio-Inspired Comput.* **2016**, *8*, 205–214. [CrossRef]

28. Cui, Z.; Li, F.; Zhang, W. Bat algorithm with principal component analysis. *Int. J. Mach. Learn. Cybern.* **2018**. [CrossRef]

29. Storn, R.; Price, K. Differential evolution: A simple and efficient heuristic for global optimization over continuous space. *J. Glob. Optim.* **1997**, *11*, 341–359. [CrossRef]

30. Wang, H.; Wang, W.; Zhou, X.; Sun, H.; Jia, Z.; Yu, X.; Cui, Z. Firefly algorithm with neighborhood attraction. *Inf. Sci.* **2017**, *382*, 374–387. [CrossRef]

31. Yu, G.; Feng, Y. Improving firefly algorithm using hybrid strategies. *Int. J. Comput. Sci. Math.* **2018**, *9*, 163–170. [CrossRef]

32. Yu, W.X.; Wang, J. A new method to solve optimization problems via fixed point of firefly algorithm. *Int. J. Bio-Inspired Comput.* **2018**, *11*, 249–256. [CrossRef]

33. Deb, S.; Deb, S.; Gao, X.Z.; et al. A new metaheuristic optimization algorithm motivated by elephant herding behaviour. *Int. J. Bio-Inspired Comput.* **2017**, *8*, 394–409.

34. Cui, Z.; Cao, Y.; Cai, X.; Cai, J.; Chen, J. Optimal LEACH protocol with modified bat algorithm for big data sensing systems in Internet of Things. *J. Parallel Distrib. Comput.* **2017**. [CrossRef]

35. Gao, M.L.; He, X.H.; Luo, D.S.; Jiang, J.; Teng, Q.Z. Object tracking with improved firefly algorithm. *Int. J. Comput. Sci. Math.* **2018**, *9*, 219–231.

36. Arloff, W.; Schmitt, K.R.B.; Venstrom, L. A parameter estimation method for stiff ordinary differential equations using particle swarm optimization. *Int. J. Comput. Sci. Math.* **2018**, *9*, 419–432. [CrossRef]

37. Cortes, P.; Guadix, J.; Muñuzuri, J.; Onoeva, L. A discrete particle swarm optimization algorithm to operate distributed energy generation networks efficiently. *Int. J. Bio-Inspired Comput.* **2018**, *12*, 226–235. [CrossRef]

38. Wang, Y.; Wang, P.; Zhang, J.; Cui, Z.; Cai, X.; Zhang, W.; Chen, J. A Novel Bat Algorithm with Multiple Strategies Coupling for Numerical Optimization. *Mathematics* **2019**, *7*, 135. [CrossRef]

39. Cui, Z.; Xue, F.; Cai, X.; Cao, Y.; Wang, G.; Chen, J. Detectin of malicious code variants based on deep learning. *IEEE Trans. Ind. Inform.* **2018**, *14*, 3187–3196. [CrossRef]

40. Niu, Y.; Tian, Z.; Zhang, M.; Cai, X.; Li, J. Adaptive two-SVM multi-objective cuckoo search algorithm for software defect prediction. *Int. J. Comput. Sci. Math.* **2018**, *11*, 282–291. [CrossRef]

41. Reyes-Sierra, M.; Coello Coello, C.A. Multi-Objective Particle Swarm Optimizers: A Survey of the State-of-the-Art. *Int. J. Comput. Intell. Res.* **2006**, *2*, 287–308.

42. Bougherara, M.; Nedjah, N.; de Macedo Mourelle, L.; Rahmoun, R.; Sadok, A.; Bennouar, D. IP assignment for efficient NoC-based system design using multi-objective particle swarm optimization. *Int. J. Bio-Inspired Comput.* **2018**, *12*, 203–213. [CrossRef]

43. Zhang, Q.; Li, H. MOEA/D: A multi-objective evolutionary algorithm based on decomposition. *IEEE Trans. Evol. Comput.* **2007**, *11*, 712–731. [CrossRef]

44. Zhang, M.; Wang, H.; Cui, Z.; Chen, J. Hybrid multi-objective cuckoo search with dynamical local search. *Memet. Comput.* **2018**, *10*, 199–208. [CrossRef]

45. Cao, Y.; Ding, Z.; Xue, F.; Rong, X. An improved twin support vector machine based on multi-objective cuckoo search for software defect prediction. *Int. J. Bio-Inspired Comput.* **2018**, *11*, 282–291. [CrossRef]

46. Cui, Z.; Zhang, J.; Wang, Y.; Cao, Y.; Cai, X.; Zhang, W.; Chen, J. A pigeon-inspired optimization algorithm for many-objective optimization problems. *Sci. China Inf. Sci.* **2019**. [CrossRef]

47. Wang, G.; Cai, X.; Cui, Z.; Min, G.; Chen, J. High Performance Computing for Cyber Physical Social Systems by Using Evolutionary Multi-Objective Optimization Algorithm. *IEEE Trans. Emerg. Top. Comput.* **2017**. [CrossRef]

48. Deb, K.; Pratap, A.; Agarwal, S.; Meyarivan, T.A.M.T. A fast and elitist multi-objective genetic algorithm: NSGA-II. *IEEE Trans. Evol. Comput.* **2002**, *6*, 182–197. [CrossRef]

49. Holland, J.H. *Adaptation in Natural and Artificial Systems*; The University of Michigan Press: Ann Arbor, MI, USA, 1975.

50. Goldberg, D.E. *Genetic Algorithms in Search, Optimization and Machine Learning*; Addison-Wesley Publishing Company: Boston, MA, USA, 1989; pp. 2104–2116.

51. Nan, G.F.; Li, M.Q.; Li, J. Estimation of Node Localization with a Real-Coded Genetic Algorithm in WSNs. In Proceedings of the 2007 International Conference on Machine Learning and Cybernetics, Hong Kong, China, 19–22 August 2007; pp. 873–878.

52. Yang, G.; Yi, Z.; Tianquan, N.; Keke, Y.; Tongtong, X. An improved genetic algorithm for wireless sensor networks localization. In Proceedings of the IEEE Fifth International Conference on Bio-Inspired Computing: Theories and Applications, Changsha, China, 23–26 September 2010; pp. 439–443.

53. Bo, P.; Lei, L. An improved localization algorithm based on genetic algorithm in wireless sensor networks. *Cogn. Neurodyn.* **2015**, *9*, 249–256.

54. Yang, X.S.; Deb, S. Cuckoo search via Levy flights. In Proceedings of the World Congress on Nature & Biologically Inspired Computing, Coimbatore, India, 9–11 December 2009; pp. 210–214.

55. Sun, B.; Cui, Z.; Dai, C.; Chen, W. DV-Hop Localization Algorithm with Cuckoo Search. *Sens. Lett.* **2014**, *12*, 444–447. [CrossRef]

56. Zhang, M.; Zhu, Z.; Cui, Z. DV-hop localization algorithm with weight-based oriented cuckoo search algorithm. In Proceedings of the Chinese Control Conference, Dalian, China, 26–28 July 2017; pp. 2534–2539.

57. Cui, Z.; Sun, B.; Wang, G.; Xue, Y.; Chen, J. A novel oriented cuckoo search algorithm to improve DV-Hop performance for cyber-physical systems. *J. Parallel Distrib. Comput.* **2017**, *103*, 42–52. [CrossRef]

58. Cui, L.; Xu, C.; Li, G.; Minga, Z.; Fenga, Y.; Lua, N. A High Accurate Localization Algorithm with DV-Hop and Differential Evolution for Wireless Sensor Network. *Appl. Soft Comput.* **2018**, *68*, 39–52. [CrossRef]

59. Chen, W.; Jiang, X.; Li, X.; Gao, J.; Xu, X.; Ding, S. Wireless Sensor Network nodes correlation method in coal mine tunnel based on Bayesian decision. *Meas. J. Int. Meas. Confed.* **2013**, *46*, 2335–2340. [CrossRef]
60. Farjow, W.; Raahemifar, K.; Fernando, X. Novel wireless channels characterization model for underground mines. *Appl. Math. Model.* **2015**, *39*, 5997–6007. [CrossRef]

mathematics

MDPI

Article

Monarch Butterfly Optimization for Facility Layout Design Based on a Single Loop Material Handling Path

Minhee Kim and Junjae Chae *

School of Air Transport, Transportation and Logistics, Korea Aerospace University, 76, Hanggonddaehak-ro, Deoyang-gu, Goyang-si, Gyeonggi-do 10540, Korea; meenykim@kau.kr
* Correspondence: jchae@kau.ac.kr

Received: 26 December 2018; Accepted: 31 January 2019; Published: 6 February 2019

Abstract: Facility layout problems (FLPs) are concerned with the non-overlapping arrangement of facilities. The objective of many FLP-based studies is to minimize the total material handling cost between facilities, which are considered as rectangular blocks of given space. However, it is important to integrate a layout design associated with continual material flow when the system uses circulating material handling equipment. The present study proposes approaches to solve the layout design and shortest single loop material handling path. Monarch butterfly optimization (MBO), a recently-announced meta-heuristic algorithm, is applied to determine the layout configuration. A loop construction method is proposed to construct a single loop material handling path for the given layout in every MBO iteration. A slicing tree structure (STS) is used to represent the layout configuration in solution form. A total of 11 instances are tested to evaluate the algorithm's performance. The proposed approach generates solutions as intended within a reasonable amount of time.

Keywords: facility layout design; single loop; monarch butterfly optimization; slicing tree structure; material handling path; integrated design

---

1. Introduction

A facility layout design has significant value to the manufacturing world. From the machine lines to the path of material handling equipment, many factors related to operation efficiency depend on the facility layout design [1]. Because a layout configuration is directly related to the material handling performance in a factory or workspace, researchers have been studying optimal facility layouts with various approaches. In academia, this came to be called the facility layout problem (FLP).

The FLP ranges over an arrangement of blocks in a given area [2]. The blocks, which are regarded as departments, can be machines, workspaces or even buildings. In an FLP, departments that are known in size have to be arranged in a given space without overlapping [3]. For cases where each department has a different size, the term unequal area facility layout problem (UAFLP) has been coined. In this study, we use the acronym FLP to mean UAFLP. Mixed integer programming (MIP) can be used to solve FLPs. However, FLPs are categorized as NP-hard because they encounter many restrictions, including location or aspect ratio limitations. These restrictions prevent FLPs from being solved in a reasonable amount of computational time [4]. Therefore, to solve large-sized problems of generally more than 12 departments, various meta-heuristic approaches have been used [5–8].

In this study, monarch butterfly optimization (MBO), a relatively new meta-heuristic, is adopted to solve the FLP. Inspired by the monarch butterfly species, MBO mimics the butterfly's group migration [9]. MBO has a simple concept and parameters relative to other meta-heuristic algorithms. The simple structure of the algorithm facilitates the implementation of complicated expressions of the layout representation in computer code. In addition, several studies have successfully demonstrated MBO's performance [10–14].

When solving an FLP using meta-heuristics, a layout representation method is necessary to construct the department sequence. Indeed, various heuristic methods concerned with layout representation have been proposed. However, pre-defined layout representations such as the flexible bay structure (FBS) or slicing tree structure (STS) have been widely used to easily represent the layout scheme. When assigning the departments to floor space, the FBS only considers one cutting direction. Therefore, the FBS is relatively easy to apply and takes less computational time. However, the STS allows both the vertical and horizontal directions. While this makes the STS more complicated and adds computational time, it also produces a greater variety of layout configurations that combine two directions than the FBS [6]. Most previous studies dealt with loop-based FLPs by using FBS representation. We adopt the STS to gain greater flexibility in solution space and to obtain better solutions.

To estimate the fitness value of the obtained layout, various evaluation methods can be applied. Typically, a rectilinear distance-based approach is used [15] to evaluate the results. However, in this study, a single loop distance is applied to evaluate the layout. This single loop can be regarded as route-of-path-based material handling equipment such as automated guided vehicles (AGVs) or power and free systems. This kind of measurement can aid the design of certain layouts operated with path-based vehicles. We define a single loop distance as the size of a single loop where the material handling equipment can access all of the departments through the path, and the path is never crossed.

To find a single loop in the obtained layout, the single loop construction method is proposed in this study. Before constructing a single loop, a department-searching procedure is conducted to determine a feasible layout. After searching every adjacent department, the single loop department is determined based on the number of adjacent departments. If there are no more departments to be included, the single loop construction method can be finished, and a single loop is obtained. This process is embedded in every MBO iteration, and single loop size minimization becomes the main objective of the layout design.

This paper proceeds as follows. The research background and the literature review are introduced in Section 2, Section 3 deals with the layout representation method and the adoption of MBO in the FLP. Section 4 defines the MBO approach with respect to the FLP. The computational results of this research and the conclusion are described in Sections 5 and 6, respectively.

2. Background and Literature Review

According to Kochhar et al. [16], because the importance of facility layout has come to the fore, the FLP has been widely studied. As mentioned, the FLP deals with the placement of departments in a given area without overlapping [17]. Typically, a good facility layout is directly related to the material handling cost. Therefore, the objective function value (OFV), which is used to evaluate the layout configuration, is also related to the movement between departments [18]. Several distance measurements exist, as shown in Figure 1. The rectilinear distance in Figure 1a is composed of the absolute distance between two datum points of the department on both the x and y axis. On the contrary, the Euclidean distance, which is the shortest diagonal distance, can be seen in Figure 1d [19]. The contour distance is described in Figure 1b. In this measurement, the movement between two departments must occur along the adjacent department's boundary. Figure 1c deals with the single loop distance. As previously defined, all departments should be accessible through a single loop. Therefore, in this case, the single loop distance can be the OFV of the layout.

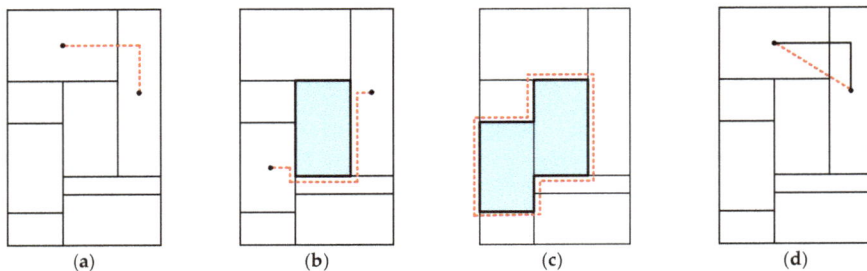

Figure 1. Examples of distance measurement. (**a**) Rectilinear distance; (**b**) Contour distance; (**c**) Single loop distance; (**d**) Euclidean distance.

However, in terms of operating circulating path-based material handling systems, the layout configured for the shortest single loop might be helpful. For instance, the derived single loop path can be regarded as a practically appropriate planned path for vehicle-based systems [20]. In this way, the FLP with the single loop can be an industrial alternative to that of general FLPs that deal with traditional distances between departments. Moreover, the single loop is much simpler than other loop networks because there are no interactions or departments along the path [21].

Two early approaches to the single loop under an FLP were conducted by Tanchoco and Sinriech [22] and Sinriech and Tanchoco [21] in the 1990s. They considered a single loop to be a more effective and cost-saving alternative for material handling equipment operation. In their papers, the optimal single-loop (OSL) guide paths for AGVs were proposed for the single closed loop guide path layout. A valid single loop problem and single loop station location problem were developed by the authors and worked well with the general FLPs. Asef-Vaziri et al. [23] defined an FLP with a single loop as the shortest loop design problem (SLDP) and introduced improved integer linear programming (ILP). To demonstrate its performance, several problems were tested. Subsequently, Ahmadi-Javid and Ramshe [24] corrected the ILP formulation and developed a cutting-plane algorithm to generate reasonable computational results. Over time, other researchers have tried to develop a variety of powerful meta-heuristics to obtain better layout solutions in less computational time. In Yang and Peters' research [25], a two-step heuristic approach was proposed for an open-field type layout with a single loop path. Their approach combined a space-filling curve with simulated annealing (SA). In the first step, a traditional block layout with a directed loop was solved with the combined heuristic approach. Next, an MIP formulation was solved using the spatial coordinate and orientation location inputs from the first stage. Hojabri et al. [26] suggested a decomposition algorithm for solving a single loop material flow. In the process of finding the solution, all departments were located along a loop and removed randomly from that loop until a feasible loop was found. This proposed algorithm was successful in solving large problems in a short amount of time. Jahandideh et al. [27] introduced a genetic algorithm (GA) to design a unidirectional loop in an FLP by considering the total loaded and empty trips. Asef-Vaziri et al. [28] proposed a single loop-based facility layout design. They used a hybrid GA and noted that the computational time could be effectively reduced even with a large number of departments.

Additional forms of single loop FLPs were studied by other researchers. Chae and Peters [29] used a single loop structure as a department placement guideline and applied SA to search for solutions. Niroomand et al. [30] built on Chae and Peters' department placement strategy and used a modified migrating birds optimization to find a better layout configuration for a closed loop-based facility layout problem (CLLP). Kang et al. [31] applied the cuckoo search algorithm to a CLLP and showed that the algorithm generated quality solutions for the CLLP. More details about loop-based FLPs can be read in the review papers provided by Asef-Vaziri and Laporte [32] and Asef-Vaziri and Kazemi [20].

To solve FLPs using various meta-heuristic approaches, a proper layout representation should be encoded to adopt the meta-heuristics. Various layout representations have been proposed. For example,

Kulturel-Konak and Konak [33] proposed location/shape representation, which employs the shape and centroid coordinates of the departments. Gonçalves and Resende [19] suggested the empty maximal spaces (EMSs) method as a layout representation. The EMSs method lists every department's minimum and maximum vertex. Instead of developing new heuristics, pre-defined and well-known layout representation methods have also been used. In essence, the FBS and STS methods are based on the cutting action and the combination of this numeric information. The FBS is a common layout representation method. It is relatively simple to adopt because it only locates the departments in one bay direction, horizontal or vertical [4]. In contrast, both horizontal and vertical directions are allowed in the STS [34]. Therefore, the STS can show more diverse layout configurations than the FBS [5]. To broaden the search space and find better solutions, we adopt the STS. The STS is explained in Section 4.1 in detail.

For the solution process, a recently introduced meta-heuristic algorithm, Wang et al.'s monarch butterfly optimization (MBO) [9], is used to solve the FLP in our study. MBO mimics the movement of the monarch butterfly species between eastern North America and Mexico. Basically, the butterflies are divided into two groups that are represented as Land 1 and Land 2. The butterflies in these two groups migrate separately according to (1) the migration operator and (2) the butterfly adjusting operator. The two operators each have a role in maintaining and developing butterfly individuals to achieve better solutions. MBO has relatively less computation time and decision parameters, and several studies have adopted MBO for various optimization problems. Wang et al. [14] proposed a new version of MBO with a self-adaptive crossover operator and greedy strategy. In its original form, MBO updates the newly generated butterfly without any restrictions. However, to improve the fitness value and convergence, a new solution can be updated when the fitness value is better than the previous one. Ghanem and Jantan [35] combined MBO with artificial bee colony (ABC) optimization to achieve global solution searching ability. To avoid the proposed algorithm falling into the local optimal solution, a modified butterfly adjusting operator is used as a mutation operator in ABC. Chen et al. [10] introduced MBO with the greedy strategy to solve the dynamic vehicle routing problem (DVRP), which is a transformed version of a VRP with dynamic customer appearance. The introduced algorithm outperformed the existing methods while drawing several new best solutions. By all accounts, MBO performed reasonably well with various optimization instances. Therefore, due to these advantages, we apply MBO to solve our layout design based on the shortest single loop path.

3. Problem Description

As mentioned, the FLP deals with the arrangement of departments in a given floor space. The general version of the FLP can be formulated mathematically as an MIP model to find the optimal solution [36]. Following this, many researchers modified MIP models to increase effectiveness and efficiency [37–39]. We introduce a conceptual model for comprehension. For the material handling equipment, which circulates along the single loop, the size of the loop is very important. Therefore, the objective function is to minimize the size of the single loop. The parameters and variables used in this study are introduced in Table 1.

Table 1. Parameters and variables.

Parameters and Variables	
n	The number of departments
L^s	The x or y size of the floor space, where $s = \{x, y\}$
l_i^s	The x or y size of the department i
lb_i^s, ub_i^s	The valid lower bound and upper bound of l_i^s
g_i^s	The centroid coordinate of department i in a direction of s
c_i^s	The south-west coordinate of department i in a direction of s
sld	The size of the single loop.
θ_{ij}^x	1, when the department i is on the left of the j. 0, otherwise
θ_{ij}^y	1, when the department i is lower than that of the j. 0, otherwise

$$\text{Minimize } sld = F^*(S) \tag{1}$$

$$\text{s.t. } a_i = l_i^x \times l_i^y, \forall i \tag{2}$$

$$lb_i^s \leq l_i^s \leq ub_i^s, \forall i, s = x, y \tag{3}$$

$$\sum_{s=x}^{y} \left(\theta_{ij}^s + \theta_{ji}^s \right) = 1, \forall i \neq j \tag{4}$$

$$g_i^s + \frac{l_i^s}{2} \leq g_j^s - \frac{l_j^s}{2} + L^s \left(1 - \theta_{ij}^s \right), \forall i \neq j, s = x, y \tag{5}$$

$$l_i^s / 2 \leq g_i^s \leq L^s - l_i^s / 2, \forall i \neq j \tag{6}$$

$$\theta_{ij}^s \in \{0, 1\}, \forall i \neq j, s = x, y \tag{7}$$

The objective function (1) deals with the size of the single loop, where S indicates the feasible layout and it is determined based on the Equations (2) to (7). The responsive single loop size is shown as the function in (1), $F(S)$. Thus, the final goal of the model is to find the layout which can provide the shortest single loop material handling path. As mentioned, these constraints are used to solve a general FLP with the rectilinear distance metric, specifically the UAFLP, after linearizing the area constraint, (2), and making the form an MIP [36–39]. Equation (3) restricts the size of each department, and (4) and (5) place the department in the given floor space without overlapping. All of the placed departments should be within a given floor space by Equation (6).

However, the layout configuration determined in the slicing tree structure, which we use in this study, is a little different from the layout generated by the MIP model. The basic structure of the slicing tree and the method of searching for the layout design are explained in Sections 4.1 and 4.2, respectively.

As mentioned, the models to find the shortest loop in a given layout were proposed by several studies [23,24,40,41]. In this study, searching for the shortest loop in the layout should proceed in every iteration of the layout design process. Thus, we introduce a simple conceptual model to explain how the shortest loop could be constructed in a given layout using the STS.

The smallest size of a loop can be obtained by evaluating possible single loops in a given layout. Thus, the objective function $F^*(S)$, as shown in Equation (8), indicates the minimum value among the possible single loop. The measure of the single loop size is calculated as follows, where T indicates the set of departments placed inside of the loop boundary.

$$F(S) \leftarrow \sum_{i \in T} 2 \left(l_i^x + l_i^y \right) - \sum_{i \in T} \sum_{j \in T} 2 \left(dl_{ij}^x + dl_{ij}^y \right) \tag{8}$$

The left term indicates the sum of the perimeter of departments that are in T, and the right term indicates the sum of the border of departments i and j, which are in T and adjacent to each other. Finally, this measure should return the minimum value to the objective function (1).

An example of this calculation is shown in Figure 2. As shown in Figure 2, the size of the loop can be calculated as the sum of the perimeter of the two departments. If we calculate separately in the x and y directions, it is equivalent to twice of length in the x direction (Figure 2a) plus twice the length in the y direction excluding the shared border of the departments (Figure 2b). The x axis contains the distances of $x(1-2)$, $x(3-4)$, $x(5-6)$, and $x(6-7)$. In this case, there are no duplicated distances between the adjacent departments. The y axis contains the distances of $y(1-2)$, $y(3-4)$, $y(4-5)$, and $y(6-7)$. In this case, the distance of $y(4-5)$ is shared by two adjacent departments. Therefore, this duplicated distance (dl_{ij}^{y}) should be subtracted. The measurement criterion of dl_{ij}^{s} is described in detail in Section 4.3.

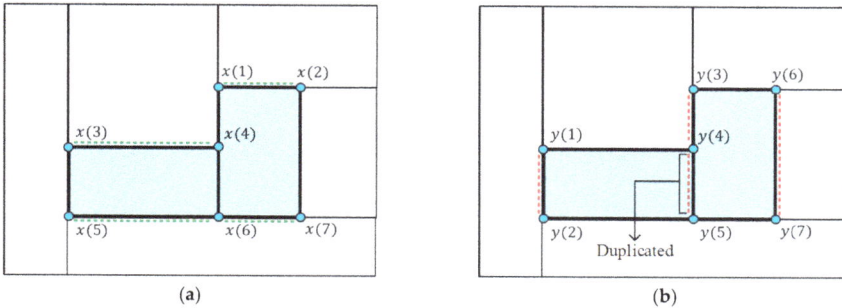

Figure 2. Loop distance measurements. (**a**) On the x axis; (**b**) On the y axis.

4. Layout Design with MBO and Single Loop Construction

As explained in Section 3, the problems we deal with are divided into two parts: the layout design and the single loop construction. In the layout design part, we use the STS to obtain the basic layout configuration, and we determine the single loop for a given layout in the single loop construction section. MBO is used to search for a better layout, which forms the shorter single loop.

4.1. Layout Representation

An FLP needs a layout representation method to be shown its layout in a certain code. The STS is adopted as a layout representation in our study. In principle, the STS has more encoding vectors than FBS, which represent components of the STS such as the department and slicing cutting sequence. Therefore, generally, the FBS has been adopted for layout representation because of its relative simplicity [27,28]. However, by allowing both directions, the STS can offer a greater variety of layouts than the FBS, as shown in Figure 3.

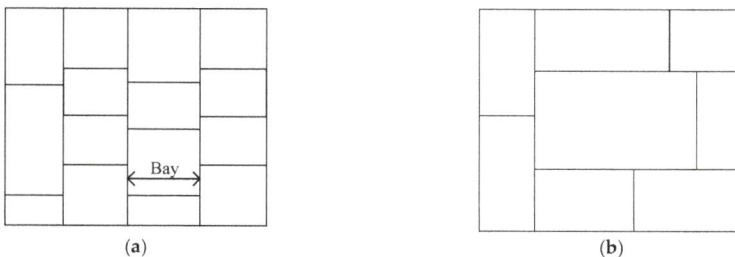

Figure 3. Layout representation. (**a**) Flexible bay structure; (**b**) Slicing tree structure.

Basically, the STS is composed of an internal and external node. The internal node represents the horizontal and vertical cutting direction. The external node, on the other hand, represents the sequence of departments in a layout [6].

As shown in Figure 4, the horizontal and vertical directions are symbolized as binary values. This type of STS with binary vectors was originally from [42,43]. In the authors' papers, the STS sequence of a layout with n number of departments can be represented with n leaves and $n - 1$ nodes. In this study, an STS with three encoding sequences is adopted, as described in Figure 5. First, the department sequence (1-3-2-5-4-6) is horizontally divided into two branches (1-3 and 2-5-4-6) because the cut orientation is 2 with a cut code of 0. Next, from the given cut orientation code of 1, the sub-branch (1-3) is separated into individual departments. At this point, the inverse vertical cut occurs because the cut code is 3 rather than 1. Accordingly, the sub-branch (1-3) is separated into 3 and 1 sequentially. The other sub-branches separate in a likewise fashion. In the third step, the sub-branch (2-5-4-6) is divided into two parts vertically, (2-5) and (4-6). Sequentially, the sub-branch (2-5) is horizontally divided into individual departments (2) and (5). The other sub-branch, (4-6), is divided into (4) and (6) by a vertical cut.

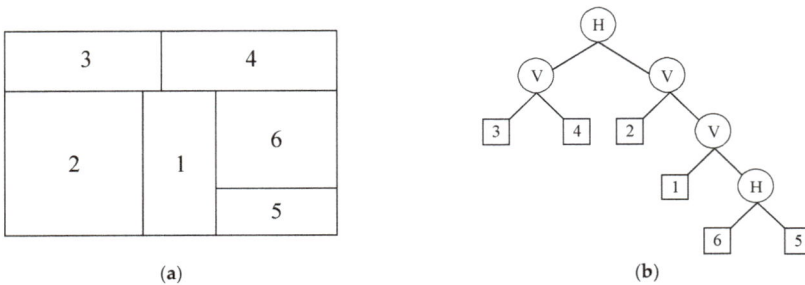

Figure 4. An example of the slicing tree structure. (**a**) Layout representation; (**b**) Tree structure.

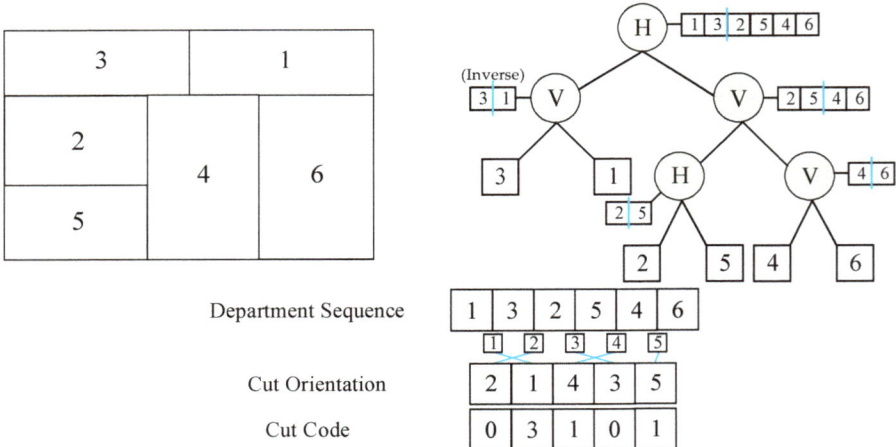

Figure 5. Slicing tree encoding scheme.

Komarudin and Wong [44] first introduced this type of STS, which is composed of the department sequence, the cut orientation, and the cut code. The department sequence indicates the order of departments in a line. The cut orientation shows the location where the horizontal or vertical cut occurs. Lastly, the cut code represents the cutting direction, and this is usually composed of binary vectors. Using this structure as a basis, Kang and Chae [5] extended the cut code from 0 to 3 to represent the inverse assignment and decrease the computation time. Likewise, in this study, the cut code is represented as 0 for the horizontal direction, 1 for the vertical direction, 2 for the inverse horizontal direction and 3 for the inverse vertical direction.

4.2. Monarch Butterfly Optimization

MBO was introduced by Wang et al. [9] for the first time. Inspired by the behavior of the monarch butterfly, MBO mimics the migration patterns of this species. The monarch butterfly migrates from the northern USA and southern Canada to Mexico during the summer and autumn. In this regard, the migrating butterfly population can be divided into two groups: Land 1 and Land 2. To simplify the migration of the monarch butterfly, the butterfly individuals stay in Land 1 for four months and Land 2 for seven months. The butterfly individuals in Land 2 move to Land 1 in April. On the other hand, the butterfly individuals in Land 1 move to Land 2 in September. Accordingly, the butterflies in Land 1 (NP_1) and Land 2 (NP_2) compose the total butterfly population (NP). This can be expressed as Equations (9) and (10).

$$NP_1 = \lceil p * NP \rceil \tag{9}$$

$$NP_2 = NP - NP_1 \tag{10}$$

p stands for the rate (%) of butterfly individuals staying in Land 1. Therefore, the subpopulation in Land 1 can be expressed as $\lceil p * NP \rceil$ where a is the nearest integer that is greater than or equal to a.

After dividing the population, the new butterfly individuals in Land 1 and Land 2 can be generated by each operator in parallel (i.e., the migration operator and butterfly adjusting operator). After the operator application, the fitness values of each generated individual are evaluated. If the newly generated individual has a better fitness value than the previous one, the old parent individual is replaced with the new one via the greedy strategy, as in Equation (11) [10].

$$x_{i,new}^{t+1} = \begin{cases} x_i^{t+1}, & \text{if } f\left(x_i^{t+1}\right) < f\left(x_i^t\right) \\ x_i^t, & \text{otherwise} \end{cases} \tag{11}$$

The overall process of the updating scheme can be seen in Figure 6.

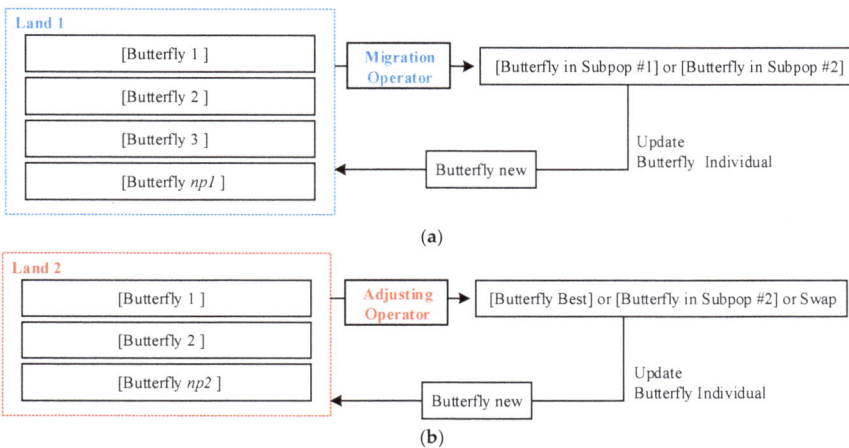

Figure 6. Updating process of MBO. (**a**) Migration operator; (**b**) Butterfly adjusting operator.

As mentioned, butterflies in Land 1 and Land 2 undergo their updating process simultaneously. The migration operator, which is the operator in Land 1, is driven by a modified random number r, which is a random decimal fraction multiplied by *peri*. According to the original author, *peri* can be regarded as a value of 1.2, which reflects the migration period of a 12-month a year.

Operator 1: Migration Operator

Begin
    for (all monarch butterfly in Subpopulation 1)
        r = random number $*$ *peri*
        if $(r \leq p)$
            $x^{t+1} = x^t_{r_1}$ in Subpopulation 1   (12)
        Else
            $x^{t+1} = x^t_{r_2}$ in Subpopulation 2   (13)
        End
    End
End

In Equation (12), x is an individual butterfly that has three separate representations: a department sequence, cut orientation and cut code. t is the current generation. Thus, x^t is a butterfly of the generation t. Similarly, $x^t_{r_1}$ indicates an individual butterfly that is randomly selected from Subpopulation 1, and $x^t_{r_2}$ indicates the one from Subpopulation 2. The new individual from Land 1, x^{t+1}, is randomly selected from Subpopulation 1 if r is less than or equal to p, as shown in Equation (12). If r is greater than p, a new butterfly is selected in Subpopulation 2, as shown in Equation (13). p, the butterfly adjusting ratio, is set to 0.41 (5/12) in accordance with the migration period in Land 1 [9].

The butterflies in Land 2 are updated via the other updating operator, the butterfly adjusting operator.

Operator 2: Butterfly Adjusting Operator

Begin
    for (all monarch butterfly in Subpopulation 2)
        rand = random number
        if $(rand \leq p)$
            $x^{t+1} = x^t_{best}$ (14)
        Else
            $x^{t+1} = x^t_{r_3}$ in Subpopulation 2   (15)
            if $(rand > bar)$
                swap two components in the   (16)
random Subpopulation
            End
        End
    End

For Subpopulation 2-oriented butterflies, if the random number *rand* is lower than or equal to p, the current best butterfly individual can exist in a new generation, as in Equation (14). This ensures that the butterfly, which provides the best fitness value, continues to the next generation. On the other hand, if *rand* is greater than p, a randomly selected butterfly in Land 2 can become a new butterfly, as in Equation (15). But in the case of Equation (16), this can be modified by a butterfly adjusting rate, *bar*. If *rand* is greater than *bar*, two components in the parent butterfly are swapped randomly and this generates a new offspring. The process of swapping two departments is described in Figure 7.

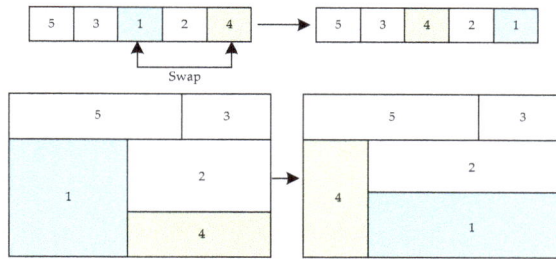

Figure 7. Swapping process.

By using this meta-heuristic approach with the STS, the evaluation method of an obtained layout is described in the next section.

4.3. Loop Construction Method

The single loop construction method is heuristically constructed in this study. The adjacent rule, which helps to find the adjacent relationship between departments, is modeled, and the loop sequence heuristic is established using the adjacent rules. For the candidate department i and the relevant department j, the adjacent rules can be defined as below.

According to Figure 8, c_i^x is a coordinate from the west corner of Department i on the x axis. In the same way, c_i^y is a coordinate from the south corner of Department i on the y axis. l_i^x and l_i^y are the horizontal and vertical length of Department i on each axis, respectively. The following description is an example of the adjacent rules between Departments i and j on the y axis. At least one adjacent rule must be satisfied. The examples of adjacent rules can be seen in Figure 9. In this process, adjacent duplication is allowed because the reverse adjacent case of the departments must be considered. As shown in Figure 10, if the departments only share a vertex, this case is not regarded as an adjacency.

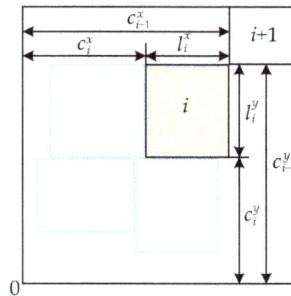

Figure 8. The example of coordinates.

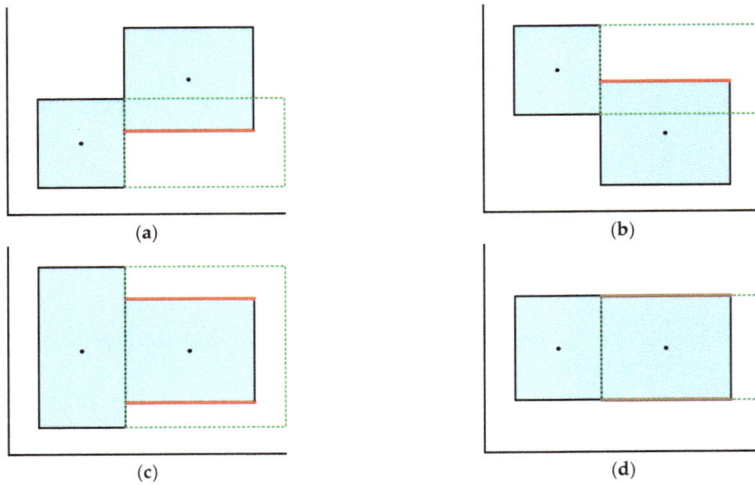

Figure 9. Examples of adjacent cases. (**a**) Upper adjacent; (**b**) Lower adjacent; (**c**) Dual adjacent-1; (**d**) Dual adjacent-2.

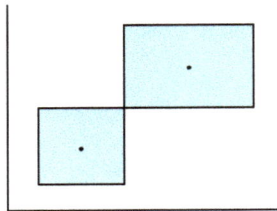

Figure 10. A case with non-contiguous departments.

- If $c_i^y \leq c_j^y < c_i^y + l_i^y$, Departments i and j have an upper adjacent relationship.
- If $c_i^y < c_j^y + l_j^y \leq c_i^y + l_i^y$, Departments i and j have a lower adjacent relationship.
- The upper two conditions are satisfied with a dual adjacent-1 relationship.
- If $c_i^y = c_j^y$ and $l_i^y = l_j^y$, Departments i and j have a dual adjacent-2 relationship.

With these adjacent cases, the duplicated distance of Departments i and j (dl_{ij}^s) can be represented as follows.

(a) Upper adjacent $\quad\quad\quad\quad\quad\quad\quad\quad\quad$ $dl_{ij}^s = \left| l_i^s - c_j^s \right|$

(b) Lower adjacent $\quad\quad\quad\quad\quad\quad\quad\quad\quad$ $dl_{ij}^s = \left| l_i^s - \left(l_j^s + c_j^s - c_i^s \right) \right|$

(c-1) Dual adjacent -1 $\quad\quad\quad\quad\quad\quad\quad$ $dl_{ij}^s = \left| l_i^s - \left((l_j^s + c_j^s) - (c_i^s + l_i^s) \right) - \left(c_j^s - c_i^s \right) \right|$

(c-2) The opposite case of dual adjacent -1 $\quad$ $dl_{ij}^s = \left| l_j^s - \left((l_j^s + c_j^s) - (c_i^s + l_i^s) \right) - \left(c_i^s - c_j^s \right) \right|$

(d) Dual adjacent -2 $\quad\quad\quad\quad\quad\quad\quad\quad$ $dl_{ij}^s = l_i^s = l_j^s$

The single loop construction method follows a number of steps. In the initial stage, all departments are regarded as loop candidate departments, and each department, except the comparison target, must be compared to determine if it is adjacent to the target or not. The illustration of a single loop construction is shown in Figure 11.

- Step 0. Initialize all loop measuring sequences.
- Step 1. For all loop candidate departments, explore every adjacent department using the adjacent rules. The adjacent departments are stored in the adjacent row of each candidate target. Once the exploration is done, eliminate the duplicated departments in each adjacent row.
- Step 2. When the exploration is finished, choose one candidate department (loop department) that has the most adjacent departments and put it in the loop sequence list. Remove the adjacent row of the candidate department from the adjacent row group.
- Step 3. Remove the adjacent departments that belong to the chosen loop department from each of the adjacent rows.
- Step 4. Remove the loop department and adjacent departments that belong to the loop department from the adjacent waiting list.
- Step 5. Iterate Steps 2 to 4 until there are no remaining departments in the adjacent waiting list. When selecting the loop candidate department, it must contain the loop department as an adjacent department.
- Step 6. Measure the loop size with the departments in the loop sequence.

All lists and rows must be initialized in Step 0. Following Figure 11a, the adjacent departments are explored and stored in the adjacent row of the corresponding target department. Afterwards, in Figure 11b, the adjacent row of Department 1 with the largest number of adjacent departments is chosen to be a loop department, and this can be stored in the loop list. The adjacent departments (2, 4, 5, 6, and 7) in the adjacent row of Department 1 are now located on the loop. Therefore, they can be erased from the adjacent waiting list. However, there is still a remaining department (3) that is not located on the loop, and the next iteration must occur. Shown in Figure 11c, the adjacent row of Departments 4 and 6 have the same number of remaining departments. At this time, to minimize the loop size, the department that makes the entire size of the loop smaller can be chosen. Therefore, Department 6 can be selected. Repeating the previous iteration, Department 6 is selected as a loop department, and the adjacent department (3) can be removed from the adjacent waiting list. Because the adjacent waiting list is empty, the loop list is completed.

The single loop construction strategy proposed in this study is relatively easy to understand. Moreover, this seems to be a reasonable method to find a loop department in the layout that is constructed by using the STS because the STS can represent complicated layouts.

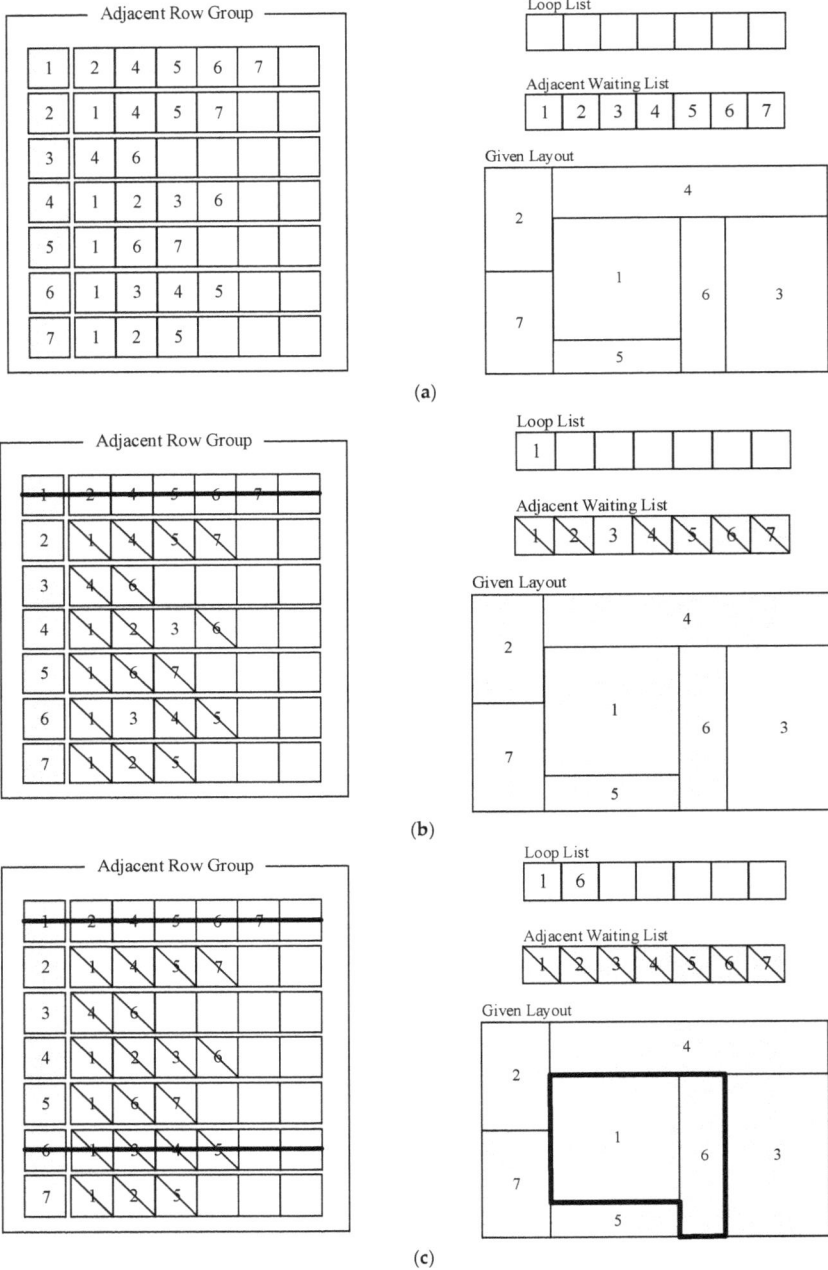

Figure 11. Single loop construction steps. (**a**) Step 1; (**b**) Steps 2–4 with Iteration 1; (**c**) Steps 2–4 with Iteration 2.

5. Computational Results

5.1. Experiment Information

The computational experiments with a set of well-known instances were tested to evaluate the performance of the proposed algorithms. The program was coded in JAVA, and the experiment was conducted on a computer with an Intel Core i5 CPU processor (3.5 GHz) and 8 GB of memory. The parameter setting for the experiment can be seen in Table 2.

Table 2. Parameters used in MBO.

Name	Rate
NP	No. of departments
p	5/12
bar	5/12
$peri$	1.2

NP, the total number of butterfly individuals, is set as equal to the number of departments of each instance. The size of NP correspondingly increases when the amount of instances increases to secure the diversification of a solution set. The remainders are set following the original parameter setting of [9]. Many studies using MBO [10–12,14,35] have adopted this same parameter setting because the original setting was based on the bio-inspired migration rate of the monarch butterfly.

Information regarding well-known instances is introduced in Table 3. From FO7 to AB20, the sum of every department's size is the same as the given floor space. For the vC10s problem, the minimum length of the department is given as 5 instead of the aspect ratio. KC15 and KC25 are introduced in this study for the first time to estimate the performance of the algorithm under the problem diversity. The area information of KC15 and KC25 can be confirmed in Tables A1 and A2. The compactness of SC30 and SC35 are less than 1.0. Simply speaking, the sum of the department's area sizes is less than the given floor size. Therefore, the empty spaces are regarded as dummy departments in this study. It is unnecessary for the dummy departments to be located along the loop, and their aspect ratio is assumed to be free.

Table 3. The instance information.

Name	Number of Departments	Floor Space (W × H)	Shape Constraint	Reference
FO7	7	8.54 × 13.00	ar = 5	Meller et al. [37]
FO8	8	11.31 × 13.00	ar = 5	Meller et al. [37]
FO9	9	12.00 × 13.00	ar = 5	Meller et al. [37]
vC10s	10	25.00 × 51.00	l^{min} = 5	van Camp et al. [45]
vC10a	10	25.00 × 51.00	ar = 5	van Camp et al. [45]
KC15	15	2.00 × 3.00	ar = 3	This study
AB20	20	2.00 × 3.00	ar = 4	Armour and Buffa [46]
KC25	25	3.00 × 3.00	ar = 4	This study
SC30	30	15.00 × 12.00	ar = 5	Liu and Meller [47]
SC35	35	16.00 × 15.00	ar = 4	Liu and Meller [47]
DU62	62	117.124 × 117.124	ar = 4	Dunker et al. [48]

5.2. Experiment Results

Several well-known instances from 7 to 62 departments are tested to evaluate the performance of the algorithm. The best OFV, the average OFV, and the corresponding computation time can be seen in Table 4. Since the STS can show the cutting style that can be obtained from both FBS and STS, as shown in Figures 12 and 13, the obtained layouts of FO7, FO8, FO9, and vC10a tend to have a bay-cutting layout form, which can also be derived by the FBS. On the other hand, the rest of the layouts are composed of both vertical and horizontal cuts that can be easily seen in the STS representation, as shown in Figures 13–16. It can be said that that as the number of departments increases, the layout tends to have an STS-like look.

Table 4. The experiment results.

Name	Best OFV	Best's CPU Time	Average OFV	Average CPU Time
FO7	12.10	10.56	13.10	11.03
FO8	12.01	9.10	16.80	13.82
FO9	12.82	12.85	15.14	14.37
vC10s	72.84	30.13	80.24	41.62
vC10a	43.81	35.03	48.16	38.74
KC15	4.89	40.45	6.12	41.82
AB20	6.00	91.42	8.97	93.52
KC25	9.18	103.59	13.05	118.40
SC30	36.78	412.38	59.13	406.67
SC35	54.50	778.52	86.24	711.49
DU62	509.41	1151.37	722.63	1104.65

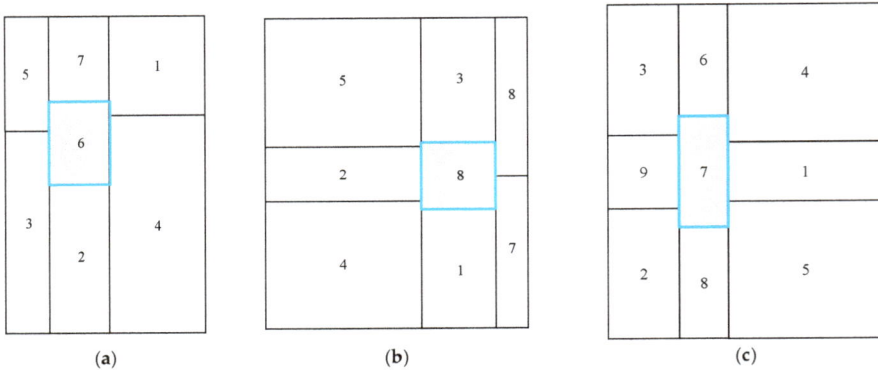

Figure 12. Layout configurations. (**a**) FO7; (**b**) FO8; (**c**) FO9.

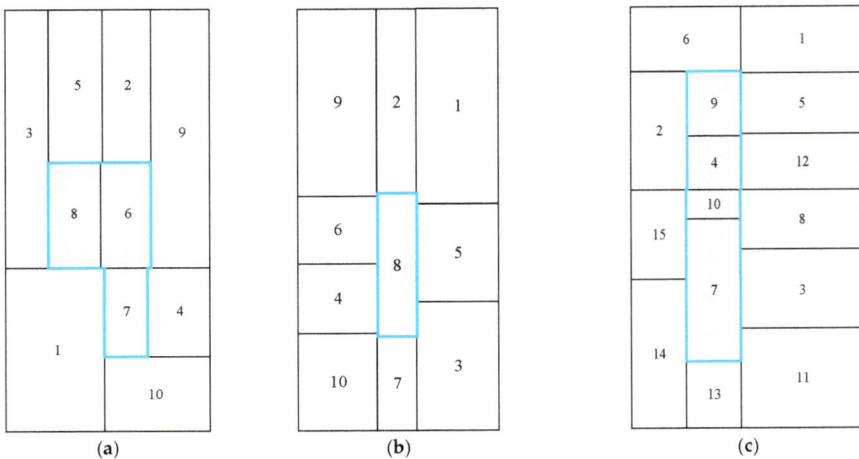

Figure 13. Layout configurations. (**a**) vC10s; (**b**) vC10a; (**c**) KC15.

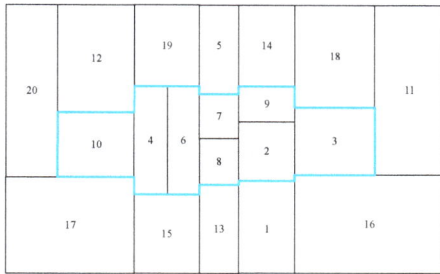
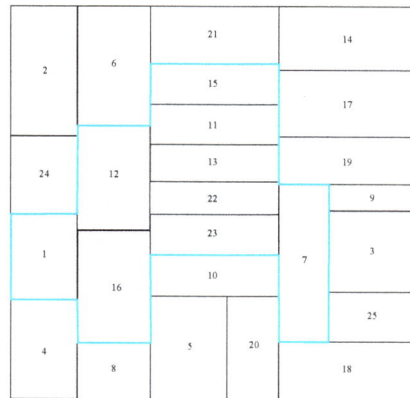

(a)   (b)

Figure 14. Layout configurations. (**a**) AB20; (**b**) KC25.

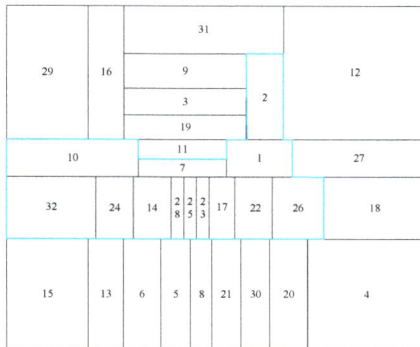
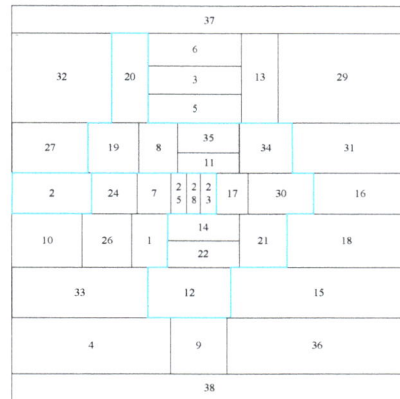

(a)   (b)

Figure 15. Layout configurations. (**a**) SC30; (**b**) SC35.

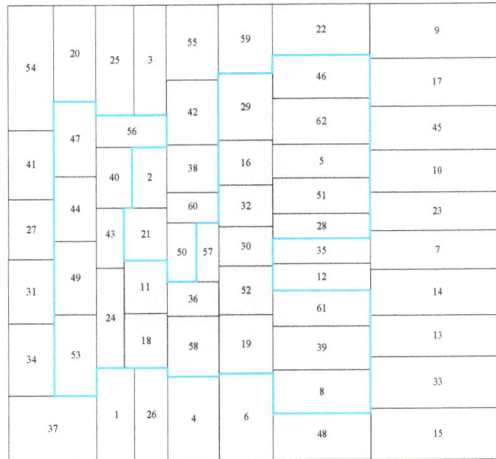

Figure 16. Layout configuration for DU62.

To the best of our knowledge, there is no previous literature that solves the UAFLP with the objective of determining the shortest single loop distance. Therefore, a direct comparison to the objective function used in other methods would not be appropriate. However, a comparison of computation times for similar instances would offer a clue in terms of the performance of the proposed MBO. Table 5 compares the present results with those provided by Asef-vaziri et al. [28]. Their approach is similar in that the pre-defined layout representation is used with a meta-heuristic to determine the layout design based on a single loop. The objective function and the values in [28] are different, as mentioned. The objective of [28] includes the loaded and empty flow between pairs of I/O points, while this study finds the shortest single loop material handling path. There are several factors that affect computation time. The computer system could be one of them. The results in [28] are generated from an 8GB RAM computer with a core-i7 CPU, while this study uses an 8GB RAM computer with a core-i5 CPU. Data structure also affects the computation time. AB20 is the only instance that uses both methods, and the proposed method provides a slightly faster CPU time. We are able to confirm that the proposed heuristic and MBO in this study can solve the problem in a relatively reasonable computation time, as shown in Table 4.

Table 5. The computation time comparison.

This Study			Asef-vaziri et al. [28]		
Problem	Average OFV	Average CPU time	Problem	Average OFV	CPU time
FO7	13.10	11.03			
FO8	16.80	13.82			
FO9	15.14	14.37			
vC10s	80.24	41.62	vC10	2.32	24.00
vC10a	48.16	38.74			
KC15	6.12	41.82	ML15	2.22	60.00
AB20	8.97	93.52	AB20	1.25	101.00
KC25	13.05	118.40	ML25	2.03	181.00
SC30	59.13	406.67	TA30	1.57	471.00
SC35	86.24	711.49	KG37	2.06	507.00
DU62	722.63	1104.65	DE62	1.38	1487.00

The layout configurations of SC30 and SC35 are shown in Figure 15. The area compactness of SC30 and SC35 is less than 1. That means the floor space is greater than the total department area. Thus, the dummy departments are included to solve the problem in the given layout structure. The departments, for which the numbers are greater than the given problem size in Figure 15, are the dummy department. For instance, department 38 in Figure 15b is one of the dummy departments, and thus, the single loop does not need to proceed through to any segment of the department. Figure 16 shows the layout configuration of DU62, and this problem does not include the dummy department.

6. Conclusions

Facility layout problems (FLPs) are placement problems that consider changes in the width and height of non-overlapping departments in a certain space. Several distance measurements, including the rectilinear distance or Euclidean distance, have been suggested to evaluate the obtained layout. However, these distance measurements are not appropriate for certain material handling systems such as circulating AVGs or power and free systems. Therefore, to determine a layout that considers the usage of circulating path-based equipment, we proposed a single loop construction method with a meta-heuristic, MBO. The MBO mimics the monarch butterfly's migration patterns. Its performance has been verified in various optimization problems. The MBO is relatively easy to apply and modify compared to other evolutionary algorithms because it has very simple operators and parameter sets.

The objective function of this study is the single loop size minimization for the circulating material handling equipment path. To evaluate the objective function values (OFVs), a single loop construction method that finds the single loop while minimizing the OFVs was introduced.

To evaluate the algorithm performance, several well-known instances were tested. MBO successfully found the single loop in a reasonable amount of time. The proposed algorithm tends to generate favorable solutions. Further studies can search for better layout placements considering better loop construction heuristics with reduced computation times. Additionally, improving MBO with other powerful methods could yield potential developments. As mentioned, to generate better solutions with MBO, some researchers took specific operators from other meta-heuristic algorithms. It might be a meaningful challenge to find different operators which fit with MBO.

Author Contributions: Conceptualization, M.K. and J.C.; methodology, M.K.; software, M.K.; validation, M.K. and J.C.; formal analysis, M.K.; investigation, M.K.; resources, J.C.; data curation, M.K.; writing—original draft preparation, M.K.; writing—review and editing, J.C.; visualization, M.K.; supervision, J.C.; project administration, J.C.; funding acquisition, J.C.

Funding: This work was supported by a 2017 Korea Aerospace University faculty research grant.

Conflicts of Interest: The authors declare no conflict of interest.

Appendix A

The instances used in this study is basically brought from other papers. However, two instances were created in this study for department size of 15 and 25. The specific area information of the new instances as follows.

Table A1. The department area of KC15.

Department number	1	2	3	4	5	6	7	8	9	10	11	12	13	14	15
Area	0.5	0.4	0.6	0.18	0.45	0.45	0.45	0.45	0.21	0.1	0.74	0.42	0.25	0.5	0.3

Table A2. The department area of KC25.

Department number	1	2	3	4	5	6	7	8	9	10	11	12	13	14	15
Area	0.33	0.5	0.4	0.38	0.45	0.49	0.45	0.24	0.14	0.3	0.3	0.42	0.25	0.5	0.3
Department number	16	17	18	19	20	21	22	23	24	25					
Area	0.45	0.51	0.44	0.34	0.3	0.43	0.24	0.3	0.3	0.24					

References

1. Tompkins, J.A.; White, J.A.; Bozer, Y.A.; Tanchoco, J.M.A. *Facilities Planning*, 4th ed.; John Wiley & Sons: Hoboken, NJ, USA, 2010.
2. Kusiak, A.; Heragu, S.S. The facility layout problem. *Eur. J. Oper. Res.* **1987**, *29*, 229–251. [CrossRef]
3. Ingole, S.; Singh, D. Unequal-area, fixed-shape facility layout problems using the firefly algorithm. *Eng. Optim.* **2017**, *49*, 1097–1115. [CrossRef]
4. Tate, D.M.; Smith, A.E. Unequal-area facility layout by genetic search. *IIE Trans.* **1995**, *27*, 465–472. [CrossRef]
5. Kang, S.; Chae, J. Harmony search for the layout design of an unequal area facility. *Expert Syst. Appl.* **2017**, *79*, 269–281. [CrossRef]
6. Drira, A.; Pierreval, H.; Hajri-Gabouj, S. Facility layout problems: A survey. *Annu. Rev. Control* **2007**, *31*, 255–267. [CrossRef]
7. Shouman, M.A.; Nawara, G.M.; Reyad, A.H.; El-darandaly, K. Facility layout problem (FLP) and intelligent techniques: A survey. In Proceedings of the 7th International Conference on Production Engineeering, Design and Control(PEDAC), Alexandria, Egypt, 13–15 February 2001; pp. 409–422.
8. Meller, R.D.; Gau, K.Y. The facility layout problem: Recent and emerging trends and perspectives. *J. Manuf. Syst.* **1996**, *15*, 351–366. [CrossRef]
9. Wang, G.-G.; Deb, S.; Cui, Z. Monarch butterfly optimization. *Neural Comput. Appl.* **2015**, 1–20. [CrossRef]
10. Chen, S.; Chen, R.; Gao, J. A Monarch Butterfly Optimization for the Dynamic Vehicle Routing Problem. *Algorithms* **2017**, *10*, 107. [CrossRef]
11. Ghetas, M.; Yong, C.H.; Sumari, P. Harmony-based monarch butterfly optimization algorithm. In *2015 IEEE International Conference on Control System, Computing and Engineering (ICCSCE)*; IEEE: George Town, Malaysia, 2015; pp. 156–161.
12. Wang, G.-G.; Hao, G.-S.; Cheng, S.; Qin, Q. A Discrete Monarch Butterfly Optimization for Chinese TSP Problem. In *International Conference in Swarm Intelligence*; Springer: Cham, Switzerland, 2016; Volume 9712, pp. 165–173.
13. Wang, G.-G.; Deb, S.; Zhao, X.; Cui, Z. A new monarch butterfly optimization with an improved crossover operator. *Oper. Res.* **2018**, *18*, 731–755. [CrossRef]
14. Wang, G.-G.; Zhao, X.; Deb, S. A Novel Monarch Butterfly Optimization with Greedy Strategy and Self-Adaptive. In *2015 Second International Conference on Soft Computing and Machine Intelligence (ISCMI)*; IEEE: Hong Kong, China, 2015; pp. 45–50.
15. Chittratanawat, S.; Noble, J.S. An integrated approach for facility layout, P/D location and material handling system design. *Int. J. Prod. Res.* **1999**, *37*, 683–706. [CrossRef]
16. Kochhar, J.S.; Foster, B.T.; Heragu, S.S. HOPE: A genetic algorithm for the unequal area facility layout problem. *Comput. Oper. Res.* **1998**, *25*, 583–594. [CrossRef]
17. Scholz, D.; Petrick, A.; Domschke, W. STaTS: A Slicing Tree and Tabu Search based heuristic for the unequal area facility layout problem. *Eur. J. Oper. Res.* **2009**, *197*, 166–178. [CrossRef]
18. Bukchin, Y.; Tzur, M. A new MILP approach for the facility process-layout design problem with rectangular and L/T shape departments. *Int. J. Prod. Res.* **2014**, *52*, 7339–7359. [CrossRef]
19. Gonçalves, J.F.; Resende, M.G.C. A biased random-key genetic algorithm for the unequal area facility layout problem. *Eur. J. Oper. Res.* **2015**, *246*, 86–107. [CrossRef]
20. Asef-Vaziri, A.; Kazemi, M. Covering and connectivity constraints in loop-based formulation of material flow network design in facility layout. *Eur. J. Oper. Res.* **2018**, *264*, 1033–1044. [CrossRef]

21. Sinriech, D.; Tanchoco, J.M.A. Solution methods for the mathematical models of single-loop AGV systems. *Int. J. Prod. Res.* **1993**, *31*, 705–725. [CrossRef]

22. Tanchoco, J.M.A.; Sinriech, D. OSL—optimal single-loop guide paths for AGVS. *Int. J. Prod. Res.* **1992**, *30*, 665–681. [CrossRef]

23. Asef-Vaziri, A.; Laporte, G.; Sriskandarajah, C. The block layout shortest loop design problem. *IIE Trans.* **2000**, *32*, 727–734. [CrossRef]

24. Ahmadi-Javid, A.; Ramshe, N. On the block layout shortest loop design problem. *IIE Trans.* **2013**, *45*, 494–501. [CrossRef]

25. Yang, T.; Peters, B.A.; Tu, M. Layout design for flexible manufacturing systems considering single-loop directional flow patterns. *Eur. J. Oper. Res.* **2005**, *164*, 440–455. [CrossRef]

26. Hojabri, H.; Hojabri, A.; Jaafari, A.A.; Farahani, L.N. A Loop Material Flow System Design. *Eng. Comput. Sci.* **2010**, *3*, 1544–1545.

27. Jahandideh, H.; Asef-Vaziri, A.; Modarres, M. Genetic Algorithm for Designing a Convenient Facility Layout for a Circular Flow Path. Available online: https://arxiv.org/vc/arxiv/papers/1211/1211.2361v1.pdf (accessed on 6 December 2018).

28. Asef-Vaziri, A.; Jahandideh, H.; Modarres, M. Loop-based facility layout design under flexible bay structures. *Int. J. Prod. Econ.* **2017**, *193*, 713–725. [CrossRef]

29. Chae, J.; Peters, B.A. A simulated annealing algorithm based on a closed loop layout for facility layout design in flexible manufacturing systems. *Int. J. Prod. Res.* **2006**, *44*, 2561–2572. [CrossRef]

30. Niroomand, S.; Hadi-Vencheh, A.; Şahin, R.; Vizvári, B. Modified migrating birds optimization algorithm for closed loop layout with exact distances in flexible manufacturing systems. *Expert Syst. Appl.* **2015**, *42*, 6586–6597. [CrossRef]

31. Kang, S.; Kim, M.; Chae, J. A closed loop based facility layout design using a cuckoo search algorithm. *Expert Syst. Appl.* **2018**, *93*, 322–335. [CrossRef]

32. Asef-Vaziri, A.; Laporte, G. Loop based facility planning and material handling. *Eur. J. Oper. Res.* **2005**, *164*, 1–11. [CrossRef]

33. Kulturel-Konak, S.; Konak, A. Linear Programming Based Genetic Algorithm for the Unequal Area Facility Layout Problem. *Int. J. Prod. Res.* **2013**, *51*, 4302–4324. [CrossRef]

34. Azadivar, F.; Wang, J. Facility layout optimization using simulation and genetic algorithms. *Int. J. Prod. Res.* **2000**, *38*, 4369–4383. [CrossRef]

35. Ghanem, W.A.H.M.; Jantan, A. Hybridizing artificial bee colony with monarch butterfly optimization for numerical optimization problems. *Neural Comput. Appl.* **2018**, *30*, 163–181. [CrossRef]

36. Montreuil, B. A Modelling Framework for Integrating Layout Design and flow Network Design. In *Material Handling '90*; Springer: Berlin, Heidelberg, 1991; Volume 2, pp. 95–115. ISBN 978-3-642-84356-3.

37. Meller, R.D.; Narayanan, V.; Vance, P.H. Optimal facility layout design. *Oper. Res. Lett.* **1998**, *23*, 117–127. [CrossRef]

38. Sherali, H.D.; Fraticelli, B.M.P.; Meller, R.D. Enhanced Model Formulations for Optimal Facility Layout. *Oper. Res.* **2003**, *51*, 629–644. [CrossRef]

39. Castillo, I.; Westerlund, T. An ε-accurate model for optimal unequal-area block layout design. *Comput. Oper. Res.* **2005**, *32*, 429–447. [CrossRef]

40. Farahani, R.Z.; Laporte, G.; Sharifyazdi, M. A practical exact algorithm for the shortest loop design problem in a block layout. *Int. J. Prod. Res.* **2005**, *43*, 1879–1887. [CrossRef]

41. Asef-Vaziri, A.; Ortiz, R.A. The value of the shortest loop covering all work centers in a manufacturing facility layout. *Int. J. Prod. Res.* **2008**, *46*, 703–722. [CrossRef]

42. Tam, K.Y. Genetic algorithms, function optimization, and facility layout design. *Eur. J. Oper. Res.* **1992**, *63*, 322–346. [CrossRef]

43. Tam, K.Y. A simulated annealing algorithm for allocating space to manufacturing cells. *Int. J. Prod. Res.* **1992**, *30*, 63–87. [CrossRef]

44. Komarudin, K.; Wong, K.Y. Applying Ant System for solving Unequal Area Facility Layout Problems. *Eur. J. Oper. Res.* **2010**, *202*, 730–746. [CrossRef]

45. van Camp, D.J.; Carter, M.W.; Vannelli, A. A nonlinear optimization approach for solving facility layout problems. *Eur. J. Oper. Res.* **1992**, *57*, 174–189. [CrossRef]

46. Armour, G.C.; Buffa, E.S. A Heuristic Algorithm and Simulation Approach to Relative Location of Facilities. *Manage. Sci.* **1963**, *9*, 294–309. [CrossRef]

47. Liu, Q.; Meller, R.D. A sequence-pair representation and MIP-model- based heuristic for the facility layout problem with rectangular departments. *IIE Trans.* **2007**, *39*, 377–394. [CrossRef]

48. Dunker, T.; Radons, G.; Westkämper, E. A coevolutionary algorithm for a facility layout problem. *Int. J. Prod. Res.* **2003**, *41*, 3479–3500. [CrossRef]

mathematics

MDPI

Article

A Novel Bat Algorithm with Multiple Strategies Coupling for Numerical Optimization

Yechuang Wang [1], Penghong Wang [1], Jiangjiang Zhang [1], Zhihua Cui [1,*], Xingjuan Cai [1], Wensheng Zhang [2] and Jinjun Chen [3]

[1] Complex System and Computational Intelligent Laboratory, Taiyuan University of Science and Technology, Taiyuan 030024, China; yechuangwang@sina.com (Y.W.); penghongwang@sina.cn (P.W.); jiangofyouth@163.com (J.Z.); xingjuancai@gmail.com (X.C.)
[2] State Key Laboratory of Intelligent Control and Management of Complex Systems, Institute of Automation Chinese Academy of Sciences, Beijing 100190, China; wensheng.zhang@ia.ac.cn
[3] Department of Computer Science and Software Engineering, Swinburne University of Technology, Melbourne 3000, Australia; jinjun.chen@gmail.com
* Correspondence: zhihua.cui@hotmail.com; Tel.: +86-138-3459-9274

Received: 10 December 2018; Accepted: 21 January 2019; Published: 1 February 2019

Abstract: A bat algorithm (BA) is a heuristic algorithm that operates by imitating the echolocation behavior of bats to perform global optimization. The BA is widely used in various optimization problems because of its excellent performance. In the bat algorithm, the global search capability is determined by the parameter loudness and frequency. However, experiments show that each operator in the algorithm can only improve the performance of the algorithm at a certain time. In this paper, a novel bat algorithm with multiple strategies coupling (mixBA) is proposed to solve this problem. To prove the effectiveness of the algorithm, we compared it with CEC2013 benchmarks test suits. Furthermore, the Wilcoxon and Friedman tests were conducted to distinguish the differences between it and other algorithms. The results prove that the proposed algorithm is significantly superior to others on the majority of benchmark functions.

Keywords: bat algorithm (BA); bat algorithm with multiple strategy coupling (mixBA); CEC2013 benchmarks; Wilcoxon test; Friedman test

---

1. Introduction

In the past ten years, many heuristic optimization algorithms, such as particle swarm optimization (PSO) [1–3], ant colony optimization (ACO) [4,5], bat algorithm (BA) with triangle-flipping strategy [6], fly algorithm (FA) [7–9], cuckoo search [10–13], pigeon-inspired optimization algorithm, and genetic algorithm (GA) [14], have been developed to solve complex computational problems. It became popular because of its superior ability, which deals with a variety of complex issues. Moreover, it has been proven that there is no heuristic algorithm that can perform generally enough to solve all optimization problems [15]. Therefore, scholars have tried to solve these problems with different bionic algorithms.

BA [16–19] is a novel heuristic optimization algorithm, inspired by the echolocation behavior of bats. This algorithm carries out the search process using artificial bats as search agents mimicking the natural pulse loudness and emission rate of real bats. To improve the performance of BA, different strategies have been proposed. We will elaborate in the following three research situations.

(I) **Parameter adjustment**

For the standard BA algorithm, four main parameters are required: frequency, emission, constants, and emission rate. The frequency is used to balance the impact of the historical optimal position on the current position. The bat individual will search far from the group historical position when the search range of frequency is large, and vice versa. In general, the choice of frequency range is determined by different issues. Hasançebi [20] set the pulse frequency range to [0–1]. Gandomi and Yang [21] sets the frequency range to [0–2] in the chaotic bat algorithm. Fister et al. [22] sets the frequency to [0–5] in their algorithm. Ali [23] sets the frequency to [0–100] in the power system. Xie et al. [24] proposed an adaptive adjustment strategy for frequency. Pérez et al. [25] designed a fuzzy controller to dynamically adjust the range of pulse frequencies, while Liu [26] replaced the frequency with a Lévy distribution. To improve the local search capability, Yilmaz and Kucuksille [27] added a random item with two randomly selected bats to explore more search space. Cai [28] introduced a linear decreasing function into the bat algorithm to enhance the global search capability.

(II) **Formula adjustment**

In terms of global search, the step size of the standard BA algorithm decreases with the increase of iterations, which causes the algorithm to be sensitive to local optimum. Focusing on this problem, Bahmani-Firouzi and Azizipanah-Abarghooee [29] proposed four different velocity updating strategies to keep a balance between exploitation and exploration. Inspired by PSO, Yilmaz and Kucuksille [30] put the inertia weight into the velocity update equation. Xie et al [24] use random parts associated with Lévy distributions instead of avoidance. To improve the local search capability, four differential evolutionary strategies were employed to replace the original local search pattern in the standard BA [31]. Xie et al. [32] also incorporated the Lévy flight in the velocity update equation, but four randomly selected bats were used to guide the search pattern. Zhu et al. [33] replace the swarm historical best position with the mean best position to enhance the convergence speed.

(III) **Application**

BA has been widely applied to various areas, including classification, wireless sensor [34], and data mining. Yang and Gandomi [35] proposed a bat algorithm to solve multi-objective problems; Bora et al. [36] proposed a bat-inspired optimization approach to solve the brushless direct current (DC) wheel motor problem; Sambariya and Prasad [37] proposed a metaheuristic bat algorithm for solving robust turning of power system stabilizer for small signal stability enhancement; Sathya and Ansari [38] highlighted the load frequency control using dual mode bat algorithm based scheduling of PI controllers for interconnected power systems; Sun and Xu [39] proposed node localization of wireless sensor networks based on a hybrid bat-quasi-newton algorithm; Cao et al. [40] improved low energy adaptive clustering hierarchy protocol based on a local centroid bat algorithm.

Furthermore, there are many applications in big data and machine learning [41,42], such as Hamidzadeh et al. [43], who proposed a novel method called chaotic bat algorithm for support vector data description (SVDD) (CBA-SVDD) to design effective descriptions of data. Alsalibi [44] proposed a novel membrane-inspired binary bat algorithm for facial feature selection. Furthermore, it outperforms recent state-of-the-art face recognition methods on three benchmark databases. Therefore, the bat algorithm has a wide range of applications. In addition, the bat algorithm has been proposed to optimize support vector machine (SVM) parameters that reduce the classification error [45]. Notably, increasing SVM prediction accuracy and avoiding local optimal trap using the bat algorithm has been very helpful in biomedical research [46,47].

The rest of this paper is organized as follows. Section 2 provides a brief description of the standard BA. In Section 3, we listed eight improvement strategies and proposed a novel bat algorithm with multiple strategy coupling. Numerical experiments on the CEC2013 benchmark set are conducted in Section 4. Finally, the discussion and future work are given in Section 5.

2. Bat Algorithm

The bat algorithm [48] was proposed by Xin-She Yang, based on the echolocation of microbats. Bats usually use echolocation to find food. During removal, bats usually send out short pulses, however, when they encounter food, their pulse send out rates increase and the frequency goes up. The increase in frequency means frequency-tuning, which shortens the echolocations' time and increases the location accuracy. In the standard bat algorithm, each individual i has a defined position $x_i(t)$ and velocity $v_i(t)$ in the search space, which will be updated as the number of iterations increases. The new positions $x_i(t)$ and velocities $v_i(t)$ can be calculated as follows:

$$x_i(t+1) = x_i(t) + v_i(t+1) \tag{1}$$

$$v_i(t+1) = v_i(t) + (x_i(t) - p(t)) \cdot f_i \tag{2}$$

$$f_i = f_{min} + (f_{max} - f_{min}) \cdot \beta \tag{3}$$

where β is a random vector with uniform distribution, the range of which is $[0, 1]$. $p(t)$ is the current global optimal solution and $f_{min} = 0$, $f_{max} = 1$.

As we also know, whether BA has global and local search capabilities depends on its parameters; therefore, it is necessary to achieve a balance between global search and local search capabilities by adopting adaptive parameters. The formula for the local search strategy is as follows:

$$x_i(t+1) = \vec{p}(t) + \varepsilon \overline{A}(t) \tag{4}$$

where ε is a random number from $[-1, 1]$, $\overline{A}(t)$ is the average loudness of population.

In addition, it achieves global search by controlling loudness $A_i(t+1)$ and pulse rate $r_i(t+1)$.

$$A_i(t+1) = \alpha A_i(t) \tag{5}$$

$$r_i(t+1) = r_i(0)[1 - \exp(-\gamma t)] \tag{6}$$

where α and γ are constants and $\alpha > 0, \gamma > 0$. $A_i(0)$ and $r_i(0)$ are initial values of loudness and pulse rate, respectively.

The following describes the execution steps of the standard bat algorithm.

Step 1: For each bat, initialize the position, velocity, and parameters and randomly generate the frequency with Equation (3).

Step 2: Update the position and velocity of each bat with Equations (1) and (2).

Step 3: For each bat, generate a random number $(0 < rand1 < 1)$. Update the temp position and calculate the fitness value for corresponding bat with Equation (4) if $rand1 < r_i(t)$.

Step 4: For each bat, generate a random number $(0 < rand2 < 1)$. Update $A_i(t)$ and $r_i(t)$ with Equations (5) and (6), respectively, if $rand2 < A_i(t)$ and $f(x_i(t)) < f(p(t))$.

Step 5: Sort each individual based on fitness values and save the best position.

Step 6: The algorithm is finished if the condition is met, otherwise, move on to Step 2.

Detailed steps about the standard bat algorithm are presented in Figure 1.

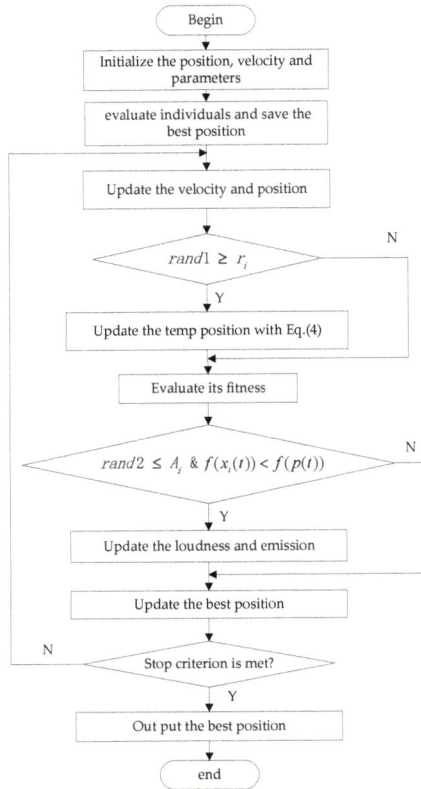

Figure 1. The flowchart of the standard bat algorithm (BA).

3. Bat Algorithm with Multiple Strategy Coupling

Through a large number of experimental studies, we found that different operators play an important role in the convergence ability of the algorithm. When the development operator increases, the global convergence ability of the algorithm becomes weaker; when the exploration operator increases, the convergence accuracy will be insufficient. Therefore, in this paper, we propose a multiple strategy autonomous selection strategy. The main idea is that different individuals choose which strategy to update the position according to the quality of fitness. In this paper, the bat algorithm with multiple strategy coupling (mixBA) formed will adopt the following eight strategies.

- (1) The velocity and position formula of the original algorithm are adopted [42]:

$$x_{ik}(t+1) = x_{ik}(t) + v_{ik}(t+1) \tag{7}$$

$$v_{ik}(t+1) = v_{ik}(t) + (x_{ik}(t) - p_k(t)) \cdot f_i \tag{8}$$

- (2) The velocity and position formula of the improved algorithm are adopted [29]:

$$x_{ik}(t+1) = x_{ik}(t) + v_{ik}(t+1) \tag{9}$$

$$v_{ik}(t+1) = v_{ik}(t) + (x_{ik}(t) - w_k(t)) \cdot f_i \tag{10}$$

where w_k is the position of the worst individual.

- (3) The position and velocity formula of Levy flight was adopted by Xie et al. [24]:

$$f_i^t = \left((f_{\max} - f_{\min}) \frac{t}{n_t} + f_{\min} \right) \beta \tag{11}$$

$$v_i^t = (\hat{x}_i - x^*) f_i^t \tag{12}$$

$$x_i^t = \hat{x}_i + usign[rand(1) - \frac{1}{2}] \oplus Levy(\lambda) \tag{13}$$

where β is a random constant, n_t is a constant, $\hat{x}_i$ is the best position of the ith bat, and x^* is the best position of those found so far. $Levy(\lambda)$ ($1 < \lambda \leq 3$) is step length of the Levy flight and $\oplus$ stands for dot product.

- (4) The position and velocity formula of Levy flight was adopted by Liu [26]:

$$x_i^{t+1} = x_i^t + Levy(\lambda) \otimes (x_i^t - x^*) \tag{14}$$

x_i^t and x_i^{t+1} are ith the position of t bat in generation and $t + 1$ generation, respectively.

- (5) The position and velocity formula with the idea of genetic algorithms was adopted:

$$x_i(t+1) = Dx_i(t) + (1 - D)x_i^*(t) \tag{15}$$

The formula is a two-point crossover operator in simulation genetic algorithm.

- (6) The position and velocity formula with the idea of PSO was adopted:

$$v_{ik}(t+1) = v_{ik}(t) + r_1(x_{ik}(t) - p_{gk}(t))_i + r_2(x_{ik}(t) - p_{ik}(t)) \tag{16}$$

$$x_{ik}(t+1) = x_{ik}(t) + v_{ik}(t+1) \tag{17}$$

where r_1 and r_2 are random constants and p_{gk} is the best position by the entire swarm.

- (7) A local disturbance strategy based on inertial parameters is adopted:

$$x_i(t+1) = x_i^*(t) + wr \tag{18}$$

$$w = w_{\max} - (w_{\max} - w_{\min}) \cdot t / T_{\max} \tag{19}$$

where $w_{\max}$ and $w_{\min}$ are the maximum and minimum values, respectively, of inertia weight and $T_{\max}$ is the maximum number of iterations.

- (8) The local search strategy of flight to optimal position is adopted.

$$x_i(t+1) = x_i^*(t) + r \cdot (p_{gk}(t) - x_{ik}(t)) \tag{20}$$

where p_{gk} is the best position by the entire swarm and r is a random constant.

The above strategies are chosen by the form of probability. Therefore, the number of bat individuals of choosing different strategies varies from generation to generation. Each strategy adjusts the probability of it being selected according to evaluation results. When the fitness value is better, the probability of the strategy will be adjusted by Equation (21).

$$P(n+1) = P(n) + (1 - \lambda) \cdot P(n) \tag{21}$$

Otherwise, it is calculated as follows:

$$P(n+1) = \lambda \cdot P(n) \tag{22}$$

where $\lambda = 0.75$, which is a parameter used to adjust the rate of probability change. The larger λ is, the slower the rate of probability reduction will be. To ensure the diversity of the population, we define the probability lower bound for each strategy as 0.01. The procedure of the bat algorithm with multiple strategy coupling is shown in Table 1.

Table 1. The procedure of the bat algorithm (BA) with multiple strategy coupling.

Algorithm 1: Bat algorithm with multiple strategy coupling
Begin
For each bat, initialize the position, velocity, parameters and probability table;
While (stop criterion is met)
Randomly generate the frequency for each bat with Equation 3;
Evaluate its fitness;
Switch num = 8
Case 1 (*rand* < *p1*)
Update the velocity and position with strategy1.
Case 2 (*p1* < *rand* < *p2*)
Update the velocity and position with strategy2
Case 3 (*p2* < *rand* < *p3*)
Update the velocity and position with strategy3.
Case 4 (*p3* < *rand* < *p4*)
Update the velocity and position with strategy4.
Case 5 (*p4* < *rand* < *p5*)
Update the velocity and position with strategy5.
Case 6 (*p5* < *rand* < *p6*)
Update the velocity and position with strategy6.
Case 7 (*p6* < *rand* < *p7*)
Update the velocity and position with strategy7.
Case 8 (*p7* < *rand* < *p8*)
Update the velocity and position with strategy8.
Evaluate its fitness;
If the position is update
Update the loudness and emission rate;
Update the probability table;
If *pi* < 0
Pi = 0.001;
End
End
Rank the bats and save the best position;
End
Output the best position;
End

4. Experimental Result

4.1. Text Functions and Parameter

The algorithm is tested on the CEC2013 benchmark set [43]. The test set can be divided into three groups, as shown in Table 2.

Table 2. The CEC2013 benchmark set.

F1–F5 belongs to uni-modal functions
F6–F20 belongs to multi-modal functions
F21–F28 belongs to composition functions

The experiment is tested on Matlab 2016a environment (2016a, MathWorks, Natick, MA, USA). For details of parameter settings for the bat algorithm with multiple strategies coupling (mixBA), please refer to Table 3. It is worth emphasizing that the parameters of the adopted strategy are not optimized in this paper. In our algorithm, we used the following indicators to evaluate the experimental results.

$$meanerror = \left| \frac{\sum_{j=1}^{51} f_j}{51} - f_{true} \right| \tag{23}$$

where f_{true} is the actual solution set of the test set.

Table 3. The CEC2013 benchmark set.

Pop size	100
Run	51
Frequency	$[0, 5]$
$A(0)$	0.95
$r(0)$	0.9
α	0.99
γ	0.9
Search Domain	$[-100, 100]^D$

4.2. Comparison of MixBA with State-of-the-Art Algorithms

In this section, we will compare mixBA with six other algorithms. The algorithms involved are presented in Table 4.

Table 4. The involved algorithm.

Bat algorithm with multiple strategy coupling (mixBA);
Standard bat algorithm (SBA);
Self-adaptive heterogeneous particle swarm optimization (PSO) [49];
Bat algorithm with Lévy distribution (LBA1) [26];
Bat algorithm with Lévy distribution (LBA2) [32];
Bat algorithm with arithmetic centroid strategy (ACBA) [40]
Oriented cuckoo search (OCS) [34]

The experimental results will be compared with the standard bat algorithm (SBA), PSO, bat algorithm with Lévy distribution LBA1, LBA2, bat algorithm with arithmetic centroid strategy (ACBA), and oriented cuckoo search (OCS) algorithms. Table A1 (in Appendix A) shows the average error obtained by different algorithms in different test functions. In the last line of the table, w refers to the number of mixBA algorithms superior to other algorithms in the test function, *t* indicates the number of performances similar to other algorithms, and L refers to the number of mixBA algorithms inferior to other algorithms. The dynamic comparison can be viewed in Figure 2.

From Table A1, mixBA won 27 functions, 28 functions, and 26 functions compared with SBA, LBA1, and LBA2, respectively. Compared with the mixBA algorithm, the PSO and OCS algorithm has seven functions and six functions, respectively, that are good.

For most of the test functions, the SBA does not find the global optimal solution. For ACBA and PSO, it only finds optimal solutions on function F1 and F5. LBA1 and LBA2 showed excellent searching ability on functions F11, F14, and F17. OCS obtains good solutions on functions F1, F4, and F5. MixBA can find reasonable solutions on the most of the functions.

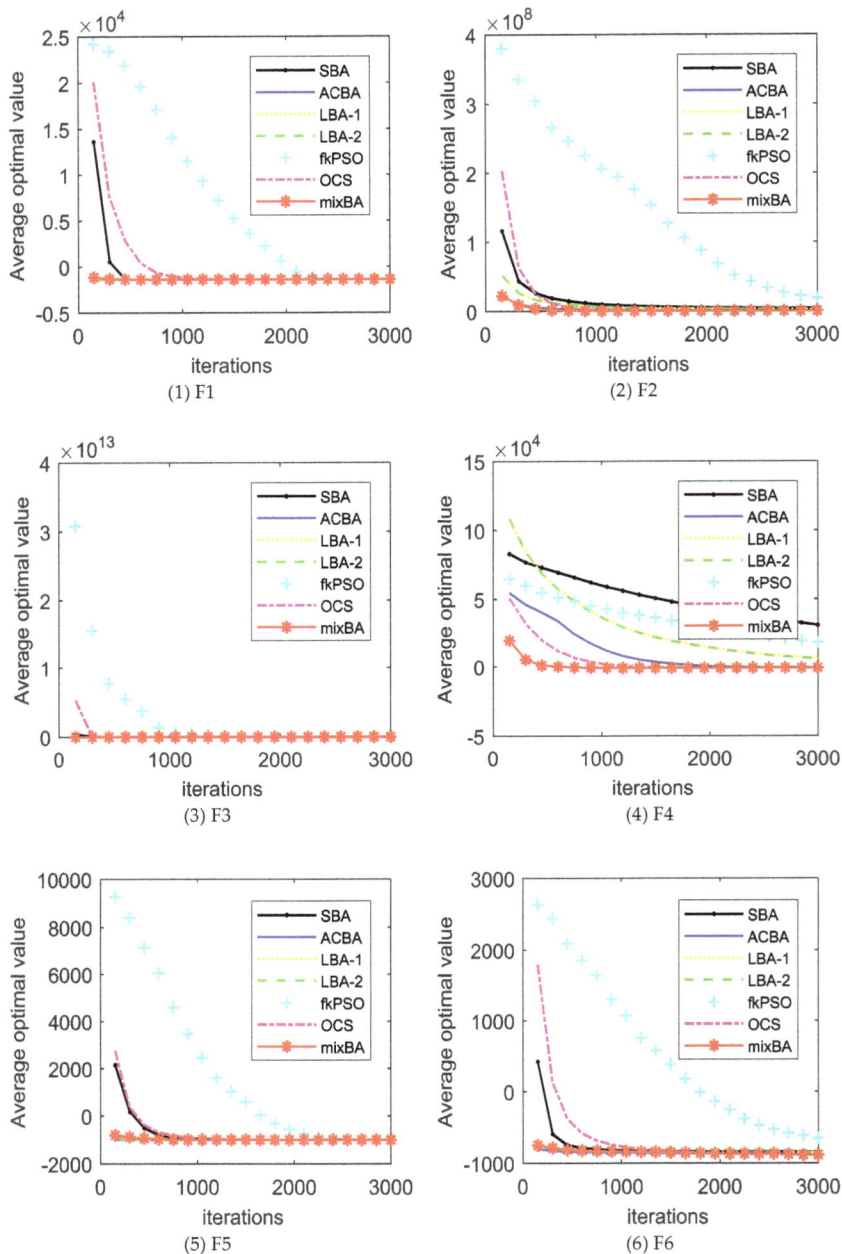

(1) F1

(2) F2

(3) F3

(4) F4

(5) F5

(6) F6

Figure 2. *Cont.*

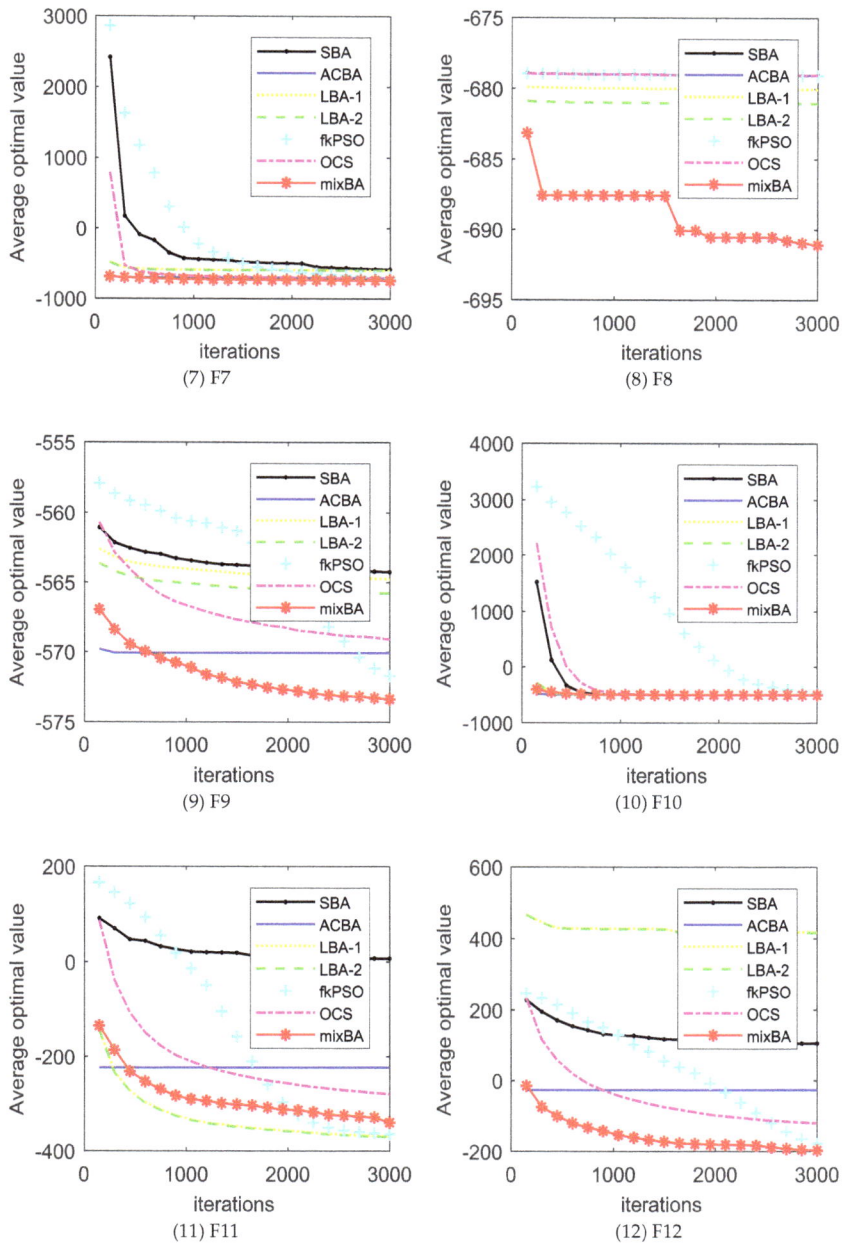

Figure 2. *Cont.*

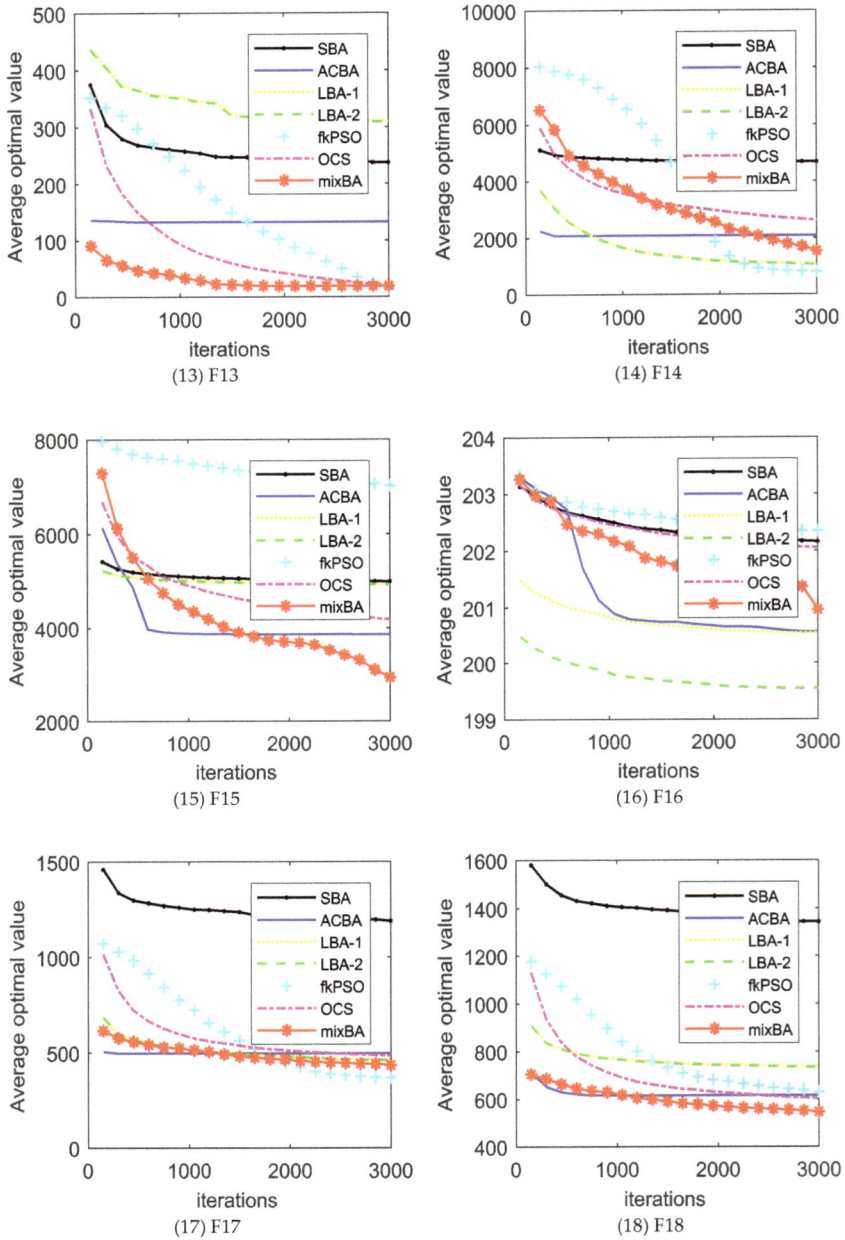

(13) F13

(14) F14

(15) F15

(16) F16

(17) F17

(18) F18

Figure 2. *Cont.*

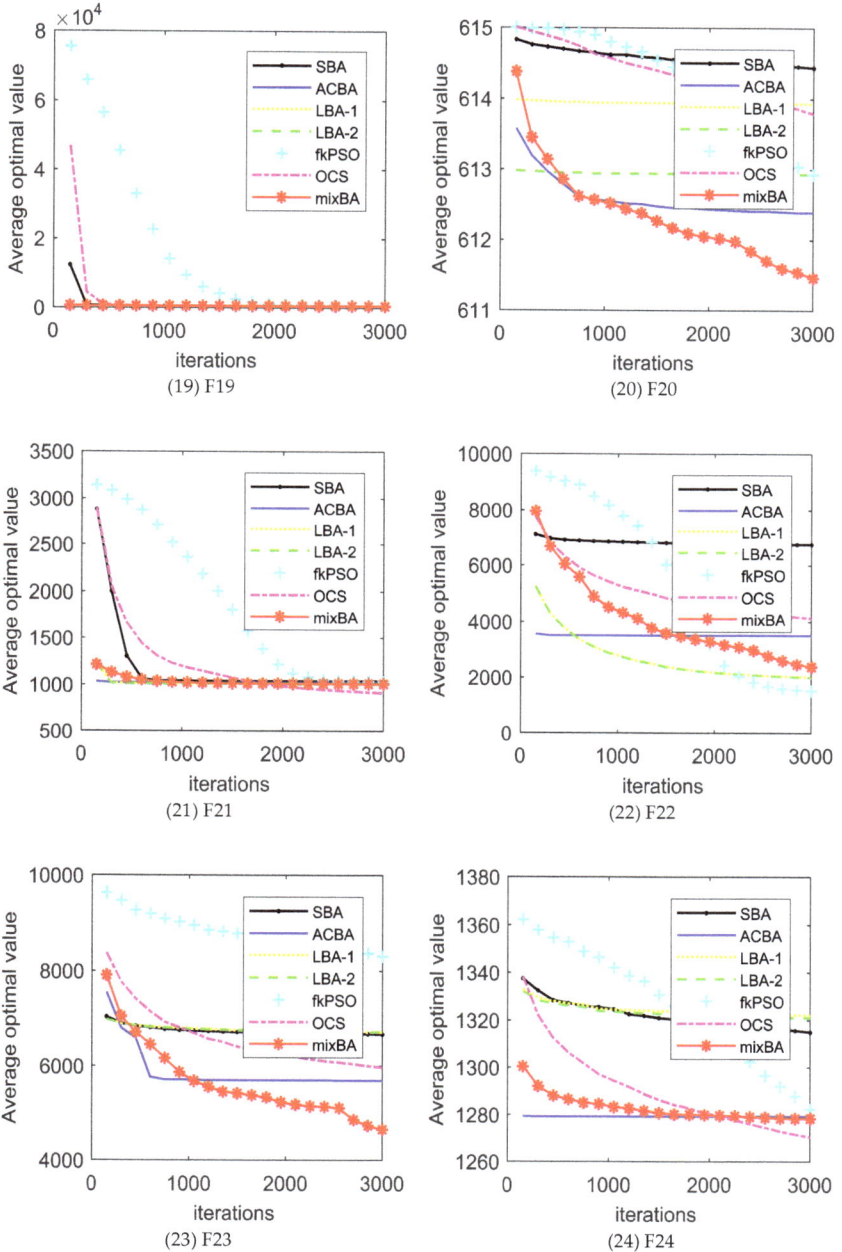

(19) F19

(20) F20

(21) F21

(22) F22

(23) F23

(24) F24

Figure 2. *Cont.*

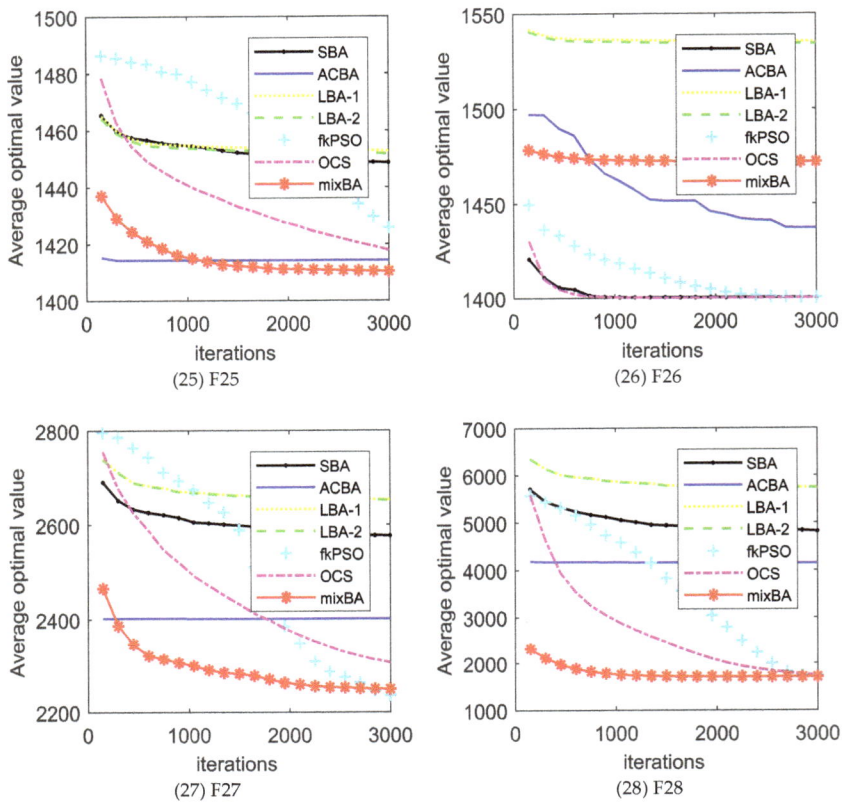

Figure 2. The convergence curves of different algorithms on the benchmark set. SBA—standard bat algorithm; ACBA—bat algorithm with arithmetic centroid strategy; LBA—bat algorithm with Lévy distribution; PSO—particle swarm optimization; OCS—oriented cuckoo search.

Table 5 presents the statistical results obtained by Friedman tests [30,50]. The smaller the ranking value, the better the performance of the algorithm. From the results, we can get the ranks of seven algorithms as follows: SBA, LBA1, LBA2, ACBA, PSO, OCS, mixBA. The highest ranking shows that mixBA is the best algorithm among the seven algorithms.

Table 5. Friedman test for the seven algorithms. SBA—standard bat algorithm; ACBA—bat algorithm with arithmetic centroid strategy; LBA—bat algorithm with Lévy distribution; PSO—particle swarm optimization; OCS—oriented cuckoo search.

Algorithm	Rankings
SBA	5.86
ACBA	3.59
LBA1	5.61
LBA2	5.05
FK-PSO	3.09
OCS	2.86
mixBA	1.95

Table 6 shows the results of the Wilcoxon test [51]. For SBA, ACBA, LBA1, LBA2 and PSO algorithm, $p < 0.05$. The results of this experiment show that the performance of mixBA is far superior to that of other algorithms. For OCS, the p-value is approximately equal to the 0.05 significance level.

Table 6. Wilcoxon test for the seven algorithms.

mixBA vs	p-Value
SBA	0
ACBA	0
LBA1	0
LBA2	0
FK-PSO	0.04
OCS	0.068

Figure 2 shows the convergence of different algorithms in different test functions. In most cases, our proposed algorithm has better results. However, it is undeniable that in a few cases, our proposed algorithm shows poor performance compared with other algorithms, such as in F11, F14, F16, F17, F22, F24, and F26. This is because the magnitude of the probability adjustment of the strategy is large, which leads to the algorithm prematurely selecting a certain strategy to make the algorithm fall into the local optimal. In addition, Table A2 shows the runtime of each algorithm on the CEC2013 benchmark set. It is obvious that the running time of the mixBA algorithm is slightly higher than that of SBA on F1 and F2, but significantly smaller than other algorithms in other test functions.

5. Conclusions

Bio-inspired computation is a collection for stochastic optimization algorithms inspired by biological phenomenon. BA is novel bio-inspired algorithm inspired by bat behaviors, and has been used to solve engineering optimization problems. However, with a single optimization strategy, it shows weakness in solving various complex optimization problems. To tackle this issue, this paper proposes a bat algorithm with multiple strategy coupling (mixBA) to improve the performance of BA. The simulation results show that the performance of the mixBA is superior to that of other algorithms. This is because of the adoption of adaptive multi-strategies coupling rules. The algorithm can adjust the probabilities of different strategies based on the evaluation index. In most cases, this manner guarantees the global convergence and local exploration ability of the algorithm. However, in some cases, this treatment causes other strategies to be ignored early and fall into the local optimum. This is the reason that the proposed algorithm performs worse than the other ones in some test functions, such as F11, F14, F16, F17, F22, F24, and F26. Therefore, we will continue to explore the impact of the probability change of the strategies on optimization performance in subsequent studies and seek better solutions.

Author Contributions: Writing—original draft preparation, Y.W.; writing—review and editing, P.W.; visualization, J.Z.; supervision, Z.C., X.C., W.Z., and J.C.

Acknowledgments: This work is supported by the National Natural Science Foundation of China under Grant No. 61806138, No. U1636220, and No. 61663028; Natural Science Foundation of Shanxi Province under Grant No. 201801D121127; Scientific and Technological innovation Team of Shanxi Province under Grant No. 201805D131007; PhD Research Startup Foundation of Taiyuan University of Science and Technology under Grant No. 20182002; and Zhejiang Provincial Natural Science Foundation of China under Grant No. Y18F030036.

Conflicts of Interest: The authors declare no conflict of interest.

Appendix A

Table A1. Comparison results for mixBA and the other six algorithms.

Function	SBA	ACBA	LBA1	LBA2	FK-PSO	OCS	mixBA
F1	1.96×10^0	0.00×10^0	8.24×10^{-1}	3.59×10^{-1}	0.00×10^0	4.51×10^{-5}	1.98×10^{-5}
F2	3.69×10^6	3.04×10^5	3.54×10^6	2.23×10^6	1.59×10^6	3.18×10^0	1.27×10^3
F3	3.44×10^8	6.47×10^7	4.78×10^8	3.58×10^8	2.40×10^8	9.96×10^6	3.38×10^7
F4	3.20×10^4	1.22×10^2	1.45×10^4	6.85×10^3	4.78×10^2	7.53×10^{-3}	6.80×10^1
F5	5.86×10^{-1}	0.00×10^0	4.74×10^{-1}	2.76×10^{-1}	0.00×10^0	4.26×10^{-3}	5.82×10^{-3}
F6	5.63×10^1	2.87×10^1	5.07×10^1	4.85×10^1	2.29×10^1	8.79×10^0	1.45×10^1
F7	2.16×10^2	9.30×10^1	1.77×10^2	2.00×10^2	6.39×10^1	6.07×10^1	5.61×10^1
F8	2.09×10^1	2.10×10^1	2.09×10^1	2.10×10^1	2.09×10^1	2.10×10^1	2.09×10^1
F9	3.57×10^1	2.99×10^1	3.40×10^1	3.62×10^1	1.85×10^1	2.79×10^1	2.55×10^1
F10	1.32×10^0	1.92×10^{-1}	1.23×10^0	1.07×10^0	2.29×10^{-1}	1.40×10^{-2}	1.77×10^{-2}
F11	4.07×10^2	1.77×10^2	1.49×10^2	3.16×10^1	2.36×10^1	6.83×10^1	6.05×10^1
F12	4.06×10^2	2.73×10^2	7.42×10^2	7.18×10^2	5.64×10^1	1.07×10^2	1.03×10^2
F13	4.37×10^2	3.33×10^2	5.59×10^2	5.11×10^2	1.23×10^2	1.84×10^2	1.14×10^2
F14	4.78×10^3	2.18×10^3	3.17×10^3	1.15×10^3	7.04×10^3	2.39×10^3	1.63×10^2
F15	4.89×10^3	3.76×10^3	4.76×10^3	4.83×10^3	3.42×10^3	3.55×10^3	2.82×10^3
F16	2.16×10^0	5.61×10^{-1}	1.33×10^0	1.54×10^0	8.48×10^{-1}	1.65×10^0	9.50×10^{-1}
F17	8.92×10^2	1.96×10^2	3.36×10^2	1.61×10^2	5.26×10^2	1.61×10^2	1.33×10^2
F18	9.44×10^2	2.16×10^2	3.28×10^2	3.35×10^2	6.81×10^2	1.86×10^2	1.45×10^2
F19	6.07×10^1	1.34×10^1	1.89×10^1	1.28×10^1	3.12×10^1	7.38×10^1	6.07×10^0
F20	1.44×10^1	1.24×10^1	1.47×10^1	1.49×10^1	1.20×10^1	1.19×10^1	1.14×10^1
F21	3.38×10^2	3.22×10^2	3.22×10^2	3.05×10^2	3.11×10^2	2.88×10^2	3.01×10^2
F22	5.94×10^3	2.70×10^3	3.32×10^3	1.20×10^3	8.59×10^3	2.81×10^3	1.58×10^3
F23	5.77×10^3	4.79×10^3	6.03×10^3	5.82×10^3	3.57×10^3	4.10×10^3	3.75×10^3
F24	3.15×10^2	2.79×10^2	3.22×10^2	3.23×10^2	2.48×10^2	2.67×10^2	2.57×10^2
F25	3.49×10^2	3.14×10^2	3.53×10^2	3.54×10^2	2.49×10^2	3.01×10^2	2.94×10^2
F26	2.00×10^2	2.37×10^2	3.54×10^2	3.36×10^2	2.95×10^2	2.00×10^2	2.72×10^2
F27	1.28×10^3	1.10×10^3	1.33×10^3	1.35×10^3	7.76×10^2	9.84×10^2	8.91×10^2
F28	3.42×10^3	2.74×10^3	4.68×10^3	4.34×10^3	4.01×10^2	3.48×10^2	3.00×10^2
w\t\l	27\1\0	27\1\0	28\0\0	26\1\1	21\7\0	22\6\0	-

Table A2. The computation time of each algorithm.

Function	SBA	ACBA	LBA1	LBA2	FK-PSO	OCS	mixBA
F1	3.23×10^1	4.82×10^1	9.22×10^1	9.07×10^1	4.01×10^1	1.10×10^2	3.29×10^1
F2	5.81×10^1	7.88×10^1	1.28×10^2	1.26×10^2	7.14×10^1	1.43×10^2	5.17×10^1
F3	6.29×10^1	8.53×10^1	1.31×10^2	1.30×10^2	7.77×10^1	1.44×10^2	5.05×10^1
F4	4.26×10^1	6.09×10^1	1.06×10^2	1.06×10^2	5.16×10^1	1.24×10^2	4.13×10^1
F5	3.32×10^1	5.05×10^1	9.42×10^1	9.40×10^1	4.23×10^1	1.12×10^2	3.40×10^1
F6	4.02×10^1	6.14×10^1	1.04×10^2	1.05×10^2	5.20×10^1	1.21×10^2	3.85×10^1
F7	1.40×10^1	1.65×10^2	2.21×10^2	2.21×10^2	1.70×10^2	2.38×10^2	8.21×10^1
F8	1.26×10^1	1.39×10^2	2.02×10^2	1.90×10^2	1.34×10^2	2.10×10^2	8.55×10^1
F9	8.54×10^1	8.99×10^2	1.09×10^3	1.10×10^3	1.03×10^3	1.13×10^3	3.59×10^2
F10	7.14×10^1	9.06×10^1	1.39×10^2	1.38×10^2	8.77×10^1	1.56×10^2	5.02×10^1
F11	8.00×10^1	1.03×10^2	1.47×10^2	1.47×10^2	9.95×10^2	1.69×10^2	5.58×10^1
F12	1.03×10^1	1.27×10^2	1.76×10^2	1.75×10^2	1.31×10^2	1.97×10^2	6.68×10^1
F13	1.05×10^1	1.20×10^2	1.75×10^2	1.76×10^2	1.33×10^2	1.99×10^2	6.67×10^1
F14	8.75×10^1	1.09×10^2	1.56×10^2	1.57×10^2	1.06×10^2	1.77×10^2	6.67×10^1
F15	9.49×10^1	1.20×10^2	1.67×10^2	1.68×10^2	1.18×10^2	1.87×10^2	6.87×10^1
F16	2.13×10^2	2.47×10^2	3.15×10^2	3.11×10^2	2.66×10^2	3.26×10^2	1.23×10^2
F17	5.90×10^1	8.09×10^1	1.29×10^2	1.27×10^2	7.07×10^2	1.41×10^2	4.89×10^1
F18	7.50×10^1	9.68×10^1	1.52×10^2	1.47×10^2	9.27×10^2	1.62×10^2	5.59×10^1
F19	5.01×10^1	6.72×10^1	1.18×10^2	1.14×10^2	6.43×10^1	1.30×10^2	4.25×10^2
F20	9.18×10^1	9.34×10^1	1.62×10^2	1.66×10^2	1.10×10^2	1.83×10^2	5.71×10^2

Table A2. *Cont.*

Function	SBA	ACBA	LBA1	LBA2	FK-PSO	OCS	mixBA
F21	2.07×10^2	2.47×10^2	3.08×10^2	2.97×10^2	2.55×10^2	3.15×10^2	1.04×10^2
F22	2.61×10^1	2.96×10^2	3.72×10^2	3.58×10^2	3.17×10^2	3.81×10^2	1.32×10^2
F23	2.85×10^1	3.22×10^2	4.02×10^2	3.90×10^2	3.50×10^2	4.09×10^2	1.42×10^2
F24	1.05×10^3	1.11×10^3	1.37×10^3	1.31×10^3	1.31×10^3	1.33×10^3	4.27×10^2
F25	1.05×10^3	1.12×10^3	1.35×10^3	1.33×10^3	1.29×10^3	1.32×10^3	4.35×10^2
F26	1.17×10^3	1.25×10^3	1.49×10^3	1.46×10^3	1.4×10^3	1.47×10^3	4.85×10^2
F27	1.14×10^3	1.17×10^3	1.44×10^3	1.42×10^3	1.40×10^3	1.42×10^3	4.62×10^2
F28	3.98×10^2	4.35×10^2	5.25×10^2	5.21×10^2	4.83×10^2	5.14×10^2	1.70×10^2
w\t\l	26\0\2	28\0\0	28\0\0	28\0\0	28\0\0	28\0\0	-

References

1. Yang, X.S. *Swarm Intelligence and Bio-Inspired Computation: Theory and Applications*; Elsevier Science Publishers B. V.: New York, NY, USA, 2013. [CrossRef]
2. Eberhart, R.; Kennedy, J. A new optimizer using particle swarm theory. In Proceedings of the International Symposium on MICRO Machine and Human Science, Nagoya, Japan, 4–6 October 1995; pp. 39–43. [CrossRef]
3. Pan, J. Diversity enhanced particle swarm optimization with neighborhood search. *Inf. Sci.* **2013**, *223*, 119–135. [CrossRef]
4. Dorigo, M.; Stützle, T. Ant Colony Optimization: Overview and Recent Advances. In *Handbook of Metaheuristics*; Springer: Cham, Switzerland, 2010. [CrossRef]
5. Stodola, P.; Mazal, J. *Applying the Ant Colony Optimization Algorithm to the Capacitated Multi-Depot Vehicle Routing Problem*; Inderscience Publishers: Geneva, Switzerland, 2016. [CrossRef]
6. Cai, X.; Wang, H.; Cui, Z.; Cai, J.; Xue, Y.; Wang, L. Bat algorithm with triangle-flipping strategy for numerical optimization. *Int. J. Mach. Learn. Cybern.* **2018**, *9*, 199–215. [CrossRef]
7. Yang, X.S.; Deb, S. Cuckoo Search via Levy Flights. *Mathematics* **2010**, 210–214. [CrossRef]
8. Cui, Z.; Li, F.; Zhang, W. Bat algorithm with principal component analysis. *Int. J. Mach. Learn. Cybern.* **2018**, 1–20. [CrossRef]
9. Zhang, M.; Wang, H.; Cui, Z.; Chen, J. Hybrid multi-objective cuckoo search with dynamical local search. *Memet. Comput.* **2018**, *10*, 199–208. [CrossRef]
10. Wang, H.; Wang, W.; Sun, H.; Rahnamayan, S. Firefly algorithm with random attraction. *Int. J. Bio-Inspired Comput.* **2016**, *8*, 33–41. [CrossRef]
11. Wang, H.; Wang, W.; Zhou, X.; Sun, H.; Zhao, J.; Yu, X.; Cui, Z. Firefly algorithm with neighborhood attraction. *Inf. Sci.* **2017**, *382*, 374–387. [CrossRef]
12. Iglesias, A.; Gálvez, A.; Collantes, M. Global-Support Rational Curve Method for Data Approximation with Bat Algorithm. In Proceedings of the IFIP International Conference on Artificial Intelligence Applications and Innovations, Bayonne, France, 14–17 September 2015. [CrossRef]
13. Iglesias, A.; Gálvez, A. Memetic electromagnetism algorithm for surface reconstruction with rational bivariate Bernstein basis functions. *Natural Comput.* **2017**, *16*, 1–15. [CrossRef]
14. Holland, J.H. *Adaptation in Natural and Artificial System*; MIT Press: Cambridge, MA, USA, 1992. [CrossRef]
15. Zhao, S.Z.; Suganthan, P.N.; Zhang, Q. Decomposition-Based Multiobjective Evolutionary Algorithm with an Ensemble of Neighborhood Sizes. *IEEE Trans. Evol. Comput.* **2012**, *16*, 442–446. [CrossRef]
16. Yang, X.S. A New Metaheuristic Bat-Inspired Algorithm. *Comput. Knowl. Technol.* **2010**, *28*, 65–74.
17. Yang, X. Bat algorithm for multi-objective optimization. *Int. J. Bio-Inspired Comput.* **2012**, *3*, 267–274. [CrossRef]
18. Tharakeshwar, T.K.; Seetharamu, K.N.; Prasad, B.D. Multi-objective optimization using bat algorithm for shell and tube heat exchangers. *Appl. Therm. Eng.* **2017**, *110*, 1029–1038. [CrossRef]
19. Damasceno, N.C.; Filho, O.G. PI controller optimization for a heat exchanger through metaheuristic Bat Algorithm, Particle Swarm Optimization, Flower Pollination Algorithm and Cuckoo Search Algorithm. *IEEE Lat. Am. Trans.* **2017**, *15*, 1801–1807. [CrossRef]
20. Hasançebi, O.; Teke, T.; Pekcan, O. A bat-inspired algorithm for structural optimization. *Comput. Struct.* **2013**, *128*, 77–90. [CrossRef]
21. Gandomi, A.H.; Yang, X.S. Chaotic bat algorithm. *J. Comput. Sci.* **2014**, *5*, 224–232. [CrossRef]

22. Fister, I.; Fong, S.; Brest, J. A novel hybrid self-adaptive bat algorithm. *Sci. World J.* **2014**, *2014*, 709–738. [CrossRef] [PubMed]

23. Ali, E.S. Optimization of Power System Stabilizers using BAT search algorithm. *Int. J. Electr. Power Energy Syst.* **2014**, *61*, 683–690. [CrossRef]

24. Xie, J.; Zhou, Y.Q.; Chen, H. A bat algorithm based on Lévy flights trajectory. *Pattern Recognit. Artif. Intell.* **2013**, *26*, 829–837. [CrossRef]

25. Pérez, J.; Valdez, F.; Castillo, O. A New Bat Algorithm Augmentation Using Fuzzy Logic for Dynamical Parameter Adaptation. In Proceedings of the Mexican International Conference on Artificial Intelligence, Cuernavaca, Mexico, 25–31 October 2015; Springer: Cham, Switzerland, 2015; pp. 433–442. [CrossRef]

26. Liu, C. Bat algorithm with Levy flight characteristics. *CAAI Trans. Intell. Syst.* **2013**, *3*, 240–246.

27. Yilmaz, S.; Kucuksille, E.U. Improved Bat Algorithm (IBA) on Continuous Optimization Problems. *Lect. Notes Softw. Eng.* **2013**, *1*, 279. [CrossRef]

28. Cai, X.; Wang, L.; Kang, Q.; Wu, Q. Adaptive bat algorithm for coverage of wireless sensor network. *Int. J. Wirel. Mob. Comput.* **2015**, *8*, 271–276. [CrossRef]

29. Bahmani-Firouzi, B.; Azizipanah-Abarghooee, R. Optimal sizing of battery energy storage for micro-grid operation management using a new improved bat algorithm. *Int. J. Electr. Power Energy Syst.* **2014**, *56*, 42–54. [CrossRef]

30. Yilmaz, S.; Kucuksille, E.U. A new modification approach on bat algorithm for solving optimization problems. *Appl. Soft Comput.* **2015**, *28*, 259–275. [CrossRef]

31. Deng, Y.; Duan, H. Chaotic mutated bat algorithm optimized edge potential function for target matching. In Proceedings of the 10th Conference on Industrial Electronics and Applications (ICIEA), Auckland, New Zealand, 15–17 June 2015. [CrossRef]

32. Xie, J.; Zhou, Y.; Chen, H. A Novel Bat Algorithm Based on Differential Operator and Lévy Flights Trajectory. *Comput. Intell. Neurosci.* **2013**, *2013*, 453–812. [CrossRef] [PubMed]

33. Zhu, B.; Zhu, W.; Liu, Z.; Duan, Q.; Cao, L. A Novel Quantum-Behaved Bat Algorithm with Mean Best Position Directed for Numerical Optimization. *Comput. Intell. Neurosci.* **2016**, *2016*, 1–17. [CrossRef]

34. Cui, Z.; Sun, B.; Wang, G.; Xue, Y.; Chen, J. A novel oriented cuckoo search algorithm to improve DV-Hop performance for cyber-physical systems. *J. Parallel Distrib. Comput.* **2017**, *103*, 42–52. [CrossRef]

35. Yang, X.S.; Gandomi, A.H. Bat Algorithm: A Novel Approach for Global Engineering Optimization. *Eng. Comput.* **2012**, *29*, 464–483. [CrossRef]

36. Bora, T.C.; Coelho, L.D.S.; Lebensztajn, L. Bat-Inspired Optimization Approach for the Brushless DC Wheel Motor Problem. *IEEE Trans. Magn.* **2012**, *48*, 947–950. [CrossRef]

37. Sambariya, D.K.; Prasad, R. Robust tuning of power system stabilizer for small signal stability enhancement using metaheuristic bat algorithm. *Int. J. Electr. Power Energy Syst.* **2014**, *61*, 229–238. [CrossRef]

38. Sathya, M.R.; Ansari, M.M.T. Load frequency control using Bat inspired algorithm based dual mode gain scheduling of PI controllers for interconnected power system. *Int. J. Electr. Power Energy Syst.* **2015**, *64*, 365–374. [CrossRef]

39. Sun, S.; Xu, B. Node localization of wireless sensor networks based on hybrid bat-quasi-Newton algorithm. *J. Comput. Appl.* **2015**, *11*, 38–42. [CrossRef]

40. Cao, Y.; Cui, Z.; Li, F.; Dai, C.; Chen, W. Improved Low Energy Adaptive Clustering Hierarchy Protocol Based on Local Centroid Bat Algorithm. *Sens. Lett.* **2014**, *12*, 1372–1377. [CrossRef]

41. Cui, Z.; Cao, Y.; Cai, X.; Cai, J.; Chen, J. Optimal LEACH protocol with modified bat algorithm for big data sensing systems in Internet of Things. *J. Parallel Distrib. Comput.* **2017**. [CrossRef]

42. Cui, Z.; Xue, F.; Cai, X.; Cao, Y.; Wang, G.G.; Chen, J. Detectin of malicious code variants based on deep learning. *IEEE Trans. Ind. Inform.* **2018**, *14*, 3187–3196. [CrossRef]

43. Hamidzadeh, J.; Sadeghi, R.; Namaei, N. Weighted Support Vector Data Description based on Chaotic Bat Algorithm. *Appl. Soft Comput.* **2017**, *60*, 540–551. [CrossRef]

44. Alsalibi, B.; Venkat, I.; Al-Betar, M.A. A membrane-inspired bat algorithm to recognize faces in unconstrained scenarios. *Eng. Appl. Artif. Intell.* **2017**, *64*, 242–260. [CrossRef]

45. Cui, Z.; Zhang, J.; Wang, Y.; Cao, Y.; Cai, X.; Zhang, W.; Chen, J. A pigeon-inspired optimization algorithm for many-objective optimization problems. *Sci. China Inf. Sci* **2019**. [CrossRef]

46. Tharwat, A.; Hassanien, A.E.; Elnaghi, B.E. A BA-based algorithm for parameter optimization of Support Vector Machine. *Pattern Recognit. Lett.* **2016**, *93*, 13–22. [CrossRef]

47. Basith, S.; Manavalan, B.; Shin, T.H.; Lee, G. iGHBP: Computational identification of growth hormone binding proteins from sequences using extremely randomised tree. *Comput. Struct. Biotechnol. J.* **2018**, *16*, 412–420. [CrossRef]

48. Zamuda, A.; Brest, J.; Mezura-Montes, E. Structured Population Size Reduction Differential Evolution with Multiple Mutation Strategies on CEC 2013 real parameter optimization. In Proceedings of the IEEE Congress on Evolutionary Computation, Cancun, Mexico, 20–23 June 2013; pp. 1925–1931. [CrossRef]

49. Manavalan, B.; Subramaniyam, S.; Shin, T.H.; Kim, M.O.; Lee, G. Machine-learning-based prediction of cell-penetrating peptides and their uptake efficiency with improved accuracy. *J. Proteome Res.* **2018**, *17*, 2715–2726. [CrossRef]

50. Friedman, J.H. Fast sparse regression and classification. *Int. J. Forecast.* **2012**, *28*, 722–738. [CrossRef]

51. Sun, H.; Wang, K.; Zhao, J.; Yu, X. Artificial bee colony algorithm with improved special centre. *Int. J. Comput. Sci. Math.* **2017**, *7*, 548–553. [CrossRef]

Article

Search Acceleration of Evolutionary Multi-Objective Optimization Using an Estimated Convergence Point

Yan Pei [1,*], Jun Yu [2] and Hideyuki Takagi [3]

[1] Computer Science Division, University of Aizu, Aizuwakamatsu 965-8580, Japan
[2] Graduate School of Design, Kyushu University, Fukuoka 815-8540, Japan; yujun@kyudai.jp
[3] Faculty of Design, Kyushu University, Fukuoka 815-8540, Japan; takagi@design.kyushu-u.ac.jp
* Correspondence: peiyan@u-aizu.ac.jp; Tel.: +81-242-37-2765

Received: 21 November 2018; Accepted: 23 January 2019; Published: 28 January 2019

Abstract: We propose a method to accelerate evolutionary multi-objective optimization (EMO) search using an estimated convergence point. Pareto improvement from the last generation to the current generation supports information of promising Pareto solution areas in both an objective space and a parameter space. We use this information to construct a set of moving vectors and estimate a non-dominated Pareto point from these moving vectors. In this work, we attempt to use different methods for constructing moving vectors, and use the convergence point estimated by using the moving vectors to accelerate EMO search. From our evaluation results, we found that the landscape of Pareto improvement has a uni-modal distribution characteristic in an objective space, and has a multi-modal distribution characteristic in a parameter space. Our proposed method can enhance EMO search when the landscape of Pareto improvement has a uni-modal distribution characteristic in a parameter space, and by chance also does that when landscape of Pareto improvement has a multi-modal distribution characteristic in a parameter space. The proposed methods can not only obtain more Pareto solutions compared with the conventional non-dominant sorting genetic algorithm (NSGA)-II algorithm, but can also increase the diversity of Pareto solutions. This indicates that our proposed method can enhance the search capability of EMO in both Pareto dominance and solution diversity. We also found that the method of constructing moving vectors is a primary issue for the success of our proposed method. We analyze and discuss this method with several evaluation metrics and statistical tests. The proposed method has potential to enhance EMO embedding deterministic learning methods in stochastic optimization algorithms.

Keywords: evolutionary multi-objective optimization; convergence point; acceleration search; evolutionary computation; optimization

---

1. Introduction

In the research area of optimization, there are single objective optimization problems and multi-objective optimization problems. The difference between these two categories of optimization problems lies in the number of fitness functions. The single objective optimization attempts to obtain only one optimal solution in one parameter space, i.e., one fitness landscape. The multi-objective optimization tries to satisfy more than one optimal condition or target, i.e., more than one fitness landscape. Usually, these optimal conditions in multi-objective optimization conflict with each other, and cannot be combined into one optimal condition. Single objective optimization and multi-objective optimization have different search targets because of the requirements of algorithm design. One tries to obtain a better optimum, the other seeks to obtain more non-dominated solutions on Pareto front.

Most multi-objective optimization research pays attention to the diversity and the number of non-dominated Pareto solutions. Evolutionary multi-objective optimization (EMO) algorithms are efficient and effective when handling multi-objective optimization problems. The EMO keeps the multiple objective functions independently and uses Pareto-based ranking schemes to maintain feasible solutions. However, the determinative programming methods use a scalarization method that needs to transfer multiple objectives into one objective. State-of-the-art studies on EMO concentrate on Pareto dominance handling in an objective space, which is tied to generating solutions to approximate the Pareto solution frontiers [1]. The primary disadvantages of EMO are its optimization capability and the non-guarantee of Pareto optimality, which cannot be perfectly solved by Pareto dominance studies in EMO.

A fitness approximation method is widely used in the evolutionary computation (EC) community to reduce the computational cost of fitness evaluations and is expected to estimate the global optimum solution area. Reference [2] investigated several approximation methods, such as polynomial models, kriging models, neural networks, and others. Reference [3] proposed a framework to manage approximation models in EC search. Reference [4] made a survey on the advances of approximating a fitness function in EC algorithms and presented some future research challenges. Inspired by the scale-spacing method [5], a uni-modal function was used to approximate a fitness function for estimating the peak point, and the EC algorithm used it as an elite individual to increase convergence speed [6]. The same method was extended into a dimension reduction space for fitness landscape approximation to enhance EC algorithms [7,8]. A fitness approximation mechanism was introduced into genetic algorithms to obtain the optimal solutions satisfactorily and quickly, and can reduce computational cost at the same time [9]. It was a unique approach to filter frequency components for approximating a fitness landscape [10], which uniformly samples in parameter space firstly, and the re-sampled EC individuals are used to obtain frequency information using the discrete Fourier transform. There are many methods to accelerate EC convergence using a fitness approximate model [11–17], and many potential subjects need to be studied further [18,19]. Reference [20] presents a comprehensive survey on the fitness approximation methods in interactive evolutionary computation (IEC). Those methods not only can enhance IEC search, but also can enhance conventional EC search [21].

The estimation method presents a novel perspective using mathematical approaches to calculate a convergence point (Convergence means of modeling the tendency for generic characteristics of populations to stabilize over time, in EC, convergence point means the global optimum in single objective problems ideally.) of a population. This is a great solution that enhances search of a stochastic optimization algorithm using a deterministic method embedded into a stochastic optimization process. Using an estimated point to accelerate EC search is one such method that implements this research philosophy [22]. The individual moving from one generation to the next supports convergence information of EC search condition. We used such information to mathematically estimate a convergence point as an elite individual to enhance single objective optimization search [23]. A clustering method was developed for bipolar tasks (Bipolar tasks mean there are two peaks in a fitness landscape in a problem, e.g., combination of two Gaussian functions $N(\mu_1, \sigma_1) + N(\mu_2, \sigma_2)$ presents two peaks in the landscape.), which proposed four improvements to increase the accuracy of estimated convergence points and was applied to a multi-modal optimization problem [24]. We attempted to combine EC algorithms with the estimation framework for bipolar tasks and analyzed the effect of proposed four improvements. From our previous studies, we found that this estimation method is effective in single objective optimization [23].

Pareto improvement information from the current generation to the next supports the promising search areas of Pareto frontier solutions in both an objective space and a parameter space. In this paper, we extend the method of estimating a convergence point into the EMO algorithm to find potential non-dominated solution areas in search spaces using the estimation information from the objective spaces. By putting an estimated point into an EMO search and deleting a dominated solution, the EMO algorithm can find more non-dominated solutions in early generations, which is a factor that motivates this study. We use the NSGA-II algorithm to evaluate our study hypothesis and verify the performance of the framework that combines the estimation method with EMO, and attempt to enhance EMO search.

This demonstrates one of original aspects of this work. We undertake a comparative study between the NSGA-II algorithm with and without the estimation method using multi-objective optimization benchmark problems and statistical tests. The advantages and disadvantages of proposed method are presented, analyzed, and discussed using experimental evaluation results.

Following this introductory section, we introduce a variety of EMO techniques and algorithms in the Section 2. We make a brief review on the estimation algorithm for a uni-modal optimization problem in the Section 3. In the Section 4, we explain how to extend such estimation methods to accelerate EMO search. There are three primary steps in this framework. The first is finding the pair information of moving vectors in objective space of EMO, the second is the estimation of a convergence point in a parameter space, and the third is the insertion of an estimated point by deleting a dominated solution to enhance EMO search. In the Section 5, we evaluate our method using the NSGA-II algorithm and some multi-objective benchmark problems, and analyze its optimization performance and characteristics of population distribution in both an objective space and a parameter space. We discuss and analyze the evaluation results using statistical tests in the Section 6. The results demonstrate that the proposed method can obtain more non-dominate Pareto solutions early. Finally, we make a conclusion of the whole work, and present some open research topics and future work in the Section 7.

2. Evolutionary Multi-Objective Optimization

Multi-objective optimization problems lie in many real-world applications and they contain multiple optimization objectives that conflict with each other. This makes conventional optimization algorithms (deterministic optimization method), e.g., linear programming method [25] and Newton-Raphson method [26], difficult to apply when solving these problems. One solution of the multi-objective optimization problems is to transform multiple objectives into a single objective by assigning different weights to each objective for a combination. This requires us to have a degree of deep understanding of multi-objective optimization problems.

Currently, more popular approaches use evolutionary multi-objective optimization algorithms because of various well-defined features and characteristics, such as strong robustness, ease of use, intelligence, and others. Almost all of these pay attention to finding a set of trade-off optimal solutions (known as Pareto optimal solutions), instead of a single optimal solution. The Pareto dominance and diversity of solutions are two primary subjects in EMO research. One attempts to obtain many non-dominated Pareto solutions, and the other tries to obtain Pareto solutions in a wide area on Pareto solution front. Here, we make a brief review of several techniques, strategies, and algorithms that solve the problems of EMO with regards to these two aspects.

2.1. Non-dominated Sorting Method

Non-dominated sorting is an elite mechanism for building a new generation of EMO algorithms for handling the non-dominated Pareto solutions. It is one of EMO selection strategies. The main motivation of this method is to find the non-dominated solutions by pairwise comparisons of all individuals. Here, a non-dominated individual means that there are no other individuals, whose all objectives are better than this one. The basic and formal calculation process of non-dominant sorting method can be implemented as the following steps.

1. Getting the first individual as a current individual;
2. Comparing all objectives of the current individual with those of all other individuals;
3. Counting the domination count N_p, which means the number of individuals that dominant the current individual;
4. Setting the individuals satisfy $N_p = 0$ as the first front, and remove these individuals from the generation temporally;
5. Repeating the above process until every individual is processed.

2.2. Crowding Distance Techniques

The Pareto solution diversity issue is also an important indicator of measurement in EMO algorithms. If the solution diversity is insufficient, it is easy to lead the Pareto optimal solutions not to be covered. Many methods attempt to maintain the solution diversity as much as possible in EMO algorithms. One of these methods uses a sharing parameter to keep the diversity during EMO search. However, it requires the preset and optimization of the parameter, and EMO performance depends on the setting. Crowding distance technique is another solution for handling this predefined parameter problems, which is calculated using a set of individuals.

The primary motivation of a crowding distance is to measure an individual density by distances. There is an aggregation level of the adjacent individuals in parameter space. Figure 1 presents a two-dimensional example of crowding distance calculation, where f_1 and f_2 are two objectives. The crowding distance can be calculated by averaging the length and width of the cuboid (marked by dash line). Averaging of the length and width of the cuboid (marked by dash line) is used to calculate the crowding distance for each individual.

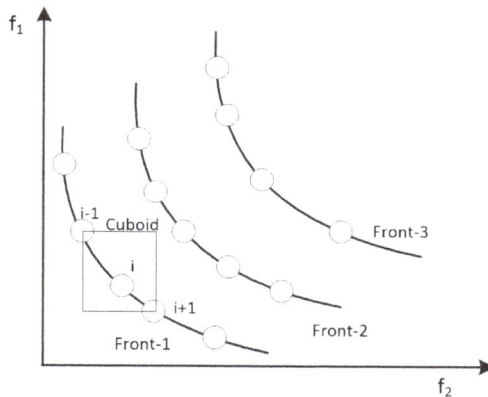

Figure 1. Two-dimensional example of crowding distance [27], the crowding distance is calculated within one Pareto front, its calculation method is $crowding\_distance = \sum_{o=1}^{P} \frac{obj_{o(i+1)} - obj_{o(i-1)}}{obj_{o_{max}} - obj_{o_{min}}}$, where o is the index of the number of objective functions, P is the number of objective functions, i is the index of individual, and obj is the value of an objective function.

Any EMO algorithms sort all individuals according to their first objective. We calculate the difference, i.e., $obj_{o(i+1)} - obj_{o(i-1)}$, between two adjacent individuals for all individuals. The first and last individual are set to infinite. This method calculates all crowding distances using multiple objectives. Finally, the final crowding distance of each individual is set to the mean of its crowding distances in all objectives.

2.3. NSGA and Its Variants

The NSGA is the first generation of EMO algorithm that uses non-dominated sorting techniques to find multiple Pareto optimal solutions with a single simulation running [28]. However, it has still suffered from several criticisms, including those relating to its high computational cost, lack of elitism, and the requirement for the setting of sharing parameter. Subsequently, its improved version, NSGA-II, was proposed to overcome all the above limitations at once by introducing fast non-dominated sorting and a tournament selection using a crowding distance to reduce computational complexity [27]. It has become one of the most popular EMO algorithms that are used to solve problems of multi-objective optimization. Recently, a more powerful version, NSGA-III [29], was also proposed, where a clustering operator replaces the crowding distance operator in NSGA-II to solve many-objective optimization problems. Actually, there are also many other EMO algorithms based on non-dominated sorting and have achieved satisfactory results, such as MOGA [30], NPGA [31], SPEA [32], SPEA2 [33], PESA [34], PESA-II [35], multi-objective chaotic evolution [36], etc.

Although various EMO algorithms have been proposed and have achieved outstanding results, most of them only focus on the study in an objective space. We therefore try to use moving vectors as a bridge between a parameter space and an objective space to analyze the landscapes of the two spaces. The motivation of this study promotes to use a mathematical method to estimate a convergence point in a parameter space using these moving vectors' information from its objective space. We expect this research to provoke the EMO researchers' attention towards the parameter space and encourage them to notice the connection between the two spaces for designing better EMO algorithms.

3. Estimating a Convergence Point

3.1. Notation Definitions

Before we explain how to estimate a convergence point from moving vectors, we offer some notations for better understanding of this section in advance. When an EC algorithm searches in a d-dimensional parameter space with n individuals ($n, d \in Z^+$), we notate the i-th individual in the current generation, a corresponding relative individual in the next generation, and their moving vector to be $\mathbf{a}_i$, $\mathbf{c}_i$, and $\mathbf{b}_i = \mathbf{c}_i - \mathbf{a}_i$, respectively, $\{\mathbf{a}_i, \mathbf{b}_i, \mathbf{c}_i \in \mathbb{R}^d; i = 1, 2, ..., n\}$ (See Figure 2). The unit vector of $\mathbf{b}_i$ is defined as $\mathbf{b}'_i = \mathbf{b}_i / \|\mathbf{b}_i\|$ ($\mathbf{b}_i^T \mathbf{b}_i = 1$). There are n moving vectors, and $\mathbf{a}_i$ is a starting point of $\mathbf{b}_i$.

We notate $\mathbf{x} \in \mathbb{R}^d$ as the estimated convergence point that has the minimal distance to the lines made by extending the line segments $\mathbf{b}'_i$. This point, $\mathbf{x}$, has a higher possibility to locate near the optimal solution in EC optimization problems. The $\mathbf{x}$ is indicated by the $\star$ mark in the Figure 2. We will explain how to obtain the $\mathbf{x}$ point by a deterministic mathematical method. In this work, all vectors are presented as column vectors.

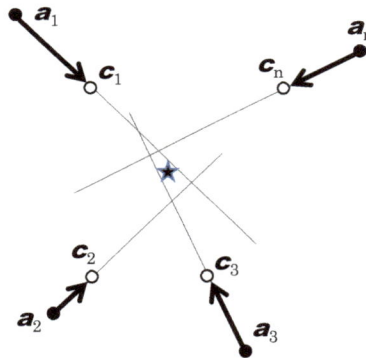

Figure 2. The convergence point ($\star$) can be estimated by the moving vectors ($\mathbf{b}_i$) between individuals ($\mathbf{a}_i$, $i = 1, 2, ..., n$) in the k-th generation and their offspring ($\mathbf{c}_i$, $i = 1, 2, ..., n$) in the ($k + 1$)-th generation.

3.2. Estimation Method of a Point from Moving Vectors

This section is primarily adopted from our previous work in reference [22]. From the principle of the law of large numbers, the estimated convergent point is the nearest one to the extension lines of these moving vectors. The lines of moving vectors can be expressed as $\mathbf{a}_i + t_i \mathbf{b}'_i$, $t_i \in \mathbb{R}$. The nearest point that is close to these extension lines can be obtained by solving the optimization problem shown in Equation (1).

$$\mathbf{x} = \min_{\mathbf{x},\{t_i\}} J(\mathbf{x},\{t_i\}) = \min_{\mathbf{x},\{t_i\}} \sum_{i=1}^{n} \|\mathbf{a}_i + t_i\mathbf{b}'_i - \mathbf{x}\|^2 \tag{1}$$

The shortest line segment from the estimated convergence point $\mathbf{x}$ to the extended moving vector $\mathbf{a}_i + t_i\mathbf{b}'_i$, $t_i \in \mathbb{R}$ is $\mathbf{a}_i + t_i\mathbf{b}'_i - \mathbf{x}$, $t_i \in \mathbb{R}$, and the relation of this line segment and the moving vector $\mathbf{b}$ or its unit vector $\mathbf{b}'$ is orthogonal. Equation (2) presents this orthogonal condition.

$$\mathbf{b}'^{\mathrm{T}}_i (\mathbf{a}_i + t_i\mathbf{b}'_i - \mathbf{x}) = 0 \quad \text{(orthogonal condition)} \tag{2}$$

From this orthogonal condition, we can use $\mathbf{x}$ to express t_i, i.e., $t_i = \frac{(\mathbf{b}'_i)^{\mathrm{T}}(\mathbf{x}-\mathbf{a}_i)}{\|\mathbf{b}'_i\|^2}$ and introduce it to Equation (1) to reduce the number of optimization parameter. The derivation process is presented in Equation (3), where I_d is an identity matrix, and $H_i = \mathbf{b}'_i\mathbf{b}'^{\mathrm{T}}_i - I_d$.

$$
\begin{aligned}
\mathbf{x} &= \min_{\mathbf{x},\{t_i\}} \sum_{i=1}^{n} \|\mathbf{a}_i + t_i\mathbf{b}'_i - \mathbf{x}\|^2 \\
&= \min_{\mathbf{x}} \sum_{i=1}^{n} \| \mathbf{b}'_i \frac{(\mathbf{b}'_i)^{\mathrm{T}}(\mathbf{x} - \mathbf{a}_i)}{\|\mathbf{b}'_i\|^2} - (\mathbf{x} - \mathbf{a}_i)\|^2 \\
&= \min_{\mathbf{x}} \sum_{i=1}^{n} \| \left\{ \frac{\mathbf{b}'_i\mathbf{b}'^{\mathrm{T}}_i}{\|\mathbf{b}'_i\|^2} - I_d \right\} (\mathbf{x} - \mathbf{a}_i)\|^2 \\
&= \min_{\mathbf{x}} \sum_{i=1}^{n} \|H_i(\mathbf{x} - \mathbf{a}_i)\|^2
\end{aligned}
\tag{3}
$$

Next, we can obtain the following objective function (Equation (4)) from Equation (1) where we have eliminated the term $\{t_i\}$.

$$J(\mathbf{x}) = \sum_{i=1}^{n} (\mathbf{x} - \mathbf{a}_i)^{\mathrm{T}} H_i^{\mathrm{T}} H_i (\mathbf{x} - \mathbf{a}_i) \tag{4}$$

Our goal is obtained by minimizing $J(\mathbf{x})$ regarding $\mathbf{x}$. Estimation of $\mathbf{x}$, i.e., $\hat{\mathbf{x}}$, is obtained by partially differentiating each element of $\mathbf{x}$ and setting them equal to 0 (shown in Equation (5)).

$$
\begin{aligned}
\frac{\partial J(\mathbf{x})}{\partial \mathbf{x}} &= 2 \sum_{i=1}^{n} H_i^{\mathrm{T}} H_i (\mathbf{x} - \mathbf{a}_i) \\
&= 2 \left\{ \left(\sum_{i=1}^{n} H_i^{\mathrm{T}} H_i \right) \mathbf{x} - \left(\sum_{i=1}^{n} H_i^{\mathrm{T}} H_i \mathbf{a}_i \right) \right\} \\
&= 0
\end{aligned}
\tag{5}
$$

Thus, the estimation of Equation (6) is obtained.

$$\hat{\mathbf{x}} = \left(\sum_{i=1}^{n} H_i^{\mathrm{T}} H_i \right)^{-1} \left(\sum_{i=1}^{n} H_i^{\mathrm{T}} H_i \mathbf{a}_i \right) \tag{6}$$

Since H_i has the characteristic of $H_i^{\mathrm{T}} H_i = H_i^2 = H_i$, i.e., a projection matrix, Equation (6) can be rewritten as in Equation (7), which we can use to estimate a convergence point using moving vectors.

$$\hat{\mathbf{x}} = \left(\sum_{i=1}^{n} H_i \right)^{-1} \left(\sum_{i=1}^{n} H_i \mathbf{a}_i \right)$$

$$= \left\{ \sum_{i=1}^{n} \left(I_d - \mathbf{b}'_i \mathbf{b}'^{\mathrm{T}}_i \right) \right\}^{-1} \left\{ \sum_{i=1}^{n} \left(I_d - \mathbf{b}'_i \mathbf{b}'^{\mathrm{T}}_i \right) \mathbf{a}_i \right\}$$

$$(7)$$

Besides these derivatives exactly estimating a convergence point, two approximated calculation methods are described in [22].

4. Accelerating EMO Search Using an Estimated Convergence Point

4.1. Philosophy of the Proposal

There are two subjects studied in the EMO algorithm research field; one study is the Pareto dominance issue, and the other one is EMO solution diversity issue. Almost all research on these two issues focuses on special handling in an objective space of an EMO algorithm, and frequently ignore the search condition in a parameter space. EMO algorithms try to find more non-dominated solutions with diversity. The solutions on the first Pareto solution frontier from the last generation to the next in an objective space supports information on how moving the variables in a parameter space can find promising Pareto solutions.

In Figure 3, we can find a set of the pairs of moving vectors in a parameter space in accord with the Pareto dominance information obtained in an objective space. We can use the moving vector information to estimate a convergence point that presents a promising area where Pareto solutions would be in a parameter space. We put such an estimated convergence point of a parameter space into EMO search and remove one of dominated solutions. EMO search should be enhanced considering such search information, and hopefully, EMO algorithm can find more non-dominated Pareto solutions quickly. This is a study hypothesis and motivation of our proposal that utilities an estimated convergence point to accelerate EMO search.

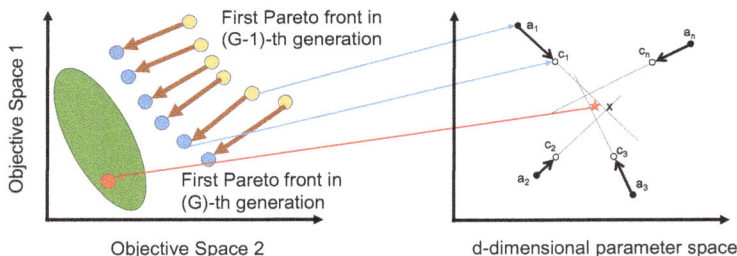

Figure 3. Estimation of promising Pareto solution area in parameter space using the dominance information from objective space to enhance EMO search.

4.2. Estimation of Pareto Solution Frontier in a Parameter Space from Pareto Improvement Information in an Objective Space

There are three primary steps and/or issues in the proposed method to enhance EMO search. The first step/issue is how to make pairs of moving vectors in a parameter space from Pareto improvement information obtained in an objective space. We make two candidate groups of non-dominated solutions in the current generation and in the last generation, so Pareto solution improvement information can be obtained from these two group individuals. Here, we design two methods to make moving vector pairs ($\mathbf{b}_i = \mathbf{c}_i - \mathbf{a}_i$ in Figure 2) .

- We pick up one of non-dominated solutions in an objective space from one group, and find the nearest non-dominated solution in the other group, and then find their corresponding individuals in a parameter space to make these two solutions form a pair. (Estimation in objective space)
- We pick up one of non-dominated solutions in an objective space from one group, and find its corresponding individual in a parameter space, and then find this individual's nearest individual in a parameter space to make these two solutions form a pair. (Estimation in parameter space)

After this, we delete these two solutions from the two groups, and we repeat this processing until one of groups becomes empty.

The second step estimates a convergence point in a parameter space using the moving vector pairs obtained from the first step. The estimation method uses Equation (7) to implement. The estimated point has high potential in the non-dominated Pareto solution frontier, and can therefore accelerate EMO search.

- Besides estimating only one estimated point, we can also estimate one point from only one single objective space each by each individually and use them together to accelerate EMO search (Estimation in each single objective space).

In the third step, we put the estimated convergence point as a search elite individual into EMO algorithms, and delete one/more of the dominated solutions in the current generation to enhance EMO search. This is the primary implementation within our proposal.

5. Experimental Evaluations

5.1. Experiment Setting

We use five multi-objective benchmark functions from the ZDT test suite [37] to evaluate our proposed methods. We embed our proposed method into conventional NSGA-II [27] with different constructing methods of moving vector, and compare our proposed method with NSGA-II. Table 1 presents the benchmark function's mathematical expressions. We examine these functions with three dimensional settings, i.e., two dimensions (2-D), 10-D, and 30-D. Table 2 shows the parameter settings of conventional NSGA-II algorithm used in the evaluation experiments.

Table 1. Multi-objective benchmark function used in evaluation [27]. All the Pareto frontier are $g(x) = 1$.

Functions	Definition		
ZDT1	$f_1(x)$	$=$	x_1
	$f_2(x)$	$=$	$g(x)[1 - \sqrt{\frac{x_1}{g(x)}}]$
	$g(x)$	$=$	$1 + 9\frac{\sum_{i=2}^n x_i}{n-1}$
ZDT2	$f_1(x)$	$=$	x_1
	$f_2(x)$	$=$	$g(x)[1 - (\frac{x_1}{g(x)})^2]$
	$g(x)$	$=$	$1 + 9\frac{\sum_{i=2}^n x_i}{n-1}$
ZDT3	$f_1(x)$	$=$	x_1
	$f_2(x)$	$=$	$g(x)[1 - \sqrt{\frac{x_1}{g(x)}} - \frac{x_1}{g(x)}\sin(10\pi x_1)]$
	$g(x)$	$=$	$1 + 9\frac{\sum_{i=2}^n x_i}{n-1}$
ZDT4	$f_1(x)$	$=$	x_1
	$f_2(x)$	$=$	$g(x)[1 - (\frac{x_1}{g(x)})^2]$
	$g(x)$	$=$	$1 + 10(n-1) + \sum_{i=2}^n [x_i^2 - 10\cos(4\pi x_i)]$
ZDT6	$f_1(x)$	$=$	$1 - \exp(-4\pi x_1)\sin^6(6\pi x_1)$
	$f_2(x)$	$=$	$g(x)[1 - (\frac{f(x_1)}{g(x)})^2]$
	$g(x)$	$=$	$1 + 9[\frac{\sum_{i=2}^n x_i}{n-1}]^{0.25}$

Three experiments are designed where different methods of constructing moving vectors, and these combined with the conventional NSGA-II algorithm. The legends displayed in figures and tables have the following meanings.

- **NSGA-II;** conventional NSGA-II algorithm;
- **Estimation in objective space;** we construct moving vectors from two subsequent non-dominated solution set in an objective space;
- **Estimation in parameter space;** we find the nearest offspring individual for each one in a parent generation, and make pairs in a parameter space; and
- **Estimation in each single objective space;** we consider each objective independently and estimation convergence point for each objective, where the estimated points may not be best on all objectives, but they have good potential in some objectives.

Table 2. NSGA-II algorithm parameter setting.

population size for 2-D, 10-D, and 30-D	20, 50, and 100
crossover rate	0.8
mutation rate	0.05
max. # of fitness evaluations, MAX_{NFC}, for 2-D, 10-D, and 30-D search	400, 1000, and 10,000
dimensions of benchmark functions, D	2, 10, and 30
# of trial runs	30

5.2. Evaluation Metrics

We set the stop conditions of each evaluation using the number of fitness calls instead of generations for fair evaluation, because our proposed methods increase additional fitness cost consumption. We set the stop conditions as 400 times, 1000 times, and 10,000 times of fitness evaluations in 2-D, 10-D, and 30-D problems, respectively. Besides, we test each benchmark function with 30 trial runs in three different dimensional settings.

Conventional NSGA-II is adopted as an example algorithm; other EMO algorithms can be also applied. Although there are many ways to generate estimated points, the greedy replacement strategy, where the estimated points will replace with the worst ranked and low diversity individuals to keep the same population size, is adopted in the proposed acceleration framework. To analyze the effect of the proposed acceleration framework, we calculate the number of non-dominated Pareto solutions in each generation shown in Figures 4–6.

Hyper volume [38] is used to evaluate the diversity and acceleration performance of our proposal. Table 3 presents the hyper volume values of our proposed method and conventional NSGA-II algorithm at the stop condition in three different dimensional settings. We apply Wilcoxon signed-rank test for 30 trail runs data to evaluate the significance of hyper volume obtained by conventional NSGA-II and our proposal. Some functions without hyper volume value is due to reference point $[-1, 1]$ setting.

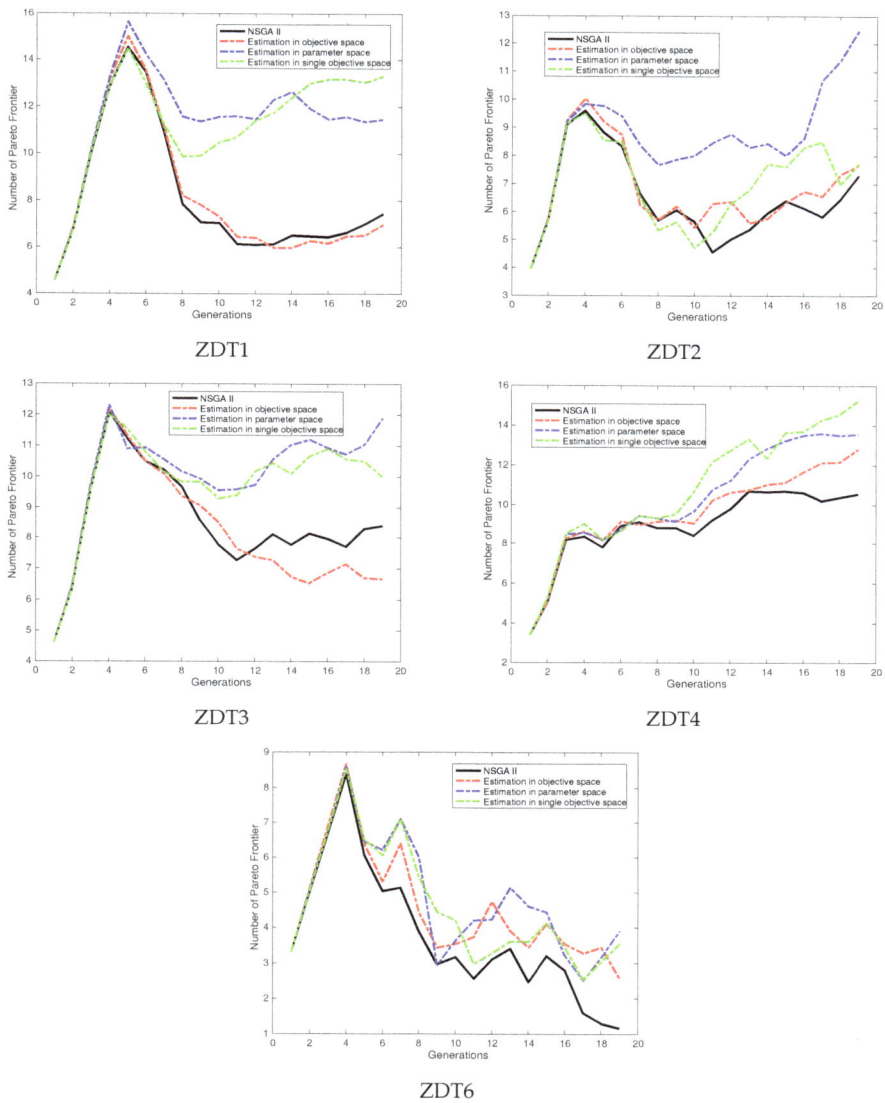

ZDT1

ZDT2

ZDT3

ZDT4

ZDT6

Figure 4. The number of Pareto solutions in every generation for 2-D benchmark problems. We can observe that proposed method can obtain more Pareto solutions for the most of cases.

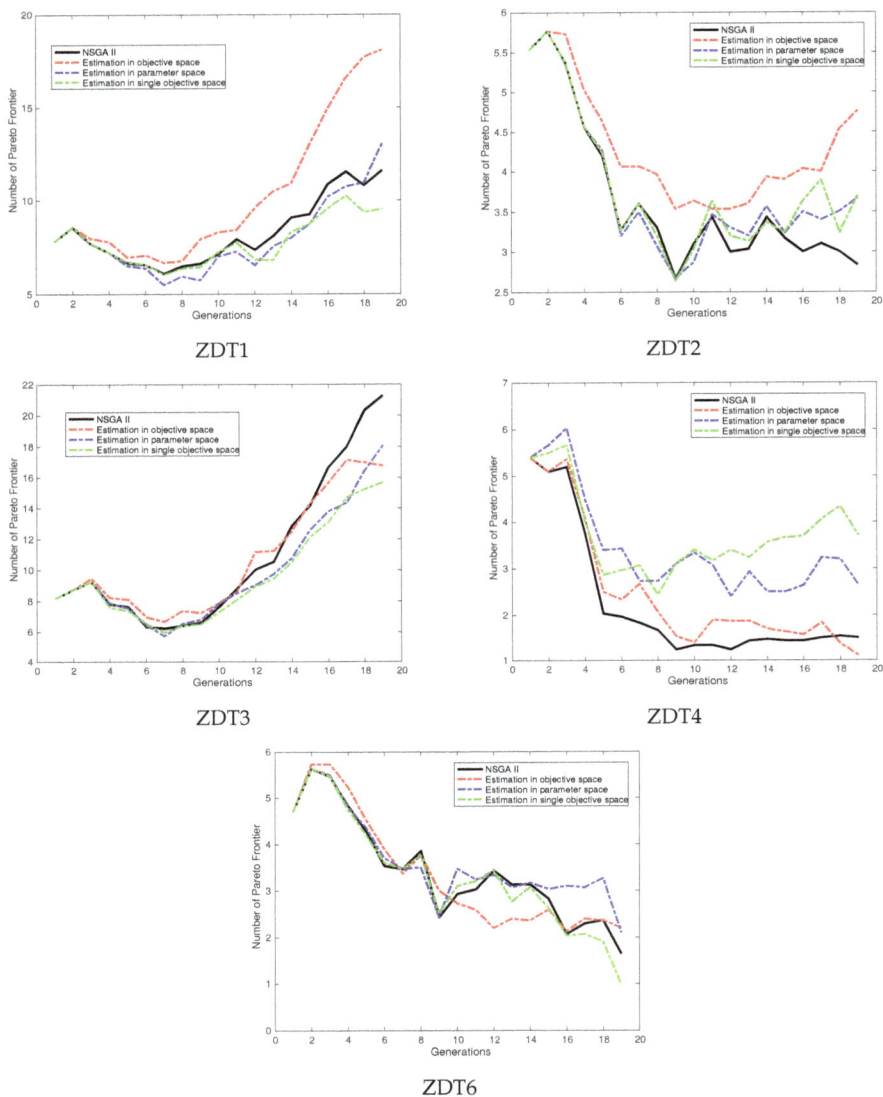

ZDT1

ZDT2

ZDT3

ZDT4

ZDT6

Figure 5. The number of Pareto solutions in every generation for 10-D benchmark problems. We can observe that proposed method can obtain more Pareto solutions for the most of cases.

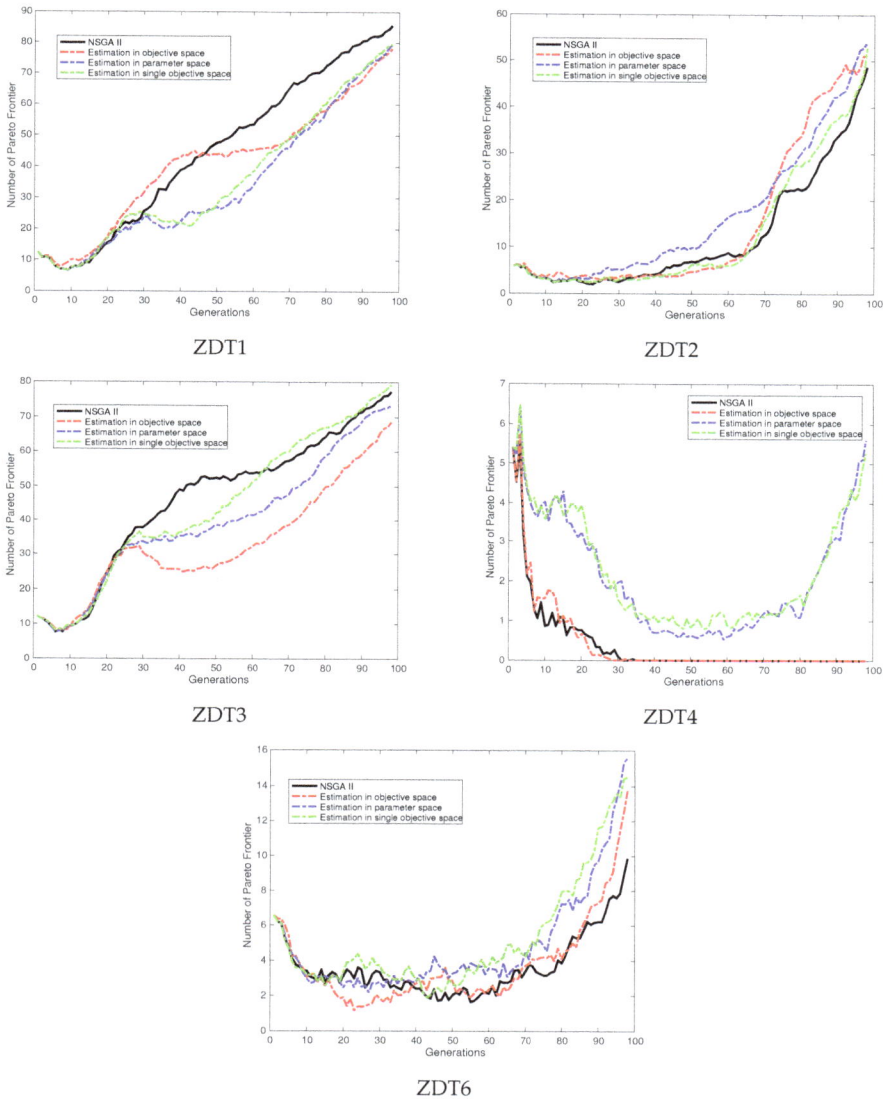

Figure 6. The number of Pareto solutions in every generation for 30-D benchmark problems. We can observe that proposed method can obtain more Pareto solutions for the most of cases.

Table 3. The average hypervolume values from 30 trials running of 4 methods in 2 dimensions (2-D), 10-D, and 30-D. Symbol † means that there is a significant difference between NSGA-II and Proposed method, i.e., NSGA-II + Estimation Point. The reference point is [1,1]. Obj., Para., and SinglePara. present objective space, parameter space, and each single parameter space, respective.

	2-D tasks			
Func.	NSGA-II	Estimation in Obj.	Estimation in Para.	Estimation in SinglePara.
ZDT1	0.414567	0.417300	0.453833 †	0.480533 †
ZDT2	0.109833	0.117367	0.124367	0.113867
ZDT3	0.556733	0.552433	0.622733 †	0.622600 †
ZDT4	0.255733	0.261600	0.284933 †	0.337167 †
ZDT6	0.000033	0.002033	0.001833	0.000967
	10-D tasks			
Func.	NSGA-II	Estimation in Obj.	Estimation in Para	Estimation in SinglePara
ZDT1	0.328033	0.337767	0.345533	0.339433
ZDT2	0.008633	0.012367	0.010100	0.008933
ZDT3	0.564067	0.545767	0.589400	0.588933
	30-D tasks			
Func.	NSGA-II	Estimation in Obj.	Estimation in Para	Estimation in SinglePara
ZDT1	0.647167	0.654000	0.651533	0.648633
ZDT2	0.183333	0.157567	0.187700	0.190433
ZDT3	0.796133	0.792600	0.791533	0.794767
ZDT6	0.066233	0.068433	0.069633	0.066733

6. Discussions

6.1. Pareto Improvement of the Proposal

Pareto dominance and Pareto solution diversity are two metrics to evaluate the performance of EMO algorithms. In this work, we calculated the average number of Pareto solutions in every generation for each benchmark problem; see Figures 4–6. This is one of evaluation metrics for Pareto dominance in EMO. We also calculated hyper volume values at the maximal number of function calls for each dimension setting, and applied Wilcoxon signed-rank test to verify the significant difference among hyper volume values in Table 3. This is a demonstration of Pareto solution diversity for each EMO algorithm. We analyze and discuss our proposed method using these results.

From Figures 4–6, we can observe that methods estimating a convergence point in a parameter space and in a single objective space can obtain more Pareto solutions from all five multi-objective benchmark problems in 2-D setting. Method estimating a convergence point in an objective space fails in two benchmark functions, i.e., ZDT1 and ZDT3 in 2-D tasks. It indicates that moving vectors constructed from information of the nearest points in an objective space cannot exactly estimate the non-dominated Pareto frontier area in a parameter space. The same case can also be found in 10-D benchmark setting for ZDT3 and ZDT4, and 30-D benchmark setting for ZDT1, ZDT3, and ZDT4. We need to further consider improving the estimation the accuracy of estimation method of a convergence point in an objective space.

That in a single objective space works well in most of cases because this method replaces more than one estimated convergence point, and increases the population diversity for EMO algorithms. This indicates that the better individuals in each objective can improve optimization performance of EMO algorithms, although there are conflicts among multi-objective functions when EMO searches for non-dominated Pareto solutions. From this viewpoint, elite strategy-based EC acceleration methods can be applied not only in single objective problems, but also have a potential to be applied in multi-objective problems.

From observation of Table 3, the values of hyper volume from our proposed method are bigger than those from conventional NSGA-II algorithm for the most of tasks in 2-D benchmark problems. The Wilcoxon signed-rank test results showed a significant difference between our proposed method and the conventional NSGA-II algorithm in estimation in a parameter space and estimation in a single objective space. These results demonstrate that our proposed method can obtain non-dominated Pareto solution with more diversities for EMO algorithms. However, it is not significant shown in 10-D and 30-D benchmark problems. It is a limitation for our proposal, and we need to improve it in our future work.

6.2. Topological Structure of Moving Vectors and Modality Characteristic of Pareto Improvement

The basic philosophy of our proposed method to accelerate EMO search lies in three hypotheses. First, we can obtain the information to improve non-dominated Pareto solutions through Pareto solution evolutions from the last generation to the current generation in an objective space. Second, after we obtain the information, the moving vectors can be made in an objective space or in a parameter space. Third, the estimated convergence point of these moving vectors has a high possibility that locals in the non-dominated Pareto solution frontier area in a parameter space. From the modality viewpoint, the distribution of Pareto solutions shows a uni-modal characteristic in the objective space. In the case of the Pareto improvement from the last generation to the current generation, do the corresponding individuals also present a uni-modal distribution characteristic in a parameter space? We examine this question here.

We present improved EMO evolutions along three generations' EMO evolution condition both in an objective space and a parameter space for these benchmark functions with 2-D setting (see Figures 7 and 8). The arrows demonstrate the Pareto improvement directions between two generations in both spaces. From Figures 7 and 8, we observe that with regards to the directions of arrows in an objective space, all of them are towards almost the same direction, i.e., their angles are less than 90 degrees. However, in the parameter space, the arrows are not towards the same direction. It displays a multi-modal distribution characteristic, e.g., in the ZDT4 and ZDT6 benchmark problems. From these observations, it indicates that Pareto improvement in the objective space presents a uni-modal characteristic, while it presents a multi-modal characteristic in a parameter space. In Figure 4, the numbers of the first Pareto frontier solution from four methods are almost the same, but their acceleration performances are not significant. This is one our discovery on the modality characteristic of Pareto improvement in both an objective space and a parameter space.

From Figures 7 and 8, there is a multi-modal characteristic in a parameter space when the Pareto improvement occurs from one generation to the next. The third hypothesis of proposed method is not always correct, therefore, the proposed method can work well in the uni-modal condition of Pareto improvement, and by chance well in the multi-modal one. From Table 3, there is not a significant difference between NSGA-II algorithm and our proposed method in ZDT6. This experiment's results verify our analysis and observations. The multi-modal characteristic of Pareto improvement in a parameter space is an issue when applying our proposal to enhance EMO search.

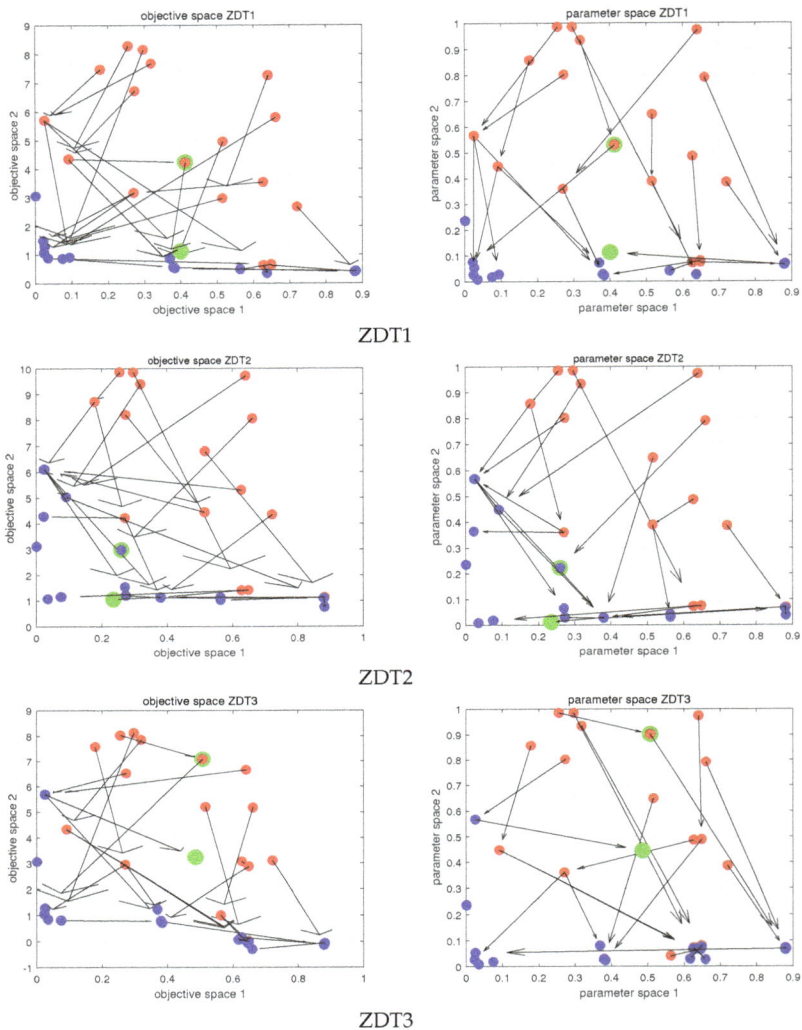

Figure 7. Two-dimensional demonstration of Pareto solution improvement in an objective space (**left**) and their corresponding individuals in parameter space (**right**) of ZDT1, ZDT2, and ZDT3. The arrows show directions of both Pareto solution improvement and moving vectors. We can observe that there is a uni-modal landscape for Pareto solution improvement in an objective space; however, it is a multi-modal landscape for Pareto improvement in a parameter space. The green point is the estimated convergence point, most of the red points and most of the blue points are in the first generation and in the third generation, respectively.

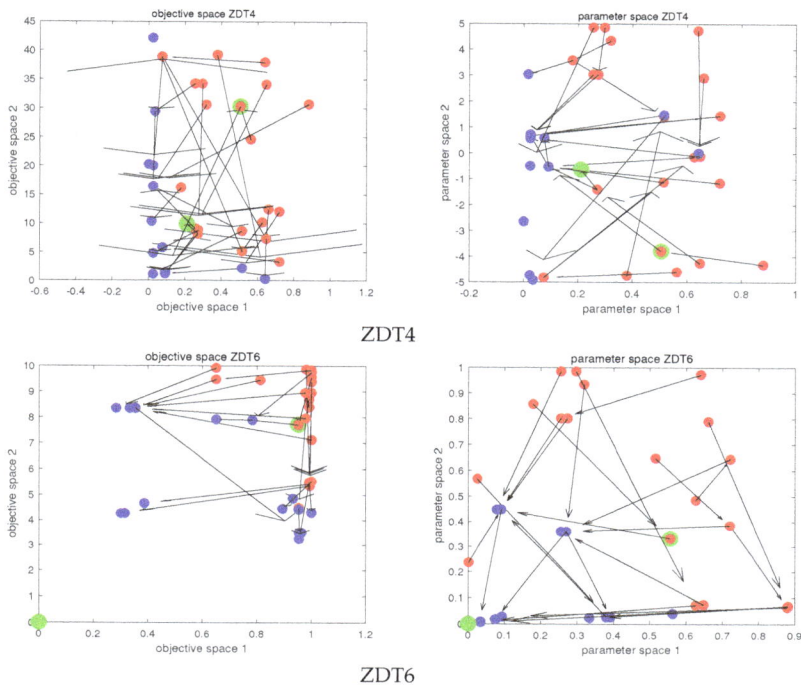

ZDT4

ZDT6

Figure 8. Two-dimensional demonstration of Pareto solution improvement in an objective space (**left**) and their corresponding individuals in parameter space (**right**) of ZDT4, and ZDT6. The arrows show directions of both Pareto solution improvement and moving vectors. We can observe that there is a uni-modal landscape for Pareto solution improvement in an objective space; however, it is a multi-modal landscape for Pareto improvement in a parameter space. The green point is the estimated convergence point, most of the red points and most of the blue points are in the first generation and in the third generation, respectively.

7. Conclusions and Future Work

In this work, we use an estimated convergence point from dominance information of Pareto solution improvement to enhance EMO search. We use NSGA-II as a test algorithm and five multi-objective functions to qualitatively evaluate our proposal. We found that our proposed method can enhance EMO search in some benchmark problems, especially for the high-dimensional and complex multi-objective problems which can obtain a greater number of Pareto solutions. We also analyzed the modality of the Pareto improvement in both an objective space and a parameter space. We found that the Pareto improvement in an objective space demonstrates a uni-modal characteristic, but a multi-modal one in parameter space. It is one of the discoveries in this work.

In the future, we will further investigate the proposed method in a variety of multi-objective problems, especially for real-world problems. How to find the exact pairs information of moving vectors is one of the potential study subjects in our method. It influences the accuracy of estimated point to make different performances of our proposed by using the point. The multi-modal characteristic of moving vectors in a parameter space is an issue for our estimation method. We will use clustering methods to find the representative moving vectors to find the estimated point in a parameter space. Another study issue is a search condition using multi-objective fitness landscape and an estimated convergence point. These and other study subjects will be involved in our future research work.

Author Contributions: Conceptualization, Y.P.; Funding acquisition, Y.P. and H.T.; Investigation, Y.P. and J.Y.; Methodology, Y.P. and H.T.; Project administration, H.T.; Software, Y.P. and J.Y.; Validation, Y.P.; Visualization, Y.P.; Writing—original draft, Y.P.

Funding: Japan Society for the Promotion of Science: JP15K00340 and 18K11470.

Acknowledgments: The work is supported by the JSPS Grant-in-Aid for Scientific Research C (JP15K00340 and 18K11470).

Conflicts of Interest: The author declares that there is no conflict of interests regarding the publication of this paper.

References

1. Li, B.; Li, J.; Tang, K.; Yao, X. Many-objective evolutionary algorithms: A survey. *ACM Comput. Surv.* **2015**, *48*, 13. [CrossRef]
2. Jin, Y. A comprehensive survey of fitness approximation in evolutionary computation. *Soft Comput.* **2005**, *9*, 3–12. [CrossRef]
3. Jin, Y.; Olhofer, M.; Sendhoff, B. A Framework for evolutionary optimization with approximate fitness functions. *IEEE Trans. Evol. Comput.* **2002**, *6*, 481–494.
4. Jin, Y. Surrogate-assisted evolutionary computation: Recent advances and future challenges. *Swarm Evol. Comput.* **2011**, *1*, 61–70. [CrossRef]
5. Witkin, A.P. Scale-space filtering. In Proceedings of the 8th International Joint Conference Artificial Intelligence, Karlsruhe, Germany, 8–12 August 1983; pp. 1019–1022.
6. Takagi, H.; Ingu, T.; Ohnishi, K. Accelerating a GA convergence by fitting a single-peak function. *J. Soft* **2003**, *15*, 219–229. [CrossRef]
7. Pei, Y.; Takagi, H. Accelerating IEC and EC searches with elite obtained by dimensionality reduction in regression spaces. *Evol. Intell.* **2013**, *6*, 27–40. [CrossRef]
8. Pei, Y.; Zheng, S.; Tan, Y.; Takagi, H. Effectiveness of approximation strategy in surrogate-assisted fireworks algorithm. *Int. J. Mach. Learn. Cybern.* **2015**, *6*, 795–810. [CrossRef]
9. Zhao, N.; Zhao, Y.; Fu, C. Genetic algorithm with fitness approximate mechanism. *J. Natl. Univ. Def. Technol.* **2014**, *36*, 116–121.
10. Pei, Y.; Takagi, H. Fourier analysis of the fitness landscape for evolutionary search acceleration. In Proceedings of the 2012 IEEE Congress on Evolutionary Computation, Brisbane, Australia, 10–15 June 2012; pp. 1–7.
11. Michael, D.S.; Hod, L. Coevolution of fitness predictors. *IEEE Trans. Evol. Comput.* **2008**, *12*, 736–749.
12. Michael, D.S.; Hod, L. Co-evolution of fitness maximizers and fitness predictors. In Proceedings of the Genetic and Evolutionary Computation Conference, Washington, DC, USA, 25–29 June 2005; pp. 1–8.
13. Michael, D.S.; Hod, L. Co-evolving fitness predictors for accelerating evaluations and reducing sampling. *Genet. Programm. Theory Pract. IV* **2006**, *5*, 113–130.
14. Michael, D.S.; Hod, L. Predicting solution rank to improve performance. In Proceedings of the 12th Annual Genetic and Evolutionary Computation Conference, Portland, OR, USA, 7–11 July 2010; pp. 949–955.
15. He, Y.; Yuen, S.Y.; Lou, Y. Exploratory landscape analysis using algorithm based sampling. In Proceedings of the 2018 Genetic and Evolutionary Computation Conference Companion, Kyoto, Japan, 15–19 July 2018; pp. 211–212.
16. Mersmann, O.; Bischl, B.; Trautmann, H.; Preuss, M.; Weihs, C.; Rudolph, G. Exploratory landscape analysis. In Proceedings of the Genetic and Evolutionary Computation Conference, Dublin, Ireland, 12–16 July 2011; pp. 829–836.
17. Wang, G.G.; Tan, Y. Improving metaheuristic algorithms with information feedback models. *IEEE Trans. Cybern.* **2017**. [CrossRef] [PubMed]
18. Wang, G.G.; Guo, L.; Gandomi, A.H.; Hao, G.S.; Wang, H. Chaotic krill herd algorithm. *Inf. Sci.* **2014**, *274*, 17–34. [CrossRef]
19. Pei, Y. Chaotic evolution: fusion of chaotic ergodicity and evolutionary iteration for optimization. *Nat. Comput.* **2014**, *13*, 79–96. [CrossRef]
20. Pei, Y.; Takagi, H. Research progress survey on interactive evolutionary computation. *J. Ambient Intell. Hum. Comput.* **2018**, 1–14. [CrossRef]

21. Pei, Y.; Takagi, H. A survey on accelerating evolutionary computation approaches. In Proceedings of the 2011 International Conference of Soft Computing and Pattern Recognition (SoCPaR), Dalian, China, 14–16 October 2011; IEEE: Piscataway, NJ, USA, 2011; pp. 201–206.

22. Murata, N.; Nishii, R.; Takagi, H.; Pei, Y. Analytical estimation of the convergence point of populations. In Proceedings of the 2015 IEEE Congress on Evolutionary Computation, Sendai, Japan, 25–28 May 2015; IEEE: Piscataway, NJ, USA, 2015; pp. 2619–2624.

23. Yu, J.; Pei, Y.; Takagi, H. Accelerating evolutionary computation using estimated convergence points. In Proceedings of the 2016 IEEE Congress on Evolutionary Computation, Vancouver, BC, Canada, 24–29 July 2016; IEEE: Piscataway, NJ, USA, 2016; pp. 1438–1444.

24. Yu, J.; Takagi, H. Clustering of moving vectors for evolutionary computation. In Proceedings of the 2015 7th International Conference of Soft Computing and Pattern Recognition, Fukuoka, Japan, 13–15 November 2015; IEEE: Piscataway, NJ, USA, 2015; pp. 169–174.

25. Dantzig, G.B. *Maximization of a Linear Function of Variables Subject to Linear Inequalities*; John Wiley & Sons: New York, NY, USA, 1951.

26. Wallis, J. A treatise of algebra, both historical and practical. *Philos. Trans.* **1685**, *15*, 1095–1106.

27. Deb, K.; Pratap, A.; Agarwal, S.; Meyarivan, T. A fast and elitist multiobjective genetic algorithm: NSGA-II. *IEEE Trans. Evol. Comput.* **2002**, *6*, 182–197. [CrossRef]

28. Srinvas, N.; Deb, K. Multi-objective function optimization using non-dominated sorting genetic algorithms. *Evol. Comput.* **1994**, *2*, 221–248. [CrossRef]

29. Deb, K.; Jain, H. An evolutionary many-objective optimization algorithm using reference-point-based nondominated sorting approach, Part I: Solving Problems With Box Constraints. *IEEE Trans. Evol. Comput.* **2013**, *18*, 577–601. [CrossRef]

30. Carlos, M.F.; Peter, F. Genetic algorithms for multiobjective optimization: Formulation discussion and generalization. In Proceedings of the 5th International Conference on Genetic Algorithms, Urbana, Champaign, IL, USA, 17–21 July 1993; pp. 416–423.

31. Rey Horn, J.; Nafpliotis, N.; Goldberg, D.E. A niched Pareto genetic algorithm for multiobjective optimization. In Proceedings of the First IEEE Conference on Evolutionary Computation, Orlando, FL, USA, 27–29 June 1994; pp. 82–87.

32. Zitzler, E.; Thiele, L. Multi-Objective evolutionary algorithms: A comparative case study and the strength Pareto approach. *IEEE Trans. Evol. Comput.* **1999**, *3*, 257–271. [CrossRef]

33. Eckart, Z.; Marco, L.; Lothar, T. SPEA2: Improving the strength Pareto evolutionary algorithm. In *Evolutionary Methods for Design, Optimization and Control with Applications to Industrial Problems, Proceedings of the EUROGEN2001 Conference, Athens, Greece, 19–21 September 2001*; International Center for Numerical Methods in Engineering: Barcelona, Spain, 2001; pp. 95–100.

34. David, W., C.; Joshua, D., K.; Martin, J., O. The Pareto-envelope based selection algorithm for multi-objective optimization. In Proceedings of the 6th International Conference on Parallel Problem Solving from Nature, Paris, France, 18–20 September 2000; pp. 839–848.

35. Corne, D.W.; Jerram, N.R.; Knowles, J.D.; Oates, M.J.; J, M. PESA-II: Region-based selection in evolutionary multiobjective optimization. In Proceedings of the Genetic and Evolutionary Computation Conference, San Francisco, CA, USA, 7–11 July 2001; pp. 283–290.

36. Pei, Y.; Hao, J. Non-dominated sorting and crowding distance based multi-objective chaotic evolution. In *International Conference in Swarm Intelligence*; Springer: Cham, Switzerland, 2017; pp. 15–22.

37. Zitzler, E.; Deb, K.; Thiele, L. Comparison of multiobjective evolutionary algorithms: Empirical results. *Evol. Comput.* **2000**, *8*, 173–195. [CrossRef] [PubMed]

38. Zitzler, E.; Brockhoff, D.; Thiele, L. The hypervolume indicator revisited: On the design of Pareto-compliant indicators via weighted integration. In *Evolutionary Multi-Criterion Optimization*; Springer: Berlin, Germany, 2007; pp. 862–876.

mathematics

Article

The Importance of Transfer Function in Solving Set-Union Knapsack Problem Based on Discrete Moth Search Algorithm

Yanhong Feng [1,2,3], Haizhong An [1,2,*] and Xiangyun Gao [1,2]

1   School of Economics and Management, China University of Geosciences, Beijing 100083, China;
    qinfyh@hgu.edu.cn (Y.F.); gaoxy@cugb.edu.cn (X.G.)
2   Key Laboratory of Carrying Capacity Assessment for Resource and Environment,
    Ministry of Natural Resources, Beijing 100083, China
3   School of Information Engineering, Hebei GEO University, Shijiazhuang 050031, China
*   Correspondence: ahz369@cugb.edu.cn

Received: 1 September 2018; Accepted: 10 December 2018; Published: 24 December 2018

Abstract: Moth search (MS) algorithm, originally proposed to solve continuous optimization problems, is a novel bio-inspired metaheuristic algorithm. At present, there seems to be little concern about using MS to solve discrete optimization problems. One of the most common and efficient ways to discretize MS is to use a transfer function, which is in charge of mapping a continuous search space to a discrete search space. In this paper, twelve transfer functions divided into three families, S-shaped (named S1, S2, S3, and S4), V-shaped (named V1, V2, V3, and V4), and other shapes (named O1, O2, O3, and O4), are combined with MS, and then twelve discrete versions MS algorithms are proposed for solving set-union knapsack problem (SUKP). Three groups of fifteen SUKP instances are employed to evaluate the importance of these transfer functions. The results show that O4 is the best transfer function when combined with MS to solve SUKP. Meanwhile, the importance of the transfer function in terms of improving the quality of solutions and convergence rate is demonstrated as well.

Keywords: set-union knapsack problem; moth search algorithm; transfer function; discrete algorithm

---

1. Introduction

The knapsack problem (KP) [1] is still considered as one of the most challenging and interesting classical combinatorial optimization problems, because it is non-deterministic polynomial hard problem and has many important applications in reality. As an extension of the standard 0–1 knapsack problem (0–1 KP) [2], the set-union knapsack problem (SUKP) [3] is a novel KP model recently introduced in [4,5]. The SUKP finds many practical applications such as financial decision making [4], data stream compression [6], flexible manufacturing machine [3], and public key prototype [7].

The classical 0–1 KP is one of the simplest KP model in which each item has a unique value and weight. However, SUKP is constructed of a set of items $S = \{U_1, U_2, U_3, \ldots, U_m\}$ and a set of elements $U = \{u_1, u_2, u_3, \ldots, u_n\}$. Each item is associated with a subset of elements. In SUKP, each item has a nonnegative profit and each element has a nonnegative weight. The goal is to maximize the total profit of a subset of items $S* \subset S$ such that the total weight of the corresponding element does not exceed the maximum capacity of knapsack C. Hence, SUKP is more complicated and more difficult to handle than the standard 0–1 KP. Thus far, only a few researchers have studied this issue despite its practical importance and NP-hard character. For example, Goldschmidt et al. applied the dynamic programming (DP) algorithm for SUKP [3]. However, when the exact algorithm is used, no satisfactory approximate solution is usually obtained in polynomial time. Afterwards, Ashwin [4] proposed an approximation algorithm A-SUKP for SUKP. Obviously, A-SUKP also has to face the inevitable problem, that is, how to compromise between achieving a high-quality solution and exponential runtime. Recently, He et al. [5] presented a binary artificial bee colony algorithm (BABC) to solve SUKP and comparative studies were conducted among BABC, A-SUKP, and binary differential evolution (DE) [8]. The results verified that BABC outperformed A-SUKP method. Ozsoydan et al. [9] proposed a swarm intelligence-based algorithm for the SUKP and designed an effective mutation procedure. Although this method does not require transfer functions, it lacks generality. Therefore, it is urgent to find an efficient metaheuristic algorithm to address SUKP whether from the perspective of academic research or practical application.

As a relatively novel nature-inspired metaheuristic algorithm, moth search (MS) algorithm was recently developed for continuous optimization by Wang [10]. Computational experiments have shown that MS is not only effective but also efficient when addressing unconstrained continuous optimization problems, compared with five state-of-the-art metaheuristic algorithms. Because of its relative novelty, extensive research on MS is relatively scarce, especially discrete version MS algorithm. Feng et al. presented a binary moth search algorithm (BMS) for discounted {0–1} knapsack problem (DKP) [11].

As we all know, the metaheuristic algorithm is usually discretized in two ways: direct discretization and indirect discretization. Direct discretization is usually achieved by modifying the evolutionary operator of the original algorithm to solve a particular discrete problem. This method depends on the algorithm used and the problem solved. Obviously, the disadvantages of direct discretization are lack of versatility and complicated operation. The latter is discretized by establishing a mapping relationship between continuous space and discrete space. Concretely speaking, indirect discretization is usually achieved by an appropriate transfer function to convert real-valued variables into discrete variables. Many discrete versions of swarm intelligence algorithms using transfer functions have been proposed to solve various optimization problems. Discrete binary particle swarm optimization [12], discrete firefly algorithm [13], and binary harmony search algorithm [14] are among the most typical algorithms. Through analyzing the literature, many kinds of transfer functions can be used, such as sigmoid function [12], tanh function [15], etc. However, most existing metaheuristics only consider one transfer function. Little research concentrates on the importance of transfer functions in solving discrete problems. In addition, a few studies [16,17] investigate the efficiency of multiple transfer functions.

In this paper, twelve principal transfer functions are used and then twelve new discrete MS algorithms are proposed to solve SUKP. These functions include four S-shaped transfer functions [16,17], named S1, S2, S3, and S4, respectively; four V-shaped transfer functions [16,17], named V1, V2, V3, and V4, respectively; and four other shapes transfer functions (Angle modulation method [18,19], Nearest integer method [20,21], Normalization method [22], and Rectified linear unit method [23]), named O1, O2, O3, and O4, respectively. Therefore, combining twelve transfer functions with MS algorithm, twelve discrete MS algorithms are naturally proposed, named as MSS1, MSS2, MSS3, MSS4, MSV1, MSV2, MSV3, MSV4, MSO1, MSO2, MSO3, and MSO4, respectively.

The remainder of the paper is organized as follows. In Section 2, we briefly introduce the SUKP problem and MS algorithm. The families of transfer functions and repair optimization mechanism are presented in Section 3. In Section 4, the twelve discrete MS algorithms are compared to shed light on how the transfer functions affect the performance of the algorithm. After that, the best algorithm (MSO4) is compared with five state-of-the-art methods on fifteen SUKP instances. Finally, we draw conclusions and suggest some directions for future research.

2. Background

To describe discrete MS algorithm for the SUKP, we first explain the mathematical model of SUKP and then introduce the MS algorithm.

2.1. Set-Union Knapsack Problem

The set-union knapsack problem (SUKP) [3,4] is a variant of the classical 0–1 knapsack problem (0–1 KP). More formally, the SUKP can be defined as follows: given a set of elements $U = \{u_1, u_2, u_3, \ldots, u_n\}$ and a set of items $S = \{U_1, U_2, U_3, \ldots, U_m\}$, such that S is the cover of U, and $U_i \neq \varnothing \wedge U_i \subset U$ ($i = 1, 2, 3, \ldots, m$) and each item U_i has a value $p_i > 0$. Each element u_j ($j = 1, 2, 3, \ldots, n$) has a weight $w_j > 0$. Suppose that set A consists of some items packed into the knapsack with capacity C, namely $A \subseteq S$. Then, the profit of A is defined as $P(A) = \sum_{U_i \in A} p_i$ and the weight of A is defined as $W(A) = \sum_{u_j \in \bigcup_{U_i \in A} U_i} w_j$. The objective of the SUKP is to find a subset A that maximizes the total value $P(A)$ on condition that the total weight $W(A) \leq C$. Then, the mathematical model of SUKP can be formulated as follows:

$$Max \ \ P(A) = \sum_{U_i \in A} p_i \tag{1}$$

$$subject \ to \ W(A) = \sum_{u_j \in \bigcup_{U_i \in A} U_i} w_j \leq C, \ \ A \subseteq S \tag{2}$$

where p_i ($i = 1, 2, 3, \ldots, m$), w_j ($j = 1, 2, 3, \ldots, n$), and C are all positive integers.

Recently, an integer programming model is proposed by He et al. [5] to solve SUKP easily by using metaheuristic algorithm; the new mathematical model of SUKP can be defined as follows:

$$Max \ \ f(Y) = \sum_{i=1}^{m} y_i p_i \tag{3}$$

$$subject \ to \ W(A_Y) = \sum_{j \in \bigcup_{U_i \in A_Y} U_i} w_j \leq C \tag{4}$$

Obviously, all the 0–1 vectors $Y = [y_1, y_2, y_3, \ldots, y_m] \in \{0, 1\}^m$ are the potential solutions of SUKP. A solution satisfying the constraint of Equation (4) is a feasible solution; otherwise, it is an infeasible solution. $A_Y = \{U_i \mid y_i \in Y, y_i = 1, 1 \leq i \leq m\} \subseteq S$. Then, $y_i = 1$ if and only if $U_i \in A_Y$.

2.2. Moth Search Algorithm

The MS algorithm [10] is a novel metaheuristic algorithm that was inspired by the phototaxis and Lévy flights of the moths in nature, which are the two most representative characteristics of moths. The MS is akin to other population-based swarm intelligence algorithms. However, MS differs from most the population-based metaheuristic algorithms, such as genetic algorithm (GA) [24,25] and particle swarm optimization algorithm (PSO) [26,27], which consist of only one population, as, in MS, the whole population is divided into two subpopulations according to the fitness, namely subpopulation1 and subpopulation2.

The MS starts its evolutionary process by first randomly generating n moth individuals. Each moth individual represents a candidate solution to the corresponding problem with a specific fitness function. In MS, two operators are considered including Lévy flights operator and straight flight operator. Correspondingly, an individual update in subpopulation1 and subpopulation2 is generated by performing Lévy flights operator and straight flight operator, respectively.

i.   Lévy flights: For each individual i in subpopulation1, it will fly around the best one in the form of Lévy flights. The resulting new solution is calculated based on Equations (5)–(7).

$$x_i^{t+1} = x_i^t + \alpha L(s) \tag{5}$$

$$\alpha = S_{max} / t^2 \tag{6}$$

$$L(s) = \frac{(\beta - 1)\Gamma(\beta - 1)\sin\left(\frac{\pi(\beta - 1)}{2}\right)}{\pi s^\beta} \tag{7}$$

where $x_i{}^t$ and $x_i{}^{t+1}$ denote the position of moth i at generation t and $t + 1$, respectively. α denotes the scale factor related to specific problem. S_{max} is the max walk step and it takes the value 1.0 in this paper. $L(s)$ represents the step drawn from Lévy flights and $\Gamma(x)$ is the gamma function. In this paper, $\beta = 1.5$ and s can be regarded as the position of moth individual in the solution space then s^β is the β power of s.

ii.   Straight flights: for each individual i in subpopulation2, it will fly towards that source of light in line. The resulting new solution is formulated as Equation (8).

$$x_i^{t+1} = \begin{cases} \lambda \times (x_i^t + \varphi \times (x_{best}^t - x_i^t)) & if\ rand > 0.5 \\ \lambda \times (x_i^t + \frac{1}{\varphi} \times (x_{best}^t - x_i^t)) & else \end{cases} \quad (8)$$

where λ and φ represent scale factor and acceleration factor, respectively. $x^t{}_{best}$ is the best individual at generation t. *Rand* is a function generating a random number uniformly distributed in (0, 1).

3. Discrete MS Optimization Method for SUKP

In this section, we describe the newly proposed discrete MS for SUKP. The main purpose of extending MS algorithm to solve the novel SUKP is to investigate the significant role of the transfer functions in terms of improving the quality of solutions and convergence rate. The basic MS algorithm was initially proposed for continuous optimization problems, while SUKP belongs to a discrete optimization problem with constraints. Therefore, the SUKP problem must contain three key elements, namely, discretization method, solution representation, and constraint handling. The three key elements are described in detail subsequently.

3.1. Transfer Functions

Transfer function is a major contributor of the discrete MS algorithm; therefore, it deserves special attention and research. In this section, 12 transfer functions are introduced. According to the shape of transfer function curve, we divide the twelve transfer functions into three groups: S-shaped transfer functions [12], V-shaped transfer functions [15], and other-shaped (O-shaped) transfer functions [19,21]. As described above, each group consists of four functions, which are named as S_i, V_i, and O_i ($i = 1, 2, 3, 4$), respectively. These transfer functions are presented in Table 1 and Figure 1.

Table 1. Twelve transfer functions.

Number	Mathematical Formula
S1 [17]	$T(x) = \frac{1}{1+e^{-2x}}$
S2 [12]	$T(x) = \frac{1}{1+e^{-x}}$
S3 [17]	$T(x) = \frac{1}{1+e^{-x/2}}$
S4 [17]	$T(x) = \frac{1}{1+e^{-x/3}}$
V1 [20]	$T(x) = \left\|erf(\frac{\sqrt{\pi}}{2}x)\right\| = \left\|\frac{\sqrt{2}}{\pi}\int_0^{\frac{\sqrt{\pi}}{2}x} e^{-t^2} dt\right\|$
V2 [12]	$T(x) = \|\tanh(x)\|$
V3 [17]	$T(x) = \left\|\frac{x}{\sqrt{1+x^2}}\right\|$
V4 [17]	$T(x) = \left\|\frac{2}{\pi}\arctan(\frac{\pi}{2}x)\right\|$
O1 [18]	$T(x) = \sin(2\pi(x-a)*b*\cos(2\pi(x-a)*c)) + d$ $(a = 0, b = 1, c = 1, d = 0)$
O2 [20]	$T(x) = \lfloor\|xmod2\|\rfloor$
O3 [22]	$T(x) = \frac{(x+x_{min})}{(\|x_{min}\|+x_{max})} \ (x_{min} \leq x \leq x_{max})$
O4 [23]	$T(x) = x$

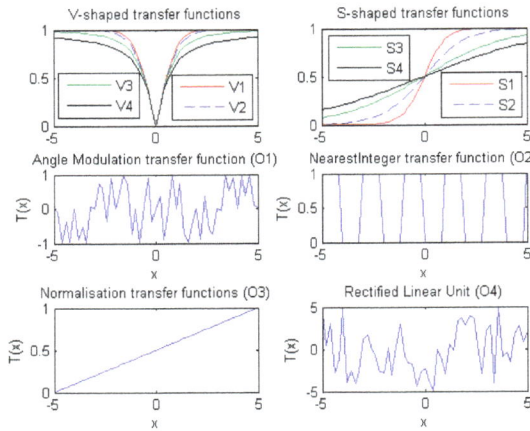

Figure 1. Twelve transfer functions.

As stated in the literature [16,17], the transfer functions define the probability that the element of position vector of each moth individual changes from 0 to 1, and vice versa. Therefore, an appropriate transfer function should ensure that a real-valued vector in a continuous search space is mapped to the value 1 in a binary search space with greater probability. Suppose applying the transfer function $T(x)$ will return a function value y ($y = 1$ or $y = 0$) through a mapping method. The probability of a transfer function with a value of 1 (PR) is displayed in Figure 2. Three groups of items, namely, 100 items, 300 items, and 500 items, were selected to count the PR value:

$$PR = \frac{\sum_{i=1}^{N}\{y_i | y_i = 1\}}{N} \times 100\% \tag{9}$$

where N represents the number of items. The value of longitudinal axis in Figure 2 is the average of PR among 100 independent runs.

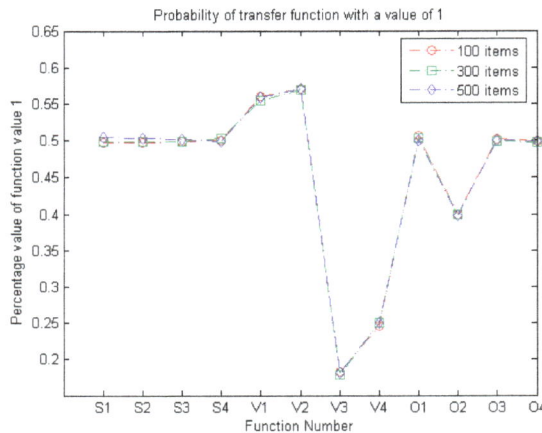

Figure 2. Probability of transfer function with a value of 1.

As shown in Figure 2, the four S-shaped transfer functions have similar PR values, which are close to 0.5. However, the PR values of the four V-shaped transfer functions differ considerably. V2 has the best *PR* value while the *PR* value of V3 is less than 0.2. It seems that V3 combining with MS should show poor performance. Similarly, V4 also demonstrates unsatisfactory performance, with a *PR* value of less than 0.25. Of the four other shapes of transfer functions, O1, O3, and O4 obtain a similar *PR* value, that is, close to 0.5. The *PR* value of O2 is slightly smaller than that of O1, O3, and O4. In sum, according to the preliminary analysis of *PR* values, it seems that V3, V4, and O2 are not suitable for combining with MS to solve binary optimization problems.

3.2. Solution Representation

The basic MS is a real-valued algorithm and each moth individual is represented as a real-valued vector. Two main operators are defined in continuous space. However, SUKP is a discrete optimization problem with constraints and the solution is a binary vector. In this paper, the most general and simplest method, mapping the real-valued vectors into binary ones by transfer functions, is opted. Concretely speaking, a real-valued vector $X = [x_1, x_2, \ldots, x_m] \in [-a, a]^m$ still evolves in continuous space. Here, m is the number of items and a is a positive real value, and $a = 5.0$ in this paper. Then, transfer function $T(x)$ is used to map X into a binary vector $Y = [y_1, y_2, \ldots, y_m] \in \{0, 1\}^m$. According to the feature of these transfer functions, three mapping methods are as follows.

The first mapping method: Choose a transfer function from S1–S4, V1–V4, and O3.

$$y_i = \begin{cases} 1 & if \ rand() \geq T(x_i) \\ 0 & else \end{cases} \tag{10}$$

where *rand*( ) is a random number in (0, 1). In Figure 1, it can be observed that S-shaped transfer functions, V-shaped transfer functions, and O3 will return a random real number between 0 and 1. Therefore, the comparison of *rand*( ) to $T(x_i)$ equals 1 or 0. Then, the mapping procedure is shown as Table 2.

Table 2. The first mapping procedure according to *S2* transfer function.

Element	x_0	x_1	x_2	x_3	x_4	x_5	x_6	x_7	x_8	x_9
X	2.96	3.32	−3.25	2.65	2.61	−1.57	−0.07	0.91	1.04	1.68
T(X)	0.95	0.97	0.04	0.93	0.93	0.17	0.48	0.71	0.74	0.84
Rand()	0.61	0.17	0.07	0.15	0.08	0.86	0.72	0.39	0.80	0.62
Y	0	0	1	0	0	1	1	0	1	0

The second mapping method: Choose the transfer function O2.

$$y_i = T(x_i) \tag{11}$$

The third mapping method: Choose either O1 or O4 as the transfer function.

$$y_i = \begin{cases} 1 & if \ T(x_i) \geq 0 \\ 0 & else \end{cases} \tag{12}$$

Then, the quality of any feasible solution Y is evaluated by the objective function f of the SUKP. Given a potential solution $Y = [y_1, y_2, \ldots, y_m]$, the objective function value $f(Y)$ is defined by

$$f(Y) = \sum_{i=1}^{m} y_i p_i \tag{13}$$

3.3. Repair Mechanism and Greedy Optimization

Clearly, SUKP is a kind of important combinatorial optimization problem with constraints. Due to the existence of constraints, the feasible search space of decision variables becomes irregular, which will increase the difficulty of finding the optimal solution. Among the many constraint processing techniques, repairing the infeasible solution is a common method to solve the combinatorial optimization problem. Michalewicz [28] introduced an evolutionary system based on repair technology. Obviously, repairing technique is dependent on specific problems and different repairing method must be designed for different problems. Consequently, He et al. [5] designed a repairing and optimization algorithm (named S-GROA) for SUKP, which can not only repair infeasible solutions but also further optimize feasible solutions. On the basis of S-GROA [5], a quadratic greedy repair and optimization strategy (QGROS) is proposed by Liu et al. [29]. In this paper, QGROS is adopted. The preprocessing phase of QGROS can be summarized as follows:

(1) Compute the frequency d_j of the element j ($j = 1, 2, 3, \ldots, n$) in the subsets $U_1, U_2, U_3, \ldots, U_m$.

(2) Calculate the unit weight R_i of the item i ($i = 1, 2, 3, \ldots, m$).

$$R_i = \sum_{j \in U_i} (w_j / d_j) \tag{14}$$

(3) Record the profit density of each item in S according to PD_i.

$$PD_i = p_i / R_i (i = 1, 2, 3, \ldots, m) \tag{15}$$

(4) Sort all the items in a non-ascending order based on PD_i ($i = 1, 2, 3, \ldots, m$) and then the index value recorded in an array H[1 ... m].

(5) Define a term $A_Y = \{U_i | y_i \in Y \wedge y_i = 1, 1 \le i \le m\}$ for any binary vector $Y = [y_1, y_2, \ldots, y_m]$ $\in \{0, 1\}^m$.

The pseudocode of QGROS [29] is outlined in Algorithm 1.

Algorithm 1. QGROS algorithm for SUKP.

Begin
  Step 1: Input: the candidate solution $Y = [y_1, y_2, \ldots, y_m] \in \{0, 1\}^m$, H[1 ... m].
  Step 2: Initialization. The m-dimensional binary vector Z = [0, 0, ..., 0].
  Step 3: Greedy repair stage
   For $i = 1$ **to** m **do**
    If ($y_{H[i]} = 1$ *and* $W(A_Z \cup \{H[i]\}) \le C$)
     $Z_{H[i]} = 1$ *and* $A_Z = A_Z \cup \{H[i]\}$.
    End if
   End for
   $Y \leftarrow Z$.
  Step 4: Quadratic greedy stage
   Do not consider the elements that have been packed into the knapsack,
   recalculate d_j ($j = 1, 2, 3, \ldots, n$), R_i ($i = 1, 2, 3, \ldots, m$), and H[1 ... m]
  Step 5: Optimization sate
   For $i = 1$ **to** m **do**
    If ($y_{H[i]} = 0$ *and* $W(A_Z \cup \{H[i]\}) \le C$)
     $y_{H[i]} = 1$ *and* $A_Y = A_Y \cup \{H[i]\}$.
    End if
   End for
  Step 6: Output:$Y = [y_1, y_2, \ldots, y_m]$ *and* $f(Y)$
 End.

In Algorithm 1, we can observe that QGROS consists of three stages. The first stage is determining whether the constraints are met for the items in the potential solution that are ready to be packed into the knapsack. At this stage, items in the potential solution that are intended to be packed into knapsack but violate constraints will be removed. Therefore, all solutions are feasible after this stage. The second stage is recalculating the frequency of each element, the unit weight of each item, and the array H[1 ... *m*]. The third stage is optimizing the remaining items by loading appropriate items into the knapsack with the aim of maximizing the use of the remaining capacity. At this stage, items in the feasible solution that are not intended to be loaded in the knapsack but satisfy the constraints will be loaded. Hence, after this stage, all solutions remain feasible and the quality of solutions is improved.

3.4. The Main Scheme of Discrete MS for SUKP

Having discussed all the components of the discrete MS algorithms in detail, the complete procedure is outlined in Algorithm 2.

3.5. Computational Complexity of the Discrete MS Algorithm

Computational complexity is the main criterion for evaluating the running time of an algorithm, which can be calculated according to its structure and implementation. In Algorithm 2, it can be seen that the computing time in each iteration is mainly dependent on the number of moths, problem dimension, and sorting of items as well as moth individual in each iteration. In addition, the computational complexity is mainly determined by Steps 1–4. In Step 1, since the Quicksort algorithm is used, the average and the worst computational costs are $O(m\log m)$ and $O(m^2)$, respectively. In Step 2, the initialization of N moth individuals costs time $O(N \times m) = O(m^2)$. In Step 3, the fitness calculation of N moth individuals costs time $O(N)$. In Step 4, Lévy flight operator has time complexity $O(N/2 \times m) = O(m^2)$, straight flight operator has time complexity $O(N/2 \times m) = O(m^2)$, QGROS has time complexity $O(m \times n)$, and sorting the population with Quicksort has average time complexity and worst time complexity of $O(N\log N)$ and $O(N^2)$, respectively. Consequently, the overall computational complexity is $O(m\log m) + O(m^2) + O(N) + O(m^2) + O(m^2) + O(m \times n) + O(N\log N) = O(m^2)$, where m is the number of items and N is the number of moths.

4. Results and Discussion

In this section, we present experimental studies on the proposed discrete MS algorithms for solving SUKP.

Algorithm 2. The main procedure of discrete MS algorithm for SUKP.

Begin

    Step 1: Sorting.

        Sort all items in S in non-increasing order according to PD_i $(0 \leq i \leq m)$, and the indexes of items are recorded in array H $[0 \ldots m]$.

    Step 2: Initialization.

        Set the maximum iteration number *MaxGen* and iteration counter $G = 1$;

        $\beta = 1.5$; the acceleration factor $\varphi = 0.618$.

        Generate N moth individuals randomly $\{X_1, X_2, \ldots, X_N\}$, $X_i \in [-a, a]^m$.

        Divide the whole population into two subpopulations with equal size: subpopulation1 and subpopulation2, according to their fitness.

        Calculate the corresponding binary vector $Y_i = T(X_i)$ by using transfer functions $(i = 1, 2, \ldots, N)$.

        Perform repair and optimization with QGROS.

    Step 3: Fitness calculation.

        Calculate the initial fitness of each individual, $f(Y_i)$, $1 \leq i \leq N$.

    Step 4: While $G < MaxGen$ **do**

        Update subpopulation 1 by using Lévy flight operator.

        Update subpopulation 2 by using fly straightly operator.

        Calculate the corresponding binary vector $Y_i = T(X_i)$ by using transfer functions $(i = 1, 2, \ldots, N)$.

        Perform repair and optimization with QGROS.

        Evaluate the fitness of the population and record the $<X_{gbest}, Y_{gbest}>$.

        $G = G + 1$.

        Recombine the two newly-generated subpopulations.

        Sort the population by fitness.

        Divide the whole population into subpopulation 1 and subpopulation 2.

    Step 5: End while

    Step 6: Output: the best results.

End.

Test instance: Three groups of thirty SUKP instances were recently presented by He et al. [5]. What needs to be specified is that the set of items $S = \{U_1, U_2, U_3, \ldots, U_m\}$ is represented as a 0–1 matrix M $= (r_{ij})$, with m rows and n columns. For each element r_{ij} in M $(i = 1, 2, \ldots, m; j = 1, 2, \ldots, n)$, $r_{ij} = 1$ if and only if $u_j = U_i$. Therefore, each instance contains four factors: (1) m denotes the number of items; (2) n denotes the number of elements; (3) density of element 1 in the matrix M$\alpha \in \{0.1, 0.15\}$; and (4) the ratio of C to the sum of all elements $\beta \in \{0.75, 0.85\}$. According to the relationship between m and n, three types of instances are generated. The first group: 10 SUKP instances with $m > n$, $m \in \{100, 200, 300, 400, 500\}$ and $n \in \{85, 185, 285, 385, 485\}$, named as F01–F10, respectively. The second group: 10 SUKP instances with $m = n$, $m \in \{100, 200, 300, 400, 500\}$ and $n \in \{100, 200, 300, 400, 500\}$, named as S01–S10, respectively. The third group: 10 SUKP instances with $m < n$, $m \in \{85, 185, 285, 385, 485\}$ and $n \in \{100, 200, 300, 400, 500\}$, named as T01–T10, respectively. We selected five instances in each group with $\alpha = 0.1$ and $\beta = 0.75$. The instances can be downloaded at http://sncet.com/ThreekindsofSUKPinstances(EAs).rar. Three categories with different relationships between m and n ($m > n$, $m = n$, and $m < n$) of 15 SUKP instances were selected for testing. The parameters and the best solution value (Best*) [5] are shown in Table 3.

Table 3. The parameters and the best solution value provided in [5] for 15 SUKP instances.

Number	Instance	m	n	Capacity	Best*
1	F01	100	85	12,015	13,251
2	F03	200	185	22,809	13,241
3	F05	300	285	36,126	10,553
4	F07	400	385	50,856	10,766
5	F09	500	485	60,351	11,031
6	S01	100	100	11,223	14,044
7	S03	200	200	25,630	11,846
8	S05	300	300	38,289	12,304
9	S07	400	400	49,822	10,626
10	S09	500	500	63,902	10,755
11	T01	85	100	12,180	11,664
12	T03	185	200	25,405	13,047
13	T05	285	300	38,922	11,158
14	T07	385	400	49,815	10,085
15	T09	485	500	62,516	10,823

Experimental environment: For fair comparisons, all proposed algorithms in this paper were coded in C++ and in the Microsoft Visual Studio 2015 environment. All the experiments were run on a PC with Intel (R) Core (TM) i7-7500 CPU (2.90 GHz and 8.00 GB RAM).

On the stopping condition, we followed the original paper [5] and set the iteration number *MaxGen* equal to *max* {m, n} for all SUKP instances. Here, m denotes the number of items and n is the number of elements in each SUKP instance. In addition, the population size of all the algorithms was set to $N = 20$. For each SUKP instance, we carried out 100 independent replications.

The parameters for the proposed discrete MS algorithms were set as follows: the max step $S_{max} = 1.0$, acceleration factor $\varphi = 0.618$, and the index $\beta = 1.5$.

4.1. The Performance of Discrete MS Algorithm with Different Transfer Functions

Computational results are summarized in Table 4, which records the results for SUKP instances with $m > n$, $m = n$, and $m < n$, respectively. For each instance, we give several criteria to evaluate the comprehensive performance of the twelve discrete MS algorithms. "Best" and "Mean" refer to the best value and the average value for each instance obtained by each algorithm among 100 independent runs. The best solution provided in [5] are given in parentheses in the first column.

Table 4. The best values and average values of twelve discrete MS algorithms on 15 SUKP instances.

Number	Criterion	MSS1	MSS2	MSS3	MSS4	MSV1	MSV2	MSV3	MSV4	MSO1	MSO2	MSO3	MSO4
F01	Best	12,698	13,283	12,861	13,057	13,003	13,003	13,044	13,044	12,973	12,678	13,283	13,283
(13251)	Mean	12,250	13,102	12,168	12,227	12,564	12,740	12,858	12,697	12,320	12,066	13,052	13,062
F03	Best	12,762	13,286	12,216	12,189	12,875	12,639	11,953	12,267	13,175	12,255	13,322	13,521
(13241)	Mean	11,777	12,860	11,465	11,261	12,176	12,100	11,321	11,193	12,371	11,007	13,101	13,193
F05	Best	10,142	10,668	9974	10,047	9966	9562	9752	9460	10,539	9656	10,643	11,127
(10553)	Mean	9588	10,196	9465	9393	9352	9120	9250	9131	9822	8987	10,381	10,302
F07	Best	10,456	11,321	9793	10,005	10,625	9539	9917	9814	10,906	9801	11,321	11,435
(10766)	Mean	9750	10,644	9349	9467	10,042	9150	9317	9265	10,141	9028	10,833	10,411
F09	Best	10,669	11,410	10,642	10,461	10,718	10,725	10,598	10,288	11,279	9808	11,172	11,031
(11031)	Mean	10,293	10,913	10,082	9965	10,420	9969	10,134	9997	10,648	9429	10,750	10,716
S01	Best	13,405	14,044	13,080	13,611	13,396	13,814	13,721	13,721	13,659	13,202	14,003	14,044
(14044)	Mean	12,725	13,478	12,418	12,607	13,211	13,569	13,540	13,503	12,899	12,339	13,583	13,649
S03	Best	11,249	11,104	10,904	11,295	11,329	10,802	10,481	10,808	11,757	11,147	11,873	12,350
(11846)	Mean	10,469	10,576	10,282	10,285	10,622	9879	10,112	10,212	10,789	9975	11,419	11,508
S05	Best	11,649	12,071	11,472	11,459	11,799	11,686	11,421	11,380	11,862	11,048	12,240	12,598
(12304)	Mean	10,979	11,650	10,753	10,787	11,199	11,206	10,898	11,165	11,272	10,153	11,721	11,541
S07	Best	10,330	10,990	10,218	10,073	10,177	9669	9957	9977	10,650	10,006	10,722	10,727
(10626)	Mean	9831	10,379	9766	9681	9968	9286	9372	9460	10,019	9426	10,327	10,343
S09	Best	10,074	10,495	9995	10,037	9938	10,025	10,043	10,052	10,199	9553	10,355	10,355
(10755)	Mean	9719	9968	9583	9675	9654	9707	9804	9713	9807	9,127	10,056	9919
T01	Best	11,034	11,573	11,158	11,332	11,427	11,027	11,151	11,076	11,195	11,159	11,519	11,735
(11664)	Mean	10,577	11,259	10,491	10,501	10,812	10,572	10,496	10,781	10,568	10,495	11,276	11,287
T03	Best	12,234	13,306	12,357	12,136	12,415	12,633	12,039	11,829	12,798	12,085	13,378	13,647
(13047)	Mean	11,549	12,621	11,624	11,522	11,745	11,743	11,535	11,400	11,980	11,267	12,948	13,000
T05	Best	11,025	11,173	11,167	10,765	10,814	10,725	10,485	11,240	11,183	10,299	11,226	11,391
(11158)	Mean	10,385	10,871	10,247	10,118	10,635	10,229	10,059	10,735	10,651	9584	10,957	10,816
T07	Best	9676	9609	9140	9169	9594	9303	9049	8965	9675	9198	9783	9739
(10085)	Mean	8987	9264	8873	8875	9161	9131	8642	8733	9154	8694	9261	9240
T09	Best	10,208	10,549	10,131	10,094	10,115	10,201	9866	10,005	10,450	9989	10,660	10,539
(10823)	Mean	9856	10,205	9753	9711	9949	9771	9506	9714	9985	9332	10,350	10,190

In Table 4, it can be easily observed that MSO4 outperforms the eleven other discrete MS algorithms and demonstrates the best comprehensive performance when solving all fifteen SUKP instances. In addition, MSS2 and MSO3 show comparable performance.

To evaluate the performance of each algorithm, the relative percentage deviation (*RPD*) was defined to represent the similarity between the best value obtained by each algorithm and the best solution 5. The *RPD* of each SUKP instance is calculated as follows.

$$RPD = (Best^* - Best)/Best^* \times 100 \tag{16}$$

where Best* is the best solution provided in [5]. Clearly, if the value of *RPD* is less than 0, the algorithm updates the best solution of the SUKP test instance in [5]. The statistical results are shown in Table 5.

Table 5. The effect of twelve transfer functions on the performance of discrete MS algorithm (*RPD* values).

Number	MSS1	MSS2	MSS3	MSS4	MSV1	MSV2	MSV3	MSV4	MSO1	MSO2	MSO3	MSO4
F01	4.17	−0.24	2.94	1.46	1.87	1.87	1.56	1.56	2.10	4.32	−0.24	−0.24
F03	3.62	−0.34	7.74	7.95	2.76	4.55	9.73	7.36	0.50	7.45	−0.61	−2.11
F05	3.89	−1.09	5.49	4.79	5.56	9.39	7.59	10.36	0.13	8.50	−0.85	−5.44
F07	2.88	−5.16	9.04	7.07	1.31	11.40	7.89	8.84	−1.30	8.96	−5.16	−6.21
F09	3.28	−3.44	3.53	5.17	2.84	2.77	3.93	6.74	−2.25	11.09	−1.28	0.00
S01	4.55	0.00	6.86	3.08	4.61	1.64	2.30	2.30	2.74	6.00	0.29	0.00
S03	5.04	6.26	7.95	4.65	4.36	8.81	11.52	8.76	0.75	5.90	−0.23	−4.25
S05	5.32	1.89	6.76	6.87	4.10	5.02	7.18	7.51	3.59	10.21	0.52	−2.39
S07	2.79	−3.43	3.84	5.20	4.23	9.01	6.30	6.11	−0.23	5.83	−0.90	−0.95
S09	6.33	2.42	7.07	6.68	7.60	6.79	6.62	6.54	5.17	11.18	3.72	3.72
T01	5.40	0.78	4.40	2.85	2.03	5.46	4.40	5.04	4.02	4.33	1.24	−0.61
T03	6.23	−1.99	5.29	6.98	4.84	3.17	7.73	9.34	1.91	7.37	−2.54	−4.60
T05	1.19	−0.13	−0.08	3.52	3.08	3.88	6.03	−0.73	−0.22	7.70	−0.61	−2.09
T07	4.06	4.72	9.37	9.08	4.87	7.75	10.27	11.11	4.07	8.80	2.99	3.43
T09	5.68	2.53	6.39	6.74	6.54	5.75	8.84	7.56	3.45	7.71	1.51	2.62
Mean	4.30	0.19	5.77	5.47	4.04	5.82	6.79	6.56	1.63	7.69	−0.14	−1.28

In Table 5, it can be seen that, in all twelve discrete MS algorithms, MSS2, MSS3, MSV4, MSO1, MSO3, and MSO4 all update the best solutions [5]. However, MSS3 and MSV4 update only one SUKP instance, T05. MSO1 updates the instances F07, F09, S07, and T05. Moreover, MSO4 still keeps the best performance because its total average *RPD* is only −1.28. The total average *RPD* of MSO3 is −0.14, which implies that MSO3 is slightly worse than MSO4 but outperforms the ten other discrete MS algorithms. Obviously, MSS2 is the third best of the twelve discrete MS algorithms. Indeed, it can also be seen that MSO4 updates and obtains the best solutions [5] ten and two times (out of 15), i.e., 66.67% and 13.33% of the whole instance set, respectively. MSO3 updates and fails to find the best solutions 5 nine and six times (out of 15), i.e., 60.00% and 40.00% of the whole instance set, respectively. MSS2 updates and obtains the best solutions 5 eight (53.33%) and one times (6.60%), respectively.

To further evaluate the comprehensive performance of twelve discrete MS algorithms in solving fifteen SUKP instances, the average ranking based on the best values are displayed in Table 6 and Figure 3, respectively. In Table 6 and Figure 3, the average ranking value of MSO4 is 1.60 and it still ranks first. In addition, MSO3 and MSS2 are the second and the third best algorithms, respectively, which is very consistent with the previous analysis. The ranking of twelve discrete MS algorithms based on the best values are as follows:

$$\begin{aligned} MSO4 \succ MSO3 \succ MSS2 \succ MSO1 \succ MSS1 \succ MSV1 \succ MSS4 \succ MSS3 \\ = MSV2 \succ MSV4 \succ MSV3 \succ MSO2 \end{aligned} \tag{17}$$

Table 6. Ranks of twelve discrete MS algorithms based on the best values.

Number	MSS1	MSS2	MSS3	MSS4	MSV1	MSV2	MSV3	MSV4	MSO1	MSO2	MSO3	MSO4
F01	11	1	10	4	7	7	5	5	9	12	1	1
F03	6	3	10	11	5	7	12	8	4	9	2	1
F05	5	2	7	6	8	11	9	12	4	10	3	1
F07	5	1	6	8	7	12	11	10	4	9	3	2
F09	7	1	8	10	6	5	9	11	2	12	3	4
S01	9	1	12	8	10	4	5	5	7	11	3	1
S03	6	8	9	5	4	11	12	10	3	7	2	1
S05	7	3	8	9	5	6	10	11	4	12	2	1
S07	5	1	6	8	7	12	11	10	4	9	3	2
S09	6	1	10	8	11	9	7	6	4	12	2	2
T01	11	2	8	5	4	12	9	10	6	7	3	1
T03	8	3	7	9	6	5	11	12	4	10	2	1
T05	7	5	6	9	8	10	11	2	4	12	3	1
T07	3	5	10	9	6	7	11	12	4	8	1	2
T09	5	2	7	9	8	6	12	10	4	11	1	3
Mean	6.67	2.60	8.27	7.87	6.80	8.27	9.67	8.93	4.47	10.07	2.27	1.60

By looking closely at Figures 2 and 3, it is not difficult to see that V3, V4, and O2 exhibit the worst performance, which is consistent in the two figures. Similar to the previous analysis in Figure 2, O1, O3, and O4 show satisfactory performance among 12 transfer functions. Thus, it can be inferred that *PR* value can be used as a criterion for selecting transfer functions.

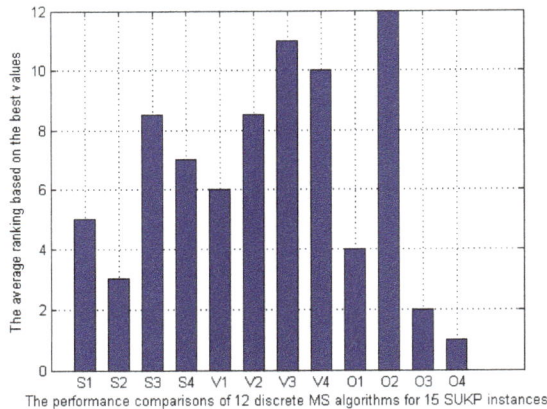

The performance comparisons of 12 discrete MS algorithms for 15 SUKP instances

Figure 3. Comparison of the average rank of 12 discrete MS algorithms for 15 SUKP instances.

To analyze the experimental results for statistical purposes, we selected three representative instances (F09, S09, and T09) and provided boxplots in Figures 4–6. In Figure 4, the boxplot of MSS2 has greater value and less height than those of other eleven algorithms. In Figures 5 and 6, MSO3 exhibits a similar phenomenon as MSS2 in Figure 4. Additionally, the performance of MSO2 is the worst. In Figures 4–6, we can also observe that MSO3 performs slightly better than MSO4 in solving large-scale instances.

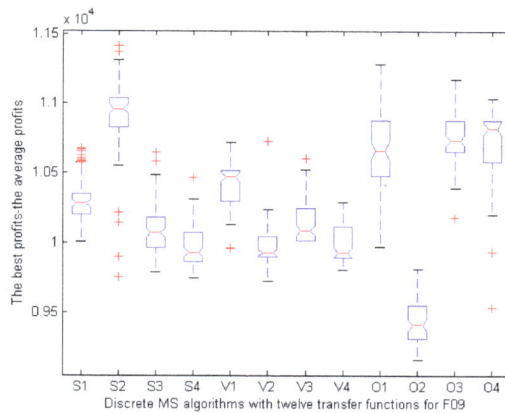

Figure 4. Boxplot of the best values on F09 in 100 runs.

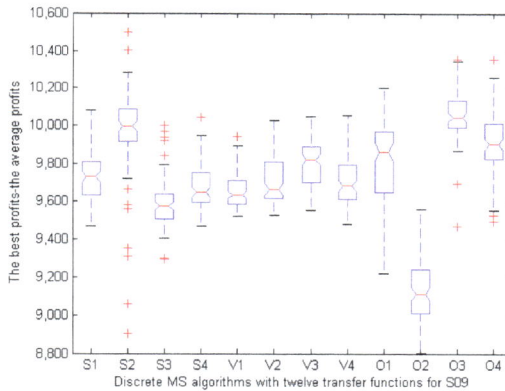

Figure 5. Boxplot of the best values on S09 in 100 runs.

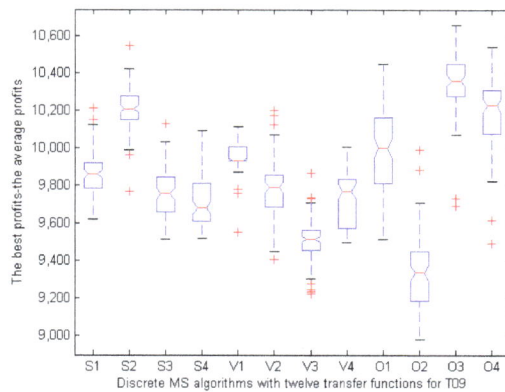

Figure 6. Boxplot of the best values on T09 in 100 runs.

Moreover, optimization process of each algorithm in solving F09, S09, and T09 instances is given in Figures 7–9, respectively. In these three figures, all the function values are the average best values achieved from 100 runs. In Figure 7, the initial value of MSS2 is greater than that of other algorithms and then it quickly converges to the global optimum. For MSO3, the same scene appears in Figures 8 and 9. Overall, MSS2 and MSO3 have stronger optimization ability and faster convergence speed than the other discrete MS algorithms.

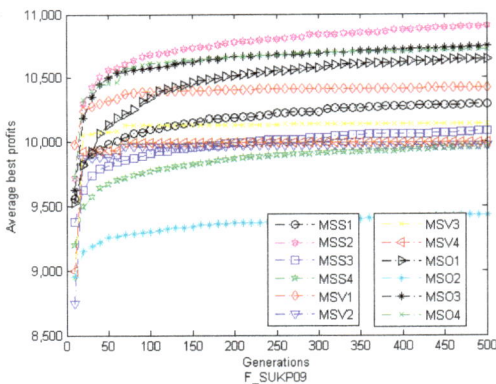

Figure 7. The convergence graph of twelve discrete MS algorithms on F09.

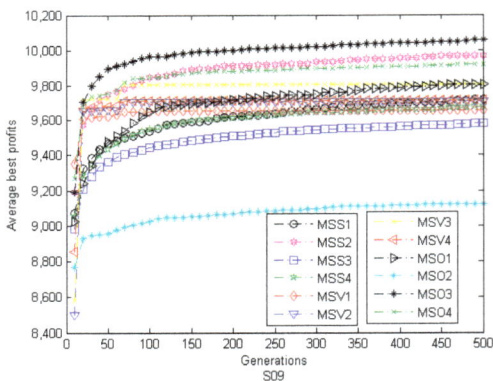

Figure 8. The convergence graph of twelve discrete MS algorithms on S09.

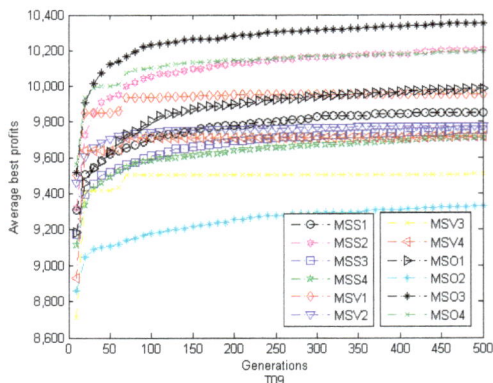

Figure 9. The convergence graph of twelve discrete MS algorithms on T09.

Through the above experimental analysis, the following conclusions can be drawn: (1) For S-shaped transfer functions, the combination of S2 and MS (MSS2) is the most effective. (2) As far as V-shaped transfer functions are concerned, the combination of V1 and MS (MSV1) shows the best performance. (3) In the case of other shapes transfer functions, the more effective algorithms are MSO4, MSO3, and MSO1. (4) By comparing the family of S-shaped transfer functions and V-shaped transfer functions, the family of S-shaped transfer functions with MS is suitable for solving SUKP problem. (5) MSO4 has advantages over other algorithms in terms of the quality of solutions. (6) As far as the stability and convergence rate are concerned, MSO3 and MSS2 perform better than other algorithms.

Overall, it is evident that MSO4 has the best results (considering *RPD* values and average ranking values) on fifteen SUKP instances. Therefore, it appears that the proposed other-shapes family of transfer functions, particularly the O4 function, has many advantages combined with other algorithms to solve binary optimization problems. Additionally, the O3 function and S2 function are also suitable functions that can be considered for selection. In brief, these results demonstrate that the transfer function plays a very important role in solving SUKP using discrete MS algorithm. Thus, by carefully selecting the appropriate transfer function, the performance of discrete MS algorithm can be improved obviously.

4.2. Estimation of the Solution Space

SUKP is a binary coded problem and the solution space can be represented as a graph $G = (V, E)$, in which vertex set $V = S$, where S is the set of solutions for a SUKP instance, $S = \{0, 1\}^n$ and edge set $E = \{(s, s') \in S \times S | d(s, s') = d_{min}\}$, where d_{min} is the minimum distance between two points in the search space. Especially, hamming distance is used to describe the similarity between individuals. Obviously, the minimum distance is 0 when all bits have the same value and the maximum distance is n, where n is the dimension of SUKP instance.

Here, MSO4 is specially selected to analyze the solution space for F01, S01, and T01 SUKP instance. The distribution of fitness at generation 0 and generation 100 is presented in Figures 10–12. The distance between each individual and the best individual is given in Figures 13–15. In Figures 10–12, we can see that, at generation 0, the fitness values are more dispersed and worse than that at generation 100. In Figure 13, it can be observed that the hamming distance varies from 0 to 35 at generation 0 while the range is 0 to 12 at generation 100. Moreover, the hamming distance can be divided into eight levels at generation 100, which demonstrates that all individuals tend to some superior individuals. However, this phenomenon is not evident in S01 and T01.

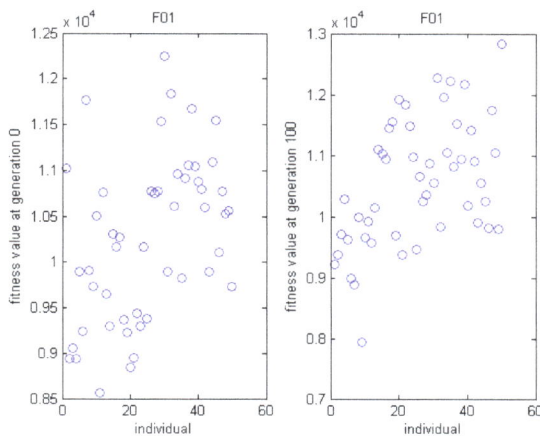

Figure 10. The distribution graph of fitness on MSO4 for F01.

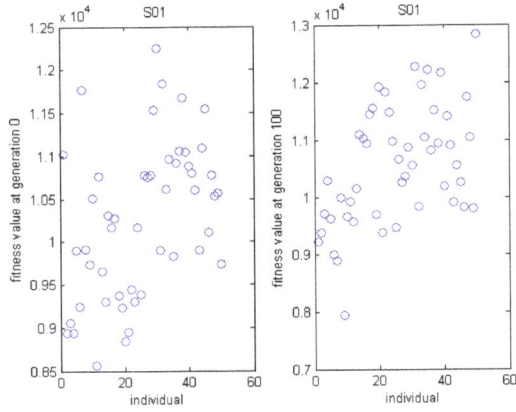

Figure 11. The distribution graph of fitness on MSO4 for S01.

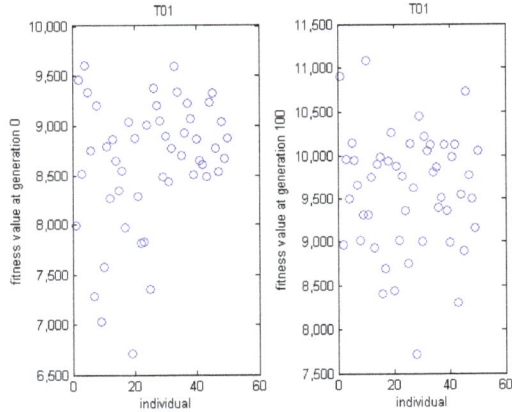

Figure 12. The distribution graph of fitness on MSO4 for T01.

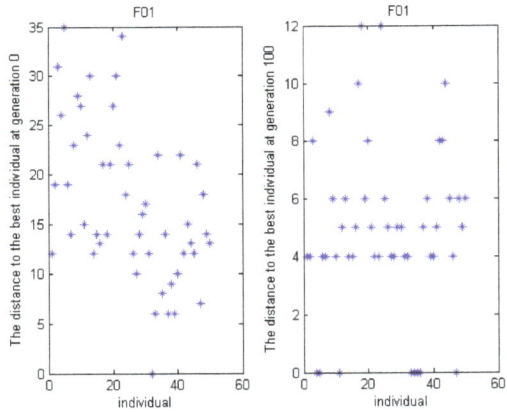

Figure 13. The distance to the best individual for F01.

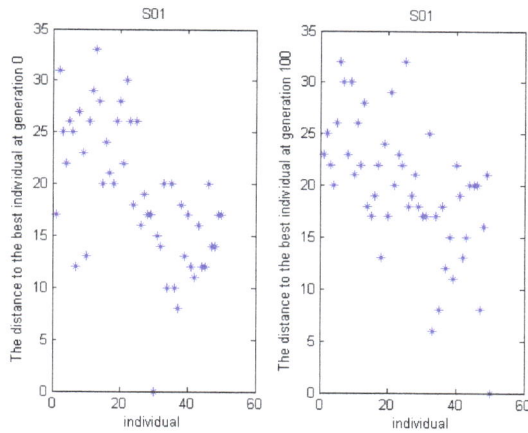

Figure 14. The distance to the best individual for S01.

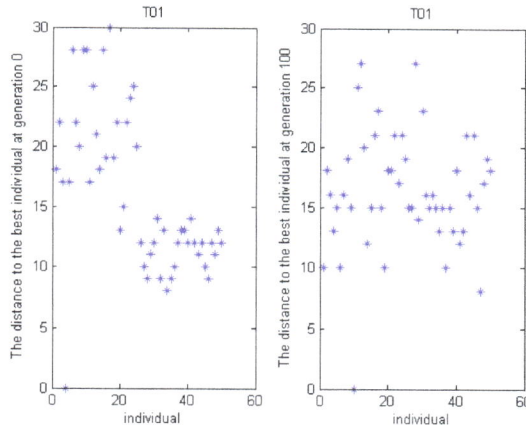

Figure 15. The distance to the best individual for T01.

To intuitively understand the similarity of the solutions, the spatial structure of the solutions at generation 100 is illustrated in Figures 16–18. In Figure 16, the first node (denoting the first individual) has the maximum degree which also shows more individuals have approached the better individual. However, the value of degree is not much different in Figures 17 and 18. This result is consistent with the previous analysis.

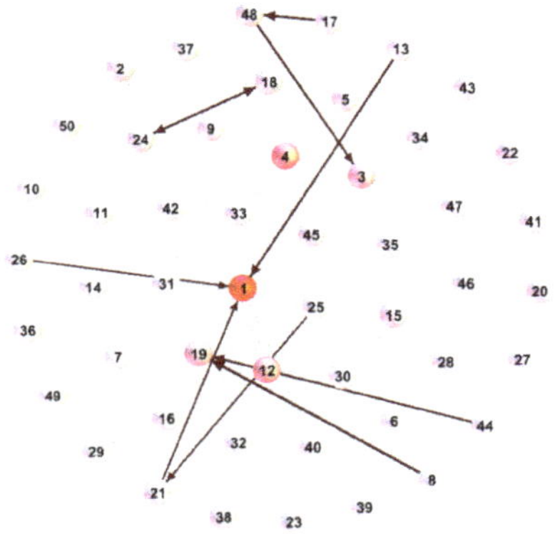

Figure 16. The spatial structure graph for F01 at generation 100.

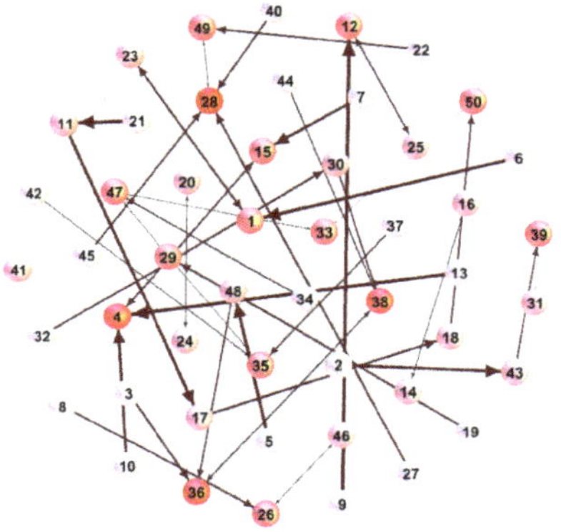

Figure 17. The spatial structure graph for S01 at generation 100.

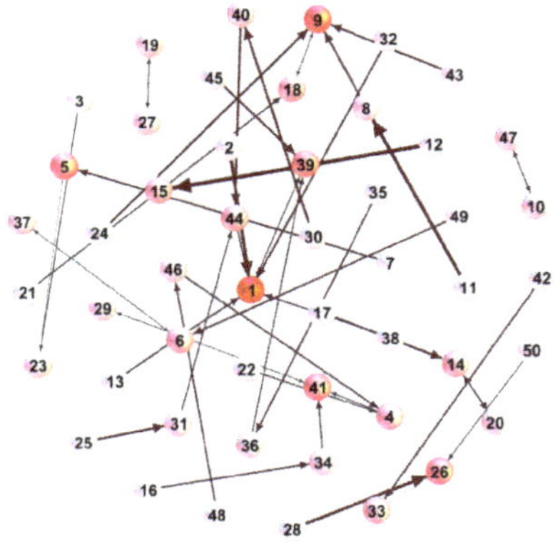

Figure 18. The spatial structure graph for T01 at generation 100.

4.3. Discrete MS Algorithm vs. Other Optimization Algorithms

To further verify the performance of discrete MS algorithm, we chose MSO4 algorithm to compare with five other optimization algorithms. These comparison algorithms include PSO [12], DE [8], global harmony search (GHS) [30], firefly algorithm (FA) [31], and monarch butterfly optimization (MBO) [32,33]. In DE, the DE/rand/1/bin scheme was adopted. PSO, FA, and MBO are classical or novel swarm intelligence algorithms that simulate the social behavior of birds, firefly, and monarch butterfly, respectively. DE is derived from evolutionary theory in nature and has been proved to be one of the most promising stochastic real-value optimization algorithms. GHS is an efficient variant of HS, which imitates the music improvisation process. It is also noteworthy that all five comparison algorithms adopt the discretization method introduced in this paper and combine with O4, respectively. The parameter setting for each algorithm are shown in Table 7.

Table 7. The parameter settings of six algorithms on SUKP.

Algorithm	Parameters	Value
PSO	Cognitive constant C1	1.0
	Social constant C2	1.0
	Inertial constant W	0.3
DE	Weighting factor F	0.9
	Crossover constant CR	0.3
GHS	Harmony memory considering rate HMCR	0.9
	Pitch adjusting rate PAR	0.3
FA	Alpha	0.2
	Beta	1.0
	Gamma	1.0
MBO	Migration ratio	3/12
	Migration period	1.4
	Butterfly adjusting rate	1/12
	Max step	1.0
MSO4	Max step S_{max}	1.0
	Acceleration factor φ	0.618
	Lévy distribution parameter β	1.5

The best results and average results obtained by six methods over 100 independent runs as well the average time cost of each computation (unit: second, represented as "time") are summarized in Table 8. The frequency (T_{Best} and T_{Mean}) and average ranking (R_{Best} and R_{Mean}) of each algorithm with the best performance based on the best values and average values are also recorded in Table 8. The average time cost of each computation for solving fifteen SUKP instances is illustrated in Figure 19. In Table 8, on best, MSO4 outperforms other methods on eight of fifteen instances (F01, F03, F05, F07, S01, S03, S05, and T03). MBO is the second most effective. In terms of average ranking, there is little difference between the performance of MSO4 and MBO. In terms of the average time cost, it can be observed in Figure 19 that DE has the slowest computing speed. However, GHS has surprisingly fast solving speed. In addition, MSO4 is second among the six algorithms. Overall, the computing speed of PSO, FA, MBO and MSO4 shows little difference.

Table 8. Computational results and comparisons on the Best and Mean on 15 SUKP instances.

Number	Criterion	PSO	DE	GHS	FA	MBO	MSO4
F01	Best	13,283	13,125	13,251	13,283	13,283	13,283
(13251)	Mean	12,981	12,923	12,492	13,041	12,941	13,062
	Time	1.297	1.923	0.330	1.627	3.684	1.398
F03	Best	13,319	13,172	12,323	13,282	13,381	13,521
(13241)	Mean	12,697	12,443	11,231	12,544	12,886	13,193
	Time	8.643	13.381	0.603	11.552	11.676	7.901
F05	Best	10,408	10,214	10,512	10,191	10,786	11,127
(10553)	Mean	9825	9420	10,179	9092	10,210	10,302
	Time	28.628	41.633	0.928	45.647	40.938	24.912
F07	Best	11,091	10,135	11,255	9740	11,142	11,435
(10766)	Mean	10,613	9573	10,642	9226	10,463	10,411
	Time	63.290	124.504	1.637	97.102	61.588	56.838
F09	Best	11,046	11,016	11,536	11,099	11,546	11,031
(11031)	Mean	10,473	10,443	11,199	10,473	10,736	10,716
	Time	138.551	225.749	3.179	170.035	157.478	124.378
S01	Best	13,814	13,519	13,522	13,814	14,044	14,044
(14044)	Mean	13,575	12,964	12,656	13,472	13,612	13,649
	Time	1.608	2.800	0.358	1.805	2.617	1.646
S03	Best	11,914	11,085	11,531	11,406	11,955	12,350
(11846)	Mean	10,978	10,408	10,925	10,833	11,056	11,508
	Time	8.437	14.543	0.476	10.753	9.371	8.112
S05	Best	12,574	12,071	12,104	11,398	12,369	12,598
(12304)	Mean	11,709	11,251	11,492	10,993	11,604	11,541
	Time	35.259	46.302	1.014	37.130	26.551	28.612
S07	Best	10,669	10,267	10,952	10,241	10,906	10,727
(10626)	Mean	10,217	9753	10,497	9827	10,237	10,343
	Time	79.622	101.118	1.718	77.458	76.049	58.433
S09	Best	10,352	10,100	10,434	10,057	10,633	10,355
(10755)	Mean	10,104	9708	10,239	9766	10,139	9919
	Time	144.377	242.428	3.013	167.492	153.835	121.622
T01	Best	11,752	11,469	11,434	11,755	11,748	11,735
(11664)	Mean	11,152	10,930	10,370	11,226	11,207	11,287
	Time	1.517	1.892	0.407	1.722	1.668	1.354
T03	Best	13,100	9624	12,618	11,487	13,008	13,647
(13047)	Mean	12,091	9,122	11,855	10,880	12,189	13,000
	Time	7.964	12.708	0.507	11.958	10.702	7.642
T05	Best	11,032	10,669	11,071	11,557	11,090	11,391
(11158)	Mean	10,656	10,490	10,722	10,983	10,686	10,816
	Time	31.753	41.176	0.822	32.175	25.044	24.539
T07	Best	9790	9250	9857	9,392	9770	9739
(10085)	Mean	9636	8897	9447	8,895	9322	9240
	Time	62.079	113.779	1.639	79.984	67.675	57.000
T09	Best	10,482	10,260	10,643	10,207	10,661	10,539
(10823)	Mean	10,111	9717	10,306	9783	10,249	10,190
	Time	121.926	195.754	2.896	144.926	136.766	114.066
T_{Best}		1	0	2	3	5	8
T_{Mean}		2	0	5	1	0	6
R_{Best}		3.27	5.47	3.40	4.40	2.20	2.27
R_{Mean}		3.17	5.47	3.27	4.43	2.53	2.13

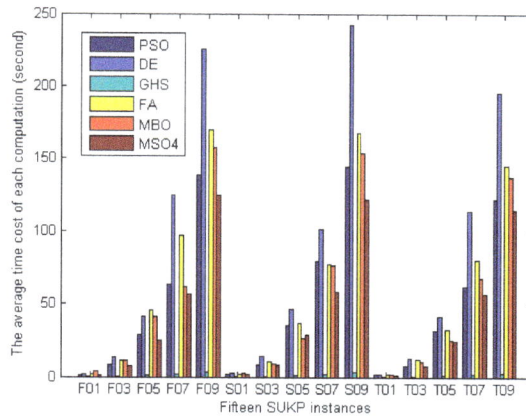

Figure 19. The average time cost of each computation for solving fifteen SUKP instances.

To investigate the difference between the results obtained by MSO4 and those by the comparison algorithm from the perspective of statistics, Wilcoxon's rank sum tests with the 5% significance level were performed. The results of rank sum tests are recorded in Table 9. In Table 9, "1" and "−1" indicate that MSO4 is superior or inferior to the corresponding comparison algorithm, respectively, while "0" shows that there is no statistical difference at 5% significance level between the two comparison algorithms. The statistical result is shown in Table 9.

In Table 8, MSO4 outperforms PSO and DE on all fifteen instances. In addition, MSO4 performs better than GHS and FA on most of the instances except for S05 and F01, respectively. Meanwhile, MSO4 is superior to MBO on eleven instances except for F07, F09, S01, and S05. Statistically, there is no difference between the performance of MSO4 and that of MBO for these four instances.

Considering the results shown in Tables 8 and 9, a conclusion can be drawn that the performance of MSO4 is superior to or at least quite competitive with the five other methods.

Table 9. Results of rank sum tests for MSO4 with the comparison algorithms.

MSO4	PSO	DE	GHS	FA	MBO
F01	1	1	1	0	1
F03	1	1	1	1	1
F05	1	1	1	1	1
F07	1	1	1	1	0
F09	1	1	1	1	0
S01	1	1	1	1	0
S03	1	1	1	1	1
S05	1	1	0	1	0
S07	1	1	1	1	1
S09	1	1	1	1	1
T01	1	1	1	1	1
T03	1	1	1	1	1
T05	1	1	1	1	1
T07	1	1	1	1	1
T09	1	1	1	1	1
1	15	15	14	14	11
0	0	0	1	1	4
−1	0	0	0	0	0

5. Conclusions

In this paper, twelve different transfer functions-based discrete MS algorithms are proposed for solving SUKP. These transfer functions can be divided into three families, S-shaped, V-shaped, and other-shaped transfer functions. To investigate the performance of twelve discrete MS algorithms, three groups of fifteen SUKP instances were employed and the experimental results were compared and analyzed comprehensively. From the experimental results, we found that MSO4 has the best performance. Furthermore, the relative percentage deviation (*RPD*) was calculated to evaluate the similarity between the best value obtained by each algorithm and the best solution provided in [5]. The results show that six algorithms update the best solutions [5] for 11 SUKP instances. The results also indicate that four other shapes transfer functions, especially the O4 function combined with MS, have merits for solving discrete optimization problems.

The comparison results on the fifteen SUKP instances among MSO4 and five state-of-the-art algorithms show that MSO4 performs competitively.

There are several possible directions for further study. First, we will investigate some new transfer functions on other algorithms such as krill herd algorithm (KH) [34–38], fruit fly optimization algorithm (FOA) [39], earthworm optimization algorithm (EWA) [40], and cuckoo search (CS) [41,42]. Second, we will study other techniques to discrete continuous optimization algorithms such as k-means framework [43]. Third, we will apply these twelve transfer functions-based discrete MS algorithms to other related and more complicated binary optimization problems including multidimensional knapsack problem (MKP) [39] and flow shop scheduling problem (FSSP) [44]. Finally, we will incorporate other strategies, namely, information feedback [45] and chaos theory [46], into MS to improve the performance of the algorithm.

Author Contributions: Writing and methodology, Y.F.; supervision, H.A.; review and editing, X.G.

Funding: This research was funded by National Natural Science Foundation of China, grant number 61806069, Key Research and Development Projects of Hebei Province, grant number 17210905.

Conflicts of Interest: The authors declare no conflict of interest.

References

1. Cormen, T.H.; Leiserson, C.E.; Rivest, R.L.; Stein, C. *Introduction to Algorithms*; MIT Press: Cambridge, MA, USA, 2009.
2. Du, D.Z.; Ko, K.I. *Theory of Computational Complexity*; John Wiley & Sons: Hoboken, NJ, USA, 2011.
3. Goldschmidt, O.; Nehme, D.; Yu, G. Note: On the set-union knapsack problem. *Naval Res. Logist. (NRL)* **1994**, *41*, 833–842. [CrossRef]
4. Arulselvan, A. A note on the set union knapsack problem. *Discret. Appl. Math.* **2014**, *169*, 214–218. [CrossRef]
5. He, Y.; Xie, H.; Wong, T.L.; Wang, X. A novel binary artificial bee colony algorithm for the set-union knapsack problem. *Future Gener. Comput. Syst.* **2017**, *78*, 77–86. [CrossRef]
6. Yang, X.; Vernitski, A.; Carrea, L. An approximate dynamic programming approach for improving accuracy of lossy data compression by Bloom filters. *Eur. J. Oper. Res.* **2016**, *252*, 985–994. [CrossRef]
7. Schneier, B. *Applied Cryptography: Protocols, Algorithms, and Source Code in C*; John Wiley & Sons: Hoboken, NJ, USA, 2007.
8. Engelbrecht, A.P.; Pampara, G. Binary differential evolution strategies. In Proceedings of the IEEE Congress on Evolutionary Computation, Singapore, 25–28 September 2007; pp. 1942–1947.
9. Ozsoydan, F.B.; Baykasoglu, A. A swarm intelligence-based algorithm for the set-union knapsack problem. *Future Gener. Comput. Syst.* **2018**, *93*, 560–569. [CrossRef]
10. Wang, G.G. Moth search algorithm: A bio-inspired metaheuristic algorithm for global optimization problems. *Memetic Comput.* **2016**. [CrossRef]
11. Feng, Y.; Wang, G.G. Binary moth search algorithm for discounted 0-1 knapsack problem. *IEEE Access* **2018**, *6*, 10708–10719. [CrossRef]

12. Kennedy, J.; Eberhart, R.C. A discrete binary version of the particle swarm algorithm. In Proceedings of the 1997 IEEE International Conference on Systems, Man, and Cybernetics—Computational Cybernetics and Simulation, Orlando, FL, USA, 12–15 October 1997; Volume 5, pp. 4104–4108.

13. Karthikeyan, S.; Asokan, P.; Nickolas, S.; Page, T. A hybrid discrete firefly algorithm for solving multi-objective flexible job shop scheduling problems. *Int. J. Bio-Inspired Comput.* **2015**, *7*, 386–401. [CrossRef]

14. Kong, X.; Gao, L.; Ouyang, H.; Li, S. A simplified binary harmony search algorithm for large scale 0-1 knapsack problems. *Expert Syst. Appl.* **2015**, *42*, 5337–5355. [CrossRef]

15. Rashedi, E.; Nezamabadi-Pour, H.; Saryazdi, S. BGSA: Binary gravitational search algorithm. *Nat. Comput.* **2010**, *9*, 727–745. [CrossRef]

16. Saremi, S.; Mirjalili, S.; Lewis, A. How important is a transfer function in discrete heuristic algorithms. *Neural Comput. Appl.* **2015**, *26*, 625–640. [CrossRef]

17. Mirjalili, S.; Lewis, A. S-shaped versus V-shaped transfer functions for binary particle swarm optimization. *Swarm Evol. Comput.* **2013**, *9*, 1–14. [CrossRef]

18. Pampara, G.; Franken, N.; Engelbrecht, A.P. Combining particle swarm optimisation with angle modulation to solve binary problems. In Proceedings of the 2005 IEEE Congress on Evolutionary Computation, Edinburgh, UK, 2–5 September 2005; Volume 1, pp. 89–96.

19. Leonard, B.J.; Engelbrecht, A.P.; Cleghorn, C.W. Critical considerations on angle modulated particle swarm optimisers. *Swarm Intell.* **2015**, *9*, 291–314. [CrossRef]

20. Costa, M.F.P.; Rocha, A.M.A.C.; Francisco, R.B.; Fernandes, E.M.G.P. Heuristic-based firefly algorithm for bound constrained nonlinear binary optimization. *Adv. Oper. Res.* **2014**, *2014*, 215182. [CrossRef]

21. Burnwal, S.; Deb, S. Scheduling optimization of flexible manufacturing system using cuckoo search-based approach. *Int. J. Adv. Manuf. Technol.* **2013**, *64*, 951–959. [CrossRef]

22. Pampará, G.; Engelbrecht, A.P. Binary artificial bee colony optimization. In Proceedings of the 2011 IEEE Symposium on Swarm Intelligence (SIS), Paris, France, 11–15 April 2011; pp. 1–8.

23. Zhu, H.; He, Y.; Wang, X.; Tsang, E.C.C. Discrete differential evolutions for the discounted {0-1} knapsack problem. *Int. J. Bio-Inspired Comput.* **2017**, *10*, 219–238. [CrossRef]

24. Changdar, C.; Mahapatra, G.S.; Pal, R.K. An improved genetic algorithm based approach to solve constrained knapsack problem in fuzzy environment. *Expert Syst. Appl.* **2015**, *42*, 2276–2286. [CrossRef]

25. Lim, T.Y.; Al-Betar, M.A.; Khader, A.T. Taming the 0/1 knapsack problem with monogamous pairs genetic algorithm. *Expert Syst. Appl.* **2016**, *54*, 241–250. [CrossRef]

26. Cao, L.; Xu, L.; Goodman, E.D. A neighbor-based learning particle swarm optimizer with short-term and long-term memory for dynamic optimization problems. *Inf. Sci.* **2018**, *453*, 463–485. [CrossRef]

27. Chih, M. Three pseudo-utility ratio-inspired particle swarm optimization with local search for multidimensional knapsack problem. *Swarm Evol. Comput.* **2017**, *39*. [CrossRef]

28. Michalewicz, Z.; Nazhiyath, G. Genocop III: A co-evolutionary algorithm for numerical optimization problems with nonlinear constraints. In Proceedings of the 1995 IEEE International Conference on Evolutionary Computation, Perth, Western Austrilia, 29 November–1 December 1995; Volume 2, pp. 647–651.

29. Liu, X.J.; He, Y.C.; Wu, C.C. Quadratic greedy mutated crow search algorithm for solving set-union knapsack problem. Microelectro. Comput. 2018, 35, 13–19.

30. Omran, M.G.H.; Mahdavi, M. Global-best harmony search. *Appl. Math. Comput.* **2008**, *198*, 643–656. [CrossRef]

31. Yang, X.S. Firefly Algorithm, Lévy Flights and Global Optimization. In *Research and Development in Intelligent Systems XXVI*; Springer: London, UK, 2010; pp. 209–218.

32. Wang, G.-G.; Deb, S.; Cui, Z. Monarch butterfly optimization. *Neural Comput. Appl.* **2015**, 1–20. [CrossRef]

33. Feng, Y.; Wang, G.G.; Li, W.; Li, N. Multi-strategy monarch butterfly optimization algorithm for discounted {0-1} knapsack problem. *Neural Comput. Appl.* **2017**, 1–18. [CrossRef]

34. Wang, G.-G.; Gandomi, A.H.; Alavi, A.H. An effective krill herd algorithm with migration operator in biogeography-based optimization. *Appl. Math. Model.* **2014**, *38*, 2454–2462. [CrossRef]

35. Wang, G.-G.; Gandomi, A.H.; Alavi, A.H. Stud krill herd algorithm. *Neurocomputing* **2014**, *128*, 363–370. [CrossRef]

36. Wang, G.; Guo, L.; Wang, H.; Duan, H.; Liu, L.; Li, J. Incorporating mutation scheme into krill herd algorithm for global numerical optimization. *Neural Comput. Appl.* **2014**, *24*, 853–871. [CrossRef]

37. Wang, H.; Yi, J.-H. An improved optimization method based on krill herd and artificial bee colony with information exchange. *Memetic Comput.* **2017**. [CrossRef]

38. Wang, G.-G.; Deb, S.; Gandomi, A.H.; Alavi, A.H. Opposition-based krill herd algorithm with Cauchy mutation and position clamping. *Neurocomputing* **2016**, *177*, 147–157. [CrossRef]

39. Wang, L.; Zheng, X.L.; Wang, S.Y. A novel binary fruit fly optimization algorithm for solving the multidimensional knapsack problem. *Knowl.-Based Syst.* **2013**, *48*, 17–23. [CrossRef]

40. Wang, G.-G.; Deb, S.; Coelho, L.D.S. Earthworm optimization algorithm: A bio-inspired metaheuristic algorithm for global optimization problems. *Int. J. Bio-Inspired Comput.* **2015**. [CrossRef]

41. Cui, Z.; Sun, B.; Wang, G.-G.; Xue, Y.; Chen, J. A novel oriented cuckoo search algorithm to improve DV-Hop performance for cyber-physical systems. *J. Parallel. Distr. Comput.* **2017**, *103*, 42–52. [CrossRef]

42. Wang, G.-G.; Gandomi, A.H.; Zhao, X.; Chu, H.E. Hybridizing harmony search algorithm with cuckoo search for global numerical optimization. *Soft Comput.* **2016**, *20*, 273–285. [CrossRef]

43. García, J.; Crawford, B.; Soto, R.; Castro, C.; Paredes, F. A k-means binarization framework applied to multidimensional knapsack problem. *Appl. Intell.* **2018**, *48*, 357–380. [CrossRef]

44. Deng, J.; Wang, L. A competitive memetic algorithm for multi-objective distributed permutation flow shop scheduling problem. *Swarm Evolut. Comput.* **2016**, *32*, 107–112. [CrossRef]

45. Wang, G.-G.; Tan, Y. Improving metaheuristic algorithms with information feedback models. *IEEE Trans. Cybern.* **2017**. [CrossRef] [PubMed]

46. Wang, G.-G.; Guo, L.; Gandomi, A.H.; Hao, G.-S.; Wang, H. Chaotic krill herd algorithm. *Inf. Sci.* **2014**, *274*, 17–34. [CrossRef]

![mathematics logo] *mathematics*

MDPI

Article

A Novel Simple Particle Swarm Optimization Algorithm for Global Optimization

Xin Zhang, Dexuan Zou * and Xin Shen

School of Electrical Engineering & Automation, Jiangsu Normal University, Xuzhou 221116, China; 2020160848@jsnu.edu.cn (X.Z.); 2020160838@jsnu.edu.cn (X.S.)
* Correspondence: 6020110007@jsnu.edu.cn; Tel.: +86-181-2003-0371

Received: 28 October 2018; Accepted: 19 November 2018; Published: 27 November 2018

Abstract: In order to overcome the several shortcomings of Particle Swarm Optimization (PSO) e.g., premature convergence, low accuracy and poor global searching ability, a novel Simple Particle Swarm Optimization based on Random weight and Confidence term (SPSORC) is proposed in this paper. The original two improvements of the algorithm are called Simple Particle Swarm Optimization (SPSO) and Simple Particle Swarm Optimization with Confidence term (SPSOC), respectively. The former has the characteristics of more simple structure and faster convergence speed, and the latter increases particle diversity. SPSORC takes into account the advantages of both and enhances exploitation capability of algorithm. Twenty-two benchmark functions and four state-of-the-art improvement strategies are introduced so as to facilitate more fair comparison. In addition, a *t*-test is used to analyze the differences in large amounts of data. The stability and the search efficiency of algorithms are evaluated by comparing the success rates and the average iteration times obtained from 50-dimensional benchmark functions. The results show that the SPSO and its improved algorithms perform well comparing with several kinds of improved PSO algorithms according to both search time and computing accuracy. SPSORC, in particular, is more competent for the optimization of complex problems. In all, it has more desirable convergence, stronger stability and higher accuracy.

Keywords: particle swarm optimization; confidence term; random weight; benchmark functions; t-test; success rates; average iteration times

1. Introduction

Since the 1950s, heuristic algorithms based on evolutionary algorithms (EAs) [1] have sprung up and been widely applied to the field of optimization control, such as moth search (MS) algorithm [2,3], genetic algorithm (GA) [4], ant colony optimization (ACO) algorithm [5], differential evolution (DE) algorithm [6], simulated annealing (SA) algorithm [7], krill herd (KH) algorithm, etc. [8–12]. Compared with traditional optimization methods such as golden section [13], Newton method [14,15], gradient method [16], heuristic algorithms have better biological characteristics and higher efficiency. It has been proved that heuristic algorithms perform well in some advanced existing fields e.g., grid computing [17], the superfluid management of 5G Networks [18], TCP/IP Mobile Cloud [19], IIR system identification [20], etc.

The Particle Swarm Optimization (PSO) algorithm proposed by Kennedy and Eberhart [21,22] in 1995 is also a member of the heuristic algorithm. Unlike other EAs, PSO does not require such steps as crossover, mutation, and selection, and it has fewer parameters. Its optimization process relies entirely on formula iteration, hence its calculation burden is low. The efficiency is very high, especially for continuous unimodal function model optimization. Due to these advantages, it has been widely used in various theoretical and practical problems such as function optimization [23], Non-Deterministic Polynomial(NP) problem [24], and multi-objective optimization [25].

PSO is a typical algorithm that relies on swarm intelligence [26–31] to optimize complex problems, and it is inspired by the foraging behavior of birds. It can be imagined that a group of gold rushers find gold in a region. They all have instruments that can detect gold mines under the stratum, and they can communicate with their nearest gold rushers. Through communication, they can know whether the person next to them finds gold. At the beginning, in order to explore this area more comprehensively, they randomly select a location to explore and maintain a certain distance. As the exploration begins, if someone finds some gold, the neighboring gold rushers can choose whether to change his position based on his own experience and whether he trusts him. This constant search may make it easier to find more gold than to be alone. In this example, a group of gold rushers and the gold are, respectively, equivalent to the particles of PSO and the optima that needs to be searched.

In actual operation, it is observed that PSO is very prone to premature convergence and falls into the local optima when faced with multimodal functions, especially some ones with traps or discontinuities. Based on this observation, a huge amount of particle swarm optimization variants have been proposed to deal with these issues. From the literature, it can be clearly observed that most of the existing PSO algorithms can be roughly divided into six categories: principle study, parameter setting, topology improvement, updating formula improvement, hybrid mechanism, practical application.

1. **Principle study:** The inertia weight factor, which adjusts the ability of PSO algorithm in local and global search was introduced by Shi and Eberhart [32], effectively avoiding falling into local optimum for PSO. Shi and Eberhart provided a way of thinking for future improvement. In 2001, Parsopoulos and Vrahatis [33]'s research showed that basic PSO can work effectively and stably in noisy environments, and in many cases, the presence of noise can also help PSO avoid falling into local optimum. The basic PSO was introduced for continuous nonlinear function [21,22]. However, because the basic PSO is easy to fall into the local optima, local PSO(LPSO) [34] was introduced in 2002. Clerc and Kennedy [35] proposed a constriction factor to enhance the explosion, stability, and convergence in a multidimensional complex space. Xu and Yu [36] used the super-martingale convergence theorem to analyze the convergence of the particle swarm optimization algorithm. The results showed that the particle swarm optimization algorithm achieves the global optima in probability and the quantum-behaved particle swarm optimization (QPSO) [37] has also been proved to have global convergence.

2. **Parameter setting:** A particle swarm optimization with fitness adjustment parameters (PSOFAB) [38], based on the fitness performance, was proposed in order to converge to an approximate optimal solution. The experimental results were analyzed by the Wilcoxon signed rank test, and its analysis showed that PSOFAP [38] is effective in increasing the convergence speed and the solution quality. It accurately adapts the parameter value without performing parametric sensitivity analysis. The inertia weight of the hybrid particle swarm optimization incorporating fuzzy reasoning and weighted particle (HPSOFW) [39] is changed based on defuzzification output. The chaotic binary PSO with time-varying acceleration coefficients (CBPSOTVAC) [40] using 116 benchmark problems from the OR-Library to test has time-varying acceleration coefficients for the multidimensional knapsack problem. A self-organizing hierarchical PSO [41] also uses time-varying acceleration coefficients.

3. **Topology improvement:** In 2006, Kennedy and Mendes [42] explained the neighborhood topologies in fully informed and best-of-neighborhood particle swarms in detail. A dynamic multiswarm particle swarm optimizer (DMSPSO) [43] was proposed, and it adopts a neighborhood topology including a random selection of small swarms with small neighborhood. Moreover, the regrouped group is dynamic and randomly assigned. In 2014, FNNPSO [44] use Fluid Neural Networks (FNNs) to create a dynamic neighborhood mechanism. The results showed that FNNPSO can outperform both the standard PSO algorithm and PCGT-PSO. Sun and Li proposed a two-swarm cooperation particle swarm optimization (TCPSO) [45] that used the slave swarm and the master swarm to exchange the information, which is beneficial for enhancing the convergence speed and maintaining the swarm diversity in TCPSO, and particles update the next particle

with information from its neighboring particles, rather than its own history best solution and current velocity. This strategy makes the particles of the subordinate group more inclined to local optimization, thus accelerating convergence. Inspired by the cluster reaction of the starlings, Netjinda et al. [46] used the collective response mechanism to influence the velocity and position of the current particle by seven adjacent ones, thereby increasing the diversity of the particles. A nonparametric particle swarm optimization (NP-PSO) [47] combines local and global topologies with two quadratic interpolation operations to enhance the PSO capability without tuning any algorithmic parameter.

4. **Updating formula improvement:** Mendes [48] changed the PSO's velocity and personal best solution updating formula and proposed a fully informed particle swarm (FIPS) algorithm to make good use of the whole entire swarm. Mendes [49] proposed a Comprehensive learning particle swarm optimizer (CLPSO) whose velocity updating formula eliminates the influence from global best solution to to suit the multimodal functions, and CLPSO uses two tournament-selected particles to help particles study better case during iteration. The results showed that CLPSO performs better than other PSO variants for multimodal problems. A learning particle swarm optimization (*LPSO) algorithm [50] was proposed with a new framework that changed the velocity updating formula so as to organically hybridize PSO with another optimization technique. *LPSO is composed of two cascading layers: exemplar generation and a basic PSO algorithm updating method. A new global particle swarm optimization (NGPSO) algorithm [51] uses a new position updating equation that relies on the global best particle to guide the searching activities of all particles. In the latter part of the NGPSO search, the random distribution based on uniform distribution is used to increase the particle swarm diversity and avoid premature convergence. Kiran proposed a PSO with a distribution-based position update rule (PSOd) [52] whose position updating formula is combined with three variables.

5. **Hybrid mechanism:** In 2014, Wang et al. [53] proposed a series of chaotic particle-swarm krill herd (CPKH) algorithms for global numerical optimization. The CPKH is a hybird Krill herd (KH) [54] algorithm with APSO [55] which has a mutation operator and chaotic theory. This hybrid algorithm, which with an appropriate chaotic map performs superiorly to the standard KH and other population-based optimization, has quick exploitation for solution. DPSO [56] is a accelerated PSO (APSO) [55] algorithm hybridized with a DE algorithm mutation operator. It has a superior performance due to combining the advantages from both APSO and DE. Wang et al., finally, studied and analyzed the effect of the DPSO parameters on convergence and performance by detailed parameter sensitivity studies. In he hybrid learning particle swarm optimizer with genetic disturbance (HLPSO-GD) [57], the genetic disturbance is used to cross the corresponding particle in the external archive, and new individuals are generated, which will improve the swarm's ability to escape from the local optima. Gong et al. proposed a genetic learning particle swarm optimization (GLPSO) algorithm that uses genetic evolution to breed promising exemplars based on *LPSO [50] enhancing the global search ability and search efficiency of PSO. PSOTD [58] namely a particle swarm optimization algorithm with two differential mutation, which has a novel structure with two swarms and two layers including bottom layer and top layer, was proposed for 44 benchmark functions. HNPPSO [59] is a novel particle swarm optimization combined with a multi-crossover operation, a vertical crossover, and an exemplar-based learning strategy. To deal with production scheduling optimization in foundry, a hybrid PSO combined the SA [7] algorithm [60] was proposed.

6. **Practical application:** Zou et al. used NGPSO [51] to solve the economic emission dispatch (EED) problems and the results showed that NGPSO is the most efficient approach for solving the economic emission dispatch (EED) problems. PS-CTPSO [61] based on the predatory search strategy was proposed to do with web service combinatorial optimization, which is an NP problem, and it improves overall ergodicity. To improve the changeability of ship inner shell, IPSO [62] was proposed for a 50,000 DWT product oil tanker. MBPSO [63] was proposed for sensor

management of LEO constellation to the problem of utilizing a low Earth orbit (LEO) infrared constellation in order to track the midcourse ballistic missile. GLNPSO [64] is for a capacitated Location-Routing Problem. The particle swarm algorithm is also applied to many other practical problems e.g., PID (Proportion Integration Differentiation) controller [65], optimal strategies of energy management integrated with transmission control for a hybrid electric vehicle [66], production scheduling optimization in foundry [60], etc.

In view of the shortcomings of PSO [21,22], three improvements are proposed in this paper. The first is Simple Particle Swarm Algorithm (SPSO). It does not use the velocity updating formula, and abandons the use of self-cognitive term. Although the speed of the algorithm has been greatly improved, some deficiencies have been found in actual tests. It is observed that the particles' difference is too small to jump out of the local optimal solution, which is not suitable for searching for multimodal problems. For this purpose, a second improvement named Simple Particle Swarm Optimization with Confidence Term (SPSOC) is proposed in this paper. That is, the confidence term is introduced in the SPSO's position updating formula. Although having a slight increase in time compared to SPSO, the results show that SPSOC is better for multi-peak function optimization. On the basis of this, the inertia weight is improved by introducing the difference between the stochastic objective function value and the worst one, and the final improvement is called Simple Particle Swarm Optimization based on Random weight and Confidence term (SPSORC). The inertial weight not only has a crucial effect on its convergence, but also plays an important role in balancing exploration and exploitation during the evolution. The strategy in this paper makes particle position movements more random. A large number of experiments suggest that all three improvements are very effective, and the combination of the three improvements has greatly improved the search efficiency of the particle swarm algorithm.

The rest of this paper is organized as follows: Section 2 introduces the basic PSO [21,22] and three recently improved PSO methods. In Section 3, three improvements are presented in detail. In Section 4, some analysis of PSO is further discussed. The experimental results are discussed and analyzed between four state-of-the-art PSOs and three improved ones proposed in this paper. Finally, this paper presents some important conclusions and the outlook of future work in Section 5.

2. Related Works

2.1. The Basic PSO

In general, the particle swarm optimization algorithm is composed of the position updating formula and the velocity updating formula. Each particle iterates with reference to its own history best solution p_{best} and the global best value g_{best} to change position and velocity information. The basic particle swarm optimization (bPSO) [21,22] algorithm iteration formula is as follows:

$$v_{in}^{t+1} = v_{in}^t + c_1 r_1 (p_{best}^t - x_{in}^t) + c_2 r_2 (g_{best}^t - x_{in}^t), \tag{1}$$

$$x_{in}^{t+1} = x_{in}^t + v_{in}^{t+1}. \tag{2}$$

As shown in the above formula, Equations (1) and (2) are the velocity updating formula and the position updating formula, respectively. The particles whose population is m search for the optima in the -dimensional space. In that process, the i-th particle's position in the n-dimensional space is x_{in} and the current velocity is v_{in}. p_{best} is the individual history best solution and g_{best} is the global one. t is the current iteration numbers. c_1 and c_2 are cognitive and social factors, and r_1 and r_2 are random numbers belonging to [0,1). Figure 1 is an optimization procedure of PSO.

From Figure 1, the area U is the solution space of a function. O is the theoretical optima that needs to be found. x_i^t is the position of the initial particle. The velocity v_i^t is the current particle velocity. v_i^{t+1} is the velocity after being affected by various aspects. Particle memory influence and swarm influence are parallel to the connecting lines from x_i^t to g_{best} and p_{best}, respectively, indicating the influence from g_{best} and p_{best}. In this generation, the particle i is affected by v_i^t first. After particle memory influence and swarm influence, i arrives at x_i^{t+1} from x_i^t at velocity v_i^{t+1}. From the next iteration, the particle will move from x_i^{t+1} towards the new position. It keeps iterating as the step above and moves to the theoretical optima more and more close.

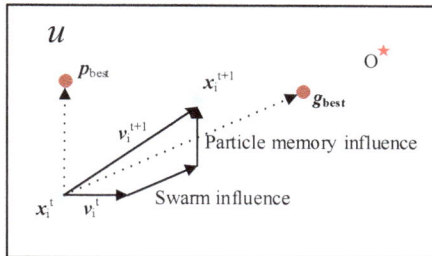

Figure 1. Optimization procedure of bPSO.

The velocity updating formula had changed to Equation (3), when Shi and Eberhart put the inertia weight ω into it, and the position updating formula remained unchanged:

$$v_{in}^{t+1} = \omega v_{in}^t + c_1 r_1 (p_{best}^t - x_{in}^t) + c_2 r_2 (g_{best}^t - x_{in}^t). \tag{3}$$

The introduction of inertia weight effectively keeps a balance between the local and global search capability. The larger the inertia weight, the stronger the global search capability of the algorithm. On the contrary, the local search capability is more prominent. This particle swarm optimization model is the most commonly used nowadays, and many scholars have improved it.

The steps to achieve it are as follows:

Step 1: Initialize the population randomly. Set the maximum number of iterations, population size, inertia weight, cognitive factors, social factors, position limits and the maximum velocity limit.

Step 2: Calculate the fitness of each particle according to fitness function or models.

Step 3: Compare the fitness of each particle with its own history best solution p_{best}. If the fitness is smaller than p_{best}, the smaller value is assigned to p_{best}, otherwise, p_{best} is reserved. Then, the fitness is compared with the global best solution g_{best}, and the method is the same as selecting p_{best}.

Step 4: Use Equations (2) and (3) to update the particle position and velocity. In addition, we must make sure that its velocity and position are, respectively, within the maximum velocity limit and position limits.

Step 5: Check if the theoretical optimum is reached, output the value and stop the operation; otherwise, return to **Step 2** (Section 2.1) until it reaches the theoretical optima or peaks the maximum number of iterations.

In this paper, a basic particle swarm optimization with decreasing linear inertia weight is used. The weight formula is as Equation (4):

$$\omega = \omega_{max} - \frac{\omega_{max} - \omega_{min}}{T} \times t. \tag{4}$$

In Equation (4), ω_{max} is starting weight. ω_{min} is final weight. t_{max} is the maximum number of iterations. PSO needs to set a still more larger starting weight ω_{max} according to the influence of inertia weight on the search capability of PSO, so as to pay more attention to the global optima. As the number of iterations increases, the weight will be decreased. The search process would be more inclined to explore the local optima, which is more conducive to the final convergence.

2.2. The PSO with a Distribution-Based Position Update Rule

In 2017, a distribution-based update rule for PSO (PSOd) [52] algorithm was proposed by Kiran. This improved strategy changed PSO's iteration formula.

$$x_{in}^{t+1} = \mu + \sigma \times Z. \tag{5}$$

Those three variables in Equation (5) work by Equations (6)–(8):

$$\mu = \frac{x_{in}^t + p_{best}^t + g_{best}^t}{3}, \tag{6}$$

$$\sigma = \sqrt{\frac{(x_{in}^t - \mu)^2 + (p_{best}^t - \mu)^2 + (p_{best}^t - \mu)^2}{3}}, \tag{7}$$

$$Z = (-2 \ln k_1)^{\frac{1}{2}} \times cos(2\pi k_2). \tag{8}$$

It works as follows:

Step 1: The population is initialized randomly.
Step 2: The fitness is calculated and compared to get the best individual history solution and the best global one.
Step 3: Equation (5) is used to update the particle position that is limited in the upper and lower limits.
Step 4: If the termination condition is met, the best solution is reported.

2.3. A Hybrid PSO with Sine Cosine Acceleration Coefficients

In order to make better use of parameters on PSO algorithm, such as inertia weight, learning factors, etc., Chen et al. proposed a hybrid PSO algorithm with the sine cosine acceleration coefficients (HPSOscac) [67].

Step 1: The population is initialized randomly.
Step 2: The reverse population of the initial population is calculated by Equation (9)

$$x_{in}' = x_{max} + x_{min} - x_{in}. \tag{9}$$

In this equation, x_{in} and x_{in}' are initial population and reverse population, respectively. x_{max} and x_{min} are combined the upper and lower limits of particles position i.e., the solution space boundary.
Step 3: Fitness values of those two populations are sorted, and the best half is used as the initial population. Then, the p_{best} and g_{best} are obtained by comparing.
Step 4: Equations (10) and (11) are used to update the inertia weight and learning factors, respectively:

$$\begin{cases} \omega^{t+1} = \frac{c}{4} \times \sin(\pi\omega^t), \\ \omega^1 = 0.4; c\epsilon(0,4], \end{cases} \tag{10}$$

$$\begin{cases} c_1 = 2 \times \sin((1 - \frac{t}{T}) \times \frac{\pi}{2}) + 0.5, \\ c_2 = 2 \times \cos((1 - \frac{t}{T}) \times \frac{\pi}{2}) + 0.5. \end{cases} \tag{11}$$

Among them, c is a constant among 0 and 4. c_1 and c_2 are cognitive and social factors, respectively.
Step 5: Updating the particle velocity and position, use Equations (1) and (12). The particle position updating formula is as follows:

$$x_{in}^{t+1} = x_{in}^t \times W_{in}^t + v_{in}^t \times W_{in}^{t'} + \rho \times g_{best}^t \times W_{in}^t. \tag{12}$$

W_{in}^t and $W_{in}^{t'}$ are the dynamic weights that control position and velocity terms. Its formula is like Equation (13). ρ is a random value between 0 and 1:

$$\begin{cases} W_{in}^t = \dfrac{\exp\frac{f_i}{f_{avg}}}{1+\exp\frac{-f_i}{f_{avg}}}t, \\ W_{in}^{t'} = 1 - W_{in}^t. \end{cases} \tag{13}$$

In this formula, f_i is the particle fitness value, and f_{avg} is the average one.

Step 6: The iteration is ended if end condition is reached. Otherwise, it comes back to **Step 2** (Section 2.3).

2.4. A Two-Swarm Cooperative PSO

A two-swarm cooperative particle swarm optimization (TCPSO) [45] was proposed who uses two particle swarms, the slave swarm and the master swarm with the clear division of their works to overcome the shortcomings such as lack of diversity, slow convergence in the later period, etc. It works like the following:

Step 1: Initialization. Initialize the slave swarm and the master swarm's velocity and position randomly.

Step 2: Calculate the fitness of these two swarms and get the g_{best}^S, p_{best}^S, g_{best} and p_{best}^M. The first two come from the slave swarm and the last two come from the master swarm.

Step 3: Reproduction and updating.

Step 3.1: Update the slave swarm by Equations (14) and (15). Ensure that velocity and position are within the limits:

$$v_{in}^{S,t+1} = c_1^S r_1 (1-r_2)(x_{kn}^{S,t} - x_{in}^{S,t}) + c_2^S (1-r_1) r_2 (g_{best} - x_{in}^{S,t}), \tag{14}$$

$$x_{in}^{S,t+1} = x_{in}^{S,t} + v_{in}^{S,t+1}. \tag{15}$$

S in these two formulas means that this variable from the slave swarm, except g_{best} in Equation (14) from the master swarm. Finally, we will get the g_{best}^S. x_k is randomly chosen from the neighberhood of the x_i according to Equation (16) [42]:

$$k\epsilon \begin{cases} [i - \frac{l}{2} + 1, i + \frac{l}{2}], if\ l\ is\ even, \\ [i - \frac{l-1}{2}, i + \frac{l-1}{2}], if\ l\ is\ odd. \end{cases} \tag{16}$$

l is the size of neighborhood. Sun and Li found that the size of neighborhood equal to 2 is best in their experiments.

Step 3.2: Update the master swarm by Equations (17) and (18). Ensure that velocity and position are within the limits:

$$\begin{aligned} v_{in}^{M,t+1} = \omega^M v_{in}^{M,t} + c_1^M r_1 (1-r_2)(1-r_3)(p_{best}^M - x_{in}^{M,t}) + c_2^M r_2 (1-r_1)(1-r_3)(g_{best}^S - x_{in}^{M,t}) \\ + c_3^M r_3 (1-r_1)(1-r_2)(g_{best} - x_{in}^{M,t}) \end{aligned} \tag{17}$$

$$x_{in}^{M,t+1} = x_{in}^{M,t} + v_{in}^{M,t+1}. \tag{18}$$

M here means that this variable is from the master swarm. In the end of **Step 3.2** (Section 2.4), g_{best} wil be obtained for the next iteration.

Step 4: Get the optima if it meets the termination condition; otherwise, go to **Step 2** (Section 2.4).

3. SPSO, SPSOC, SPSORC

3.1. Simple PSO

Zou et al. proposed a novel harmony search algorithm [68] that used the optimal harmony and worst harmony in the harmony memory to guide the configuration of the harmony vector. It obtained very suitable results. Inspired by its thoughts, we try to round off the velocity formula and cognitive term of PSO and directly use the social term to control the algorithm optimization, so that the formula Equation (21) is the most simplified, namely the Simple Particle Swarm Optimization (SPSO) algorithm. According to the results of the literature [69], the influence of the velocity term on the performance of the particle swarm algorithm can be neglected. Drawing on literature [69], we can simply do the following derivation. Before abandoning the velocity updating formula, SPSO velocity updating formula is shown as follows. Particles' positions are updated according to Equation(2):

$$v_{in}^{t+1} = \omega v_{in}^t + cr(g_{best}^t - x_{in}^t). \tag{19}$$

According to Equations (19) and (2), we make the following assumptions:

Hypothesis 1. *The update of particles per dimension is independent from each other, except that g_{best} is the one that connects the information to the other dimensions.*

Hypothesis 2. *When particle i is updated, the other particles' velocities and positions are not changed.*

Hypothesis 3. *The particles' positions are moving continuously.*

According to the above assumptions, it is only necessary to prove a certain dimension of a certain particle search process that can be universal. Iterating over Equations (19) and (2) yields a second-order differential equation:

$$x^{t+2} + (rc - \omega - 1)x^{t+1} + \omega x^t = rcg_{best}^t. \tag{20}$$

We can observe that there is no velocity updating in Equation (20). This result can be applied to each dimension update of other particles. Now, we get SPSO's updating formula:

$$x_{in}^{t+1} = \omega x_{in}^t + cr(g_{best}^t - x_{in}^t). \tag{21}$$

SPSO only uses this formula to iterate. The experimental results show that this strategy improves the search efficiency and stability of the bPSO.

It works like the following:

Step 1: The maximum generation, population number, inertia weight, learning factor are set up. Population is initialized.

Step 2: Fitness is calculated according to the function.

Step 3: Every particle compares with its history best solution to get the p_{best} and compares with the global best one to get the g_{best}.

Step 4: Particle position is updated by Equation (21).

Step 5: If the theoretical optimal value is not found, the program returns to **Step 2** (Section 3.1); otherwise, the program stops.

After changing, the particle direction is only affected by the global optima. Graphical display of one of the particle optimization process is shown in Figure 2.

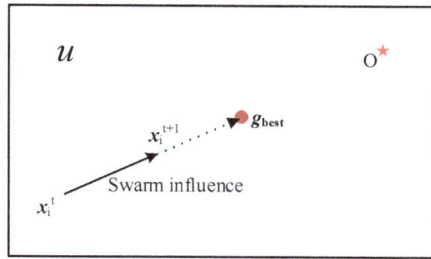

Figure 2. Optimization procedure of SPSO.

As shown in the figure, compared with the bPSO optimization process diagram in Section 2.1, in the optimization of SPSO, the particles are only affected by g_{best} and the direction of the particles always faces g_{best}. This feature also brings some drawbacks. For example, whether the algorithm can or cannot search for the theoretical optima depends entirely on the selection position of the global optima, which makes it likely for particles develop in a certain local optimal direction. It is possible to reach the current g_{best} value directly if the movement is fast enough. This is a search trajectory of one, while when all particles are only optimized in one direction, it obviously reduces the difference between the population. The lack of diversity directly leads to the fact that SPSO are easily trapped in local optimal solutions.

What is gratifying is that SPSO is very fast because of the simplification. This is very suitable for single-peak problems. This advantage can be clearly reflected in the experimental results in Section 4. However, the unconstrained functions, especially single-peak problems, are a minority after all. In order to make this improvement apply into more functions or environment, we propose adding a confidence term so that some part of the particles can determine the distance to advance based on its own level of trust to g_{best}, so as to get rid of the defects that all particles are looking for at one point.

3.2. SPSO with Confidence Term

In order to better solve the multimodal problem and make the improvement universal, we decided to add a confidence item(SPSOC) that rewrites Equation (21) into Equation (22):

$$x_{in}^{t+1} = \omega_1 x_{in}^t + cr_1(g_{best}^t - x_{in}^t) - \omega_2 r_2 g_{best}^t. \tag{22}$$

Compared with SPSO, the algorithm formula adds one item, namely the confidence term. ω_2 is the inertia weight of the confidence term. r_2 is the random value between [0, 1].

Referring to Figure 3, the principle of the item can be understood as: at a certain iteration, the position calculated by the SPSO moves a distance suffered from confidence influence. The effect is equivalent to the particle being optimized from x_i^t to $x_i'^{t+1}$. Then, the particle retreats a distance from the beginning in the opposite direction to the g_{best} direction. Finally, this particle reaches the position of x_i^{t+1}. Using the inertia weight ω_2 and the random number r_2, the distance of the particles retreating in the opposite direction would be uncertain. It can be imagined that the degree of particles trust at different generation is different, that is, the influence of g_{best} is different. This improvement can effectively slow the convergence of particles, so that the particles are not too dense, thus maintaining particle diversity.

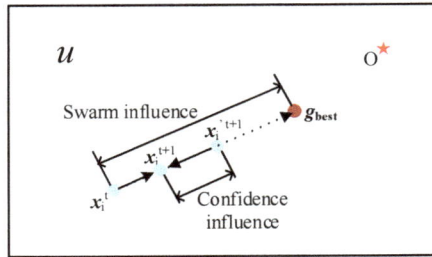

Figure 3. Optimization procedure of SPSOC.

A discussion of the impact of this improved algorithm using a combination of different weights will be explained in the experiment of Section 4.4. In order to minimize the program running time and ensure that the program structure is simple and the effect is optimal, this paper makes $\omega_1 = \omega_2$. SPSOC's iteration process is the same as SPSO.

3.3. SPSOC Based on Random Weight

Adding a confidence item to the SPSO does significantly enhance the search ability of the algorithm, but it does not achieve theoretical optimization when searching for most of the benchmark functions. Compared with many improved PSOs proposed recently, SPSOC has no big advantage except for the short amount of time. Therefore, we think about randomization improvement of inertia weight named SPSOC based on random weight (SPSORC). The improved inertia weight formula is shown in Equation (23):

$$\omega = \begin{cases} \frac{p_{best}^r - f_{best}}{f_{worst} - f_{best}}, & if\ set\ minimum\ as\ target, \\ \frac{f_{best} - p_{best}^r}{f_{best} - f_{worst}}, & if\ set\ maximum\ as\ target. \end{cases} \tag{23}$$

In this formula, if we set the minimum as the target we want to find, f_{best} is the minimum fitness in the current iteration, f_{worst} is the worst fitness target value in the current iteration, and p_{best}^r is one of the most p_{best} that a random particle has searched for from total swarm.

The use of Equation (23) allows the weights to be generated randomly, which effectively reduces the possibility that the algorithm falls into a local solution and enhance the exploitation capability. This strategy will at least make algorithms better for some multimodel problems. The random weight, however, also increases the risk of finding non-optimal solutions. This will be reflected in the large amount of experimental data in Section 4, but the experimental results show that the overall search ability of SPSOC has been very significantly improved.

It is more concise that the flow of SPSORC is similar to that of the bPSO, which just calculates the random weight. Its procedure is shown in Table 1.

Table 1. The procedure of SPSORC.

Line	Procedure of SPSORC
1	Initialize parameters: dimension N, population size m, iteration number T, weight ω, learing factors c_1, c_2, etc; % Step 1
2	Initialize and reserve matrix space: $\boldsymbol{p}_{best} = [\text{Inf}_{11} \cdots \text{Inf}_{mN}]$, $\boldsymbol{g}_{best} = [\text{Inf}_1 \cdots \text{Inf}_m]$, x_{min} = lower limits of position, x_{max} = upper limits;
3	**For** $i = 1:m$
4	**For** $j = 1:N$
5	Randomly initialize velosity and position: v_{in}, x_{in}; % Step 2
6	**End For**
7	**End For**
8	**For** $i = 1:m$
9	Calculate the fitness. Compared to get the p_{best}^1 and g_{best}^1
10	**End For**
11	**While** the optima is not found or the termination condition is not met
12	Calculate the f_{best} and f_{worst}. Then, get the ω by Equation (23); % Step 3
13	**For** $i = 1:m$
14	**For** $j = 1:N$
15	Update the particle positon according to Equation (22); % Step 4
16	**If** $x_{in}^t > x_{max}$
17	$x_{in}^t = x_{max}$;
18	**ElseIf** $x_{in}^t < x_{min}$
19	$x_{in}^t = x_{min}$;
20	**End If**
21	**End For**
22	Substitute the current particle into the fitness formula to calculate the fitness value of the current particle;
23	Compare to get the p_{best} and g_{best};
24	**End For**
25	**End While**
26	**Return** Results. % Step 5

4. Experimental Study and Results Analysis

4.1. Benchmark Functions

The aim of this improved strategy is to solve the problem of unconstrained optimization better. In order to demonstrate the effectiveness of the algorithm more fully, this experiment will use 22 commonly used benchmark functions to simulate and contrast, including the unimodal benchmark functions represented by Sphere Function, the complex multimodel solution functions such as Rastrigrin Problem, the ill-conditioned quadratic Rosenbrock function, Xin–She Yang 3 with discontinuity and trap near the optimal solution, noise-containing functions like Quartic Function and other functions which is hard to find the best solution. Of course, these 22 functions also contain four test functions (f_7, f_{15}, f_{20} and f_{21}) with negative optima.

These 22 benchmark functions arranged in alphabetical order are shown in Table A1. The following test functions may change slightly in form for consistency or convenience because of the large number of types for test function versions, but the test results will not be affected. The last column, 'Accuracy (50)', is the convergence accuracy we want to reach for the test function in the 50-dimensional case, which will be used in Section 4.5.3, 'Success Rate and Average Iteration Times'.

4.2. Parameters Setting and Simulation Environment

One of the reasons why particle swarm optimization algorithm was proposed late but had a relatively wide range of use is that it needs fewer parameters and is set up simply. When dealing with general problems, its requirements on population numbers, the maximum iteration numbers and other parameters are not high, which also determines that the algorithm has the advantages of small size and fast searching speed when it is implemented. Under normal circumstances, the population set at

40 can get a good solution for most problems. More complex problems can be solved by increasing the population number and the maximum iteration times.

Table 2 is about the specific parameter settings. NR is the number of times each algorithm searches for the benchmark functions. m is population number. T is the maximum iteration times per search. ω_{max} and ω_{min} are the maximum weight and the minimum weight. c_1 and c_2 are the acceleration factors.

Table 2. Parameters for candidates.

	NR	m	T	ω_{max}	ω_{min}	c_1	c_2	c_3
bPSO	30	40	100	0.9	0.4	2	2	-
PSOd	30	40	100	-	-	-	-	-
HPSOscac	30	40	100	Equation (10)	-	Equation (11)	Equation (11)	-
TCPSO	30	80	100	0.9	-	1.6	1.6	1.6
SPSO	30	40	100	0.9	0.4	-	2	-
SPSOC	30	40	100	0.9	0.4	-	2	-
SPSORC	30	40	100	Equation (23)	-	-	2	-

Simulation environment is shown in Table 3.

Table 3. Simulation environment.

Operation System	**Windows 7 Professional ($\times$ 32)**
CPU	Core 2 Duo 2.26 GHz
Memory	4.00 GB
Platform	Matlab R2014a
Network	Gigabit Ethernet

4.3. Discussion on Improvement Necessity For SPSO

The search speed is a great advantage of SPSO because of a simple structure. However, its advantages are its disadvantages. The over-simplified structure makes the SPSO's population lack of diversity, which makes it converge to local optima quickly, so the further improvement of SPSO becomes indispensable. Therefore, in Section 3, we present two improvements to SPSO. In this section, we will let SPSO, SPSOC and SPSORC solve the high dimension benchmark functions. Then, we discuss the necessity of those two improvement steps in Section 3 by analyzing its results.

In this experiment, the function dimension is set to 200 dimensions. The other parameters are consistent with the parameter setting table in Table 2 of Section 4.2. Table 4 shows the optimal results of the experiment. The minimum number in each set of data (min and mean) is represented in bold in the following table.

From the experimental results for the 200-dimensional benchmark functions in Table 4, we can see that SPSORC can search the other 21 functions for the theoretical optimal solution or a better solution than SPSO and SPSOC except for searching Quartic Function with noises. The optimization results of SPSO and SPSOC, however, are in straitened circumstances compared to SPSORC. SPSO gets better solutions four times, while SPSOC gets better solutions six times. Compared with the 30 search average solutions of SPSO and SPSOC, the optima of SPSOC is also smaller. it is indicated that the solution searched by SPSO after adding the confidence item can be kept smaller and its performance is greatly improved and the optimization capability is more enhanced after using random inertia weight. Thus, it can be seen that the two improvements to the SPSO are very necessary. SPSO is more inclined to exploration, which is more conducive to the local search of particles. Confidence term change the trajectory of some particles, which increases particle diversity. Meanwhile, the random inertia weight balances the exploitation capabilities of the algorithm so that it improves the search range and robustness of the algorithm significantly.

Table 4. Discussion on the necessity of improving SPSO.

Instance	SPSO		SPSOC		SPSORC	
	min	mean	min	mean	min	max
f_1	4.44×10^{-15}	4.44×10^{-15}	$\mathbf{8.88 \times 10^{-16}}$	$\mathbf{8.88 \times 10^{-16}}$	$\mathbf{8.88 \times 10^{-16}}$	1.48×10^{-15}
f_2	1.53×10^{-18}	5.91×10^{-3}	6.94×10^{-56}	6.93×10^{-50}	0	$\mathbf{6.65 \times 10^{-268}}$
f_3	2.16×10^{-35}	2.68×10^{-35}	3.84×10^{-106}	1.33×10^{-97}	0	0
f_4	5.40×10^{-76}	7.42×10^{-76}	2.27×10^{-216}	8.96×10^{-195}	0	0
f_5	0	8.32×10^{-3}	0	0	0	3.37×10^{-16}
f_6	1.61×10^{-30}	2.63×10^{-30}	2.88×10^{-102}	4.12×10^{-93}	0	0
f_7	-1.12×10^{1}	7.87×10^{-2}	$\mathbf{-1.49 \times 10^{2}}$	$\mathbf{-1.49 \times 10^{2}}$	$\mathbf{-1.49 \times 10^{2}}$	$\mathbf{-1.49 \times 10^{2}}$
f_8	9.51×10^{-1}	6.61×10^{1}	0	0	0	1.87×10^{1}
f_9	$\mathbf{2.33 \times 10^{-4}}$	3.15×10^{-2}	9.12×10^{-4}	2.30×10^{-2}	5.75×10^{-3}	3.63×10^{-1}
f_{10}	0	4.52×10^{1}	0	0	0	6.51×10^{-16}
f_{11}	1.49×10^{2}	$\mathbf{1.49 \times 10^{2}}$	1.49×10^{2}	1.49×10^{2}	$\mathbf{1.48 \times 10^{2}}$	1.49×10^{2}
f_{12}	7.97×10^{-33}	$\mathbf{2.79 \times 10^{-32}}$	2.95×10^{-82}	8.39×10^{-21}	0	1.77×10^{-04}
f_{13}	6.25×10^{-49}	1.07×10^{-45}	1.89×10^{-75}	1.24×10^{-60}	0	0
f_{14}	1.20×10^{-17}	1.36×10^{-17}	5.46×10^{-55}	3.75×10^{-47}	0	1.48×10^{-270}
f_{15}	-5.26×10^{3}	-3.37×10^{3}	-4.31×10^{3}	-2.61×10^{3}	$\mathbf{-5.99 \times 10^{3}}$	$\mathbf{-3.80 \times 10^{3}}$
f_{16}	1.21×10^{-34}	1.47×10^{-34}	1.91×10^{-106}	1.81×10^{-96}	0	1.90×10^{-321}
f_{17}	0	1.64×10^{-315}	0	0	0	1.36×10^{-256}
f_{18}	1.76×10^{-29}	2.81×10^{-6}	2.34×10^{-60}	1.85×10^{-18}	0	0
f_{19}	7.48×10^{-25}	1.02×10^{-16}	1.75×10^{-21}	1.04×10^{-14}	0	$\mathbf{3.96 \times 10^{-20}}$
f_{20}	3.63×10^{-55}	3.95×10^{-43}	7.93×10^{-46}	1.69×10^{-33}	-1	-1
f_{21}	1.75×10^{-44}	1.78×10^{-42}	1.59×10^{-36}	4.71×10^{-32}	-1	-1
f_{22}	2.61×10^{-36}	$\mathbf{3.15 \times 10^{-36}}$	2.20×10^{-33}	6.20×10^{-14}	0	2.14×10^{-8}

4.4. Discussion on Weight Selection for SPSOC

The proposed SPSOC has two inertia weights. The first inertia weight balances the search ability to global optima and the local one, while the second weight determines the degree to which the particle converges to the global optima in current generation. Obviously, whether these two weights are set properly or not has a significant impact on the performance of the algorithm. Then, the discussion of how these two inertia weights should be selected becomes quite necessary. The experiment comparing the optimal solution and the average solution found by the algorithm with different weights introduces three kinds of inertia weight strategies, which are divided into six kinds of situations. Those three kinds of inertia weights used in the experiment are as follows:

1. Linear decreasing inertia weight, i.e., Equation (4);
2. Classic nonlinear dynamic inertia weight, i.e., Equation (24);

$$\omega = \begin{cases} \omega_{max}, x_{in} > f_{avg}, \\ \omega_{min} - (\omega_{max} - \omega_{min}) \times \frac{x_{in} - f_{min}}{f_{avg} - f_{min}}, x_{in} \leq f_{avg}. \end{cases} \tag{24}$$

3. Random inertia weight proposed in this paper, i.e., Equation (23).

Table 5 reports the results of this experiment. Taking $\omega_{2,1}$ for example, the first subscript 2 indicates that ω_1 in Equation (22) uses the second kind of weight formula i.e., Equation (24), and the second subscript 1 indicates that ω_2 uses the first kind of weight formula i.e., Equation (4). The others are similar. Experimental benchmark functions' upper dimensions are set at 100. We represnt the minimum value for min and mean in bold in the following table.

Table 5. Discussion on the weights selection of SPSOC.

Instance		Different Weight Matching					
		ω_{21}	ω_{31}	ω_{32}	ω_{11}	ω_{22}	ω_{22}
f_1	min	$\mathbf{8.88 \times 10^{-16}}$	$\mathbf{8.88 \times 10^{-16}}$	$\mathbf{8.88 \times 10^{-16}}$	$\mathbf{8.88 \times 10^{-16}}$	$\mathbf{8.88 \times 10^{-16}}$	8.88×10^{-16}
	mean	1.80×10^1	3.85×10^{-15}	1.27×10^1	$\mathbf{8.88 \times 10^{-16}}$	1.53×10^1	1.36×10^{-15}
f_2	min	1.54×10^{-85}	4.67×10^{-65}	9.25×10^{-83}	8.09×10^{-61}	2.10×10^{-80}	0
	mean	4.34×10^{-16}	1.53×10^{-57}	7.90×10^{-75}	2.73×10^{-53}	4.14×10^{-63}	0
f_3	min	7.53×10^{-163}	3.54×10^{-125}	1.81×10^{-169}	8.92×10^{-122}	2.07×10^{-156}	0
	mean	1.75×10^{-38}	4.85×10^{-109}	4.84×10^{-91}	1.37×10^{-101}	4.97×10^{-50}	0
f_4	min	1.78×10^{-240}	8.32×10^{-258}	$\mathbf{0}$	4.48×10^{-235}	2.22×10^{-304}	0
	mean	1.68×10^{-34}	7.32×10^{-214}	5.03×10^{-106}	1.99×10^{-205}	2.65×10^{-21}	0
f_5	min	$\mathbf{0}$	$\mathbf{0}$	2.17×10^1	$\mathbf{0}$	$\mathbf{0}$	0
	mean	1.20×10^2	$\mathbf{0}$	2.17×10^1	$\mathbf{0}$	2.17×10^{-2}	0
f_6	min	2.20×10^{-161}	2.59×10^{-121}	1.45×10^{-155}	6.26×10^{-114}	1.41×10^{-151}	0
	mean	3.58×10^{-67}	4.98×10^{-104}	7.88×10^{-142}	2.66×10^{-92}	9.16×10^{-72}	0
f_7	min	$\mathbf{-9}$	$\mathbf{-9}$	$\mathbf{-9}$	$\mathbf{-9}$	$\mathbf{-9}$	-9
	mean	$\mathbf{-9}$	$\mathbf{-9}$	$\mathbf{-9}$	$\mathbf{-9}$	$\mathbf{-9}$	-9
f_8	min	1.30	$\mathbf{0}$	9.02×10^{-1}	$\mathbf{0}$	9.02×10^{-1}	0
	mean	1.50	1.54	1.60	$\mathbf{7.29 \times 10^{-2}}$	1.25	3.25×10^{-1}
f_9	min	9.13×10^{-4}	1.37×10^{-3}	1.51×10^{-3}	1.05×10^{-3}	$\mathbf{5.92 \times 10^{-4}}$	2.71×10^{-3}
	mean	2.01×10^{-2}	3.16×10^{-2}	4.61×10^{-2}	$\mathbf{1.51 \times 10^{-2}}$	3.11×10^{-2}	4.44×10^{-2}
f_{10}	min	$\mathbf{0}$	$\mathbf{0}$	$\mathbf{0}$	$\mathbf{0}$	$\mathbf{0}$	0
	mean	2.89×10^1	$\mathbf{0}$	$\mathbf{0}$	$\mathbf{0}$	5.67×10^{-7}	0
f_{11}	min	$\mathbf{7.28}$	8.03	7.69	8.04	7.86	7.97
	mean	$\mathbf{8.09}$	8.83	8.81	8.45	8.22	8.54
f_{12}	min	1.60×10^{-157}	1.76×10^{-116}	8.56×10^{-158}	5.08×10^{-105}	2.32×10^{-148}	0
	mean	3.04×10^{-22}	7.01×10^{-8}	9.69×10^1	6.71×10^{-67}	3.90×10^{-35}	0
f_{13}	min	1.35×10^{-96}	2.36×10^{-73}	1.39×10^{-93}	2.91×10^{-77}	1.92×10^{-92}	0
	mean	9.62×10^{-26}	3.02×10^{-59}	3.63×10^{-43}	7.25×10^{-58}	2.42×10^{-41}	0
f_{14}	min	1.56×10^{-84}	1.14×10^{-64}	2.91×10^{-85}	1.66×10^{-60}	6.95×10^{-84}	0
	mean	4.04×10^{-56}	1.89×10^{-55}	1.94×10^{-67}	3.04×10^{-52}	5.80×10^{-35}	0
f_{15}	min	$\mathbf{-1.60 \times 10^3}$	-1.57×10^3	-1.52×10^3	-1.36×10^3	-1.38×10^3	-1.34×10^3
	mean	-8.99×10^2	-9.81×10^2	$\mathbf{-1.05 \times 10^3}$	-8.06×10^2	-8.09×10^2	-8.84×10^2
f_{16}	min	9.16×10^{-164}	1.56×10^{-121}	3.15×10^{-164}	3.83×10^{-117}	1.96×10^{-150}	0
	mean	2.09×10^{-27}	9.99×10^{-101}	1.05×10^{-144}	1.05×10^{-97}	1.04×10^{-62}	0
f_{17}	min	1.78×10^{-199}	2.74×10^{-154}	1.09×10^{-192}	8.20×10^{-148}	2.80×10^{-158}	0
	mean	6.33×10^{-1}	1.43×10^{-116}	1.63×10^{-33}	2.37×10^{-118}	6.53×10^{-24}	0
f_{18}	min	5.61×10^{-84}	1.49×10^{-52}	1.97×10^{-82}	2.24×10^{-59}	7.48×10^{-80}	0
	mean	3.89×10^{-13}	7.36×10^{-19}	1.94×10^{-19}	4.60×10^{-29}	1.48×10^{-19}	0
f_{19}	min	3.54×10^{-3}	1.01×10^{-2}	9.08×10^{-3}	1.47×10^{-2}	1.09×10^{-2}	$\mathbf{2.24 \times 10^{-2}}$
	mean	6.15×10^{-2}	4.33×10^{-2}	4.10×10^{-2}	9.36×10^{-2}	7.27×10^{-2}	2.24×10^{-2}
f_{20}	min	3.97×10^{-25}	7.05×10^{-17}	3.97×10^{-25}	1.81×10^{-12}	3.97×10^{-25}	-1
	mean	3.68×10^{-14}	2.75×10^{-11}	4.58×10^{-10}	1.01×10^{-7}	3.97×10^{-25}	-1
f_{21}	min	9.28×10^{-4}	5.44×10^{-4}	1.88×10^{-3}	1.61×10^{-3}	2.95	-1
	mean	2.85	2.35×10^{-3}	1.07×10^{-1}	3.14×10^{-3}	2.95	-1
f_{22}	min	7.70×10^{-155}	7.35×10^{-115}	3.27×10^{-159}	1.75×10^{-113}	1.61×10^{-146}	0
	mean	3.08×10^{-30}	2.71×10^{-45}	5.07×10^{-100}	2.32×10^{-88}	1.26×10^{-40}	0

As can be seen from the experimental data in Table 5, when the ω_1 and ω_2 take the random weights proposed in this paper, the obtained optima are satisfactory. It has 19 times to find the best solution, but only dominated by other algorithms when searching for the three functions (f_9, f_{11} and f_{15})—followed by ω_1 using the second kind of weight improvement strategy and ω_2 using the first strategy with the way. This method has six times to search for smaller results. The conclusion of this discussion is that the optimization of the algorithm is better when ω_1 is equal to ω_2. If they all use the randomized weights proposed in this paper at the same time, the capability of the SPSOC will be the best and it can easily do this with most of the benchmark functions. Comparing with the 30-times average values, we can find that, when the weight ω_1 is equal to ω_2, the average value is smaller and the randomization strategy proposed in this paper is the best among them. If they use the same weight equation, the algorithm will be simpler and faster because only one weight needs to be calculated.

However, this paper uses only six kinds of collocation ways which are combined into three kinds of improvement strategies to carry on the simulation experiment. Whether there is a better weight improvement strategy to make SPSOC have a better performance needs to be further developed and improved.

4.5. Comparison and Analysis with Other PSOs

The most commonly used method which better reflects that the improved algorithm is excellent is bound to be compared with other classical improvement methods. In this section, we have a comparative test between three improved strategies proposed in this paper and bPSO and its three representative improved ones namely, bPSO, PSOd, HPSO-SCAC and TCPSO. The experiment consists of three parts mainy. The first part is to test the seven kinds of particle swarm algorithms separately for 0-dimensional, 50-dimensional and 100-dimensional functions. Each function is searched for 30 times. A *t*-test is used to analyze the large amount of experimental data obtained. Twenty-two distinct evolution curves of fitness from optimizng 100-dimensional functions will be analyzed briefly. Here, all the experiments were conducted on the same conditions as Zhang et al. [70,71]. The second part is to analyze the complexity by the Big O notation [72] and the actual running time for search for the optima in 50-dimensional problems. The third part is to calculate the success rate and the average iteration times of seven algorithms in solving twenty-two 50-dimensional problems, respectively. The stability and effectiveness of the algorithm will be analyzed by these two indices. More details of those three parts will be elaborated in sequence in the following subsections.

4.5.1. Different Dimensional Experiments and *t*-Test Analysis

Students' *t*-test (*t*-test) is a frequently used method of data analysis in statistics to compare whether two sets of data is in one solution space or not, that is, the comparasion for data differences. In this paper, a two-independent-samples *t*-test as the following formulas is used to analyze the difference between the 30 optima searched by SPSORC and the 30 ones by others:

$$t = \frac{(\bar{X}_1 - \bar{X}_2) - (\mu_1 - \mu_2)}{S_{\bar{X}_1 - \bar{X}_2}} = \frac{(\bar{X}_1 - \bar{X}_2)}{S_{\bar{X}_1 - \bar{X}_2}}, \tag{25}$$

$$S_{\bar{X}_1 - \bar{X}_2} = \sqrt{\frac{S_c^2}{n_1} + \frac{S_c^2}{n_2}}, \tag{26}$$

where $\bar{X}_1$ and $\bar{X}_2$ are, respectively, the average of two sets of data; S_c^2 is the combined variance; The sample size is 30; the two-tailed test level is taken as 0.05. The Matlab R2014a test2 function (MathWorks, Natick, MA, USA) instruction is used to calculate directly so as to avoid unnecessary calculation error in the paper.Table 6 shows the optima of the seven improved algorithms for the 10-dimensional, 50-dimensional and 100-dimensional benchmark functions from Table A1, respectively. In Table 6, each algorithm solves the specified function 30 times separately and minimum value(min), average values(mean) and standard deviation values(std) of them are calculated. The minimum one of this three sets of data are highlighted in boldface. '+', '−' and '=' respectively indicate that the SPSORC results are *'better'* than, *'worse'* than and *'same'* as the improved algorithm. To calculate the SPSORC's net score for convenience, '1', '−1', and '0' corresponding to the three symbols here indicate the score of the SPSORC.

Table 6. Optimization results for the function in 3 kinds of dimension.

f		10				50				100			
		min	mean	std	ttest	min	mean	std	ttest	min	mean	std	ttest
f_1	bPSO	3.16×10^{-1}	1.29	6.68×10^{-1}	+(1)	1.26×10^{1}	1.77×10^{1}	1.82	+(1)	1.95×10^{1}	2.02×10^{1}	3.23×10^{-1}	+(1)
	PSOd	1.16	3.32	1.64	+(1)	1.09×10^{1}	1.36×10^{1}	1.16	+(1)	1.39×10^{1}	1.54×10^{1}	6.67×10^{-1}	+(1)
	HPSOscac	2.08×10^{-10}	2.20	5.69	+(1)	1.68×10^{-11}	1.13	2.98	+(1)	6.66×10^{-10}	6.94×10^{-1}	2.65	=(0)
	TCPSO	8.92×10^{-01}	2.08	6.35×10^{-1}	+(1)	1.03×10^{1}	1.39×10^{1}	2.08	+(1)	1.72×10^{1}	1.85×10^{1}	7.74×10^{-1}	+(1)
	SPSO	$\mathbf{8.88 \times 10^{-16}}$	3.38×10^{-15}	1.66×10^{-15}	+(1)	$\mathbf{8.88 \times 10^{-16}}$	3.73×10^{-15}	1.45×10^{-15}	+(1)	$\mathbf{8.88 \times 10^{-16}}$	3.85×10^{-15}	1.35×10^{-15}	+(1)
	SPSOC	$\mathbf{8.88 \times 10^{-16}}$	$\mathbf{8.88 \times 10^{-16}}$	0	=(0)	$\mathbf{8.88 \times 10^{-16}}$	$\mathbf{8.88 \times 10^{-16}}$	0	=(0)	$\mathbf{8.88 \times 10^{-16}}$	$\mathbf{8.88 \times 10^{-16}}$	0	−(−1)
	SPSORC	$\mathbf{8.88 \times 10^{-16}}$	$\mathbf{8.88 \times 10^{-16}}$	0		$\mathbf{8.88 \times 10^{-16}}$	$\mathbf{8.88 \times 10^{-16}}$	0		$\mathbf{8.88 \times 10^{-16}}$	2.19×10^{-15}	4.01×10^{-15}	
f_2	bPSO	3.46×10^{-2}	6.10×10^{-1}	6.44×10^{-1}	+(1)	3.97×10^{1}	5.60×10^{1}	8.63	+(1)	1.43×10^{2}	1.68×10^{2}	1.25×10^{1}	+(1)
	PSOd	4.28×10^{-3}	1.29×10^{-1}	1.79×10^{-1}	+(1)	1.53×10^{1}	2.12×10^{1}	4.77	+(1)	5.40×10^{1}	6.94×10^{1}	1.07×10^{1}	+(1)
	HPSOscac	0	6.88×10^{-77}	3.77×10^{-76}	=(0)	0	1.27×10^{-68}	6.95×10^{-68}	=(0)	0	8.67×10^{-49}	4.75×10^{-48}	=(0)
	TCPSO	6.42×10^{-2}	1.46	1.69	+(1)	2.24×10^{1}	4.92×10^{1}	1.33×10^{1}	+(1)	1.05×10^{2}	1.35×10^{2}	2.11×10^{1}	+(1)
	SPSO	2.58×10^{-24}	2.81×10^{-20}	1.49×10^{-1}	=(0)	4.28×10^{-20}	2.40×10^{-4}	1.25×10^{-3}	=(0)	3.06×10^{-19}	1.94×10^{-3}	1.06×10^{-2}	=(0)
	SPSOC	7.18×10^{-59}	1.01×10^{-45}	5.54×10^{-45}	=(0)	2.12×10^{-56}	5.22×10^{-41}	2.73×10^{-40}	=(0)	1.07×10^{-53}	5.71×10^{-42}	2.35×10^{-41}	=(0)
	SPSORC	0	1.46	0		0	5.61×10^{-281}	0			0	0	
f_3	bPSO	1.31×10^{-3}	8.95×10^{-1}	4.78	=(0)	7.69×10^{2}	1.37×10^{3}	4.27×10^{2}	+(1)	1.12×10^{4}	1.37×10^{4}	1.33×10^{3}	+(1)
	PSOd	6.53×10^{-4}	2.91×10^{-1}	3.50×10^{-1}	+(1)	2.95×10^{2}	5.68×10^{2}	1.66×10^{2}	+(1)	2.88×10^{3}	4.32×10^{3}	1.02×10^{3}	+(1)
	HPSOscac	0	8.09×10^{-132}	4.43×10^{-131}	=(0)	0	1.62×10^{-155}	8.88×10^{-155}	=(0)	0	3.60×10^{-120}	1.97×10^{-119}	=(0)
	TCPSO	1.66×10^{-2}	7.68×10^{-2}	6.09×10^{-2}	+(1)	1.70×10^{2}	4.56×10^{2}	2.85×10^{2}	+(1)	3.40×10^{3}	5.24×10^{3}	1.15×10^{3}	+(1)
	SPSO	1.80×10^{-52}	2.39×10^{-46}	6.91×10^{-46}	=(0)	2.97×10^{-38}	1.11×10^{-32}	4.16×10^{-32}	=(0)	3.39×10^{-35}	2.39×10^{-29}	1.31×10^{-28}	=(0)
	SPSOC	4.92×10^{-112}	3.50×10^{-90}	1.89×10^{-89}	=(0)	1.91×10^{-108}	9.54×10^{-73}	5.22×10^{-72}	=(0)	7.38×10^{-104}	5.55×10^{-69}	3.04×10^{-68}	=(0)
	SPSORC	0				0				0	0	0	
f_4	bPSO	2.10×10^{-8}	5.60×10^{-7}	5.78×10^{-7}	+(1)	7.19	3.05×10^{1}	1.35×10^{1}	+(1)	2.32×10^{2}	4.93×10^{2}	1.54×10^{2}	+(1)
	PSOd	3.81×10^{-7}	4.00×10^{-4}	9.29×10^{-4}	+(1)	1.69	4.84	2.59	+(1)	1.94×10^{1}	6.03×10^{1}	2.18×10^{1}	+(1)
	HPSOscac	0	9.26×10^{-285}	0	+(1)	0	6.20×10^{-224}	0	+(1)	0	1.36×10^{-274}	0	+(1)
	TCPSO	2.30×10^{-8}	3.68×10^{-6}	4.79×10^{-6}	+(1)	4.50×10^{-1}	4.35	6.46	+(1)	4.11×10^{1}	1.07×10^{2}	5.20×10^{1}	+(1)
	SPSO	6.17×10^{-108}	4.10×10^{-94}	2.23×10^{-93}	=(0)	9.65×10^{-83}	3.38×10^{-72}	1.77×10^{-71}	=(0)	4.66×10^{-78}	1.25×10^{-69}	4.73×10^{-69}	=(0)
	SPSOC	5.88×10^{-234}	1.19×10^{-183}	0	=(0)	2.87×10^{-218}	1.52×10^{-152}	8.31×10^{-152}	=(0)	1.56×10^{-222}	4.13×10^{-155}	2.26×10^{-154}	=(0)
	SPSORC	0	$\mathbf{1.73 \times 10^{-321}}$	0		0	$\mathbf{2.96 \times 10^{-323}}$	0		0	$\mathbf{2.02 \times 10^{-320}}$	0	
f_5	bPSO	4.22×10^{-1}	9.36×10^{-1}	1.64×10^{-1}	+(1)	6.46×10^{1}	2.11×10^{2}	6.99×10^{1}	+(1)	8.91×10^{2}	1.15×10^{3}	1.48×10^{2}	+(1)
	PSOd	1.37×10^{-1}	8.46×10^{-1}	6.67×10^{-1}	+(1)	4.36×10^{1}	8.91×10^{1}	2.69×10^{1}	+(1)	2.38×10^{1}	3.18×10^{2}	4.27×10^{1}	+(1)
	HPSOscac	3.38×10^{-6}	6.26×10^{1}	1.14×10^{2}	+(1)	7.84×10^{-2}	4.02×10^{2}	5.68×10^{2}	+(1)	1.57×10^{-5}	1.37×10^{3}	1.67×10^{3}	+(1)
	TCPSO	1.04	1.16	9.72×10^{-2}	+(1)	1.90×10^{1}	6.24×10^{1}	3.63×10^{1}	+(1)	2.46×10^{2}	4.14×10^{2}	7.97×10^{1}	+(1)
	SPSO	0	3.87×10^{-1}	3.41×10^{-1}	−(−1)	0	5.40×10^{-2}	1.49×10^{-1}	=(0)	0	7.33×10^{-3}	2.37×10^{-2}	=(0)
	SPSOC	0	0	0		0	0	0	=(0)	0	0	0	=(0)
	SPSORC	0	3.70×10^{-18}	2.03×10^{-17}		0	0	0		0	5.18×10^{-17}	1.23×10^{-16}	

Table 6. Cont.

		10				50				100			
		min	mean	std	ttest	min	mean	std	ttest	min	mean	std	ttest
f_6	bPSO	4.16×10^{3}	6.97×10^{5}	1.22×10^{6}	+(1)	1.38×10^{8}	3.70×10^{8}	1.82×10^{8}	+(1)	7.95×10^{8}	1.90×10^{9}	6.34×10^{8}	+(1)
	PSOd	2.60×10^{3}	5.58×10^{4}	7.67×10^{4}	+(1)	1.80×10^{7}	8.68×10^{8}	4.80×10^{8}	+(1)	2.99×10^{8}	5.38×10^{8}	1.37×10^{8}	+(1)
	HPSOscac	0	3.69×10^{-149}	2.02×10^{-148}	=(0)	0	3.77×10^{-157}	2.06×10^{-156}	=(0)	3.40×10^{-124}	3.40×10^{-124}	1.86×10^{-123}	=(0)
	TCPSO	5.51×10^{4}	8.43×10^{5}	1.26×10^{6}	+(1)	4.79×10^{7}	1.91×10^{8}	1.33×10^{8}	+(1)	3.51×10^{8}	1.04×10^{9}	6.43×10^{8}	+(1)
	SPSO	1.44×10^{-45}	5.73×10^{-41}	1.56×10^{-40}	+(1)	1.71×10^{-33}	1.72×10^{-28}	7.68×10^{-28}	=(0)	8.51×10^{-33}	2.42×10^{-27}	5.20×10^{-27}	=(0)
	SPSOC	3.77×10^{-113}	8.59×10^{-84}	4.71×10^{-83}	+(1)	1.15×10^{-101}	1.94×10^{-76}	7.40×10^{-76}	+(1)	1.47×10^{-102}	1.67×10^{-76}	9.16×10^{-76}	+(1)
	SPSORC	**0**	**0**	**0**		**0**	**0**	**0**		**0**	**0**	**0**	
f_7	bPSO	-6.76	-5.32	6.63×10^{-1}	+(1)	-1.20×10^{1}	-9.16	1.64	+(1)	-1.66×10^{1}	-1.23×10^{1}	2.52	+(1)
	PSOd	-7.93	-6.49	6.48×10^{-1}	+(1)	-2.64×10^{1}	-2.13×10^{1}	2.01	+(1)	-3.78×10^{1}	-3.27×10^{1}	2.57	+(1)
	HPSOscac	-2.46	-2.46	4.73×10^{-1}	+(1)	-3.51	-4.39	4.35	+(1)	-1.48×10^{1}	-5.40	4.02	+(1)
	TCPSO	-6.83	-4.91	9.48×10^{-1}	+(1)	-1.43×10^{1}	-8.66	3.01	+(1)	-1.58×10^{1}	-9.52	3.65	+(1)
	SPSO	-9	-5.47	2.81	+(1)	-4.90×10^{1}	-1.20×10^{1}	1.79×10^{1}	+(1)	-3.31×10^{1}	-3.96	8.44	+(1)
	SPSOC	-9	**-9**	**0**	=(0)	-4.90×10^{1}	**-4.90×10^{1}**	**0**	=(0)	-9.90×10^{1}	**-9.90×10^{1}**	**0**	−(−1)
	SPSORC	-9	-9	0		-4.90×10^{1}	-4.90×10^{1}	0		-9.90×10^{1}	-9.90×10^{1}	2.64×10^{-15}	
f_8	bPSO	9.02×10^{-1}	1.71	5.13×10^{-1}	+(1)	1.04×10^{1}	1.24×10^{1}	1.36	+(1)	2.39×10^{1}	2.69×10^{1}	1.36	+(1)
	PSOd	1.65	2.41	3.26×10^{-1}	+(1)	1.94×10^{1}	2.11×10^{1}	3.01	+(1)	4.29×10^{1}	4.53×10^{1}	9.08×10^{-1}	+(1)
	HPSOscac	2.22×10^{-16}	1.82	1.47	+(1)	6.75×10^{-12}	1.43×10^{1}	9.26	+(1)	9.03×10^{-6}	2.68×10^{1}	2.09×10^{1}	+(1)
	TCPSO	7.05×10^{-1}	1.68	7.04×10^{-1}	+(1)	9.91	1.19×10^{1}	1.05	+(1)	2.33×10^{1}	2.56×10^{1}	1.65	+(1)
	SPSO	1.47	2.81	5.56×10^{-1}	+(1)	2.11×10^{-4}	1.87×10^{1}	6.17	+(1)	6.62×10^{-1}	4.30×10^{1}	8.94	+(1)
	SPSOC	0	**2.87×10^{-2}**	**1.57×10^{-1}**	+(1)	0	**6.58×10^{-1}**	**6.65×10^{-1}**	−(−1)	0	1.50	8.22	+(1)
	SPSORC	0	4.43×10^{-1}	1.16		0	1.47	5.59		0	**0**	**0**	
f_9	bPSO	1.97×10^{-2}	9.62×10^{-2}	5.02×10^{-2}	+(1)	4.21	2.94×10^{1}	1.63×10^{1}	+(1)	2.64×10^{2}	4.90×10^{2}	1.22×10^{2}	+(1)
	PSOd	1.23×10^{-2}	5.86×10^{-2}	3.47×10^{-2}	+(1)	3.09	6.44	3.01	+(1)	3.35×10^{1}	6.07×10^{1}	2.06×10^{1}	+(1)
	HPSOscac	4.17×10^{-1}	1.27×10^{-2}	1.70×10^{-2}	+(1)	6.76	8.72×10^{2}	6.17×10^{2}	+(1)	7.34×10^{2}	3.51×10^{3}	2.21×10^{3}	+(1)
	TCPSO	2.68×10^{-2}	7.16×10^{-2}	4.05×10^{-2}	+(1)	2.39	5.04	3.05	+(1)	6.29×10^{1}	1.07×10^{2}	4.38×10^{1}	+(1)
	SPSO	2.16×10^{-4}	2.27×10^{-2}	2.16×10^{-2}	=(0)	2.39×10^{-3}	**2.13×10^{-2}**	1.70×10^{-2}	=(0)	2.52×10^{-3}	**3.08×10^{-2}**	2.90×10^{-2}	−(−1)
	SPSOC	9.79×10^{-4}	**2.07×10^{-2}**	2.35×10^{-2}	=(0)	**1.00×10^{-3}**	2.15×10^{-2}	**1.65×10^{-2}**	=(0)	**9.11×10^{-2}**	2.15×10^{-2}	**1.66×10^{-2}**	−(−1)
	SPSORC	**7.03×10^{-4}**	3.65×10^{-2}	3.08×10^{-2}		4.43×10^{-3}	6.27×10^{-2}	5.32×10^{-2}		5.05×10^{-3}	1.72×10^{-1}	2.82×10^{-1}	
f_{10}	bPSO	8.03	2.76×10^{1}	1.21×10^{1}	+(1)	4.55×10^{2}	5.47×10^{2}	5.13×10^{1}	+(1)	1.05×10^{3}	1.31×10^{3}	9.25×10^{1}	+(1)
	PSOd	1.05×10^{-1}	8.86	4.17	+(1)	1.65×10^{2}	2.06×10^{2}	2.38×10^{1}	+(1)	5.15×10^{2}	6.36×10^{2}	5.58×10^{1}	+(1)
	HPSOscac	9.35×10^{-5}	5.13×10^{1}	5.40×10^{1}	+(1)	2.50×10^{-1}	3.47×10^{2}	3.03×10^{2}	+(1)	3.08×10^{-1}	6.13×10^{2}	5.71×10^{2}	+(1)
	TCPSO	1.12×10^{1}	4.68×10^{1}	2.10×10^{1}	+(1)	4.02×10^{2}	5.42×10^{2}	7.35×10^{1}	+(1)	1.03×10^{3}	1.22×10^{3}	1.15×10^{2}	+(1)
	SPSO	0	1.78×10^{1}	2.31×10^{1}	+(1)	0	1.09	3.67	=(0)	0	3.24×10^{-1}	1.24	=(0)
	SPSOC	**0**	**0**	**0**	=(0)	**0**	**0**	**0**	=(0)	**0**	**0**	**0**	=(0)
	SPSORC	0	0	0		0	0	0		0	2.96×10^{-16}	9.43×10^{-16}	

Table 6. *Cont.*

		10				50				100			
		min	mean	std	ttest	min	mean	std	ttest	min	mean	std	ttest
f_{11}	bPSO	1.31×10^1	6.35×10^3	2.28×10^{04}	=(0)	5.06×10^6	1.71×10^7	8.00×10^6	+(1)	1.80×10^8	2.98×10^8	7.48×10^7	+(1)
	PSOd	**6.50**	8.95×10^2	2.03×10^3	+(1)	1.89×10^6	6.49×10^6	3.13×10^8	+(1)	2.08×10^7	4.22×10^7	1.46×10^7	+(1)
	HPSOscac	9.00×10^3	8.68×10^7	8.12×10^7	+(1)	5.64×10^5	1.01×10^9	6.13×10^8	+(1)	2.42×10^8	2.41×10^9	1.33×10^9	+(1)
	TCPSO	4.24×10^1	8.01×10^2	1.04×10^3	+(1)	5.18×10^5	2.45×10^6	1.01×10^6	+(1)	3.84×10^7	7.88×10^7	3.19×10^7	+(1)
	SPSO	7.74	**8.15**	$\mathbf{1.28 \times 10^1}$	−(−1)	4.81×10^1	4.87×10^1	3.08×10^{-1}	−(−1)	9.81×10^1	$\mathbf{9.88 \times 10^1}$	2.02×10^{-1}	=(0)
	SPSOC	8.11	8.32	1.99×10^{-1}	+(1)	4.81×10^1	4.87×10^1	3.05×10^{-1}	+(1)	9.82×10^1	9.89×10^1	$\mathbf{1.48 \times 10^{-1}}$	=(0)
	SPSORC	8.00	8.64	6.52×10^{-1}		4.86×10^1	4.89×10^1	$\mathbf{1.05 \times 10^{-1}}$	+(1)	9.82×10^1	9.89×10^1	1.54×10^{-1}	=(0)
f_{12}	bPSO	6.41×10^2	3.89×10^3	1.89×10^3	+(1)	3.09×10^5	1.53×10^6	1.12×10^6	+(1)	5.44×10^6	1.94×10^7	1.18×10^7	+(1)
	PSOd	1.66×10^2	1.73×10^3	7.79×10^2	+(1)	1.23×10^5	3.16×10^5	1.43×10^5	+(1)	1.26×10^6	4.46×10^6	2.47×10^6	+(1)
	HPSOscac	0	7.30×10^{-131}	4.00×10^{-130}	=(0)	0	1.56×10^{-144}	8.56×10^{-144}	=(0)		1.10×10^{-120}	6.01×10^{-120}	=(0)
	TCPSO	4.44×10^2	3.89×10^3	1.64×10^3	+(1)	5.48×10^5	2.50×10^6	1.72×10^6	+(1)	1.05×10^7	3.86×10^7	3.14×10^7	+(1)
	SPSO	2.39×10^{-45}	9.73×10^{-42}	3.27×10^{-41}	=(0)	1.84×10^{-38}	1.73×10^{-36}	3.39×10^{-36}	=(0)	1.09×10^{-36}	5.55×10^{-35}	8.02×10^{-35}	+(1)
	SPSOC	3.41×10^{-117}	3.44×10^{-52}	1.56×10^{-51}	=(0)	8.83×10^{-74}	2.38×10^{-24}	9.28×10^{-24}	=(0)	8.79×10^{-66}	2.06	1.13×10^1	+(1)
	SPSORC	0	0	0		0	0	0					
f_{13}	bPSO	5.56×10^{-9}	1.23×10^{-6}	2.22×10^{-6}	+(1)	7.47×10^{-10}	1.01×10^{-6}	1.96×10^{-6}	+(1)	1.25×10^{-8}	9.08×10^{-7}	1.18×10^{-6}	+(1)
	PSOd	1.42×10^{-34}	1.88×10^{-30}	4.49×10^{-30}	+(1)	7.15×10^{-36}	6.07×10^{-31}	1.23×10^{-30}	+(1)	6.10×10^{-35}	4.29×10^{-30}	1.65×10^{-29}	=(0)
	HPSOscac	0	1.06×10^{-83}	5.79×10^{-83}	=(0)	0	1.13×10^{-73}	6.15×10^{-73}	=(0)	3.62×10^{-321}	3.14×10^{-86}	1.25×10^{-85}	=(0)
	TCPSO	3.23×10^{-4}	2.43×10^{-2}	2.73×10^{-2}	+(1)	4.45×10^{-04}	1.88×10^{-2}	1.83×10^{-2}	+(1)	2.17×10^{-4}	2.45×10^{-2}	2.75×10^{-2}	+(1)
	SPSO	3.11×10^{-51}	5.41×10^{-47}	1.16×10^{-46}	+(1)	3.93×10^{-53}	6.62×10^{-47}	1.57×10^{-46}	+(1)	8.32×10^{-52}	1.34×10^{-45}	5.05×10^{-45}	=(0)
	SPSOC	7.15×10^{-77}	4.92×10^{-60}	2.68×10^{-59}	+(1)	2.13×10^{-77}	7.72×10^{-60}	4.23×10^{-59}	+(1)	3.79×10^{-79}	1.10×10^{-60}	6.01×10^{-60}	=(0)
	SPSORC	0	0	0		0	0	0		0	0	0	
f_{14}	bPSO	6.81×10^{-2}	2.34×10^{-1}	1.12×10^{-1}	+(1)	8.67×10^1	3.08×10^2	9.07×10^2	+(1)	2.94×10^2	3.55×10^2	2.47×10^1	+(1)
	PSOd	2.74×10^{-2}	6.15×10^{-1}	5.46×10^{-1}	+(1)	3.43×10^1	5.81×10^1	1.55×10^1	+(1)	1.19×10^2	1.62×10^2	2.39×10^1	+(1)
	HPSOscac	0	1.04×10^{-59}	5.71×10^{-59}	=(0)	1.79×10^{-238}	3.77×10^{-61}	1.45×10^{-60}	=(0)	0	5.56×10^{-63}	2.57×10^{-62}	=(0)
	TCPSO	2.30×10^{-1}	4.88×10^{-1}	1.66×10^{-1}	+(1)	9.95×10^1	8.94×10^{13}	4.83×10^{14}	+(1)	3.25×10^2	1.60×10^{37}	7.61×10^{37}	=(0)
	SPSO	1.38×10^{-25}	1.28×10^{-23}	2.21×10^{-23}	+(1)	6.30×10^{-20}	3.34×10^{-17}	6.71×10^{-17}	+(1)	1.70×10^{-19}	2.24×10^{-15}	5.95×10^{-15}	+(1)
	SPSOC	8.49×10^{-58}	9.72×10^{-48}	3.86×10^{-47}	+(1)	7.86×10^{-55}	1.25×10^{-34}	6.82×10^{-34}	+(1)	2.92×10^{-56}	3.42×10^{-37}	1.85×10^{-36}	+(1)
	SPSORC	0	0	0		0	0	0					
f_{15}	bPSO	-3.83×10^3	-3.41×10^3	3.02×10^2	+(1)	-1.31×10^4	$\mathbf{-1.12 \times 10^4}$	1.03×10^3	+(1)	-1.91×10^4	-1.57×10^4	1.40×10^3	−(−1)
	PSOd	-3.83×10^3	-3.18×10^3	3.54×10^2	+(1)	-1.10×10^4	-9.40×10^3	7.22×10^2	+(1)	-1.77×10^4	-1.46×10^4	1.31×10^3	−(−1)
	HPSOscac	-1.91×10^3	-1.28×10^3	3.96×10^2	+(1)	-5.62×10^3	-3.34×10^3	1.02×10^3	+(1)	-7.58×10^3	-4.87×10^3	1.20×10^3	+(1)
	TCPSO	$\mathbf{-4.06 \times 10^3}$	$\mathbf{-3.41 \times 10^3}$	3.01×10^2	−(−1)	$\mathbf{-1.34 \times 10^4}$	-1.11×10^4	9.24×10^2	−(−1)	$\mathbf{-2.04 \times 10^4}$	$\mathbf{-1.72 \times 10^4}$	1.78×10^3	−(−1)
	SPSO	-1.51×10^3	-9.53×10^2	2.22×10^2	=(0)	-3.05×10^3	-1.91×10^3	5.22×10^2	=(0)	-4.12×10^3	-2.69×10^3	6.78×10^2	=(0)
	SPSOC	-1.19×10^3	-7.02×10^2	$\mathbf{1.89 \times 10^2}$	=(0)	-2.61×10^3	-1.54×10^3	4.49×10^2	=(0)	-3.62×10^3	-2.09×10^3	6.72×10^2	=(0)
	SPSORC	-1.35×10^3	-8.69×10^2	2.55×10^2		-2.62×10^3	-1.87×10^3	$\mathbf{4.16 \times 10^2}$		-4.26×10^3	-2.82×10^3	$\mathbf{6.11 \times 10^2}$	=(0)

Table 6. *Cont.*

		10				50				100			
		min	mean	std	ttest	min	mean	std	ttest	min	mean	std	ttest
f_{16}	bPSO	2.15×10^{-1}	2.48	2.76	$+(1)$	7.73×10^{3}	1.87×10^{4}	6.62×10^{3}	$+(1)$	9.11×10^{4}	1.15×10^{5}	1.47×10^{4}	$+(1)$
	PSOd	9.32×10^{-1}	1.02×10^{1}	1.21×10^{1}	$+(1)$	4.26×10^{3}	9.67×10^{3}	2.81×10^{3}	$+(1)$	2.34×10^{4}	3.40×10^{4}	6.67×10^{3}	$+(1)$
	HPSOscac	0	6.95×10^{-126}	3.81×10^{-125}	$=(0)$	0	5.30×10^{-144}	2.90×10^{-143}	$=(0)$		1.15×10^{-92}	6.31×10^{-92}	$=(0)$
	TCPSO	1.41	7.52	5.62	$+(1)$	1.74×10^{3}	6.23×10^{3}	4.27×10^{3}	$+(1)$	3.12×10^{4}	5.11×10^{4}	1.05×10^{4}	$+(1)$
	SPSO	1.38×10^{-50}	3.20×10^{-44}	1.11×10^{-43}	$=(0)$	2.69×10^{-36}	5.43×10^{-32}	1.61×10^{-31}	$=(0)$	3.40×10^{-35}	8.65×10^{-29}	2.53×10^{-29}	$=(0)$
	SPSOC	2.11×10^{-114}	3.46×10^{-89}	1.66×10^{-88}	$=(0)$	8.39×10^{-107}	5.65×10^{-70}	3.10×10^{-69}	$=(0)$	2.21×10^{-103}	5.27×10^{-79}	2.85×10^{-78}	$=(0)$
	SPSORC	0	0	0		0	0	0		0	0	0	
f_{17}	bPSO	2.52×10^{-12}	1.09×10^{-8}	2.37×10^{-8}	$+(1)$	7.44×10^{-55}	2.75×10^{-3}	2.89×10^{-3}	$+(1)$	1.48×10^{-2}	2.36×10^{-1}	2.58×10^{-1}	$+(1)$
	PSOd	2.07×10^{-11}	3.03×10^{-6}	9.79×10^{-6}	$+(1)$	5.86×10^{-8}	4.53×10^{-5}	7.70×10^{-5}	$=(0)$	7.17×10^{-7}	1.85×10^{-4}	3.29×10^{-4}	$+(1)$
	HPSOscac	0	5.38×10^{-135}	2.95×10^{-134}	$=(0)$	0	2.62×10^{-148}	1.34×10^{-147}	$=(0)$		2.87×10^{-154}	1.57×10^{-153}	$=(0)$
	TCPSO	1.51×10^{-8}	6.10×10^{-7}	6.72×10^{-7}	$+(1)$	6.45×10^{-6}	5.69×10^{-4}	1.10×10^{-3}	$+(1)$	1.48×10^{-3}	1.27×10^{-1}	3.65×10^{-1}	$+(1)$
	SPSO	1.16×10^{-94}	4.96×10^{-86}	1.50×10^{-85}	$=(0)$	6.17×10^{-94}	1.10×10^{-84}	5.95×10^{-84}	$=(0)$	1.78×10^{-92}	8.69×10^{-87}	1.71×10^{-86}	$+(1)$
	SPSOC	8.84×10^{-141}	1.97×10^{-115}	1.08×10^{-114}	$=(0)$	6.16×10^{-158}	1.01×10^{-121}	5.52×10^{-121}	$=(0)$	2.52×10^{-146}	1.22×10^{-118}	6.19×10^{-118}	$+(1)$
	SPSORC	0	0	0		0	0	0			$\mathbf{4.94 \times 10^{-324}}$	0	
f_{18}	bPSO	5.10×10^{-4}	6.20×10^{-1}	2.15	$=(0)$	1.82×10^{13}	2.27×10^{19}	1.11×10^{20}	$=(0)$	2.83×10^{34}	6.97×10^{46}	3.76×10^{47}	$=(0)$
	PSOd	5.79×10^{-4}	3.73×10^{-1}	6.15×10^{-1}	$+(1)$	2.48×10^{4}	8.59×10^{10}	4.26×10^{11}	$+(1)$	8.86×10^{20}	5.66×10^{32}	2.80×10^{33}	$=(0)$
	HPSOscac	5.26×10^{-319}	1.20×10^{3}	3.72×10^{3}	$+(1)$	4.63×10^{-237}	3.53×10^{28}	1.66×10^{29}	$+(1)$		2.88×10^{54}	1.34×10^{55}	$=(0)$
	TCPSO	2.75×10^{-3}	3.15×10^{-4}	4.89×10^{-1}	$+(1)$	1.10×10^{8}	4.18×10^{16}	2.20×10^{17}	$=(0)$	4.98×10^{30}	1.22×10^{42}	6.28×10^{42}	$=(0)$
	SPSO	2.25×10^{-3}	1.23×10^{-4}	6.69×10^{-4}	$=(0)$	1.65×10^{-33}	3.98×10^{-6}	2.12×10^{-5}	$=(0)$	1.97×10^{-34}	1.67×10^{-5}	6.35×10^{-5}	$=(0)$
	SPSOC	4.34×10^{-70}	2.51×10^{-21}	1.37×10^{-20}	$=(0)$	5.00×10^{-64}	2.23×10^{-8}	1.22×10^{-7}	$=(0)$	7.22×10^{-63}	3.73×10^{-20}	2.04×10^{-19}	$=(0)$
	SPSORC	0	0	0		0	0	0		0	0	0	
f_{19}	bPSO	9.08×10^{-4}	2.66×10^{-3}	4.25×10^{-4}	$=(0)$	1.59×10^{-19}	$\mathbf{1.68 \times 10^{-19}}$	3.35×10^{-21}	$=(0)$	1.88×10^{-40}	$\mathbf{1.94 \times 10^{-40}}$	3.45×10^{-42}	$-(-1)$
	PSOd	5.66×10^{-4}	$\mathbf{9.64 \times 10^{-4}}$	$\mathbf{3.49 \times 10^{-4}}$	$-(-1)$	8.35×10^{-18}	1.41×10^{-16}	2.64×10^{-16}	$-(-1)$	3.03×10^{-31}	1.19×10^{-28}	3.03×10^{-28}	$-(-1)$
	HPSOscac	1.59×10^{-3}	3.43×10^{-3}	3.74×10^{-3}	$=(0)$	1.79×10^{-19}	1.79×10^{-19}	$\mathbf{4.90 \times 10^{-35}}$	$=(0)$	2.04×10^{-40}	2.04×10^{-40}	$\mathbf{4.15 \times 10^{-56}}$	$-(-1)$
	TCPSO	2.12×10^{-3}	3.50×10^{-3}	2.34×10^{-3}	$+(1)$	1.45×10^{-19}	9.59×10^{-16}	5.21×10^{-15}	$+(1)$	$\mathbf{1.69 \times 10^{-40}}$	4.24×10^{-12}	2.32×10^{-33}	$-(-1)$
	SPSO	7.91×10^{-4}	5.49×10^{-2}	3.83×10^{-2}	$=(0)$	1.32×10^{-7}	4.32×10^{-7}	8.07×10^{-7}	$+(1)$	2.94×10^{-13}	2.36×10^{-12}	1.22×10^{-11}	$=(0)$
	SPSOC	2.34×10^{-2}	9.80×10^{-2}	5.49×10^{-2}	$+(1)$	1.79×10^{-8}	3.53×10^{-5}	4.99×10^{-5}	$=(0)$	3.57×10^{-18}	1.58×10^{-09}	4.91×10^{-9}	
	SPSORC	-1.00	9.35	2.22×10^{-2}	$+(1)$	0	3.81×10^{-7}	6.51×10^{-7}	$+(1)$		1.13×10^{-13}	2.35×10^{-13}	
f_{20}	bPSO	3.97×10^{-25}	3.97×10^{-25}	1.87×10^{-40}	$+(1)$	9.83×10^{-123}	9.83×10^{-123}	0	$+(1)$	9.66×10^{-245}	9.66×10^{-245}	0	$+(1)$
	PSOd	4.66×10^{-25}	2.90×10^{-19}	1.51×10^{-18}	$+(1)$	4.66×10^{-63}	1.10×10^{-51}	5.13×10^{-51}	$+(1)$	7.68×10^{-90}	1.59×10^{-73}	8.73×10^{-73}	$+(1)$
	HPSOscac	3.97×10^{-25}	3.97×10^{-25}	1.87×10^{-40}	$+(1)$	9.83×10^{-123}	9.83×10^{-123}	0	$=(0)$	9.66×10^{-245}	9.66×10^{-245}	0	$+(1)$
	TCPSO	3.97×10^{-25}	3.97×10^{-25}	1.87×10^{-40}	$+(1)$	9.83×10^{-123}	9.83×10^{-123}	0	$+(1)$	9.66×10^{-245}	9.66×10^{-245}	0	$+(1)$
	SPSO	9.48×10^{-14}	2.11×10^{-8}	9.45×10^{-8}	$+(1)$	8.01×10^{-31}	1.32×10^{-21}	4.61×10^{-21}	$=(0)$	1.95×10^{-49}	3.65×10^{-34}	1.91×10^{-33}	$+(1)$
	SPSOC	1.15×10^{-12}	4.77×10^{-7}	1.37×10^{-6}	$+(1)$	1.19×10^{-22}	2.22×10^{-16}	4.89×10^{-16}	$=(0)$	1.69×10^{-41}	5.40×10^{-22}	2.89×10^{-21}	$+(1)$
	SPSORC	-1.00	-1.00	0		-1.00	-9.33×10^{-1}	2.54×10^{-1}	$=(0)$	-1.00	$\mathbf{-9.00 \times 10^{-1}}$	3.05×10^{-1}	

Table 6. *Cont.*

		10				50				100			
		min	mean	std	ttest	min	mean	std	ttest	min	mean	std	ttest
f_{21}	bPSO	3.73×10^{-7}	3.57×10^{-6}	3.53×10^{-6}	+(1)	5.81×10^{-19}	9.30×10^{-17}	1.28×10^{-16}	+(1)	1.04×10^{-32}	4.59×10^{-29}	9.83×10^{-29}	+(1)
	PSOd	2.54×10^{-8}	8.53×10^{-6}	1.06×10^{-5}	+(1)	1.35×10^{-20}	6.03×10^{-20}	$\mathbf{5.19 \times 10^{-20}}$	+(1)	3.97×10^{-39}	7.51×10^{-38}	$\mathbf{9.24 \times 10^{-38}}$	+(1)
	HPSOscac	$-\mathbf{1.00}$	-5.27×10^{-1}	5.13×10^{-1}	+(1)	$-\mathbf{1.00}$	-3.33×10^{-2}	1.83×10^{-1}	+(1)	3.14×10^{-31}	1.87×10^{-14}	5.55×10^{-14}	+(1)
	TCPSO	8.71×10^{-7}	1.33×10^{-5}	1.67×10^{-5}	+(1)	6.93×10^{-20}	2.42×10^{-18}	3.80×10^{-18}	+(1)	1.30×10^{-38}	1.17×10^{-32}	5.38×10^{-32}	+(1)
	SPSO	4.39×10^{-4}	1.04×10^{-3}	3.19×10^{-4}	+(1)	4.00×10^{-16}	2.15×10^{-14}	1.99×10^{-14}	+(1)	2.51×10^{-29}	6.77×10^{-27}	1.10×10^{-26}	+(1)
	SPSOC	6.88×10^{-4}	3.24×10^{-3}	1.69×10^{-3}	+(1)	9.35×10^{-14}	7.28×10^{-12}	1.12×10^{-11}	+(1)	1.13×10^{-25}	3.10×10^{-22}	6.75×10^{-22}	+(1)
	SPSORC	$-\mathbf{1.00}$	$-\mathbf{1.00}$	0		$-\mathbf{1.00}$	$-\mathbf{1.00}$	2.92×10^{-17}	+(1)	$-\mathbf{1.00}$	$-\mathbf{1.00}$	2.92×10^{-17}	+(1)
f_{22}	bPSO	2.26	2.62×10^{1}	2.76×10^{1}	+(1)	1.16×10^{3}	2.76×10^{3}	4.53×10^{3}	+(1)	3.39×10^{3}	1.70×10^{5}	5.90×10^{5}	=(0)
	PSOd	5.39×10^{-1}	4.98	2.55	+(1)	1.80×10^{2}	3.28×10^{2}	9.49×10^{1}	+(1)	7.23×10^{2}	9.14×10^{2}	1.20×10^{2}	+(1)
	HPSOscac	0	1.93×10^{-174}	0	+(1)	0	1.17×10^{-167}	0	+(1)	0	$\mathbf{1.89 \times 10^{-143}}$	$\mathbf{1.04 \times 10^{-142}}$	=(0)
	TCPSO	5.77×10^{-1}	1.15×10^{1}	1.52×10^{1}	+(1)	1.03×10^{3}	1.60×10^{4}	5.36×10^{4}	=(0)	2.51×10^{3}	6.23×10^{5}	2.09×10^{6}	=(0)
	SPSO	1.63×10^{-48}	1.59×10^{-43}	6.83×10^{-43}	=(0)	1.69×10^{-40}	3.13×10^{-38}	6.42×10^{-38}	+(1)	1.89×10^{-40}	3.88×10^{-38}	8.35×10^{-38}	=(0)
	SPSOC	1.29×10^{-113}	1.69×10^{-81}	8.46×10^{-81}	=(0)	3.35×10^{-61}	2.71×10^{-29}	1.08×10^{-28}	+(1)	8.00×10^{-61}	3.24×10^{-20}	1.76×10^{-19}	=(0)
	SPSORC	0	0	0		0	0	0		0	1.64×10^{-4}	8.99×10^{-4}	=(0)

The search results of the heuristic algorithm are random, so the average value (mean) of the results after multiple searches is the most valuable data. Observing the mean values in Table 6, when searching for 10-dimensional functions, we can see that SPSORC outperforms the other six PSO on 16 functions (f_1, f_2, f_3, f_4, f_6, f_7, f_{10}, f_{12}, f_{13}, f_{14}, f_{16}, f_{17}, f_{18}, f_{20}, f_{21} and f_{22}) in terms of the criteria 'mean', and 15 out of 16 were theoretical optimal solutions. Secondly, the number of that SPSOC find the minimum mean solutions for 10-dimensional functions is 6 times (f_1, f_5, f_7, f_8, f_9, f_{10}). PSOd, TCPSO and SPSO only find the minimum mean once. bPSO and HPSOscac are unable to search for the minimum mean at one time. Regarding the three functions (f_1, f_7, f_{10}), SPSOC and SPSORC can obtain the same mean.

On the other hand, both SPSORC and HPSOscac can find the same values in many functions including f_2, f_3, f_4, f_6, f_{12}, f_{13}, f_{14}, f_{16}, f_{17}, f_{21} and f_{22}. In addition, SPSORC has achieved the best results for ninth, indicating that SPSORC and HPSOscac have the opportunity to yield a better solution than the average one, but the results are volatility, especially for HPSOscac.

The use of standard deviation (std) can observe the volatility of the algorithm results. The standard deviation is a measure of the degree to which a set of data averages is dispersed. A larger standard deviation represents a larger difference between most of the values and their average values; a smaller standard deviation means that these values are closer to the average. It is clear that the standard deviation of SPSORC is almost the smallest of all algorithms.

The comparison of 'min', 'mean' and 'std' from 50-dimensional and 100-dimensional dimensional functions between SPSORC and other six PSO variants is also illustrated in Table 6. It is clearly seen that, for both 50-dimensional and 100-dimensional functions except for f_9, f_{11}, f_{15}, f_{19}, SPSORC is able to obtain better 'mean' than the most of the other improved strategies. In 50-dimensional optima values, SPSORC outperforms the other six PSOs on 17 functions (f_1, f_2, f_3, f_4, f_5, f_6, f_7, f_{10}, f_{12}, f_{13}, f_{14}, f_{16}, f_{17}, f_{18}, f_{20}, f_{21} and f_{22}), in which 14 out of 17 were searched for theoretical optima. SPSOC has searched for the best solution five times (f_1, f_5, f_7, f_8, f_{10}). Then, PSOd and SPSO has two times (f_{15}, f_{19}) and once (f_9), respectively. Regrettably, other algorithms have no chance. Almost the same situation also appears in 100-dimensional results. In addition, as for 'std' of SPSORC, it should be noted that the 'std' without any fluctuations many times is markedly superior to that of the others. Second, with regard to the experimental results comparison in 100-dimensional results of Table 6, SPSOC is also able to achieve good performance with smaller mean value such as for f_1, f_5, f_7, f_9 and f_{10} out of the twenty-two 100-dimensional test functions.

To sum up, from Table 6, it has been identified experimentally that SPSORC is superior or highly competitive with several improved PSO variants, and this improving strategy is shown to be able to find fairly good solutions for most of the well-known benchmark functions.

Table 7 is a summary of the scores based on the *t*-test analysis of search results in three kinds of dimensions. *Score* is the net score of SPSORC, which is better than the score obtained by the comparison function minus the number of comparison functions. Take the comparison result of SPSOC and RSMPSOc in Table 7 as an example, that is, the A6 algorithm in 100-dimensional in Table 7. The SPSORC result is seven times better than SPSOC and three times worse than SPSOC. Therefore, the net score of SPSORC is: $Score = 7 - 3 = 4$, and the calculation process of other net scores is the same.

Observed from Table 7, SPSORC has a stable net score for these six algorithms, all greater than 0. Careful observation can show us that the performance is slightly higher in the 50-dimensional scores compared to 10-and-100 dimensions. It is again proven that the capability of SPSORC is better than bPSO, PSOd, HPSOscac, TCPSO, SPSO and SPSOC, especially when solving the 50-dimensional problem. The convergence curves of seven improved PSO algorithms on twenty-two benchmark functions with 100 dimensions are plotted in Section 4.5.1 Figure 4, respectively.

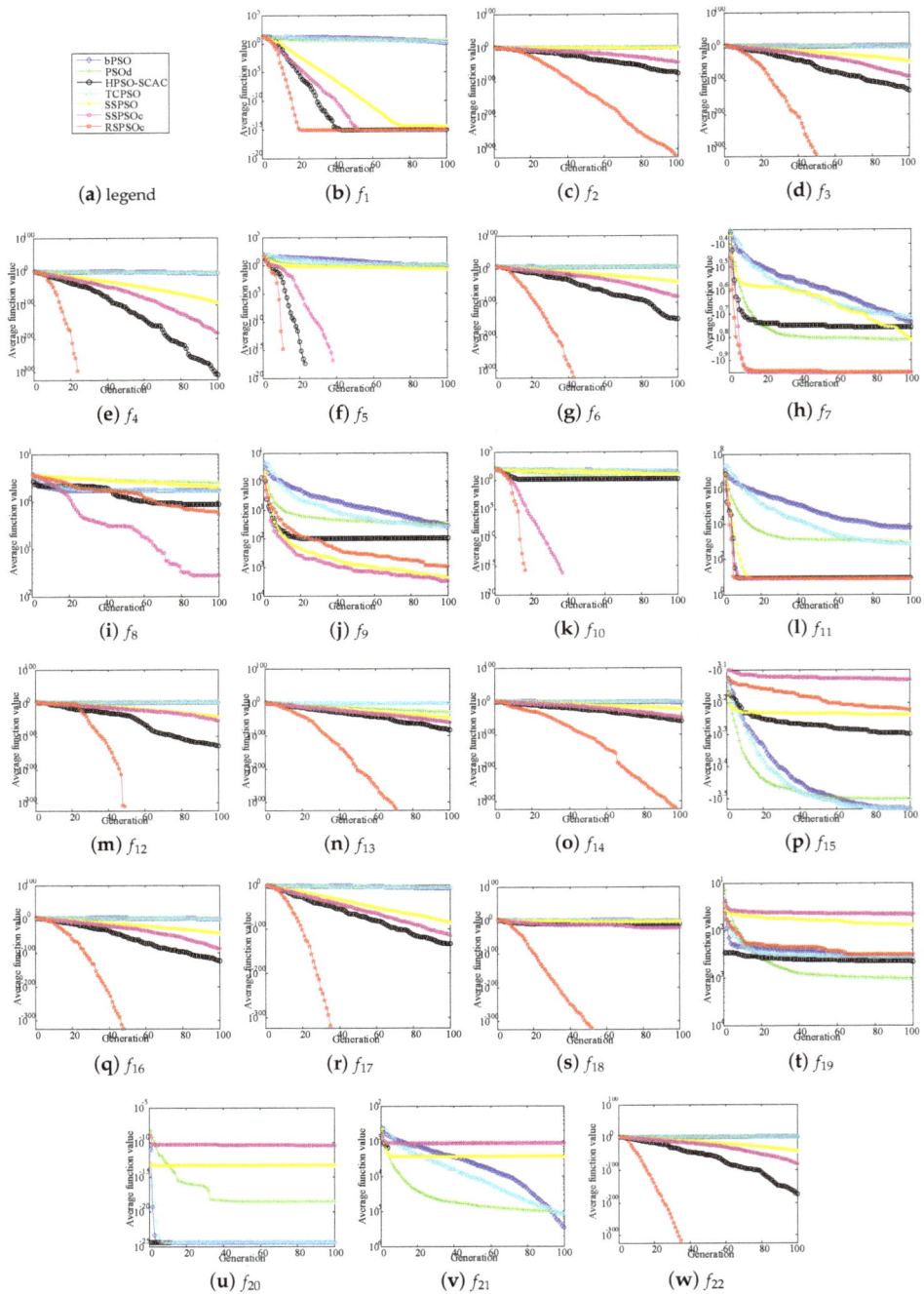

Figure 4. Average convergence curves of seven improved PSO algorithms for twenty-two functions in 10 dimensions.

Table 7. Simulation environment.

N	Results	bPSO	PSOd	HPSOscac	SPSO	SPSOC	SPSORC
10	+	18	20	12	20	11	6
	=	4	1	10	1	10	14
	−	0	1	0	1	1	2
	Score	18	19	12	19	10	4
50	+	19	21	13	19	9	8
	=	2	1	9	3	11	13
	−	1	0	0	0	2	1
	Score	18	21	13	19	7	7
100	+	18	18	10	16	9	7
	=	2	2	11	4	12	12
	−	2	2	1	2	1	3
	Score	16	16	9	14	8	4

Figure 4a is the legend for the other twenty-two convergence curves. Figure 4b indicates that, on f_1, PSPSOC converges the fastest in the early stage among the seven improvements. HPSOscac converges relatively slowly compared to SPSORC. The order of performance on f_1 is SPSORC, HPSOscac, SPSOC, SPSO, bPSO, PSOd, TCPSO. Almost the same situation occurs simultaneously on the other 11 function convergence curves. This should be the effect of algorithm simplifying so that the algorithm can converge very quickly in the early stage. On f_3, f_4, f_5, et al., RSPSO has found the best solution within the maximum generation. Figure 4 l,m, on f_{11} and f_{12}, show that SPSORC converges relatively slowly compared to HPSOscac in the beginning, but it surpasses HPSOscac in about 20th generation.

Next, it is further analyzed by the box diagram in Figure 5. Box diagram is mainly used to reflect the characteristics of the original data distribution, and can also compare the distribution characteristics of multiple sets of data. On the same number of axes, the box plots of several sets of data are arranged in parallel. Shape information such as median, tail length, outliers, and distribution intervals of several batches of data can be seen at a glance. + indicates an abnormal point.

From Figure 5a, the order of these boxes from high to low is bPSO, TCPSO, PSOd, HPSOscac, SPSO, SPSORC and SPSOC, respectively. The upper quartile and median values of bPSO and TCPSO are closer to the upper edge, which indicates that the data of the two algorithms are more biased toward larger values. In comparison, the box of PSOd is more symmetrical and the data distribution is relatively uniform. Unfortunately, the distribution of the boxes of these three algorithms is too high, and the search results are not good. The box of HPSOscac is at the bottom of the coordinate system. However, we can clearly see that there are many outliers in its data. Some even have exceeded the median of PSOd. Its skewed nature tends to be smaller, but the distribution of data is more scattered. Compared with the above four algorithms, the distribution of the boxes of the three algorithms proposed in this paper is obviously more optimistic. The optimization results of the three algorithms are almost neat, concentrated and smaller. The situation of other box diagrams is not much different from that of Figure 5a. Throughout the 22 box diagrams in Figure 5, the bPSO, PSOd, HPSOscac and TCPSO seem to have more difficulty locating the solution than the SPSORC for from the box diagrams. The boxes of PSOd are mostly too top, followed by TCPSO. HPSOscac has a lot of outliers. Its maximum and minimum span is large, and distribution is extremely non-uniform and decentral. It is observed that the results of HPSOscac are highly volatile and the improvement of the algorithm is unstable. This may be related to its weight mixed with the trigonometric function. For the above reasons, the results of SPSORC and SPSOC are not obvious in the box diagram, almost all posted at the bottom. Combining the results of Table 6, we can roughly know that SPSORC has better performance, and the more oblate box can show that the 30 search results have little differences and the performance is very stable.

To sum up, the test results indicate that: Both confidence term and random weight can enhance diversity. The former can yield a significant improvement in performance, while the latter can preserve much more diversity. The aforementioned two methods are compatible. Combining both of them with SPSO can preserve the highest diversity and achieve the best overall performance among the six improved strategies.

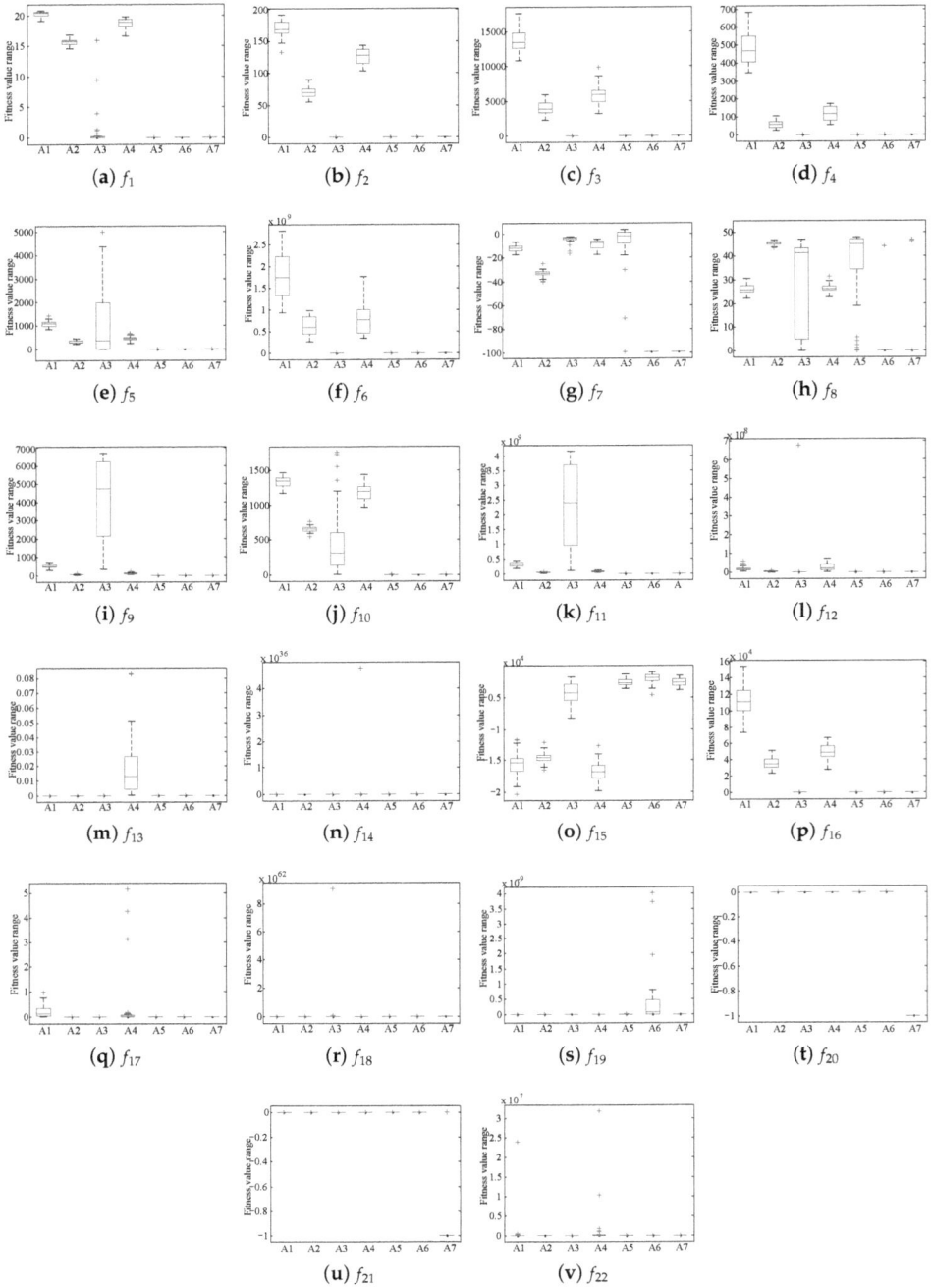

Figure 5. Box diagram of thirty results, each function with 100 dimensions.

4.5.2. Algorithm Complexity Analysis

Comparing the steps of the bPSO algorithm, the time complexity of SPSO's two improvements mainly depends on two aspects: (1) random initialization, and (2) particle velocity and position updating. These two parts can all be expressed as $O(m \times N)$ by the Big O notation [72], in that, m is the population and N is the problem dimensions. In this paper, we haven't changed the algorithm's initialization method, so we only compare the time complexity from particle velocity and position updating. The SPSO, which does not consider the inertia weight updating and confidence term calculating, has a reduced computational complexity compared to the basic particle swarm optimization algorithm, but the Big O notation can also be represented by $O(m \times N)$. Compared with the bPSO and the SPSO, the most complex algorithm we proposed is named SPSORC, which has increased weight, and the confidence term has surely increased in computational complexity, but we can see from Table 1 that its loop body has not changed. According to the Big O notation, its time complexity is still $O(m \times N)$. Overall, the complexity of SPSO and its two improved ones are not increased by orders of magnitude.

Then, we analyze the real computational time from Table 8. In Table 8, we show the computational time for three kinds of dimension functions.

Table 8. Real computational time.

N	Alg	bPSO	PSOd	HPSOscac	TCPSO	SPSO	SPSOC	SPSORC
10	f_1	0.0346	0.0408	0.0681	0.0755	0.0193	0.0293	0.0298
	f_2	0.0324	0.0396	0.0651	0.0731	0.0188	0.0295	0.0296
	f_3	0.0309	0.0374	0.0606	0.0673	0.0178	0.0280	0.0274
	f_4	0.0378	0.0465	0.0700	0.0857	0.0269	0.0374	0.0367
	f_5	0.0334	0.0421	0.0655	0.0783	0.0196	0.0298	0.0304
	f_6	0.0358	0.0444	0.0676	0.0825	0.0245	0.0349	0.0348
	f_7	0.0380	0.0449	0.1343	0.0856	0.0244	0.0328	0.0332
	f_8	0.0542	0.0626	0.0867	0.1214	0.0423	0.0517	0.0561
	f_9	0.0484	0.0559	0.0815	0.1066	0.0362	0.0470	0.0475
	f_{10}	0.0329	0.0410	0.0654	0.0758	0.0199	0.0294	0.0311
	f_{11}	0.0303	0.0388	0.0623	0.0704	0.0174	0.0278	0.0286
	f_{12}	0.0428	0.0517	0.0760	0.0967	0.0313	0.0421	0.0423
	f_{13}	0.0288	0.0378	0.0622	0.0676	0.0180	0.0285	0.0289
	f_{14}	0.0290	0.0381	0.0608	0.0690	0.0181	0.0286	0.0280
	f_{15}	0.0355	0.0435	0.0736	0.0803	0.0215	0.0325	0.0324
	f_{16}	0.0288	0.0379	0.0624	0.0681	0.0181	0.0289	0.0280
	f_{17}	0.0365	0.0462	0.0681	0.0852	0.0265	0.0368	0.0363
	f_{18}	0.0471	0.0555	0.0800	0.1043	0.0347	0.0466	0.0466
	f_{19}	0.0358	0.0428	0.0676	0.0812	0.0220	0.0319	0.0319
	f_{20}	0.0447	0.0543	0.0769	0.1006	0.0339	0.0439	0.0407
	f_{21}	0.0406	0.0488	0.0745	0.0914	0.0269	0.0369	0.0347
	f_{22}	0.0294	0.0381	0.0617	0.0695	0.0181	0.0292	0.0287
50	f_1	0.1515	0.1912	0.2889	0.3502	0.0813	0.1311	0.1325
	f_2	0.1515	0.1874	0.2897	0.3465	0.0815	0.1329	0.1329
	f_3	0.1323	0.1718	0.2726	0.3145	0.0732	0.1247	0.1241
	f_4	0.1798	0.2172	0.3179	0.4042	0.1174	0.1713	0.1704
	f_5	0.1589	0.1976	0.2964	0.3633	0.0868	0.1378	0.1383
	f_6	0.1761	0.2157	0.3130	0.3969	0.1163	0.1687	0.1678
	f_7	0.1820	0.2190	0.5396	0.4080	0.1150	0.1543	0.1552
	f_8	0.2847	0.3195	0.4206	0.5978	0.2137	0.2620	0.2682
	f_9	0.2330	0.2727	0.3809	0.5160	0.1716	0.2246	0.2250
	f_{10}	0.1554	0.1933	0.2921	0.3546	0.0851	0.1343	0.1356
	f_{11}	0.1380	0.1775	0.2819	0.3236	0.0765	0.1289	0.1295
	f_{12}	0.2696	0.3012	0.4074	0.5677	0.1996	0.2533	0.2527
	f_{13}	0.1306	0.1709	0.2713	0.3072	0.0702	0.1236	0.1246
	f_{14}	0.1324	0.1733	0.2629	0.3136	0.0744	0.1287	0.1276
	f_{15}	0.1569	0.1965	0.3368	0.3581	0.0913	0.1459	0.1466
	f_{16}	0.1348	0.1727	0.2703	0.3113	0.0727	0.1262	0.1244
	f_{17}	0.1738	0.2125	0.3016	0.3952	0.1126	0.1623	0.1649
	f_{18}	0.2277	0.2688	0.3719	0.5035	0.1682	0.2176	0.2197
	f_{19}	0.1644	0.2027	0.2985	0.3724	0.0985	0.1469	0.1495
	f_{20}	0.2089	0.2507	0.3380	0.4681	0.1501	0.1996	0.1881
	f_{21}	0.1892	0.2241	0.3240	0.4226	0.1209	0.1707	0.1541
	f_{22}	0.1308	0.1680	0.2694	0.3077	0.0700	0.1220	0.1222

Table 8. Cont.

N	Alg	bPSO	PSOd	HPSOscac	TCPSO	SPSO	SPSOC	SPSORC
	f_1	0.2977	0.3719	0.5642	0.6830	0.1561	0.2571	0.2612
	f_2	0.3023	0.3793	0.5775	0.6944	0.1595	0.2655	0.2660
	f_3	0.2665	0.3431	0.5417	0.6247	0.1416	0.2485	0.2491
	f_4	0.3588	0.4369	0.6381	0.8083	0.2379	0.3400	0.3387
	f_5	0.3158	0.3929	0.5894	0.7232	0.1716	0.2729	0.2756
	f_6	0.3555	0.4288	0.6287	0.7992	0.2284	0.3352	0.3343
	f_7	0.3655	0.4419	1.3028	0.8107	0.2323	0.3049	0.3061
	f_8	0.5580	0.6271	0.8395	1.2018	0.4311	0.5230	0.5311
	f_9	0.4651	0.5443	0.7505	1.0211	0.3380	0.4447	0.4465
	f_{10}	0.3098	0.3841	0.5771	0.7101	0.1667	0.2646	0.2687
100	f_{11}	0.2744	0.3521	0.5583	0.6413	0.1477	0.2529	0.2522
	f_{12}	0.6752	0.7524	0.9544	1.4332	0.5468	0.6518	0.6535
	f_{13}	0.2651	0.3426	0.5400	0.6193	0.1385	0.2442	0.2468
	f_{14}	0.3217	0.3646	0.5121	0.6165	0.1431	0.2478	0.2477
	f_{15}	0.3154	0.3918	0.6178	0.7155	0.1852	0.2899	0.2911
	f_{16}	0.2632	0.3413	0.5416	0.6519	0.1466	0.2518	0.2498
	f_{17}	0.3599	0.4450	0.6326	0.8335	0.2255	0.3279	0.3377
	f_{18}	0.4758	0.5581	0.7631	1.0589	0.3491	0.4483	0.4592
	f_{19}	0.3406	0.4093	0.6189	0.7589	0.1980	0.2975	0.3253
	f_{20}	0.4381	0.5238	0.6735	0.9385	0.2970	0.4000	0.3777
	f_{21}	0.3812	0.4604	0.6413	0.8513	0.2433	0.3474	0.3103
	f_{22}	0.2658	0.3424	0.5421	0.6204	0.1367	0.2440	0.2443

The time in Table 8 is the average time required to run 30 times independently. The average length of time varies slightly depending on the problem. Observing the running time of the seven algorithms, it is certain that the running time of SPSORC is similar to the computational time of other algorithms. It is clear that the lowest running time is obtained by SPSO, since it greatly simplifies bPSO. SPSOC has increased slightly over time due to confidence term. HPSOscac's trigonometric function improvement strategy makes the algorithm better applicable to multimodal problems. However, because the regularity distribution of the trigonometric function increases the particle diversity, the particle is difficult to converge at the later stage, and the actual calculation time is longer. TCPSO uses the dual population to optimize problems through information exchange. Thus, SPSO and its improved strategies do not simply consume runtime to improve algorithm performance. The real computational time is basically distributed as Figure 6. A1–A7 are namely bPSO, PSOd, HPSOscac, TCPSO, SPSO, SPSOC and SPSORC.

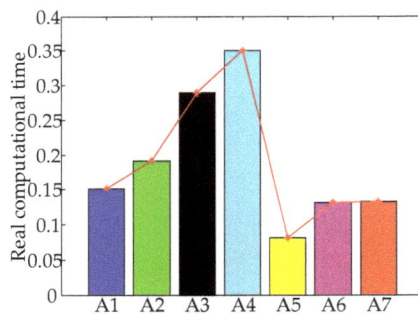

Figure 6. Real computational time of f_{11} in 50 dimensions.

4.5.3. Success Rate and Average Iteration Times

The success rate (SR) is the percentage between the times that each algorithm can successfully achieve convergence accuracy for function optimization and the total number of times. The average iterations times (AIT) is the average iteration numbers required by the algorithm to find the convergence accuracy. The former can examine the stability and accuracy of the algorithm, while the latter mainly examines the efficiency of the algorithm. The convergence accuracy used for the success rate and the average iteration times in this paper is the accuracy of the 50-dimensional test functions we want to meet. The specific values can refer to the last column of Table A1. The other parameters are set according to Table 2. Figure 7 shows the Radar charts with average iteration times of eight functions. We can surely, in Figure 7, find that the point of SPSORC is always close to the center origin.

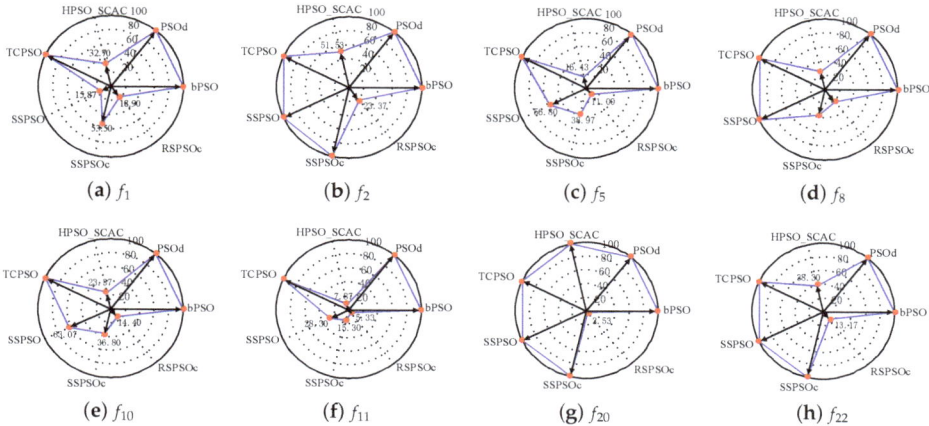

Figure 7. Radar charts of average iteration times.

Specific data for the average iteration times and the success rate of the seven algorithms in solving 50-dimensional problems are referred to in Table 'AIT' is the average iteration times. 'SR' represents the success rate. In order to facilitate the use of differentials, 'SR' is expressed in percentage form. The results round off to retain two digits after the decimal point. '−' means that the algorithm fails to search this function when convergence accuracy is reached within the maximum generations. For the convenience of observation, we will show the minimum AIT and the maximum SR for each function in the bold format.

The results can be analyzed from Table 9. bPSO, PSOd, TCPSO and SPSO have higher success rates three times. SPSOC and HPSOscac gets six and seven times, respectively. SPSORC, surprisingly, has 15 times, and seven of them (f_{12}, f_{13}, f_{14}, f_{17}, f_{18}, f_{20}, f_{21}) have the best success rate that none of the other six algorithms have achieved. One can see that the SPSORC strategy has a wider range of types, high precision and stability. Comparing the average iteration times, it is clearly shown that TCPSO, SPSO and SPSOC do not get the lowest average iteration times.

The above two data comparisons reveal that SPSORC has a large advantage compared with the other six algorithms. Not only is it more stable, but the search efficiency is also faster. When faced with a unimodal function, SPSORC can converge to effective precision more quickly. For other multi-peak complex problems, it is not to be outdone, except for the extremely difficult functions such as Rosenbrock Function and Schwefel's Problem 2.26, which show weak stability. It can be stated that the improved method can be adapted to a variety of test environments, and the results are quite excellent.

Table 9. Success rate and average iteration times.

	bPSO		PSOd		HPSOscac		TCPSO		SPSO		SPSOC		SPSORC	
	AIT	SR	AIT	SR	AIT	SR	AIT	SR	AIT	SR	AIT	SR	AIT	SR
f_1	-	0.00%	-	0.00%	32.70	0.00%	-	0.00%	15.87	3.33%	53.50	100.00%	18.90	93.33%
f_2	-	0.00%	-	0.00%	51.53	100.00%	-	0.00%	-	0.00%	-	0.00%	23.37	100.00%
f_3	-	0.00%	-	0.00%	35.33	100.00%	-	0.00%	-	100.00%	77.33	100.00%	17.50	100.00%
f_4	-	0.00%	-	0.00%	58.60	100.00%	-	0.00%	-	0.00%	-	0.00%	16.47	100.00%
f_5	-	0.00%	-	0.00%	16.43	0.00%	-	0.00%	56.80	76.67%	38.97	100.00%	11.00	100.00%
f_6	-	0.00%	-	0.00%	54.73	100.00%	-	0.00%	-	0.00%	-	0.00%	20.70	100.00%
f_7	-	100.00%	-	100.00%	-	36.67%	-	100.00%	-	100.00%	34.50	100.00%	13.77	100.00%
f_8	-	0.00%	-	0.00%	25.00	0.00%	-	0.00%	-	0.00%	37.03	96.67%	21.80	96.67%
f_9	-	0.00%	-	0.00%	6.27	0.00%	-	0.00%	12.30	96.67%	7.40	100.00%	11.80	66.67%
f_{10}	-	0.00%	-	0.00%	23.37	0.00%	-	0.00%	63.07	53.33%	36.80	100.00%	14.40	100.00%
f_{11}	-	0.00%	-	0.00%	7.87	0.00%	-	0.00%	28.30	100.00%	15.30	100.00%	5.33	100.00%
f_{12}	-	0.00%	-	0.00%	56.93	100.00%	-	0.00%	-	0.00%	-	0.00%	20.97	100.00%
f_{13}	2.83	0.00%	-	0.00%	62.93	93.33%	4.67	0.00%	-	0.00%	-	0.00%	19.57	100.00%
f_{14}	-	0.00%	1.10	0.00%	66.70	96.67%	-	0.00%	-	0.00%	-	0.00%	16.50	100.00%
f_{15}	1.20	100.00%	-	100.00%	1.20	60.00%	1.37	100.00%	1.23	16.67%	4.27	0.00%	2.30	10.00%
f_{16}	-	0.00%	-	0.00%	57.73	100.00%	-	0.00%	-	0.00%	-	0.00%	24.50	100.00%
f_{17}	-	0.00%	-	0.00%	26.63	43.33%	-	0.00%	-	0.00%	-	0.00%	20.07	100.00%
f_{18}	-	0.00%	-	0.00%	14.63	36.67%	-	0.00%	-	0.00%	-	0.00%	8.20	100.00%
f_{19}	1.67	100.00%	6.30	100.00%	1.00	100.00%	1.60	100.00%	9.20	3.33%	4.07	0.00%	1.20	16.67%
f_{20}	-	0.00%	-	0.00%	-	0.00%	-	0.00%	-	0.00%	-	0.00%	2.53	90.00%
f_{21}	-	0.00%	-	0.00%	0.63	0.00%	-	0.00%	-	0.00%	-	0.00%	16.73	96.67%
f_{22}	-	0.00%	-	0.00%	38.30	100.00%	-	0.00%	-	0.00%	-	0.00%	13.17	100.00%

5. Conclusions

Due to the effect on particle swarm optimization, in this paper, a Simple Particle Swarm Optimization based on Confidence term and Random inertia weight namely SPSORC has been proposed. SPSORC adopts three different improving strategies—first, particle updating formulas only use positional items and social items to enhance the exploration capability; second, the confidence term is introduced to increase particle diversity and avoid excessive particle convergence. Finally, a random inertia weight is formulated to keep the balance between exploration and exploitation. Extensive experiments in Section 4 on twenty-two benchmark functions validate and discuss SPSO and its further improvements' effectiveness, efficiency, robustness and scalability. It has been demonstrated that, in most cases, SPSORC performs a better capability of exploitation and exploration than, or at least highly competitively with, basic PSO and its state-of-the-art improved ones introduced in this paper.

In our future work, we intend to incorporate different initialization strategies, multi-swarm and hybrid algorithms into SPSORC. This may result in very competitive algorithms. Obviously, many adaptive methods for PSO have been proposed. In order to improve the performance of the proposed approach and its application, the research on particle swarm optimization algorithm and its improvements is a promising research direction. Furthermore, we will apply the proposed approach to solve some other practical existing engineering optimization problems, e.g., machine-tool spindle design, logistics distribution region partitioning problem, economic load dispatch problem, etc. With these evolutionary algorithms, it is unnecessary to know the computing environment and to calculate the gradient and other information. Thus, it is helpful to save on the cost of computing. Even better, we can calculate the problem with more dimensions and goals at once, including some discontinuous problems.

Author Contributions: X.Z. suggested the improving strategy and wrote the original draft preparation. D.Z. was responsible for checking this paper. X.S. provided a provincial project.

Funding: The National Natural Science Foundation of China (No. 61403174) and the Postgraduate Research and Practice Innovation Program of Jiangsu Province (No. KYCX17_1575, No. KYCX17_1576).

Acknowledgments: This work was supported by the National Natural Science Foundation of China (No. 61403174) and the Postgraduate Research and Practice Innovation Program of Jiangsu Province (No. KYCX17_1575, No. KYCX17_1576).

Conflicts of Interest: The authors declare no conflict of interest.

Abbreviations

The following abbreviations are used in this manuscript:

PSO	Particle Swarm Optimization
bPSO	The basic PSO [21,22]
PSOd	A distribution-based update rule for PSO [52]
HPSOscac	A hybrid PSO with sine cosine acceleration coefficients [67]
TCPSO	A two-swarm cooperative PSO [45]
SPSO	Simple PSO
SPSOC	Simple PSO with Confidence Term
SPSORC	Simple PSO based on Random weight and Confidence term
v	Particle velocity
x	Particle position
p_{best}	Personal historical best solution
g_{best}	Global best solution
ω	Inertia weight
ω_{max}	The maximum weight
ω_{min}	The minimum wight
c_1	Self-cognitive factor
c_2	Social communication factor
U (in Figures 1–3)	Solution space of a function
O (in Figures 1–3)	The theoretical optima of a function
i	The current particle
n	The current dimension

N	The maximum dimension
t	The current generation
T	The upper limit of generation
NR	The number of times that algorithm search for problem
m	Population size
min	The minimum values from the optima in which algorithms search for the problem 30 times
mean	The average values for the optima in which algorithms search for the problem 30 times
std	The average values for the optima in which algorithms search for the problem 30 times
ttest (in Table 6)	t-test results
AIT (in Section 4.5.2)	Average iteration times
SR (in Section 4.5.2)	Success rate
A1–A7 (in Figures 5 and 6)	bPSO, PSOd, HPSOscac, TCPSO, SPSO, SPSOC and SPSORC, respectively

Appendix A. Benchmark Function Appendix

Table A1. Benchmark functions.

Instance	Expression	Domain	Analytical Solution	Accuracy (50)
Ackley's Path Function	$f_1(x) = -20e^{-0.2\sqrt{\frac{1}{30}\sum_{i=1}^{N}x_i^2}} - e^{\frac{1}{30}\sum_{i=1}^{N}\cos 2\pi x_i} + 20 + e$	$[-32, 32]$	$f_1(0,\cdots,0) = 8.88 \times 10^{-16}$	1×10^{-15}
Alpine Function	$f_2(x) = \sum_{i=1}^{N}|x_i \sin(x_i) + 0.1x_i|$	$[-10, 10]$	$f_2(0,\cdots,0) = 0$	1×10^{-60}
Axis Parallel Hyperellipsoid	$f_3(x) = \sum_{i=1}^{N}ix_i^2$	$[-5.12, 5.12]$	$f_3(0,\cdots,0) = 0$	1×10^{-15}
De Jong's Function 4 (no noise)	$f_4(x) = \sum_{i=1}^{N}ix_i^4$	$[-1.28, 1.28]$	$f_4(0,\cdots,0) = 0$	1×10^{-240}
Girewank Problem	$f_5(x) = \frac{1}{4000}\sum_{i=1}^{N}x_i^2 - \prod_{i=1}^{N}(\frac{x_i}{\sqrt{i}}) + 1$	$[-600, 600]$	$f_5(0,\cdots,0) = 0$	1×10^{-15}
High Conditioned Elliptic Function	$f_6(x) = \sum_{i=1}^{N}(10^6)^{\frac{i-1}{n-1}}x_i^2$	$[-100, 100]$	$f_6(0,\cdots,0) = 0$	1×10^{-110}
Inverted Cosine Wave Function	$f_7(x) = -\sum_{i=1}^{N-1}(e^{\frac{-x_i^2-x_{i+1}^2-0.5x_ix_{i+1}}{8}}) \times \cos(4\sqrt{x_i^2 + x_{i+1}^2 + 0.5x_ix_{i+1}})$	$[-5, 5]$	$f_7(0,\cdots,0) = -N+1$	-4.9×10^{-1}
Pathological Function	$f_8(x) = \sum_{i=1}^{N-1}(0.5 + \frac{\sin^2\sqrt{100x_i^2+x_{i+1}^2}-0.5}{1+\frac{1}{1000}(x_i^2-2x_ix_{i+1}+x_{i+1}^2)})$	$[-100, 100]$	$f_8(0,\cdots,0) = 0$	1×10^{-5}
Quartic Function, i.e, noise	$f_9(x) = \sum_{i=1}^{N}ix_i^4 + \text{rand}[0, 1)$	$[-10, 10]$	$f_9(0,\cdots,0) = 0$	1×10^{-1}
Rastrigin Problem	$f_{10}(x) = \sum_{i=1}^{N}[x_i^2 - 10\cos(2\pi x_i) + 10]$	$[-5.12, 5.12]$	$f_{10}(0,\cdots,0) = 0$	1×10^{-20}
Rosenbrock Problem	$f_{11}(x) = \sum_{i=1}^{N-1}[100(x_{i+1} - x_i^2)^2 + (x_i - 1)^2]$	$[-30, 30]$	$f_{11}(0,\cdots,0) = 0$	5×10^{1}
Schwefel's Problem 1.2	$f_{12}(x) = \sum_{i=1}^{N}(\sum_{j=1}^{m}x_j)^2$	$[-100, 100]$	$f_{12}(0,\cdots,0) = 0$	1×10^{-100}
Schwefel's Problem 2.21	$f_{13}(x) = \max_i|x_i|, 1 \le i \le 30$	$[-100, 100]$	$f_{13}(0,\cdots,0) = 0$	1×10^{-80}
Schwefel's Problem 2.22	$f_{14}(x) = \sum_{i=1}^{N}|x_i| + \prod_{i=1}^{N}|x_i|$	$[-10, 10]$	$f_{14}(0,\cdots,0) = 0$	1×10^{-60}
Schwefel's Problem 2.26	$f_{15}(x) = \sum_{i=1}^{N}|x_i \sin(x_i) + 0.1x_i|$	$[-500, 500]$	$f_{15}(s,\cdots,s) = -419\,N$ $s \approx 420.97$	-2.5×10^3
Sphere Function	$f_{16}(x) = \sum_{i=1}^{N}x_i^2$	$[-100, 100]$	$f_{16}(0,\cdots,0) = 0$	1×10^{-120}
Sum of Different Power Function	$f_{17}(x) = \sum_{i=1}^{N}|x_i|^{i+1}$	$[-1, 1]$	$f_{17}(0,\cdots,0) = 0$	1×10^{-300}
Xin–She Yang 1	$f_{18}(x) = \sum_{i=1}^{N}\text{rand}[0, 1) \times |x_i|^i$	$[-5, 5]$	$f_{18}(0,\cdots,0) = 0$	1×10^{-60}
Xin–She Yang 2	$f_{19}(x) = \frac{\sum_{i=1}^{N}|x_i|}{e^{\sum_{i=1}^{N}\sin x_i^2}}$	$[-2\pi, 2\pi]$	$f_{19}(0,\cdots,0) = 0$	1×10^{-8}

Table A1. *Cont.*

Instance	Expression	Domain	Analytical Solution	Accuracy (50)
Xin–She Yang 3	$f_{20}(x) = e^{-\sum_{i=1}^{N}(\frac{x_i}{\beta})^{2a}} - 2e^{-\sum_{i=1}^{N}x_i^2}\prod_{i=1}^{N}\cos^2 x_i$ $\beta = 15, a = 3$	$[-20,20]$	$f_{20}(0,\cdots,0) = -1$	-1
Xin–She Yang 4	$f_{21}(x) = [\sum_{i=1}^{N}\sin^2 x_i - e^{-\sum_{i=1}^{N}x_i^2}]e^{\sum_{i=1}^{N}\sin^2\sqrt{|x_i|}}$	$[-10,10]$	$f_{21}(0,\cdots,0) = -1$	-1
Zakharov Function	$f_{22}(x) = \sum_{i=1}^{N}x_i^2 + (\sum_{i=1}^{N}0.5ix_i^2)^2 + (\sum_{i=1}^{N}0.5ix_i^2)^4$	$[-5,10]$	$f_{22}(0,\cdots,0) = 0$	1×10^{-80}

References

1. Denysiuk, R.; Gaspar-Cunha, A. Multiobjective Evolutionary Algorithm Based on Vector Angle Neighborhood. *Swarm Evol. Comput.* **2017**, *37*, 663–670. [CrossRef]
2. Wang, G.G. Moth search algorithm: A bio-inspired metaheuristic algorithm for global optimization problems. *Memet. Comput.* **2016**, *10*, 1–14. [CrossRef]
3. Feng ,Y.H.; Wang, G.G. Binary moth search algorithm for discounted 0–1 knapsack problem. *IEEE Access* **2018**, *6*, 10708–10719. [CrossRef]
4. Grefenstette, J.J. Genetic algorithms and machine learning. *Mach. Learn.* **1988**, *3*, 95–99. [CrossRef]
5. Dorigo, M.; Gambardella, L.M. Ant colony system: A cooperative learning approach to the traveling salesman problem. *IEEE Trans. Evol. Comput.* **1997**, *1*, 53–66. [CrossRef]
6. Storn, R.; Price, K. Differential Evolution—A Simple and Efficient Heuristic for global Optimization over Continuous Spaces. *J. Glob. Optim.* **1997**, *11*, 341–359. [CrossRef]
7. Kirkpatrick, S.; Vecchi, M.P. Optimization by simulated annealing. *Science* **1983**, *220*, 671–680._0035. [CrossRef] [PubMed]
8. Wang, G.G.; Gandomi, A.H.; Alavi, A.H.; Deb, S. A multi-stage krill herd algorithm for global numerical optimization. *Int. J. Artif. Intell. Tools* **2016**, *25*, 1550030. [CrossRef]
9. Wang, G.G.; Deb, S.; Gandomi, A.H.; Alavi, A.H. Opposition-based krill herd algorithm with Cauchy mutation and position clamping. *Neurocomputing* **2016**, *177*, 147–157. [CrossRef]
10. Wang, G.G.; Gandomi, A.H.; Alavi, A.H. An effective krill herd algorithm with migration operator in biogeography-based optimization. *Appl. Math. Model.* **2014**, *38*, 2454–2462. [CrossRef]
11. Wang, G.G.; Guo, L.H.; Wang, H.Q.; Duan, H.; Liu, L.; Li, J. Incorporating mutation scheme into krill herd algorithm for global numerical optimization. *Neural Comput. Appl.* **2014**, *24*, 853–871. [CrossRef]
12. Wang, G.G.; Gandomi, A.H.; Hao, G.S. Hybrid krill herd algorithm with differential evolution for global numerical optimization. *Neural Comput. Appl.* **2014**, *25*, 297–308. [CrossRef]
13. Ding, X.; Guo, H.; Guo, S. Efficiency Enhancement of Traction System Based on Loss Models and Golden Section Search in Electric Vehicle. *Energy Procedia* **2017**, *105*, 2923–2928. [CrossRef]
14. Ramos, H.; Monteiro, M.T.T. A new approach based on the Newton's method to solve systems of nonlinear equations. *J. Comput. Appl. Math.* **2016**, *318*, 3–13. [CrossRef]
15. Fazio, A.R.D.; Russo, M.; Valeri, S.; Santis, M.D. Linear method for steady-state analysis of radial distribution systems. *Int. J. Electr. Power Energy Syst.* **2018**, *99*, 744–755. [CrossRef]
16. Du, X.; Zhang, P.; Ma, W. Some modified conjugate gradient methods for unconstrained optimization. *J. Comput. Appl. Math.* **2016**, *305*, 92–114. [CrossRef]
17. Pooranian, Z.; Shojafar, M.; Abawajy, J.H.; Abraham, A. An efficient meta-heuristic algorithm for grid computing. *J. Comb. Optim.* **2015**, *30*, 413–434. [CrossRef]
18. Shojafar, M.; Chiaraviglio, L.; Blefari-Melazzi, N.; Salsano, S. P5G: A Bio-Inspired Algorithm for the Superfluid Management of 5G Networks. In Proceedings of the GLOBECOM 2017: 2017 IEEE Global Communications Conference, Singapore, 4–8 December 2017; pp. 1–7. [CrossRef]
19. Shojafar, M.; Cordeschi, N.; Abawajy, J.H.; Baccarelli, E. Adaptive Energy-Efficient QoS-Aware Scheduling Algorithm for TCP/IP Mobile Cloud. In Proceedings of the IEEE Globecom Workshops, San Diego, CA, USA, 6–10 December 2015; pp. 1–6. [CrossRef]
20. Zou, D.X.; Deb, S.; Wang, G.G. Solving IIR system identification by a variant of particle swarm optimization. *Neural Comput. Appl.* **2018**, *30*, 685–698. [CrossRef]

21. Kennedy, J. Particle Swarm Optimization. In Proceedings of the 1995 International Conference on Neural Networks, Perth, WA, Australia, 27 November–1 December 1995; Volume 4, pp. 1942–1948. [CrossRef]

22. Shi, Y.H.; Eberhart, R.C. Empirical study of particle swarm optimization. In Proceedings of the 1999 Congress on Evolutionary Computation-CEC99(Cat. No. 99TH8406), Washington, DC, USA, 6–9 July 1999; pp. 1945–1950. [CrossRef]

23. Chen, Y.; Li, L.; Xiao, J.; Yang, Y.; Liang, J.; Li, T. Particle swarm optimizer with crossover operation. *Eng. Appl. Artif. Intell.* **2018**, *70*, 159–169. [CrossRef]

24. Zou, D.; Gao, L.; Li, S.; Wu, J.; Wang, X. A novel global harmony search algorithm for task assignment problem. *J. Syst. Softw.* **2010**, *83*, 1678–1688. [CrossRef]

25. Niu, W.J.; Feng, Z.K.; Cheng, C.T.; Wu, X.Y. A parallel multi-objective particle swarm optimization for cascade hydropower reservoir operation in southwest China. *Appl. Soft Comput.* **2018**, *70*, 562–575. [CrossRef]

26. Feng, Y.; Wang, G.G.; Deb, S.; Lu, M.; Zhao, X.J. Solving 0–1 knapsack problem by a novel binary monarch butterfly optimization. *Neural Comput. Appl.* **2017**, *28*, 1619–1634. [CrossRef]

27. Wang, G.G.; Deb, S.; Coelho, L.D.S. Elephant Herding Optimization. In Proceedings of the International Symposium on Computational and Business Intelligence, Bali, Indonesia, 7–9 December 2015; pp. 1–5. [CrossRef]

28. Wang, G.; Guo, L.; Gandomi, A.H.; Cao, L.; Alavi, A.H.; Duan, H.; Li, J. Levy-flight krill herd algorithm. *Math. Probl. Eng.* **2013**, *2013*, 682073. [CrossRef]

29. Wang, G.G.; Gandomi, A.H.; Alavi, A.H. Stud krill herd algorithm. *Neurocomputing* **2014**, *128*, 363–370. [CrossRef]

30. Wang, G.G.; Deb, S.; Coelho, L. Earthworm optimization algorithm: A bio-inspired metaheuristic algorithm for global optimization problems. *Int. J. Bio-Inspired Comput.* **2018**, *12*, 1–22. [CrossRef]

31. Wang, G.G.; Chu, H.; Mirjalili, S. Three-dimensional path planning for UCAV using an improved bat algorithm. *Aerosp. Sci. Technol.* **2016**, *49*, 231–238. [CrossRef]

32. Shi, Y.; Eberhart, R. A modified particle swarm optimizer. In Proceedings of the 1998 IEEE International Conference on Evolutionary Computation Proceedings, IEEE World Congress on Computational Intelligence (Cat. No.98TH8360), Anchorage, AK, USA, 4–9 May 1998; pp. 69–73. [CrossRef]

33. Parsopoulos, K.E.; Vrahatis, M.N. Particle swarm optimizer in noisy and continuously changing environments. In *Artificial Intelligence and Soft Computing*; Hamza, M.H., Ed.; IASTED/ACTA Press: Anaheim, CA, USA, 2001; pp. 289–294.

34. Kennedy, J.; Mendes, R. Population structure and particle swarm performance. In Proceedings of the IEEE Congress on Evolutionary Computation, Honolulu, HI, USA, 12–17 May 2002; Volume 2, pp. 1671–1676. [CrossRef]

35. Clerc, M.; Kennedy, J. The particle swarm-explosion, stability, and convergence in a multidimensional complex space. *IEEE Trans. Evol. Comput.* **2002**, *6*, 58–73. [CrossRef]

36. Xu, G.; Yu, G. Reprint of: On convergence analysis of particle swarm optimization algorithm. *J. Shanxi Norm. Univ.* **2018**, *4*, 25–32. [CrossRef]

37. Sun, J.; Wu, X.; Palade, V.; Fang, W.; Lai, C.H.; Xu, W.B. Convergence analysis and improvements of quantum-behaved particle swarm optimization. *Inf. Sci.* **2012**, *193*, 81–103. [CrossRef]

38. Li, S.F.; Cheng, C.Y. Particle Swarm Optimization with Fitness Adjustment Parameters. *Comput. Ind. Eng.* **2017**, *113*, 831–841. [CrossRef]

39. Li, N.J.; Wang, W.; Hsu, C.C.J. Hybrid particle swarm optimization incorporating fuzzy reasoning and weighted particle. *Neurocomputing* **2015**, *167*, 488–501. [CrossRef]

40. Chih, M.; Lin, C.J.; Chern, M.S.; Ou, T.Y. Particle swarm optimization with time-varying acceleration coefficients for the multidimensional knapsack problem. *J. Chin. Inst. Ind. Eng.* **2014**, *33*, 77–102. [CrossRef]

41. Ratnaweera, A.; Halgamuge, S.K.; Watson, H.C. Self-organizing hierarchical particle swarm optimizer with time-varying acceleration coefficients. *IEEE Trans. Evol. Comput.* **2004**, *8*, 240–255. [CrossRef]

42. Kennedy, J.; Mendes, R. Neighborhood topologies in fully informed and best-of-neighborhood particle swarms. *IEEE Trans. Syst. Man Cybern. Part C* **2006**, *36*, 515–519. [CrossRef]

43. Zhao, S.Z.; Suganthan, P.N.; Pan, Q.K.; Tasgetiren, M.F. Dynamic multi-swarm particle swarm optimizer with harmony search. *Expert Syst. Appl.* **2011**, *38*, 3735–3742. [CrossRef]

44. Majercik, S.M. Using Fluid Neural Networks to Create Dynamic Neighborhood Topologies in Particle Swarm Optimization. In Proceedings of the International Conference on Swarm Intelligence, Brussels, Belgium, 10–12 September 2014; Springer: Cham, Switzerland; New York, NY, USA, 2014; Volume 8667, pp. 270–277. [CrossRef]

45. Sun, S.; Li, J. A two-swarm cooperative particle swarms optimization. *Swarm Evol. Comput.* **2014**, *15*, 1–18. [CrossRef]

46. Netjinda, N.; Achalakul, T.; Sirinaovakul, B. Particle Swarm Optimization inspired by starling flock behavior. *Appl. Soft Comput.* **2015**, *35*, 411–422. [CrossRef]

47. Beheshti, Z.; Shamsuddin, S.M. Non-parametric particle swarm optimization for global optimization. *Appl. Soft Comput.* **2015**, *28*, 345–359. [CrossRef]

48. Mendes, R.; Kennedy, J.; Neves, J. The fully informed particle swarm: Simpler, maybe better. *IEEE Trans. Evol. Comput.* **2004**, *8*, 204–210. [CrossRef]

49. Liang, J.J.; Qin, A.K.; Suganthan, P.N.; Subramanian, B. Comprehensive learning particle swarm optimizer for global optimization of multimodal functions. *IEEE Trans. Evol. Comput.* **2006**, *10*, 281–295. [CrossRef]

50. Gong, Y.J.; Li, J.J.; Zhou, Y.; Li, Y.; Chung, H.S.H.; Shi, Y.H.; Zhang, J. Genetic Learning Particle Swarm Optimization. *IEEE Trans. Cybern.* **2017**, *46*, 2277–2290. [CrossRef]

51. Zou, D.; Li, S.; Li, Z.; Kong, X. A new global particle swarm optimization for the economic emission dispatch with or without transmission losses. *Energy Convers. Manag.* **2017**, *139*, 45–70. [CrossRef]

52. Kiran, M.S. Particle Swarm Optimization with a New Update Mechanism. *Appl. Soft Comput.* **2017**, *60*, 607–680. [CrossRef]

53. Wang, G.G.; Gandomi, A.H.; Alavi, A.H. A chaotic particle-swarm krill herd algorithm for global numerical optimization. *Kybernetes* **2013**, *42*, 962–978. [CrossRef]

54. Gandomi, A.H.; Alavi, A.H. Krill herd: A new bio-inspired optimization algorithm. *Commun. Nonlinear Sci. Numer. Simul.* **2012**, *17*, 4831–4845. [CrossRef]

55. Yang, X.S. *Nature-Inspired Metaheuristic Algorithm*; Luniver Press: Beckington, UK, 2008; ISBN 1905986106, 9781905986101.

56. Wang, G.G.; Gandomi, A.H.; Yang, X.S.; Alavi, A.H. A novel improved accelerated particle swarm optimization algorithm for global numerical optimization. *Eng. Comput.* **2014**, *31*, 1198–1220. [CrossRef]

57. Liu, Y.; Niu, B.; Luo, Y. Hybrid learning particle swarm optimizer with genetic disturbance. *Neurocomputing* **2015**, *151*, 1237–1247. [CrossRef]

58. Chen, Y.; Li, L.; Peng, H.; Xiao, J.; Yang, Y.; Shi Y. Particle Swarm Optimizer with two differential mutation. *Appl. Soft Comput.* **2017**, *61*, 314–330. [CrossRef]

59. Liu, Z.G.; Ji, X.H.; Liu, Y.X. Hybrid Non-parametric Particle Swarm Optimization and its Stability Analysis. *Expert Syst. Appl.* **2017**, *92*, 256–275. [CrossRef]

60. Bewoor, L.A.; Prakash, V.C.; Sapkal, S.U. Production scheduling optimization in foundry using hybrid Particle Swarm Optimization algorithm. *Procedia Manuf.* **2018**, *22*, 57–64. [CrossRef]

61. Xu, X.; Rong, H.; Pereira, E.; Trovati, M.W. Predatory Search-based Chaos Turbo Particle Swarm Optimization (PS-CTPSO): A new particle swarm optimisation algorithm for Web service combination problems. *Future Gener. Comput. Syst.* **2018**, *89*, 375–386. [CrossRef]

62. Guan, G.; Yang, Q.; Gu, W.W.; Jiang W.; Lin, Y. Ship inner shell optimization based on the improved particle swarm optimization algorithm. *Adv. Eng. Softw.* **2018**, *123*, 104–116. [CrossRef]

63. Qin, Z.; Liang, Y.G. Sensor Management of LEO Constellation Using Modified Binary Particle Swarm Optimization. *Optik* **2018**, *172*, 879–891. [CrossRef]

64. Peng, Z.; Manier, H.; Manier, M.A. Particle Swarm Optimization for Capacitated Location-Routing Problem. *IFAC PapersOnLine* **2017**, *50*, 14668–14673. [CrossRef]

65. Copot, C.; Thi, T.M.; Ionescu, C. PID based Particle Swarm Optimization in Offices Light Control. *IFAC PapersOnLine* **2018**, *51*, 382–387. [CrossRef]

66. Chen, S.Y.; Wu, C.H.; Hung, Y.H.; Chung, C.T. Optimal Strategies of Energy Management Integrated with Transmission Control for a Hybrid Electric Vehicle using Dynamic Particle Swarm Optimization. *Energy* **2018**, *160*, 154–170. [CrossRef]

67. Chen, K.; Zhou, F.; Yin, L.; Wang, S.; Wang, Y.; Wan, F. A Hybrid Particle Swarm Optimizer with Sine Cosine Acceleration Coefficients. *Inf. Sci.* **2017**, *422*, 218–241. [CrossRef]

68. Zou, D.; Gao, L.; Wu. J.; Li, S. Novel global harmony search algorithm for unconstrained problems. *Neurocomputing* **2010**, *73*, 3308–3318. [CrossRef]

69. Hu, W.; Li, Z.S. A Simpler and More Effective Particle Swarm Optimization Algorithm. *J. Softw.* **2007**, *18*, 861–868. [CrossRef]

70. Zhang, X.; Zou, D.X.; Kong, Z.; Shen, X. A Hybrid Gravitational Search Algorithm for Unconstrained Problems. In Proceedings of the 30th Chinese Control and Decision Conference, Shenyang, China, 9–11 June 2018; pp. 5277–5284. [CrossRef]

71. Zhang, X.; Zou, D.X.; Shen, X. A Simplified and Efficient Gravitational Search Algorithm for Unconstrained Optimization Problems. In Proceedings of the 2017 International Conference on Vision, Image and Signal Processing, Osaka, Japan, 22–24 September 2017 ; pp. 11–17. [CrossRef]

72. Müller, P. Analytische Zahlentheorie. In *Funktionentheorie 1*; Springer: Berlin/Heidelberg, Germany, 2006; pp. 386–456. [CrossRef]

![mathematics logo](Σ mathematics)

MDPI

Article

Energy-Efficient Scheduling for a Job Shop Using an Improved Whale Optimization Algorithm

Tianhua Jiang [1,*], Chao Zhang [2], Huiqi Zhu [1], Jiuchun Gu [1] and Guanlong Deng [3]

[1]   School of Transportation, Ludong University, Yantai 264025, China; zhuhuiqi0505@126.com (H.Z.);
     gujiuchun@163.com (J.G.)
[2]   Department of Computer Science and Technology, Henan Institute of Technology, Xinxiang 453003, China;
     zhangchao915@foxmail.com
[3]   School of Information and Electrical Engineering, Ludong University, Yantai 264025, China; dglag@163.com
*   Correspondence: jth1127@163.com

Received: 28 September 2018; Accepted: 26 October 2018; Published: 28 October 2018

Abstract: Under the current environmental pressure, many manufacturing enterprises are urged or forced to adopt effective energy-saving measures. However, environmental metrics, such as energy consumption and CO_2 emission, are seldom considered in the traditional production scheduling problems. Recently, the energy-related scheduling problem has been paid increasingly more attention by researchers. In this paper, an energy-efficient job shop scheduling problem (EJSP) is investigated with the objective of minimizing the sum of the energy consumption cost and the completion-time cost. As the classical JSP is well known as a non-deterministic polynomial-time hard (NP-hard) problem, an improved whale optimization algorithm (IWOA) is presented to solve the energy-efficient scheduling problem. The improvement is performed using dispatching rules (DR), a nonlinear convergence factor (NCF), and a mutation operation (MO). The DR is used to enhance the initial solution quality and overcome the drawbacks of the random population. The NCF is adopted to balance the abilities of exploration and exploitation of the algorithm. The MO is employed to reduce the possibility of falling into local optimum to avoid the premature convergence. To validate the effectiveness of the proposed algorithm, extensive simulations have been performed in the experiment section. The computational data demonstrate the promising advantages of the proposed IWOA for the energy-efficient job shop scheduling problem.

Keywords: energy-efficient job shop scheduling; dispatching rule; nonlinear convergence factor; mutation operation; whale optimization algorithm

---

1. Introduction

Nowadays, manufacturing enterprises are facing not only the economic pressure, but also environmental challenges. With the consideration of sustainable development, reducing energy consumption becomes an important target for manufacturing companies. To implement such measures, some researchers focused on developing more energy-efficient machines or machining processes [1,2]. However, it has been indicated that a significant energy-saving opportunity may be missed by focusing solely on the machines or processes, and the operational method can be adopted from the manufacturing system-level perspective [3]. In recent years, increasingly more attention has been paid to production scheduling problems with the consideration of energy efficiency. Compared with the investment in new energy-saving machines and production redesign, the optimization of scheduling scheme requires a modest investment and is more easily applied to existing production systems.

One of the earliest energy-efficient production scheduling methods was investigated by Mouzon et al. [4]. They developed a turn-on/off scheduling strategy of machines to control the overall energy consumption. Some dispatching rules are proposed to solve the multi-objective mathematical model with the objective of minimizing the energy consumption and total completion time. Since then, energy-efficient production scheduling has become a new research spot in the manufacturing field. Mouzon and Yildirim [5] proposed a greedy randomized multi-objective adaptive searching algorithm to minimize total energy consumption and total tardiness in a single-machine system. Yildirim and Mouzon [6] developed a multi-objective genetic algorithm to deal with a single-machine scheduling problem with the objective of minimizing the energy consumption and completion time. Shrouf et al. [7] designed a genetic algorithm for a single-machine scheduling with the consideration of variable energy prices. Che et al. [8] considered a single-machine scheduling problem under the time-of-use (TOU) strategy. Li et al. [9] presented an energy-aware multi-objective optimization algorithm for the hybrid flow shop scheduling problem with the setup energy consumption. Liu et al. [10] proposed a branch-and-bound algorithm for the permutation flow shop scheduling problem in order to minimize the wasted energy consumption. Dai et al. [11] proposed a genetic-simulated annealing algorithm to optimize makespan and energy consumption in a flexible flow shop. Ding et al. [12] considered an energy-efficient permutation flow shop scheduling problem to minimize total carbon emissions and makespan. Mansouri and Aktas [13] addressed multi-objective genetic algorithms for a two-machine flow shop scheduling problem to optimize energy consumption and makespan. Luo et al. [14] presented an ant colony optimization algorithm with the criterion to minimize production efficiency and electric power cost in a hybrid flow shop. Regarding the above literature, most of the corresponding studies are oriented to a single machine [5–8] and flow shop [9–14]. By contrast, the energy-efficient job shop scheduling problem is not fully studied. However, many real-life problems can be taken as a job shop scheduling problem (JSP), such as production scheduling in the industry, departure and arrival times of logistic problems, the delivery times of orders in a company, and so on [15]. Therefore, in this paper, the job shop is selected as the research object, and the scheduling problem is investigated from the perspective of energy consumption reduction.

In some actual manufacturing systems, machines can not be frequently turned on/off because restarting action may consume a large amount of additional energy or damage the machine tools [16]. Under this situation, the framework of machine speed scaling is an alternative method to control the energy consumption in the workshop, by which machines are allowed to work at different speeds when processing different jobs. When the machine works at a higher speed, the processing time decreases but the amount of energy consumption increases, and when the machine works at a lower speed, the processing time increases while the amount of energy consumption decreases. Compared with the classical JSP, the complexity of the EJSP under study mainly lies in the addition of an energy-related objective and machine speed selection. As the classical JSP is well known as an NP-hard problem, it is clear that the EJSP is difficult to solve using exact methods. Although meta-heuristic algorithms have been paid increasingly more attention by researchers in the manufacturing field, its application to the energy-efficient job shop scheduling problem with machine speed scaling framework appears to be limited. Salido et al. [15] presented a multi-objective genetic algorithm with the objective of minimizing the makespan and the energy consumption. Zhang and Chiong [16] proposed a multi-objective genetic algorithm and two local search strategies to optimize the total energy consumption and total weighted tardiness. Tang and Dai [17] proposed a genetic-simulated annealing algorithm to minimize the makespan and energy consumption. Escamilla et al. [18] presented a genetic algorithm to minimize the makespan and the energy consumption. Yin et al. [19] developed a multi-objective genetic algorithm based on simplex lattice design to optimize the productivity, energy efficiency, and noise. In view of the complexity of the problem under study, meta-heuristic algorithms can achieve satisfactory scheduling within a reasonable time. The application of a meta-heuristic algorithm on the EJSP will be a research hot spot in the manufacturing field.

Because of their promising advantage, meta-heuristic algorithms have received increasing interest in solving complex optimization problems [20]. Recently, many meta-heuristic algorithms have been presented and improved to solve various problems [21–32]. Whale optimization algorithm (WOA) is a new swarm-based algorithm, which mimics the hunting behavior of humpback whales in nature [33]. It is well-known that the cooperation between exploration and exploitation is very important for the searching ability of a meta-heuristic algorithm. For the existing meta-heuristics, some algorithms have better global search abilities, while others have better local search abilities. In general, a well-designed hybrid method is used to balance the capacity of exploration and exploitation. By contrast, the main unique feature of the WOA algorithm is that it can maintain a good relationship between exploration and exploitation by self-tuning some parameters in the iteration process. At present, WOA has been adopted to deal with a variety of optimization problems in different fields; for example, global optimization [34], feature selection [35], content-based image retrieval [36], 0–1 knapsack problem [37], permutation flow shop scheduling problem [38], and so on. By considering its efficiency, an improved whale optimization algorithm (IWOA) is developed for solving the energy-efficient job shop scheduling problem with machine speed scaling framework. However, to the best of the authors' knowledge, the research on WOA to solve the EJSP has not yet been attempted. This paper aims to develop the IWOA in solving the EJSP with the objective of minimizing the energy consumption cost and completion-time cost. The main improvement of the algorithm lies on the introduction of dispatching rules (DR), nonlinear convergence factor (NCF), and mutation operation (MO), where DR is used to create the suitable initial population, NCF is adopted to balance the capacities of exploration and exploitation, and MO is employed to maintain the diversity of the population and avoid the premature convergence. Extensive experiments are conducted to validate the effectiveness of the proposed algorithm.

The rest of this paper is organized as follows. Section 2 introduces the description of the problem. Section 3 addresses the original whale optimization algorithm. Section 4 describes the implementation of the proposed IWOA algorithm. Section 5 shows the experimental results of IWOA and Section 6 provides conclusions and future works.

2. Problem Description

The EJSP can be described as follows: n jobs need to be processed on m machines in the workshop. The main difference between the EJSP and the classical JSP is that each machine can adjust its speed for different jobs in a finite and discrete speed set $v = \{v_1, v_2, \ldots, v_d\}$. The higher the speed of machine, the shorter the processing time of the operation assigned to it. When job i is assigned to machine k, there is a basic processing time represented by q_{ik}. If v_d is selected, the processing time of job i can be measured by p_{ikd}, that is, $p_{ikd} = q_{ik}/v_d$, and the energy consumption cost per unit time is defined as E_{kd}. For job i on machine k, if $v_{d'} > v_d$, $E_{kd'} \times p_{ikd'} > E_{kd} \times p_{ikd}$ holds. In other words, a machine working at a higher speed will decrease the processing time, but increase the energy consumption cost. Some additional constraints are involved as follows:

(1) Any job can not be processed on more than one machine simultaneously.
(2) Each machine can only process one operation at a time.
(3) Preemption is not allowed once a job starts to be processed on a machine.
(4) Setup time is not considered in this paper.
(5) The speed of a machine can not be changed during the processing of an operation.
(6) Each machine can not be turned off completely until all jobs assigned to it are finished. During the idle periods, each machine will be on a stand-by mode with energy consumption cost per unit time SE_k.

For such a problem, two sub-problems should be considered, that is, operation permutation and speed-level selection. For the classical JSP, the complete time or its related cost is very important for optimization decision-making. However, under the current environmental pressure, environmental metrics can not be ignored in the energy-efficient scheduling problem; for example, energy-consumption cost. Therefore, the optimization objective of the problem under study is aiming to obtain an optimal scheduling scheme to minimize the sum of energy-consumption cost and completion-time cost.

$$\min F = \sum_{i=1}^{n}\sum_{k=1}^{m}\sum_{d=1}^{D_k} E_{kd}p_{ikd}x_{ikd} + \sum_{k=1}^{m} SE_k(C_k - W_k) + \lambda C_{\max} \tag{1}$$

$$\text{s.t.} \sum_{d=1}^{D_k} x_{ikd} = 1, i = 1, 2, \ldots, n; k = 1, 2, \ldots, m \tag{2}$$

$$C_{ik} - \sum_{d=1}^{D_k} p_{ikd}x_{ikd} + L(1 - y_{ijk}) \geq C_{ij}, \ldots i = 1, 2, \ldots, n; j, k = 1, 2, \ldots, m \tag{3}$$

$$C_{lk} - C_{ik} + L(1 - z_{ilk}) \geq \sum_{d=1}^{D_k} p_{lkd}x_{lkd}, i, l = 1, 2, \ldots, n; k = 1, 2, \ldots, m \tag{4}$$

$$C_{ik} \geq 0, i = 1, 2, \ldots, n; k = 1, 2, \ldots, m \tag{5}$$

$$x_{ikd} \in \{0, 1\}, i = 1, 2, \ldots, n; k = 1, 2, \ldots, m; d = 1, 2, \ldots, D_k \tag{6}$$

$$y_{ijk} \in \{0, 1\}, i = 1, 2, \ldots, n; j, k = 1, 2, \ldots, m \tag{7}$$

$$s_{ilk} \in \{0, 1\}, i, l = 1, 2, \ldots, n; k = 1, 2, \ldots, m \tag{8}$$

where F means the sum of energy-consumption cost and completion-time cost. E_{kd} denotes the energy consumption cost per unit time of machine k with speed-level d. p_{ikd} is the processing time of job i on machine k with speed-level d. x_{ikd} is a 0–1 variable, if job i is processed on machine k with speed-level d, $x_{ikd} = 1$; otherwise, $x_{ikd} = 0$. D_k means the number of adjustable speed levels of machine k. SE_k is the energy consumption cost per unit time of machine k in the stand-by mode. C_k denotes the final completion time of machine k. W_k represents the total workload of machine k. $C_{\max}$ defines the final completion time of jobs (makespan) in the workshop. λ is the cost coefficient relevant to final completion-time. C_{ik} is the completion time of job i processing on machine k. L is a big position number. y_{ijk} is a 0–1 variable, if machine j performs job i prior to machine k, $y_{ijk} = 1$; otherwise, $y_{ijk} = 0$. s_{ilk} is a 0–1 variable, if job i is processed on machine k prior to job l, $s_{ilk} = 1$; otherwise, $s_{ilk} = 0$.

Equation (1) defines the optimization objective of the problem; constraint (2) ensures that the machine's speed can not be adjusted when an operation is processing on it; constraint (3) represents the precedence constraints between operations of a job; constraint (4) means that each machine only processes one operation at the same time; constraint (5) denotes the nonnegative feature of completion time; constraints (6)–(8) show the relevant 0–1 variables.

3. Overview of the Original WOA

The whale optimization algorithm (WOA) is a new population-based optimization algorithm, which mimics the hunting behavior of humpback whales in nature [33]. In this algorithm, two searching phases are involved for exploitation and exploration. In the exploitation phase, the position of each whale is updated by bubble-net attacking strategy, which is conducted by shrinking encircling the prey and the spiral shape movement based on the best solution discovered so far. In the exploration phase, the position of each whale is updated according to a randomly selected search agent rather than the best solution discovered so far. Because of the space limit, the overview of the original WOA is briefly shown below. The more detailed introduction can be easily found in the literature [33].

3.1. Exploitation Phase

(1)   Shrinking encircling mechanism

Humpback whales can observe the location of prey and encircle them in the hunting process. To model the algorithm, the current best search agent is assumed to be the target prey or close to the optimal solution. When the best search agent is discovered, other whales will update their positions towards the best whale, which can be represented as follows:

$$D = |C \cdot X^*(t) - X(t)| \tag{9}$$

$$X(t+1) = X^*(t) - A \cdot D \tag{10}$$

$$A = 2ar - a \tag{11}$$

$$C = 2r \tag{12}$$

where t is the current iteration number, X^* indicates the position vector of the best search agent found so far, and X defines the position vector of an individual whale. A and C are coefficient vectors. | | means the absolute value, and $\cdot$ is an element-by-element multiplication. r denotes a random vector inside $[0, 1]$. The elements of a are linearly decreased from 2 to 0 over the course of iterations. The shrinking encircling mechanism is implemented by decreasing the element value of a according to Equation (13), where t_{max} is the maximum of the iteration.

$$a = 2 - \frac{2t}{t_{max}} \tag{13}$$

(2)   Spiral updating mechanism

In addition to the shrinking encircling behavior, a spiral path is created to simulate the helix-shaped movement of whales, which can be defined as follows:

$$X(t+1) = D' \cdot e^{bl} \cdot \cos(2\pi l) + X^*(t) \tag{14}$$

$$D' = |X^*(t) - X(t)| \tag{15}$$

where b is a constant used to determine the shape of the logarithmic spiral and l is a random number inside $[-1, 1]$.

In the exploitation phase, whales move around the prey in a shrinking circle and along a spiral path simultaneously, which are chosen according to a probability of 50%. This can be represented by Equation (16), where h is a random number in the range $[0, 1]$.

$$X(t+1) = \begin{cases} X^*(t) - A \cdot D & \text{if } h < 0.5 \\ D' \cdot e^{bl} \cdot \cos(2\pi l) + X^*(t) & \text{if } h \geq 0.5 \end{cases} \tag{16}$$

3.2. Exploration Phase

Except for the bubble-net attacking mechanism, the humpback whales search for prey randomly in the exploration phase. The mechanism is also conducted based on the variation of the vector A. When $|A| < 1$, the exploitation is utilized by updating the positions towards the current best search agent, when $|A| \geq 1$, the exploration is adopted to search the global optimum. The mathematical model can be represented as follows:

$$D'' = |C \cdot X_{rand}(t) - X(t)| \tag{17}$$

$$X(t+1) = X_{rand}(t) - A \cdot D'' \tag{18}$$

where X_{rand} is a position vector selected from the current population at random.

3.3. Pseudo Code of WOA

The pseudo code of the original WOA algorithm can be shown in Figure 1.

Randomly initialize the whale population.
Evaluate the fitness values of whales and find out the best search agent X^*.
while $t < t_{max}$
    Calculate the value of a according to Equation (13)
    for each search agent
      if $h < 0.5$ **then**
            if $|A| < 1$ **then** $X(t+1) = X^*(t)\text{-}A \cdot D$
            else if $|A| \geq 1$ **then** $X(t+1) = X_{rand}(t)\text{-}A \cdot D''$
            end if
      else if $h \geq 0.5$ **then**
            $X(t+1) = D' \cdot e^{bl} \cdot \cos(2\pi l) + X^*(t)$
      end if
    end for
    Evaluate the fitness of $X(t+1)$ and update X^*
end while

Figure 1. The pseudo code of the original whale optimization algorithm (WOA).

4. Implement of the Proposed IWOA

4.1. Scheduling Solution Representation

As mentioned above, the energy-efficient job shop scheduling problem consists of two sub-problems, that is, operation permutation and speed-level selection. To obtain a feasible scheduling scheme, it needs to choose suitable processing speeds for jobs and arrange them on each machine. Therefore, the scheduling solution can be represented by a two-segment string with the size of $2mn$. The first segment tries to arrange the processing sequence of operations on each machine, and the second aims to choose an appropriate speed level for each job.

Taking a 3×2 (three jobs, two machines) EJSP, for example, five speed levels are considered for each job on machines. The scheduling solution can be shown by Figure 2. For the first segment, each element means the job code, where the elements with the same values represent different operations of the same job. For the second segment, each element represents the speed-level selected for the relevant job, which is stored in a given order. O_{ik} represents the kth operation of job i.

O_{11}	O_{31}	O_{21}	O_{22}	O_{32}	O_{12}	O_{11}	O_{12}	O_{21}	O_{22}	O_{31}	O_{32}
1	3	2	2	3	1	4	4	2	4	2	5
Operation permutation						Speed-level selection					

Figure 2. Scheduling solution representation.

4.2. Individual Position Vector

In the proposed IWOA, the individual position is still a multi-dimensional real vector, which is also made up by two segments, that is, $X = \{x(1), x(2), \ldots, x(mn), x(mn+1), \ldots, x(2mn)\}$, where $x(j) \in [x_{min}, x_{max}], j = 1, 2, \ldots, 2mn$. The first segment $X_1 = \{x(1), x(2), \ldots, x(mn)\}$ presents the information of operation permutation, and the second segment $X_2 = \{x(mn+1), \ldots, x(2mn)\}$ gives the information of speed-level selection. For the above 3×2 EJSP (three jobs, two machines), the individual position vector can be shown by Figure 3, where element values are stored according to the given order.

O_{11}	O_{12}	O_{21}	O_{22}	O_{31}	O_{32}	O_{11}	O_{12}	O_{21}	O_{22}	O_{31}	O_{32}
-0.5	2.4	-2.8	0.3	-2.5	1.9	1.1	0.9	-1.6	1.5	-1.5	2.7

| Operation permutation | Speed-level selection |

Figure 3. Individual position vector.

4.3. Conversion between Individual Position Vector and Scheduling Solution

The original WOA was proposed to deal with the continuous optimization problem. However, considering the discrete characteristics of the EJSP, it is very important to find a method to establish the mapping relationship between the individual position vector and the discrete scheduling solution. In the previous study, a method is proposed to implement the conversion between the continuous individual vector and the discrete scheduling solution for the classical flexible job shop (FJSP) [39]. It is known that FJSP is made up by operation permutation and machine selection. Seen from Figure 2, the structure of the solution is similar to that of the FJSP, where the speed-level selection is taken place of the machine selection vector. Therefore, the conversion method in the literature [39] is modified for the EJSP in this study. To facilitate the expression, the intervals $[x_{\min}, x_{\max}]$ are all set as $[-\varepsilon, \varepsilon], \varepsilon > 0$ in the proposed method.

4.3.1. Conversion from Individual Position Vector to Scheduling Solution

For the operation permutation segment, the ranked-order-value (ROV) rule is used to implement the conversion from the individual position to the scheduling solution. In the rule, position values in X_1 are first ranked in an increasing order, then the operation permutation can be acquired according to the new order, which is shown in Figure 4.

Job code	1	1	2	2	3	3
X_1	-0.5	2.4	-2.8	0.3	-2.5	1.9

Ranked order	-2.8	-2.5	-0.5	0.3	1.9	2.4

Operation permutation	2	3	1	2	3	1

Figure 4. The conversion process from individual position vector to operation permutation.

For the speed-level selection segment, the conversion process can be modified from the method proposed by Yuan and Xu [40], which can be represented by Equation (19). $z(j)$ denotes the size of alternative speed-level set for the operation corresponding to the jth element, $u(j)$ means the selected speed level, $u(j) \in [1, z(j)]$. In the procedure, $x(j)$ is first converted to a real number belonging to $[1, z(j)]$, then $u(j)$ is given the nearest integer value for the converted real number. For the above example, $\varepsilon = 3$ and $z(j) = 5$. The conversion process is shown in Figure 5.

$$u(j) = round(\frac{x(j) + \varepsilon}{2\varepsilon}(z(j) - 1) + 1), 1 \leq j \leq mn \tag{19}$$

| X2 | 1.1 | 0.9 | -1.6 | 1.5 | -1.5 | 2.7 |

⇩

| converted real number | 3.7 | 3.6 | 1.9 | 4.0 | 2.0 | 4.8 |

⇩

| Speed selection | 4 | 4 | 2 | 4 | 2 | 5 |

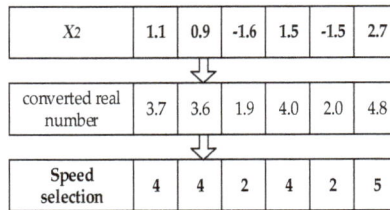

Figure 5. The conversion process from individual position vector to speed-level selection.

4.3.2. Conversion from Scheduling Solution to Individual Position Vector

For the operation permutation segment, mn real numbers are first randomly generated between $[-\varepsilon, \varepsilon]$ and then ranked in an increasing order. According to the ranked order and the scheduling solution, the individual position vector X_1 can be obtained according to the conversion process in Figure 6.

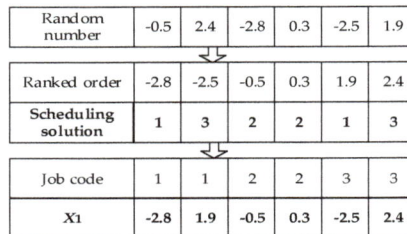

| Random number | -0.5 | 2.4 | -2.8 | 0.3 | -2.5 | 1.9 |

⇩

| Ranked order | -2.8 | -2.5 | -0.5 | 0.3 | 1.9 | 2.4 |
| Scheduling solution | 1 | 3 | 2 | 2 | 1 | 3 |

⇩

| Job code | 1 | 1 | 2 | 2 | 3 | 3 |
| X1 | -2.8 | 1.9 | -0.5 | 0.3 | -2.5 | 2.4 |

Figure 6. The conversion process from operation permutation to individual position vector.

For the speed-level selection segment, the conversion is generally an inverse process of Equation (19). However, there is a special case, that is, $z(j) = 1$, $x(j)$ is obtained by choosing a random value between $[-\varepsilon, \varepsilon]$. The conversion process is shown in Figure 7.

$$x(j) = \begin{cases} \frac{2\varepsilon}{z(j)-1}(u(j)-1) - \varepsilon, & z(j) \neq 1 \\ x(j) \in [-\varepsilon, \varepsilon], & z(j) = 1 \end{cases} \tag{20}$$

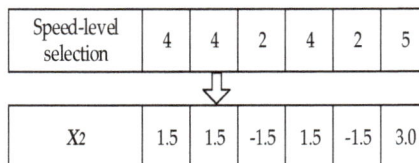

| Speed-level selection | 4 | 4 | 2 | 4 | 2 | 5 |

⇩

| X2 | 1.5 | 1.5 | -1.5 | 1.5 | -1.5 | 3.0 |

Figure 7. The conversion process from speed-level selection to individual position vector.

4.4. Initial Scheduling Generation

For a population-based optimization algorithm, the quality of initial solutions is very important for the computational performance. According to the characteristics of the scheduling solution, the population initialization process can be divided into two phases. In the speed-level selection phase, a random generation rule is employed to obtain the initial speeds for jobs. In the operation permutation phase, five dispatching rules are adopted as follows: the Most Work Remaining (MWR), the Most Operation Remaining (MOR), the Shortest Processing Time (SPT), the Longest Processing

Time (LPT), and the Random Rule (RR). Each dispatching rule is randomly selected to generate a scheduling solution.

MWR: give the highest priority to the job with the most amount of remaining work.
MOR: give the highest priority to the job with the most number of remaining operations.
SPT: give the highest priority to the job with the shortest processing time.
LPT: give the highest priority to the job with the longest processing time.
RR: select jobs for the permutation at random.

4.5. Nonlinear Convergence Factor

Like other population-based optimization algorithms, the cooperation between the abilities of exploitation and exploration is crucial for the performance of the algorithm. As seen from Figure 1, a suitable adjustment of search vector A can allow the WOA algorithm to smoothly transition between exploration and exploitation. By decreasing the value of A, some iterations are focused on exploration ($|A| \geq 1$) and the others are devoted to exploitation ($|A| < 1$). According to Equation (11), the value of A is determined by the variation of a. However, a are linearly decreased from 2 to 0 over the course of iterations, which can not well reflect the nonlinear search process of the algorithm. Therefore, a nonlinear adjustment curve of a is adopted in (21), where a_{max} and a_{min} define the maximum and minimum values of a, respectively.

$$a = a_{max} - (a_{max} - a_{min}) \sin(t\pi/(2t_{max})) \tag{21}$$

4.6. Mutation Operation

According to the characteristics of the WOA, whales cluster around the local optimum at the latter stage of the optimization, which will decrease the population diversity and lead to premature convergence. In this study, a kind of adaptive mutation operator is presented to overcome this drawback, where the mutation rate of each individual can be calculate according to Equation (15). p_g defines the mutation rate of the gth individual and f represents the fitness function $\frac{Q}{T}$, where Q is a constant. f_{max} and f_{min} are the maximum and minimum values of f in the current generation, respectively. In Equation (15), the mutation rate of each whale is changed along with the fitness in the evolution process. If a random number is smaller than p_g, the mutation operations for the two segments are conducted.

For the operation permutation segment, an inverse mutation operator is used to inverse the order of elements between two randomly selected positions in a candidate individual. For the speed-level selection segment, a single-point mutation is employed to select a new speed level to take place of the original one, and then a new individual position is obtained by Equation (20).

$$p_g(t) = 1 - \frac{f_{max}(t) - f_g(t)}{f_{max}(t) - f_{min}(t)} \tag{22}$$

4.7. Pseudo Code of IWOA

The pseudo code of the IWOA algorithm can be shown in Figure 8.

Create the initial population by the method in Section 4.4.
Evaluate the fitness values for whales, and obtain the best search agent X^*.
while $t < t_{max}$
  Calculate the value of a according to Equation (21)
  for each search agent
    if $h < 0.5$ **then**
        if $|A| < 1$ **then** $X(t+1) = X^*(t) - A \cdot D$
        else if $|A| \geq 1$ **then** $X(t+1) = X_{rand}(t) - A \cdot D'$
        end if
    else if $h \geq 0.5$ **then**
        $X(t+1) = D' \cdot e^{bl} \cdot \cos(2\pi l) + X^*(t)$
    end if
  end for
  Perform the adaptive mutation operation to individuals
  Evaluate the fitness of $X(t+1)$ and update X^*
end while

Figure 8. The pseudo code of the improved WOA (IWOA).

5. Computational Results

5.1. Experimental Settings

The implementation of the proposed IWOA algorithm is coded by using Fortran language and run on VMware Workstation with 2GB main memory under WinXP. In this section, 38 instances modified from those of the classical JSP (FT06, FT10, and FT20 in the work of [41] and LA01–LA35 in the work of [42]) are used to validate the efficiency of the IWOA. For each instance, ten independent replications are run by different algorithms. Here, the processing times in the classical JSP are taken as the basic processing times. The speed for processing operations can be selected from $v = \{v_1, v_2, v_3, v_4, v_5\} = \{1.0, 1.2, 1.5, 2.0, 2.5\}$. The energy consumption cost per unit time of machine k can be calculated according to $E_{kd} = \xi_k \times v_d^2, d = 1, 2, 3, 4, 5$, where ξ_k is randomly generated following a discrete uniform distribution in the literature [2,4]. The stand-by energy consumption cost per unit time of machine k is calculated by $SE_k = \xi_k / 4$. In addition, the completion time cost per unit time λ is set to be 15.0.

5.2. Effectiveness of the Improvement Strategies

In this paper, three strategies are adopted to improve the performance of the IWOA algorithm, namely, DR, NCF, and MO. In this subsection, the effectiveness of the three strategies are first tested. In Table 1, instance names are listed in the first column, and computational data are reported in the following columns. 'IWOA' is the proposed algorithm in this study. 'IWOA-R' represents the algorithm where the initial solutions are only created by the random rule. 'IWOA-L' represents the algorithm where the elements of a are linearly decreased by Equation (13). 'IWOA-NMO' represents the algorithm where the mutation operation is excluded from the IWOA algorithm. In addition, 'Best' represents the best value in the ten runs. 'Avg' means the average results of the ten runs. 'SD' is the standard deviation of total cost obtained by ten runs. 'Time' is the average computational time (in seconds) in the ten runs. Boldface denotes the optimal values obtained by algorithms. To facilitate the comparison, the same parameters are set for the compared algorithms, for example, population size is 200 and maximum iteration is 2000.

Table 1. Effectiveness analysis of improvement strategy.

Instance	IWOA-R				IWOA-L				IWOA-NMO				IWOA			
	Best	Avg	SD	Time	Best	Avg	SD	Time	Best	Avg	SD	Time	Best	Avg	SD	Time
FT06	**1364.4**	1381.5	13.1	27.0	1374.4	1384.3	6.7	27.0	1383.3	1389.8	**6.9**	15.9	1370.4	1382.8	**6.2**	26.9
FT10	31,756.9	32,689.8	652.5	108.2	31,809.4	32,564.2	569.2	109.6	31,720.1	32,394.9	**322.9**	64.0	**31,619.3**	32,726.4	649.4	107.8
FT20	35,225.1	35,925.9	**337.0**	111.5	35,138.0	35,888.1	425.9	113.7	35,005.8	35,525.3	370.1	66.2	**34,544.2**	**35,155.7**	376.8	111.4
LA01	18,053.5	18,278.0	164.1	40.8	**18,031.3**	18,281.8	157.6	39.2	18,033.6	18,408.3	243.7	23.6	18,096.0	**18,204.8**	**88.8**	39.0
LA02	17,663.3	17,893.5	**131.0**	38.5	17,699.8	17,935.7	160.5	38.7	17,911.5	18,253.6	218.7	23.3	**17,661.0**	17,967.5	157.8	38.2
LA03	**15,679.5**	**15,877.2**	142.5	38.5	15,765.6	15,989.5	189.1	38.8	15,873.5	16,153.8	183.0	23.2	15,679.6	15,913.4	219.2	38.7
LA04	16,114.6	16,327.7	119.2	38.0	16,223.0	16,384.8	**107.1**	40.0	16,129.9	16,467.2	215.8	23.5	**16,039.7**	**16,209.3**	136.1	38.0
LA05	14,506.1	14,667.6	**96.0**	38.6	14,447.2	14,657.9	133.4	39.7	14,613.5	14,813.4	138.4	23.6	**14,442.6**	**14,608.0**	97.8	38.2
LA06	**25,145.6**	25,529.1	206.2	68.7	25,248.1	25,654.4	204.1	69.4	25,464.8	25,758.7	196.6	40.8	25,198.4	**25,501.0**	**134.8**	68.2
LA07	24,510.7	24,770.1	**183.8**	67.0	24,534.1	24,868.4	227.2	67.1	24,513.2	24,867.2	238.4	39.9	**24,172.9**	**24,619.2**	203.1	66.3
LA08	24,511.2	**24,765.4**	213.7	73.6	24,717.8	24,889.7	**132.6**	73.9	24,700.3	24,922.0	140.0	39.5	**24,422.8**	24,775.7	202.1	67.6
LA09	27,349.1	**27,538.6**	136.5	72.3	27,401.1	27,692.2	**122.0**	74.0	27,610.5	27,798.6	149.9	44.0	**27,329.5**	27,608.6	122.5	73.5
LA10	**24,959.0**	25,349.7	216.4	73.6	25,210.1	25,455.5	**164.4**	73.5	25,206.4	25,405.9	164.9	43.9	25,081.3	**25,315.5**	178.9	73.2
LA11	34,287.2	34,668.2	**213.6**	112.6	34,406.6	34,697.8	220.4	113.9	34,227.4	34,706.6	233.3	64.6	**34,221.7**	**34,659.8**	258.5	110.1
LA12	29,927.1	30,282.2	178.2	110.8	30,213.3	30,432.0	163.0	111.9	30,088.7	30,287.8	**159.4**	64.3	**29,905.6**	30,221.5	173.9	111.6
LA13	**32,974.4**	**33,316.7**	232.8	110.4	33,379.4	33,617.1	143.1	117.0	33,192.5	33,401.8	**124.9**	67.5	33,198.1	33,490.7	216.0	110.2
LA14	33,991.7	**34,116.7**	**114.7**	113.3	34,084.4	34,478.8	187.5	114.7	33,979.4	34,398.0	249.7	67.2	**33,794.5**	34,214.3	242.8	114.0
LA15	35,817.5	**36,102.2**	**159.3**	113.9	36,017.9	36,327.0	209.0	115.1	35,979.6	36,315.0	258.1	66.2	**35,802.4**	36,199.8	203.0	113.1
LA16	31,572.7	32,171.2	549.2	109.1	31,746.8	32,480.8	453.5	111.4	32,215.1	32,856.2	**313.0**	64.5	**31,516.7**	**32,118.2**	397.7	108.9
LA17	27,734.0	28,259.8	**249.5**	110.7	28,258.8	28,710.7	265.3	111.4	28,103.6	28,570.6	277.9	64.4	**27,606.6**	**28,166.0**	377.8	110.1
LA18	**30,631.3**	**31,300.5**	291.1	106.2	31,147.3	31,530.1	**188.4**	108.3	31,192.0	31,465.4	243.4	63.0	30,826.6	31,480.9	525.1	107.5
LA19	30,938.3	**31,432.7**	**270.4**	108.1	30,951.9	31,608.5	343.4	109.3	**30,626.6**	31,684.9	491.6	63.3	31,167.9	**31,653.8**	352.9	108.5
LA20	32,959.8	33,481.9	379.0	108.3	33,121.5	33,816.8	501.0	108.0	33,213.6	33,863.7	**365.5**	61.6	**32,675.2**	**33,359.4**	384.3	104.1
LA21	46,010.4	46,775.4	473.7	199.4	46,475.3	47,252.7	665.3	196.7	**45,814.5**	**46,567.8**	707.6	115.4	46,139.7	46,595.8	**256.6**	190.7
LA22	41,008.2	42,285.9	721.5	191.3	41,528.4	42,418.6	**430.6**	197.2	**40,960.6**	**42,078.4**	640.5	116.3	41,378.6	42,185.1	556.5	193.9
LA23	44,957.1	45,687.4	619.0	191.6	45,303.6	46,203.4	602.4	203.2	45,527.4	46,142.1	722.9	116.1	**44,848.3**	45,655.5	626.4	195.5
LA24	42,921.1	44,018.8	442.6	191.0	43,142.3	44,258.3	614.3	185.4	42,587.5	43,819.4	775.2	108.0	**41,747.3**	**43,487.9**	869.3	191.3
LA25	**42,397.1**	**42,925.9**	332.0	184.0	42,805.1	43,693.6	470.3	186.2	42,602.1	43,175.7	**310.0**	106.4	43,328.0	43,686.9	364.9	185.7
LA26	59,010.9	59,783.1	426.7	289.8	60,087.4	60,912.2	569.9	299.8	**58,381.8**	**59,645.3**	745.5	171.7	58,956.5	60,361.9	792.4	292.0
LA27	60,258.6	62,040.3	1017.7	288.8	60,815.2	62,214.4	871.5	286.9	60,106.7	61,706.2	867.3	166.3	60,345.2	61,834.3	**701.7**	287.3
LA28	59,776.6	**60,419.2**	606.8	268.6	59,157.5	61,859.0	1448.4	292.5	**59,131.6**	60,600.6	777.6	162.5	59,361.1	60,933.3	**580.5**	282.0
LA29	57,139.0	58,076.0	**596.4**	278.2	**56,908.8**	58,345.2	1148.2	290.1	56,927.2	58,145.1	705.9	169.8	56,971.4	**58,028.9**	934.3	289.8
LA30	**60,689.9**	**61,396.5**	**696.2**	279.7	61,680.8	62,924.5	1226.9	280.1	60,856.4	62,305.2	1064.5	156.9	60,849.4	61,970.9	753.8	280.0
LA31	85,498.9	87,045.9	1149.1	532.6	85,708.0	87,871.1	1597.1	587.8	85,180.3	86,549.2	1422.7	320.8	**85,121.3**	**86,301.1**	**551.2**	588.0
LA32	90,282.2	93,019.1	**1049.8**	572.0	92,704.6	94,834.7	1647.9	572.5	89,949.9	**92,833.3**	1937.3	321.6	91,552.0	92,848.3	1208.2	570.5
LA33	**83,128.1**	84,780.9	1376.6	592.1	85,535.2	86,504.5	712.2	563.3	83,408.3	84,776.2	1063.2	314.6	83,406.5	84,898.0	**667.4**	557.6
LA34	85,278.7	87,079.6	**1172.8**	596.9	86,813.6	88,850.1	1395.4	590.0	**84,950.7**	**86,471.5**	1615.4	334.5	85,921.4	87,369.6	1379.5	578.7
LA35	**86,418.3**	88,466.7	1606.2	578.2	88,908.5	90,462.3	1168.2	582.1	87,212.1	87,976.7	**590.0**	327.6	87,327.1	89,208.7	1336.9	563.3

From the experimental data in Table 1, the following can be observed: (1) In comparisons of the *'Best'* value, the IWOA algorithm yields 18 optimal values, which is significantly better than the other three algorithms. The second best algorithm, namely IWOA-R, only obtains 10 optimal values. (2) In comparisons of the *'Avg'* value, the IWOA algorithm yields 16 optimal values, which is more than those of the other three algorithms. The second best algorithm, namely IWOA-R, can obtain 13 optimal values. (3) In comparisons of the *'SD'* value, IWOA-R performs better than other three algorithms. The second best algorithm, the proposed IWOA algorithm, yields 8 optimal values, which is more than those of IWOA-L and IWOA-NMO. (4) In comparisons of the *'Time'* value, IWOA-NMO spends a shorter time than other three algorithms. Compared with the IWOA-NMO, the increase of computation time is mainly the result of the addition of the mutation operation in IWOA.

5.3. Effectiveness of the Proposed IWOA

To demonstrate the effectiveness of the proposed IWOA algorithm, the proposed algorithm is compared with genetic algorithm 1 (GA1), genetic algorithm 2 (GA2), and the teaching-learning based optimization (TLBO) algorithm. For GA1, the initial population is generated by the proposed DR. The precedence preserving order-based crossover (POX) and the two-point crossover (TPX) are adopted as the crossover operators for the operation permutation and the speed selection, respectively. In addition, the swap mutation and one-point mutation are adopted for the operation permutation and the speed selection, respectively. For GA2, the algorithm in the work of [18] is used for solving the problem under study. For the TLBO, the algorithm in the work of [43] is modified with the addition of speed selection. Parameters are set as follows: In GA1 and GA2, the population size is 200, the maximum of iteration is 2000, the crossover rate is 0.8, and the mutation rate is 0.1. In the TLBO, the population size is 200 and the maximum of iteration is 2000. Ten independent replications are conducted for each instance.

From the experimental data in Table 2, the following can be easily observed: (1) In comparisons of the *'Best'* value, the proposed IWOA algorithm can obtain all the optimal values. (2) In comparisons of the *'Avg'* value, the proposed IWOA algorithm can also obtain all the optimal values. (3) In comparisons of the *'SD'* value, the IWOA algorithm yields 15 optimal values, which is better than GA1 and GA2. (4) In comparisons of the *'Time'* value, GA1 spends a shorter time than other two algorithms. The proposed IWOA spends more time because it contains the conversion process between the individual position vector and the scheduling solution. However, by comparison, the IWOA can yield the best values in an acceptable time.

Table 2. Comparison between different algorithms.

Instance	GA1				GA2				TLBO				IWOA			
	Best	Avg	SD	Time	Best	Avg	SD	Time	Best	Avg	SD	Time	Best	Avg	SD	Time
FT06	1528.0	1540.6	10.0	6.2	1550.6	1582.9	20.0	7.3	1639.3	1658.4	13.5	81.9	1370.4	1382.8	6.2	26.9
FT10	37,145.2	38,322.0	819.8	22.8	38,638.7	39,772.5	620.2	30.1	38,719.6	39,618.0	423.7	370.4	31,619.3	32,726.4	649.4	107.8
FT20	41,573.8	42,212.4	496.4	25.7	42,777.3	43,877.3	514.1	32.7	41,716.3	42,161.3	245.1	772.6	34,544.2	35,155.7	376.8	111.4
LA01	20,929.0	21,239.9	166.4	9.9	21,845.8	22,302.7	279.7	11.6	22,766.9	23,052.0	181.5	163.8	18,096.0	18,204.8	88.8	39.0
LA02	20,321.9	20,801.4	219.9	9.8	21,223.0	21,626.4	330.1	11.3	21,480.3	21,875.7	200.7	164.8	17,661.0	17,967.5	157.8	38.2
LA03	18,436.8	18,895.8	299.5	9.9	19,269.9	19,624.5	227.7	11.5	19,640.7	20,034.8	158.8	163.2	15,679.6	15,913.4	219.2	38.7
LA04	18,390.9	19,037.8	277.2	9.9	19,085.3	19,620.1	216.7	11.5	19,983.2	20,447.2	265.1	163.1	16,039.7	16,209.3	136.1	38.0
LA05	16,501.3	16,829.2	185.7	9.8	17,424.2	17,732.9	228.1	11.6	18,111.9	18,372.4	121.5	163.4	14,442.6	14,608.0	97.8	38.2
LA06	29,645.5	29,832.5	156.9	16.7	30,332.6	31,156.2	428.6	21.1	30,925.3	31,757.5	351.9	390.8	25,198.4	25,501.0	134.8	68.2
LA07	28,288.9	28,821.7	306.8	16.7	29,614.2	30,278.9	422.5	21.2	30,293.6	30,494.5	156.5	409.5	24,172.9	24,619.2	203.1	66.3
LA08	28,418.7	28,959.5	288.3	17.0	29,442.4	30,591.2	463.1	21.3	30,574.5	30,829.1	168.8	416.2	24,422.8	24,775.7	202.1	67.6
LA09	31,487.4	32,028.0	300.4	16.9	32,652.8	33,383.0	476.8	21.3	33,563.3	34,137.6	391.2	412.3	27,329.5	27,608.6	122.5	73.5
LA10	29,021.5	29,703.9	356.9	16.8	30,753.2	31,272.7	276.3	21.2	32,018.4	32,311.7	201.3	416.1	25,081.3	25,315.5	178.9	73.2
LA11	39,735.1	40,732.2	563.2	25.6	40,691.1	42,380.5	694.1	29.9	42,098.2	42,568.8	250.3	831.8	34,221.7	34,659.8	258.5	101.1
LA12	34,360.3	35,109.6	567.1	25.2	36,116.2	36,705.9	364.6	30.2	36,685.3	37,024.0	210.6	767.1	29,905.6	30,221.5	173.9	111.6
LA13	37,641.6	38,759.5	902.9	25.5	40,046.0	41,036.5	377.0	30.8	40,866.4	41,199.9	168.7	775.3	33,198.1	33,490.7	216.0	110.2
LA14	40,079.5	40,657.0	407.2	25.5	42,035.5	42,668.4	307.0	30.3	42,644.1	43,063.7	235.5	804.6	33,794.5	34,214.3	242.8	114.0
LA15	42,138.3	42,630.5	355.8	25.6	44,265.0	44,865.7	406.6	30.6	43,729.4	44,087.5	250.3	800.1	35,802.4	36,199.8	203.0	113.1
LA16	37,245.6	38,160.1	775.0	23.2	39,352.3	40,191.7	505.5	28.0	39,637.1	40,341.5	366.2	406.5	31,516.7	32,118.2	397.7	108.9
LA17	32,693.6	33,180.5	450.0	23.0	34,319.1	35,011.7	475.3	27.9	34,707.3	35,280.3	375.9	400.3	27,606.6	28,166.0	377.8	110.1
LA18	35,469.3	36,751.3	648.4	23.2	37,326.8	38,536.1	611.9	27.9	38,204.9	39,038.9	368.0	392.1	30,826.6	31,480.9	525.1	107.5
LA19	36,099.4	37,003.0	649.9	23.1	38,359.7	39,177.2	405.7	27.7	38,186.7	38,962.7	391.6	404.3	31,167.9	31,653.8	352.9	108.5
LA20	38,179.1	39,348.4	546.1	22.8	40,626.2	41,834.9	596.4	28.1	41,519.6	41,982.8	278.8	400.4	32,675.2	33,359.4	384.3	104.1
LA21	54,607.7	55,786.4	641.3	40.5	57,276.3	58,907.5	656.6	52.5	56,961.4	57,707.3	389.4	949.1	46,139.7	46,595.8	256.6	190.7
LA22	49,559.5	51,284.0	755.0	40.6	51,457.2	53,373.3	924.2	50.9	52,280.6	52,541.4	183.0	893.2	41,378.6	42,185.1	556.5	193.9
LA23	53,917.7	54,612.9	561.0	40.6	56,499.0	57,683.4	682.8	52.0	56,709.3	57,164.8	259.8	955.8	44,848.3	45,655.5	626.4	195.5
LA24	51,056.0	52,393.9	826.4	40.7	54,503.0	55,457.0	716.7	53.2	53,032.8	54,086.9	559.4	936.2	41,747.3	43,487.9	869.3	191.3
LA25	51,583.8	52,236.1	389.8	40.5	54,128.9	51,823.4	496.6	51.8	52,744.1	53,644.7	485.7	940.4	43,328.0	43,686.9	364.9	185.7
LA26	70,951.3	72,097.3	895.9	60.7	75,180.8	75,817.0	547.6	80.9	73,634.3	73,967.6	288.3	1807.0	58,956.5	60,361.9	792.4	292.0
LA27	72,413.8	74,974.8	1004.8	61.3	76,673.0	78,085.6	807.9	86.4	75,245.7	75,683.7	322.5	1845.4	60,345.2	61,834.3	701.7	287.3
LA28	71,365.7	72,774.3	782.7	61.3	74,918.6	76,472.9	834.1	85.3	73,722.1	74,669.0	608.7	1780.7	59,361.1	60,933.3	580.5	282.0
LA29	67,354.4	69,077.6	1235.2	60.9	70,744.8	72,150.4	1041.7	84.8	69,980.3	70,706.9	482.7	1777.3	56,971.4	58,028.9	934.3	289.8
LA30	72,707.5	73,966.6	589.3	59.8	76,737.0	77,547.9	659.0	85.1	74,160.6	75,329.5	711.6	1792.8	60,849.4	61,970.9	753.8	280.0
LA31	99,556.7	104,319.6	2041.1	113.8	106,691.4	108,275.1	979.2	168.9	113,707.5	114,566.3	703.8	4807.2	85,121.3	86,301.9	551.2	588.0
LA32	110,241.3	112,299.6	1208.0	113.8	115,254.6	117,103.4	859.6	165.9	113,250.8	113,450.3	220.1	4958.3	91,552.0	92,848.3	1208.2	570.5
LA33	100,960.7	102,467.9	1171.0	111.3	104,826.8	106,660.0	957.1	172.8	102,476.4	103,051.6	428.6	5273.3	83,406.5	84,898.0	667.4	557.6
LA34	102,234.6	104,521.0	1902.4	111.2	106,811.5	108,573.6	734.5	172.7	105,440.0	105,768.8	275.5	5064.8	85,921.4	87,369.6	1379.5	578.7
LA35	104,497.8	105,599.2	1097.5	112.0	107,670.7	110,468.2	1210.6	190.6	106,133.5	106,803.1	720.4	4814.7	87,327.1	89,208.7	1336.9	563.3

6. Conclusions

In this study, an improved whale optimization algorithm (IWOA) is proposed to solve an energy-efficient job shop scheduling problem. The conversion method between the scheduling solution and the individual position vector is first designed. After that, three improvement strategies are adopted in the algorithm, namely, the dispatching rules (DR), the nonlinear convergence factor (NCF), and the mutation operation (MO). The DR is adopted to generate the initial solutions. The NCF is used to coordinate the capacity of exploration and exploitation. The MO is employed to avoid the premature convergence.

Extensive experiments are conducted for testing the performance of the IWOA algorithm. From the experimental results, it can be seen that the proposed improvement strategies can improve the computational result of the algorithm. In addition, compared with GA1, GA2, and TLBO, the proposed IWOA algorithm can obtain better results in an acceptable time.

In future work, the energy-efficient scheduling will be further studied by considering some more practical constraints, for example, flexible processing routing, time-of-use electricity policy, and renewable energy, among others. In addition, the energy-efficient scheduling problem will be extended to some complex workshop, such as flexible job shop and assembly job shop, among others.

Author Contributions: T.J. and C.Z. developed the improved whale optimization algorithm. H.Z. and J.G. designed the experiments and tested the effectiveness of the IWOA. G.D. provided theoretical knowledge and refined the paper.

Funding: This research was funded by the Training Foundation of Shandong Natural Science Foundation of China under Grant ZR2016GP02, the National Natural Science Foundation Project of China under Grant 61403180, the Special Research and Promotion Program of Henan Province under Grant 182102210257, the Project of Henan Province Higher Educational Key Research Program under Grant 16A120011, the Project of Shandong Province Higher Educational Science and Technology Program under Grant J17KA199, and the Talent Introduction Research Program of Ludong University under Grant 32860301.

Conflicts of Interest: The authors declare no conflict of interest.

References

1. Duflou, J.R.; Sutherland, J.W.; Dornfeld, D.; Herrmann, C.; Jeswiet, J.; Kara, S.; Hauschild, M.; Kellens, K. Towards energy and resource efficient manufacturing: A processes and systems approach. *CIRP Ann.- Manuf. Technol.* **2012**, *61*, 587–609. [CrossRef]
2. Fang, K.; Uhan, N.; Zhao, F.; Sutherland, J.W. *A New Shop Scheduling Approach in Support of Sustainable Manufacturing*; Glocalized solutions for sustainability in manufacturing; Springer: Berlin/Heidelberg, Germany, 2011; pp. 305–310.
3. Liu, Y.; Dong, H.; Lohse, N.; Petrovic, S.; Gindy, V. An investigation into minimising total energy consumption and total weighted tardiness in job shops. *J. Clean. Prod.* **2014**, *65*, 87–96. [CrossRef]
4. Mouzon, G.; Yildirim, M.B.; Twomey, J. Operational methods for minimization of energy consumption of manufacturing equipment. *Int. J. Prod. Res.* **2007**, *45*, 4247–4271. [CrossRef]
5. Mouzon, G.; Yildirim, M.B. A framework to minimise total energy consumption and total tardiness on a single machine. *Int. J. Sustain. Eng.* **2008**, *1*, 105–116. [CrossRef]
6. Yildirim, M.B.; Mouzon, G. Single-machine sustainable production planning to minimize total energy consumption and total completion time using a multiple objective genetic algorithm. *IEEE. Trans. Eng. Manag.* **2012**, *59*, 585–597. [CrossRef]
7. Shrouf, F.; Ordieres-Meré, J.; García-Sánchez, A.; Ortega-Mier, M. Optimizing the production scheduling of a single machine to minimize total energy consumption costs. *J. Clean. Prod.* **2014**, *67*, 197–207. [CrossRef]
8. Che, A.; Zeng, Y.; Lyu, K. An efficient greedy insertion heuristic for energy-conscious single machine scheduling problem under time-of-use electricity tariffs. *J. Clean. Prod.* **2016**, *129*, 565–577. [CrossRef]
9. Li, J.Q.; Sang, H.Y.; Han, Y.Y.; Wang, C.G.; Gao, K.Z. Efficient multi-objective optimization algorithm for hybrid flow shop scheduling problems with setup energy consumptions. *J. Clean. Prod.* **2018**, *181*, 584–598. [CrossRef]

10. Liu, G.S.; Zhang, B.X.; Yang, H.D.; Chen, X.; Huang, G.Q. A branch-and-bound algorithm for minimizing the energy consumption in the PFS problem. *Math. Probl. Eng.* **2013**, *2013*, 546810. [CrossRef]

11. Dai, M.; Tang, D.; Giret, A.; Salido, M.A.; Li, W.D. Energy-efficient scheduling for a flexible flow shop using an improved genetic-simulated annealing algorithm. *Robot. Comput.-Int. Manuf.* **2013**, *29*, 418–429. [CrossRef]

12. Ding, J.Y.; Song, S.; Wu, C. Carbon-efficient scheduling of flow shops by multi-objective optimization. *Eur. J. Oper. Res.* **2016**, *248*, 758–771. [CrossRef]

13. Mansouri, S.A.; Aktas, E.; Besikci, U. Green scheduling of a two-machine flowshop: Trade-off between makespan and energy consumption. *Eur. J. Oper. Res.* **2016**, *248*, 772–788. [CrossRef]

14. Luo, H.; Du, B.; Huang, G.Q.; Chen, H.; Li, X. Hybrid flow shop scheduling considering machine electricity consumption cost. *Int. J. Prod. Econ.* **2013**, *146*, 423–439. [CrossRef]

15. Salido, M.A.; Escamilla, J.; Giret, A.; Barber, F. A genetic algorithm for energy-efficiency in job-shop scheduling. *Int. J. Adv. Manuf. Technol.* **2016**, *85*, 1303–1314. [CrossRef]

16. Zhang, R.; Chiong, R. Solving the energy-efficient job shop scheduling problem: A multi-objective genetic algorithm with enhanced local search for minimizing the total weighted tardiness and total energy consumption. *J. Clean. Prod.* **2016**, *112*, 3361–3375. [CrossRef]

17. Tang, D.; Dai, M. Energy-efficient approach to minimizing the energy consumption in an extended job-shop scheduling problem. *Chin. J. Mech. Eng.* **2015**, *28*, 1048–1055. [CrossRef]

18. Escamilla, J.; Salido, M.A.; Giret, A.; Barber, F. A metaheuristic technique for energy-efficiency in job-shop scheduling. *Knowl. Eng. Rev.* **2016**, *31*, 475–485. [CrossRef]

19. Yin, L.; Li, X.; Gao, L.; Lu, C.; Zhang, Z. Energy-efficient job shop scheduling problem with variable spindle speed using a novel multi-objective algorithm. *Adv. Mech. Eng.* **2017**, *9*. [CrossRef]

20. Wang, G.G.; Tan, Y. Improving metaheuristic algorithms with information feedback models. *IEEE. Trans. Cybern.* **2017**. [CrossRef] [PubMed]

21. Wang, G.G.; Gandomi, A.H.; Alavi, A.H. An effective krill herd algorithm with migration operator in biogeography-based optimization. *Appl. Math. Model.* **2014**, *38*, 2454–2462. [CrossRef]

22. Wang, G.G.; Gandomi, A.H.; Alavi, A.H. Stud krill herd algorithm. *Neurocomputing* **2014**, *128*, 363–370. [CrossRef]

23. Wang, G.G.; Guo, L.; Gandomi, A.H.; Hao, G.S.; Wang, H. Chaotic krill herd algorithm. *Inform. Sci.* **2014**, *274*, 17–34. [CrossRef]

24. Wang, G.; Guo, L.; Wang, H.; Duan, H.; Liu, L.; Li, J. Incorporating mutation scheme into krill herd algorithm for global numerical optimization. *Neural Comput. Appl.* **2014**, *24*, 853–871. [CrossRef]

25. Feng, Y.; Wang, G.G.; Dong, J.; Wang, L. Opposition-based learning monarch butterfly optimization with Gaussian perturbation for large-scale 0-1 knapsack problem. *Comput. Electr. Eng.* **2018**, *67*, 454–468. [CrossRef]

26. Rizk-Allah, R.M.; El-Sehiemy, R.A.; Wang, G.G. A novel parallel hurricane optimization algorithm for secure emission/economic load dispatch solution. *Appl. Soft. Comput.* **2018**, *63*, 206–222. [CrossRef]

27. Jiang, T.H.; Zhang, C. Application of Grey Wolf Optimization for solving combinatorial problems: Job shop and flexible job shop scheduling cases. *IEEE. Access.* **2018**, *6*, 26231–26240. [CrossRef]

28. Jiang, T.H.; Deng, G.L. Optimizing the low-carbon flexible job shop scheduling problem considering energy consumption. *IEEE. Access.* **2018**, *6*, 46346–46355. [CrossRef]

29. Han, Y.Y.; Gong, D.W.; Jin, Y.C.; Pan, Q.K. Evolutionary Multi-objective Blocking Lot-streaming Flow Shop Scheduling with Machine Breakdowns. *IEEE. Trans. Cybern.* **2018**. [CrossRef]

30. Han, Y.Y.; Liang, J.; Pan, Q.K.; Li, J.Q. Effective hybrid discrete artificial bee colony algorithms for the total flow time minimization in the blocking flow shop problem. *Int. J. Adv. Manuf. Technol.* **2013**, *67*, 397–414. [CrossRef]

31. Han, Y.Y.; Pan, Q.K.; Li, J.Q.; Sang, H.Y. An improved artificial bee colony algorithm for the blocking flow shop scheduling problem. *Int. J. Adv. Manuf. Technol.* **2012**, *60*, 1149–1159. [CrossRef]

32. Li, J.Q.; Duan, P.Y.; Sang, H.Y.; Wang, S.; Liu, Z.M.; Duan, P. An efficient optimization algorithm for resource-constrained steelmaking scheduling problems. *IEEE. Access.* **2018**, *6*, 33883–33894. [CrossRef]

33. Mirjalili, S.; Lewis, A. The whale optimization algorithm. *Adv. Eng. Softw.* **2016**, *95*, 51–67. [CrossRef]

34. Ling, Y.; Zhou, Y.; Luo, Q. Lévy flight trajectory-based whale optimization algorithm for global optimization. *IEEE. Access.* **2017**, *5*, 6168–6186. [CrossRef]

35. Mafarja, M.; Mirjalili, S. Whale optimization approaches for wrapper feature selection. *Appl. Soft. Comput.* **2018**, *62*, 441–453. [CrossRef]

36. El Aziz, M.A.; Ewees, A.A.; Hassanien, A.E. Multi-objective whale optimization algorithm for content-based image retrieval. *Multimed. Tools Appl.* **2018**, *77*, 26135–26172. [CrossRef]

37. Abdel-Basset, M.; El-Shahat, D.; Sangaiah, A.K. A modified nature inspired meta-heuristic whale optimization algorithm for solving 0–1 knapsack problem. *Int. J. Mach. Learn. Cybern.* **2017**, 1–20. [CrossRef]

38. Abdel-Basset, M.; Manogaran, G.; El-Shahat, D.; Mirjalili, S. A hybrid whale optimization algorithm based on local search strategy for the permutation flow shop scheduling problem. *Future Gener. Comput. Syst.* **2018**, *85*, 129–145. [CrossRef]

39. Jiang, T.H. Flexible job shop scheduling problem with hybrid grey wolf optimization algorithm. *Control Decis.* **2018**, *33*, 503–508. (In Chinese)

40. Yuan, Y.; Xu, H. Flexible job shop scheduling using hybrid differential evolution algorithms. *Comput. Ind. Eng.* **2013**, *65*, 246–260. [CrossRef]

41. Fisher, H.; Thompson, G.L. Probabilistic learning combinations of local job-shop scheduling rules. *Ind. Sched.* **1963**, *3*, 225–251.

42. Lawrence, S. *Resource Constrained Project Scheduling: An Experimental Investigation of Heuristic Scheduling Techniques*; Graduate School of Industrial Administration (GSIA), Carnegie Mellon University: Pittsburgh, PA, USA, 1984.

43. Baykasoglu, A.; Hamzadayi, A.; Kose, S.Y. Testing the performance of teaching-learning based optimization (TLBO) algorithm on combinatorial problems: Flow shop and job shop scheduling cases. *Inform. Sci.* **2014**, *276*, 204–218. [CrossRef]

mathematics

MDPI

Article

Urban-Tissue Optimization through Evolutionary Computation

Diego Navarro-Mateu [1], Mohammed Makki [2] and Ana Cocho-Bermejo [1,*]

[1] UIC Barcelona School of Architecture, Universitat Internacional de Catalunya, c/ Immaculada 22, 08017 Barcelona, Spain; navarro@uic.es

[2] Architectural Association, 36 Bedford Square, Bloomsbury, London WC1B 3ES, UK; mohammed.makki@aaschool.ac.uk

* Correspondence: acocho@uic.es

Received: 1 August 2018; Accepted: 29 September 2018; Published: 2 October 2018

Abstract: The experiments analyzed in this paper focus their research on the use of Evolutionary Computation (EC) applied to a parametrized urban tissue. Through the application of EC, it is possible to develop a design under a single model that addresses multiple conflicting objectives. The experiments presented are based on Cerdà's master plan in Barcelona, specifically on the iconic Eixample block which is grouped into a 4 × 4 urban Superblock. The proposal aims to reach the existing high density of the city while reclaiming the block relations proposed by Cerdà's original plan. Generating and ranking multiple individuals in a population through several generations ensures a flexible solution rather than a single "optimal" one. Final results in the Pareto front show a successful and diverse set of solutions that approximate Cerdà's and the existing Barcelona's Eixample states. Further analysis proposes different methodologies and considerations to choose appropriate individuals within the front depending on design requirements.

Keywords: evolution; computation; urban design; biology; shape grammar; architecture; SPEA 2

1. Introduction

1.1. Relevance of the Interdisciplinary Experiment

Jane Jacobs set the grounds of cities as problems of organized complexity in 1961 [1]. Until then, academics had defended the idea that any urban-planning problem could be perfectly described with a clear definition of all of its variables, classifying it as a problem of disorganized complexity or even as a problem of simplicity. Just recently, the tools for studying cities as the complex systems Jacobs described have become available for experts within the architectural discipline.

Along his exploration of complexity and the science of design, Simon Herbert [2] in "The sciences of the artificial" defended the science of the artificial as the science of engineering, but not engineering-science; understanding the complex system of the city is as an Artifact acting as an interface between the inner and the outer environment. The science of design is then the science of creating the artificial. Understanding that, with just minimal assumptions about the inner environment, we can predict behavior from knowledge of the system goals and its outer environment. So understanding cities as artificial and adaptable systems that have certain rules makes them particularly susceptible to simulation via simplified models.

Apart from the quality of the data then, as he stated, adaptation to the environment can be improved by combining predictive control and homeostatic and feedback methods.

Precedence for the application of an evolutionary model as a problem-solving strategy dates back to the early 20th century. It has since been developed into a model that has been applied in a multitude of different fields to provide solutions to problems that required objectivity, optimality, and efficiency.

1.2. Significance of Variation

Variation of blocks and Superblocks increases the potential for the urban fabric in which they are embedded to adapt to changes in environmental and climatic conditions. It also helps to construct patterns of spatial differentiation that are identified with the perception of urban culture and qualities that make a city a good place to live. The Universal city, beloved in the early 20th century by Modernists, has been built everywhere, and is all too frequently simply comprised of a uniform array of a single-block type distributed across a grid, with little, if any, adjustment to specific ecological or environmental contexts. Their attempts to generate substance and quality within the urban landscape through copious amounts of noncontextualized repetition have proven to be unsuccessful.

The attempt of predicting how a city will grow, either morphologically or temporally, may have been the modernist's biggest challenge. Although it may be possible to make short-term predictions, driven by rules inherent to strategies of urban design, political influences, economic patterns and social impacts; long-term predictions are the ones that are usually impossible to make [3] (pp. 109–127).

Today, rapidly changing climatic conditions and the exponential growth and mobility of populations are accelerating changes to the environmental context of many cities across the world. The stresses on future cities demand an approach that enables the urban fabric to accommodate rapid change, allowing territories for the freedom to communicate and overlap with one another in response to internal and external stimuli within the city's environment [4].

Moreover, the successive random and subjective choices made by each inhabitant of the city amplifies the city's unpredictability, imposing a shift in mindset from understanding a problem to having a single solution, to one that requires multiple solutions, each unique in its own way.

In an urban context, this variation is explained as a "formal diversity of solutions responding to the same situations" [3] (p. 112), and although the system cannot be predicted and designed in advance, it can be addressed through the application of multiple simulations, each generating a population of solutions, thus bypassing the demand for prediction (which is usually associated with generating a single solution).

This brings forward the need to clearly differentiate between 'the solution' and 'the population'. This is best described within biology, where there is a clear boundary between the 'typologist' and the 'populationist'. Leading evolutionary biologist, Ernst Mayr, highlights their distinction in his essay, Typological versus Population Thinking, where he states, "For the typologist, the type (tidos) is real and the variation an illusion, while for the populationist, the type (average) is an abstraction and only the variation is real" [5] (p. 28).

The populationist believes that each solution is unique, and by attempting to define a collection of unique solutions through a single representative they lose the individual characteristics that defined each solution within the population. In doing so, an assumption is made: the 'statistical average' solution is the best suited to adapt to the stresses of its environment.

However, nature contradicts this, as individuals within a species show significant variation and display unique traits that have evolved differently in response to the same environmental stresses.

As such, the populationist's approach of signifying importance to variation between solutions rather than an average representative serves as an optimal model for generating variation of design solutions to a design problem that cannot be addressed through a single 'average' design solution; as Mayr states, "An individual that will show in all of its characters the precise mean value for the population as a whole does not exist" [5] (p. 29).

1.3. Challenge and Hypothesis

The challenge of this research lies in developing a computational process that is capable of generating adequate variation of urban morphology that is optimal for multiple conflicting objectives. One widely used approach to multiobjective computation is for the designer to give greater weight to one objective over the others, or to vary the reactive importance of the objectives in a cascading rank; however, this makes the process deterministic on the initial conditions and decisions of ranking.

1.4. Barcelona's Urban Model

1.4.1. Urban Growth

In 1859, Ildefons Cerdà proposed 'L'Eixample', an urban solution aimed towards accommodating Barcelona's growing population through extending the city's urban fabric beyond its walls.

The distribution of functions within the urban plan would later be the primary cause in transforming Barcelona into one of the highest population-density cities of Europe [6]. Through his new plan, Cerdà aimed to address issues of population growth, building density, unsanitary conditions, illnesses and high mortality rates that were impacting the city's development during the 19th century. As such, Cerdà engaged three primary domains: sanitation, circulation, and social equality. For these reasons, the original project only built on two sides of each block, thus enabling greater views and better ventilation (Figure 1).

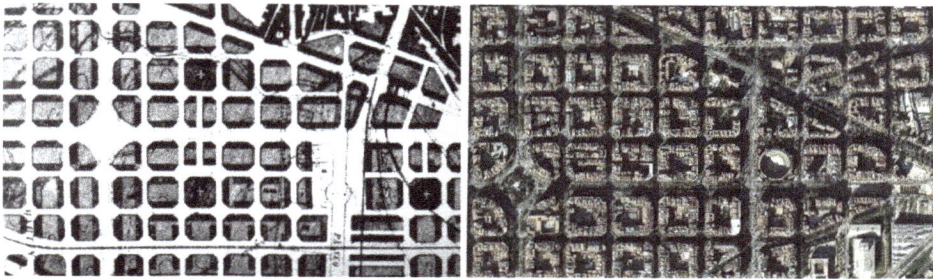

Figure 1. Comparison. (**Left**) Fragment of Cerdà's original plan: block types and orientations [7]. (**Right**) Current state of Barcelona, where all sides and inside parts are built.

1.4.2. Existing Urban Setting

Cerdà's initial plans attempted to provide a solution to a problem with multiple conflicting criteria. The primary conflicting criteria were the requirement of accommodating a high-density ratio yet maintaining a high number of street-accessible green spaces. However, rather than generating a solution that accommodated both criteria, a trade-off strategy directed the city towards a solution that prioritized population density over green spaces.

Although unknown at the time, Barcelona was following a preference-based approach that found it necessary to "convert the task of finding multiple trade-off solutions in a multiobjective optimization (problem) to one of finding a single solution of a transformed single-objective optimization problem" [8] (p. 7).

Although Cerda's original plan engaged a balanced relationship between open space and liveable space, several changes to the plan were imposed after Cerda's proposal. Moreover, political and investment opportunities transformed the original two-sided block with an open courtyard into a four-sided chamfered block with an enclosed courtyard, thus giving rise to the iconic Barcelona's eight-sided block. In doing so, Cerdà's intention to maintain a high percentage of open spaces as well as visual connectivity throughout the city was disregarded by the decision to completely modify the original two-sided block [9] (Figure 2). As such, the green area/inhabitant ratio of Barcelona is currently recorded as 6.5 m^2 per person, which is more than half the ratio recommended by the World Health Organization, W.H.O. [10].

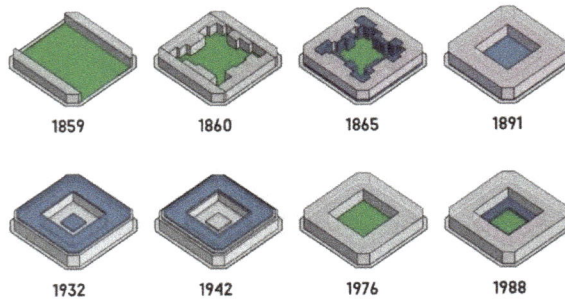

Figure 2. Development of a typical block in the Eixample throughout history.

Modifications to Cerdà's original plan have been in a continuous state of development (Figure 2). Most prominently, Barcelona's driving change factors are: the geographic limitations (Cornella mountains, Besós river, Llobregat river, and the maritime front), the hierarchical and relational changes in specific areas such as Barcelona's future center (Les Glories), and the rethinking of L'Eixample—which is engaged in this article.

Approved by the Barcelona Town Hall in 2012, the Superblock project ('super illes') (Figure 3) aims to introduce improved and sustainable mobility, public-space rehabilitation, biodiversity and green areas, accessibility, social cohesion, and energetic self-sufficiency within the city's fabric. The Superblock becomes an intermediate unit (smaller than a neighborhood yet larger than block) to allow for the development of new relationships between blocks and streets [11].

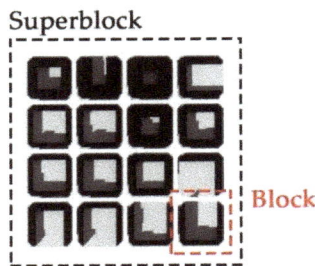

Figure 3. A superblock is composed of 16 blocks arranged in a square grid (4 × 4).

Nonetheless, the existing density of Barcelona constrains attempts to revert the city closer to Cerda's original proposal, so only minor changes have been addressed.

Thus, rather than attempting to restructure the existing city, the experiments carried out in the following chapters apply an evolutionary-design strategy that aims to generate an urban patch that incorporates Cerdàs's original design objectives while taking into account Barcelona's current population density.

The use of evolutionary population-based solvers empowers the possibility to modify, evaluate, and select a set of candidate solutions per iteration, rather than a single optimal solution. Such a process allows all objectives to be considered without the requisite of employing a trade-off strategy during simulation. More importantly, it allows for the emergence of morphological variation of different solutions, each suitable for a specific function, thus moving away from the homogeneity of 20th-century urban-planning strategies towards a more bottom–up approach of urban form.

2. Materials and Methods

2.1. Evolutionary Strategy

It is concluded that it is an appropriate approach to implement an evolutionary algorithm (EA) for our particular case of study. Evolutionary computation, based on genes and chromosomes containing the code for nature's designs, uses solution populations competing/co-operating to improve over time through interactions with the environment.

In a comparison between EA vs. derivative-based methods we can clearly state:

- EAs can be much slower (but they are any-time algorithms).
- EAs are less dependent on initial conditions (still need several runs).
- EAs can use alternative error functions: not continuous or differentiable, including structural terms.
- EAs are not easily stuck in local optima.
- EAs are better "scouters" (global searchers).

The experiments in Section 4 employ an evolutionary solver as the underlying driver for the design process. However, the methods by which different evolutionary strategies apply the principles of selection and variation are notably diverse in different evolutionary algorithms.

The most progressive multiobjective evolutionary algorithms (e.g., NSGA-2, Strength Pareto Evolutionary Algorithm 2 (SPEA-2)) excelled through their ability to achieve the most diverse Pareto optimal set in both an efficient timeframe, as well as a reasonable computational environment [12]. As such, the selected algorithm in which the presented experiments were run is the SPEA-2 [13] within the Octopus software, an evolutionary solver plugin for the design modeling platform Rhino 3D.

The EA is embedded in a 3D parametric model of the architectural topology capable of reproducing all variables related to Cerdà's original plan strategy and Eixample's current situation (Figure 4).

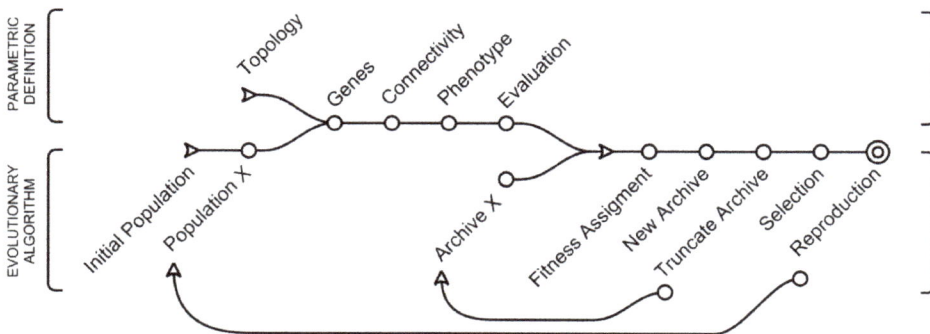

Figure 4. Workflow diagram explaining the combination between parametric modeling and evolutionary algorithm (EA).

2.2. Experimental Setup

By Definition, the geometric, environmental, and social relationships constitute the basis of urban design. These relationships become a complex system with a behavior and efficiency that are difficult to evaluate and predict. For this reason, being able to establish geometric relationships within urban patterns in 3D modeling software allows us to manipulate geometry's mathematical definition. Once the definition is established, analyses and manipulations of geometrical variables together with social and environmental ones can be developed through a range of existing plugins. These plugins allow researchers from the architectural and engineering fields to manipulate such a complex mathematical set of relationships.

In this particular scenario, the experiment was run through the use of multiple plugins within the 3D-NURBS modeling software Rhinoceros3D. Grasshopper3D (visual algorithmic modeling) serves as the primary platform for Octopus (multiobjective EA, developed by Robert Vierlinger and Bollinger + Grohmann Engineers) and Wallacei (analytic engine for data outputted by the EA, developed by Mohammed Makki and Milad Showkatbakhsh).

Considering the multiple objectives and goals originally aimed by Cerdà, the experiment sets out to generate an urban patch that optimizes for four primary objectives:

- CY—Larger courtyards for open public spaces (number of mesh faces exposed).
- B—High solar exposure on the building façades (number of mesh faces exposed).
- C—Greater block connectivity (numerical value based on Figure 5).
- DE—High population density (one that is close to the current state) (hab/km^2).

All objectives must be maximized, understanding that greater amount of light and open spaces are always positive characteristics in an overpopulated city.

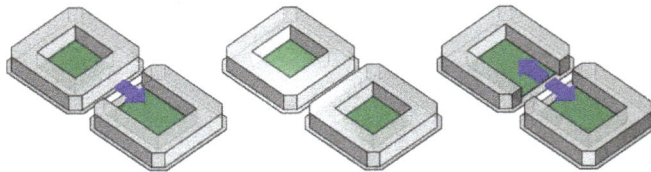

Figure 5. Connectivity possibilities between the blocks. In order (value): block to opening (0), block to block (1), and opening to opening (2). Courtyard connectivity is ranked to encourage larger courtyards between blocks and generate wide fields of view. A low ranking discourages blocks that have courtyards with one-sided access.

To address the connectivity objective, the phenotype is made up of several blocks. Therefore, individuals/chromosomes from the experiment are Superblocks composed from 16 blocks (a 4 × 4 grid) (Figure 3). Each of the blocks inside of the phenotype is governed by a gene pool of variables that transform the block's morphology. The variables are:

- D—Block Depth (0.3%–0.7%) of the block side.
- Sd—Subdivisions (2–6 parts/side).
- O—Two-sided block's organization (parallel vs. corner).
- A—Block orientation (0°, 90°, 180°, 270°).
- Fa—Deletion (amount of blocks' façades deleted: 0, 1, 2, 3, 4).
- Fn—Minimum and maximum floors (2–6).
- Fex—Minimum and maximum extra floors (2–6).

In the context of the developed experiment, the possible values for the variables are:

$$D \in (0.3, 0.4, 0.5, 0.6, 0.7)$$

$$Sd \in (2, 3, 4, 5, 6)$$

$$O \in (p, c)$$

$$A \in (0, 90, 180, 270)$$

$$Fa \in (0, 1, 2, 3, 4)$$

$$Fn \in (2, 3, 4, 5, 6)$$

$$Fex \in (2, 3, 4, 5, 6)$$

Genes Fn and Fex were programmed to generate random numbers inside the definition of every individual.

Therefore, the design space is defined by the bounds of the genes' ranges. The number of k-element variations (V) of n-elements with repetition allowed, is:

$$Vn,k = n^k \tag{1}$$

Based on the described genes, the number of possible variations for the variables within Superblocks and blocks is:

$$\text{Superblocks variations: } (5^{16})(5^{16})(2^{16})(4^{16})(5^{16})(5^{16})(5^{16}) = 2.32 \cdot 10^{70}$$

$$\text{Blocks variations: } (5)(5)(2)(4)(5)(5)(5)(5) = 25000$$

Because of the multiobjective nature of the software, objectives are not merged as a single objective. The population-based evolutionary solver addresses every individual independently for each of the fitness criteria. Therefore, the optimal solutions within the Pareto front that achieve a high fitness value regarding one criterion might also be significantly low in another criterion, resulting in "multiple optimal solutions in its final population" [8] (p. 8).

The Pareto front as a result of a 4-objective (as mentioned before CY, B, C, DE) optimization problem that is also a 4-dimensional geometry. Figure 6 shows the Pareto front represented by a 3 spatial axis (DE, CY, C), while the 4th (B) is shown as a gradient color.

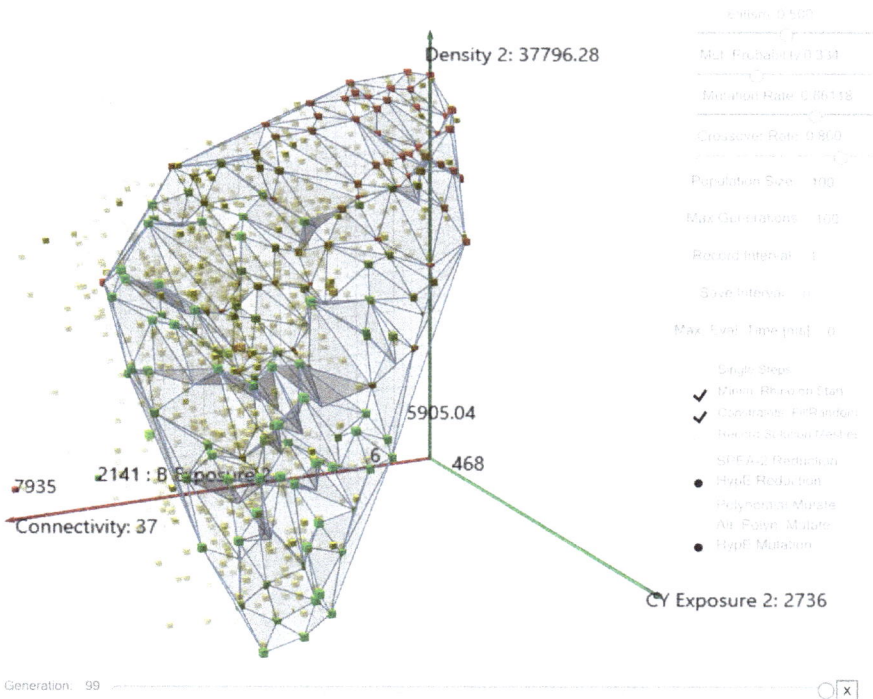

Figure 6. Octopus plugin screenshot. Since 4 fitness criteria have been set, the Pareto front is also composed of 4 dimensions. Therefore, a color gradient is represented on top of a 3-dimensional mesh to represent the 4th criteria (from green to red).

The simulation settings should balance a search and optimization strategy that is both explorative and employs an efficient selection and variation strategy that directs the algorithm towards an optimal solution set within a feasible number of generations [7]; as such, the following settings were employed:

- Generation size: 100.
- Generation count: 100 (2 + 98).
- Selection method: Elitism 50% (method fixed by the plugin, percentage set by the researcher).
- Mutation Probability: 33% (initial value 10%).
- Mutation Rate: 66% (initial value 50%).
- Crossover Rate: 80% (default 1-point crossover).

The fitness function is defined by geometrical analysis of the resulting phenotypes. Such a relationship would be complex to evaluate through a pure mathematical model. Therefore, it is necessary to include and evaluate the geometrical properties of the phenotypes.

- CY—Larger courtyards: courtyard is converted into a mesh with 4425 faces. Each mesh has a vector attached related to a virtual sun that will validate intersecting operations with the block itself.
- B—High solar exposure: calculated with vectors in a subdivided mesh to check self-shadowing or shadows from neighbor buildings. Number of faces in mesh depends on the phenotype.
- C—Greater block connectivity: A network of lines is drawn through proximity operations. The definition checks intersections with this network to establish its relationship with neighboring buildings.
- DE—Density objective: based on density in Barcelona's current Eixample, considering number of floors and total area built by the phenotype.

Solver parameters have been modified in comparison to previous experiments [14]. Based on early attempts, the following measures were taken (Figure 6):

1. Because of the low generation size (100 individuals), the probability and strength of mutations have been increased to 0.33 and 0.66, respectively. Although slower in the process, mutations should compensate for a low initial population, producing results outside of the original genes.
2. With the same purpose, the amount of genes has been reduced, deleting those that had little or no effect on the overall shape of the block. The simplification of the phenotype helps to lighten the computational load and reduces the amount of permutations.

3. Experiment Results

Unlike other single-objective design experiments, the multiobjective evolutionary solver tends to produce significant geometric variety. Because of the conflicting objectives, rather radical individuals can be spotted in the Pareto front. The variety of phenotypes throughout the simulation reflected an appropriate balance of exploration vs. exploitation within the algorithm, thus reducing the risk for the premature convergence of the population towards a local optimum (Figure 7).

Analysis through the Wallacei plugin demonstrates a successful evolutionary run through presenting increased diversity within the population accompanied with increased fitness levels. Figure 8 depicts both higher variance levels for the last generations and increasing trend lines in the standard deviation for each generation. Standard Deviation Value (Equation (2)) has been calculated for each generation (x is the solution's fitness value and μ is the generation's mean fitness value).

$$\sigma = \sqrt{\frac{1}{N} \sum_{i=1}^{n} (x_i - \mu)^2} \tag{2}$$

Consequently, the Normal Distribution (Equation (3)) curve for each generation has been plotted through the following calculation (to three standard Deviations): (x is the solution's fitness value, μ is the generation's mean fitness value, and σ is the standard deviation value.)

$$f(x) = \frac{1}{\sqrt{2\pi}\sigma} \, e^{-\left(\frac{(x-\mu)^2}{2\sigma^2}\right)}$$ (3)

On the other hand, mean values have either remained stable or decreased (which, in the context of the experiment, translates to higher fitness). Moreover, the results in the Mean Values Trend line charts values (Figure 8) present a noticeable increase on the average fitness per generation, mainly credited to the connectivity and density objectives.

Further checks on the relationship between the different fitness objectives provide further indication to the success of the evolutionary simulation. Figure 9 clearly shows an expanding Pareto front for conflicting criteria. On the other hand, converging objectives generate greater distribution with a narrower front.

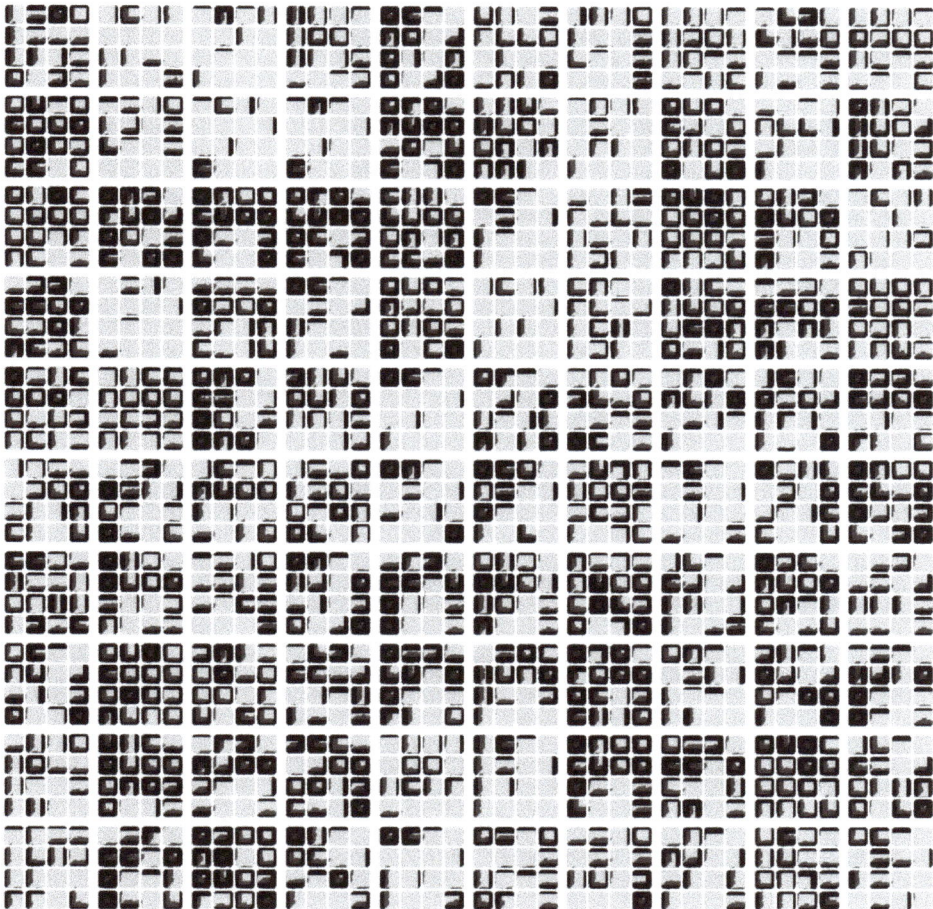

Figure 7. Render with shadow analysis from the last generation (num. 100) that contains 100 individuals (Superblocks). Individuals have been arranged in a square grid for visual purposes. Each individual is composed by 16 blocks.

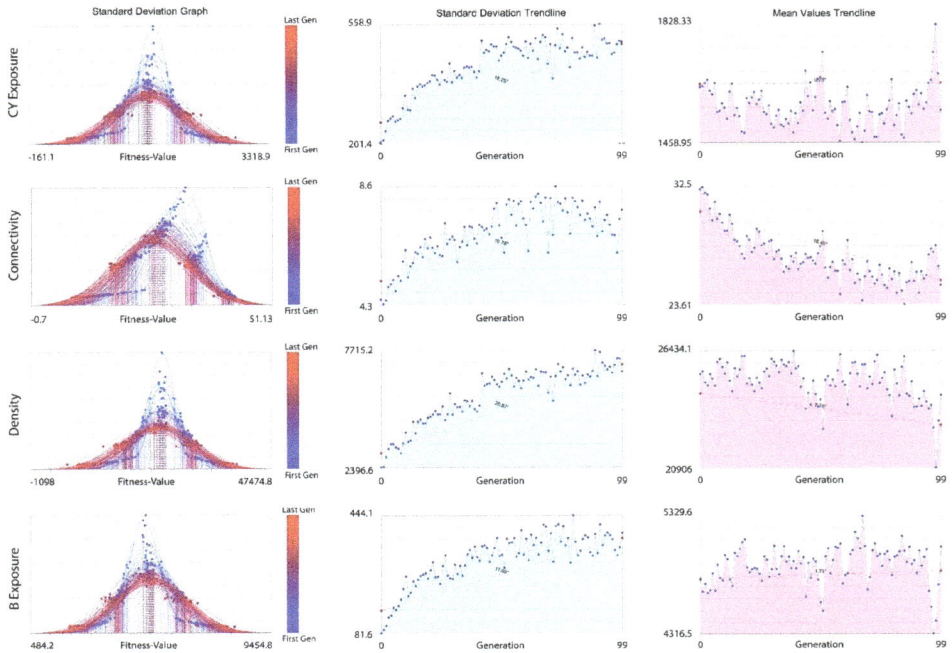

Figure 8. Standard deviation graph, standard deviation trend line, and mean values trend line for all of the fitness criteria: CV Exposure (number of mesh faces), Connectivity (numeric value based on Figure 5, Density (hab/km^2)), and B Exposure (number of mesh faces). Comparison of the objective fitness values in all generations, from blue (first generation) to red (last generation).

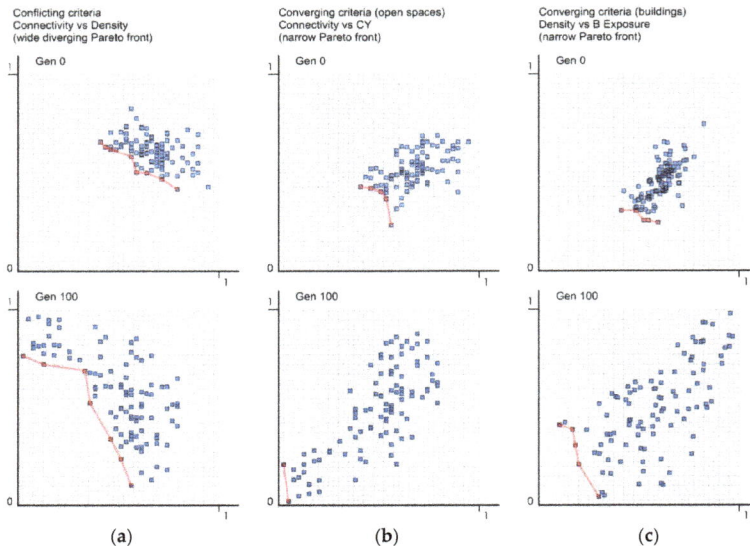

Figure 9. Pareto front analysis for criteria comparison. (**a**) Conflicting criteria for Connectivity and Density; (**b**) converging criteria for Connectivity and CY Exposure; (**c**) converging criteria for Density and B Exposure. Values normalized to the 0–1 range for all objectives.

As a result of the strategy employed within the evolutionary solver, a significant number of solutions are outputted as a final result in each iteration. Even reducing the selection to the Pareto front makes it inefficient to visually analyze each individual in the simulation. Thus, statistical analysis of the generated solutions plays a pivotal role in the selection and modification of optimal solutions. For this reason, phenotypes carefully need to be translated into remapped data in order to approach significant correlation.

For this reason, the addon Wallacei allows for a better evaluation of the individuals through Formulas (4) and (5) (the fitness values for each solution are exported through text files from Octopus, which are then read by Wallacei).

Relative Difference: (x_n is the solution's Ranking for specific fitness criteria).

$$RD = (|x_2 - x_1|) + (|x_3 - x_2|) + (|x_4 - x_3|\ldots + |x_n - x_{n-1}|) \tag{4}$$

Fitness Average: (x_n is the solution's Ranking for specific fitness criteria).

$$FA = \frac{x_1 + x_2 + x_3 + x_4 \ldots + x_n}{n} \tag{5}$$

Two selection strategies were defined in order to sort the Pareto front individuals using the Parallel Coordinate Plot: Fitness average ranking and relative difference between ranking. In both strategies, the top three ranked individuals (26-57-81 and 02-54-52) were selected; results in Figure 10 show great differences between them. Fitness average has the possibility to introduce extreme individuals that are specialized in one criterion. This specialization lets the individual reach high rankings by weakening the other criteria. Meanwhile, Relative Difference individuals tend to find an equilibrium between all the fitness criteria.

As explained above, the experiment aims to generate an urban tissue that is able to reach high-density ratios while simultaneously introducing open spaces and incorporating greater courtyard relationships. Due to the impossibility in reaching the existing Eixample density and Cerdà's connectivity at the same time, none of the Pareto solutions was a 'perfect' solution. However, multiple individuals reached a successful equilibrium that would meet W.H.O. requirements without excessively sacrificing current density. Results provided a largely successful and diverse set of solutions, allowing the designer to choose a solution (or solutions) that best fit the design objectives.

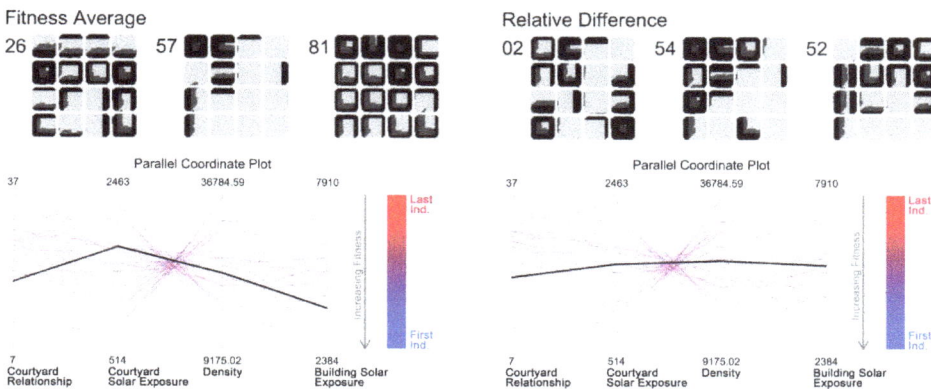

Figure 10. (**Top**) First three ranked individuals for Fitness Average (individuals num.: 26, 57, and 81) and Relative Difference (individuals num.: 02, 54, and 52). (**Below**) Parallel Coordinate Plot shows fitness relation for the first ranked individual (26 and 02). Individuals 57 and 81 show high specialization in connectivity and CY exposure, and density and B exposure, respectively.

The genome of each phenotype is comprised of all of the individual genes that define the phenotype's morphology. In this case, each gene is represented through the numerical parameter that controls how much morphological change is imposed on the phenotype. Moreover, each genome is divided into multiple gene sequences, each of which is defined by highlighting the part of the phenotype to which the gene sequence is applied. The genomes for each of the six selected phenotypes are presented in Tables 1 and 2 below. Additionally, each phenotype's genome is plotted as a polyline and compared to other selected genomes (Figures 11 and 12):

Table 1. The genomes of the phenotypes selected through the analysis of the mean fitness rank (the body part of the phenotype onto which each gene sequence is applied to is presented in italics).

Phenotype	Genome
26	['*MainCourtyard*', '0.6', '0.7', '0.4', '0.3', '0.7', '0.6', '0.6', '0.6', '0.4', '0.4', '0.7', '0.6', '0.6', '0.5', '0.5', '0.6', '*SubDivisions*', '4.0', '4.0', '4.0', '4.0', '4.0', '4.0', '4.0', '4.0', '4.0', '4.0', '4.0', '4.0', '4.0', '4.0', '4.0', '4.0', '*Organization*', '5.0', '6.0', '6.0', '5.0', '5.0', '5.0', '5.0', '6.0', '6.0', '6.0', '6.0', '6.0', '5.0', '6.0', '5.0', '5.0', '*Angle*', '2.0', '3.0', '2.0', '2.0', '2.0', '1.0', '0.0', '3.0', '0.0', '0.0', '3.0', '3.0', '3.0', '2.0', '3.0', '2.0', '*Connectivity*', '1.0', '2.0', '0.0', '3.0', '3.0', '2.0', '1.0', '2.0', '3.0', '3.0', '0.0', '2.0', '1.0', '1.0', '0.0', '3.0', '*min_floors*', '3.0', '6.0', '6.0', '5.0', '5.0', '6.0', '5.0', '5.0', '4.0', '4.0', '4.0', '5.0', '5.0', '3.0', '4.0', '5.0', '*max_floors*', '2.0', '2.0', '6.0', '3.0', '2.0', '4.0', '4.0', '4.0', '2.0', '6.0', '6.0', '4.0', '6.0', '5.0', '6.0', '4.0', '*min_extra_floors*', '4.0', '4.0', '5.0', '3.0', '3.0', '1.0', '2.0', '1.0', '3.0', '1.0', '2.0', '5.0', '2.0', '4.0', '5.0', '3.0', '*max_extra_floors*', '3.0', '2.0', '4.0', '5.0', '0.0', '1.0', '1.0', '4.0', '2.0', '4.0', '1.0', '4.0', '4.0', '2.0', '4.0', '0.0']
57	['*MainCourtyard*', '0.5', '0.4', '0.6', '0.3', '0.3', '0.5', '0.4', '0.3', '0.5', '0.5', '0.6', '0.6', '0.3', '0.3', '0.4', '0.5', '*SubDivisions*', '4.0', '4.0', '4.0', '4.0', '4.0', '4.0', '4.0', '4.0', '4.0', '4.0', '4.0', '4.0', '4.0', '4.0', '4.0', '4.0', '*Organization*', '6.0', '6.0', '6.0', '6.0', '5.0', '6.0', '6.0', '5.0', '5.0', '6.0', '6.0', '5.0', '6.0', '5.0', '5.0', '6.0', '*Angle*', '2.0', '2.0', '2.0', '0.0', '1.0', '1.0', '3.0', '2.0', '2.0', '0.0', '1.0', '0.0', '1.0', '1.0', '3.0', '0.0', '*Connectivity*', '3.0', '3.0', '4.0', '0.0', '4.0', '3.0', '2.0', '1.0', '4.0', '4.0', '4.0', '3.0', '4.0', '4.0', '3.0', '4.0', '*min_floors*', '4.0', '6.0', '2.0', '5.0', '2.0', '4.0', '3.0', '5.0', '3.0', '6.0', '3.0', '3.0', '5.0', '5.0', '4.0', '4.0', '*max_floors*', '3.0', '4.0', '6.0', '5.0', '3.0', '3.0', '4.0', '5.0', '2.0', '4.0', '3.0', '6.0', '6.0', '5.0', '4.0', '3.0', '*min_extra_floors*', '2.0', '5.0', '5.0', '4.0', '1.0', '4.0', '3.0', '4.0', '2.0', '5.0', '3.0', '0.0', '3.0', '4.0', '3.0', '1.0', '*max_extra_floors*', '4.0', '2.0', '5.0', '4.0', '0.0', '2.0', '3.0', '2.0', '5.0', '3.0', '2.0', '4.0', '5.0', '4.0', '3.0', '4.0']
81	['*MainCourtyard*', '0.6', '0.7', '0.7', '0.5', '0.7', '0.7', '0.7', '0.3', '0.7', '0.6', '0.4', '0.3', '0.7', '0.6', '0.6', '0.6', '*SubDivisions*', '4.0', '4.0', '4.0', '4.0', '4.0', '4.0', '4.0', '4.0', '4.0', '4.0', '4.0', '4.0', '4.0', '4.0', '4.0', '4.0', '*Organization*', '5.0', '6.0', '5.0', '5.0', '5.0', '6.0', '5.0', '6.0', '6.0', '5.0', '5.0', '5.0', '6.0', '5.0', '6.0', '6.0', '*Angle*', '1.0', '1.0', '2.0', '2.0', '1.0', '3.0', '3.0', '0.0', '0.0', '0.0', '2.0', '3.0', '0.0', '0.0', '2.0', '3.0', '*Connectivity*', '1.0', '0.0', '0.0', '0.0', '1.0', '0.0', '0.0', '1.0', '1.0', '0.0', '0.0', '0.0', '1.0', '2.0', '0.0', '1.0', '*min_floors*', '5.0', '5.0', '4.0', '6.0', '6.0', '4.0', '5.0', '5.0', '3.0', '4.0', '5.0', '4.0', '5.0', '6.0', '5.0', '3.0', '*max_floors*', '4.0', '3.0', '3.0', '4.0', '3.0', '5.0', '3.0', '6.0', '3.0', '3.0', '6.0', '6.0', '6.0', '6.0', '3.0', '3.0', '*min_extra_floors*', '1.0', '5.0', '4.0', '4.0', '2.0', '5.0', '2.0', '1.0', '2.0', '1.0', '4.0', '5.0', '5.0', '5.0', '3.0', '1.0', '*max_extra_floors*', '3.0', '3.0', '5.0', '5.0', '4.0', '3.0', '5.0', '4.0', '4.0', '4.0', '1.0', '5.0', '5.0', '3.0', '5.0', '1.0']

Table 2. The genomes of the phenotypes selected through the analysis of the relative difference rank (the body part of the phenotype onto which each gene sequence is applied to is presented in italics).

Phenotype	Genome
2	['*MainCourtyard*', '0.5', '0.3', '0.6', '0.7', '0.7', '0.3', '0.5', '0.3', '0.7', '0.3', '0.3', '0.4', '0.3', '0.5', '0.4', '0.6', '*SubDivisions*', '4.0', '4.0', '4.0', '4.0', '4.0', '4.0', '4.0', '4.0', '4.0', '4.0', '4.0', '4.0', '4.0', '4.0', '4.0', '4.0', '*Organization*', '5.0', '5.0', '6.0', '6.0', '5.0', '6.0', '5.0', '5.0', '5.0', '5.0', '6.0', '5.0', '6.0', '5.0', '5.0', '6.0', '*Angle*', '2.0', '2.0', '2.0', '3.0', '1.0', '0.0', '3.0', '2.0', '0.0', '0.0', '1.0', '0.0', '1.0', '0.0', '3.0', '1.0', '*Connectivity*', '0.0', '3.0', '1.0', '0.0', '4.0', '3.0', '1.0', '1.0', '2.0', '4.0', '4.0', '3.0', '0.0', '1.0', '2.0', '4.0', '*min_floors*', '5.0', '5.0', '3.0', '4.0', '6.0', '4.0', '3.0', '5.0', '4.0', '4.0', '3.0', '4.0', '6.0', '4.0', '2.0', '4.0', '*max_floors*', '6.0', '5.0', '2.0', '3.0', '2.0', '4.0', '4.0', '5.0', '3.0', '4.0', '4.0', '2.0', '4.0', '2.0', '4.0', '3.0', '*min_extra_floors*', '2.0', '2.0', '5.0', '4.0', '1.0', '3.0', '1.0', '1.0', '1.0', '2.0', '1.0', '0.0', '4.0', '4.0', '3.0', '1.0', '*max_extra_floors*', '3.0', '2.0', '5.0', '0.0', '4.0', '3.0', '4.0', '0.0', '4.0', '1.0', '5.0', '1.0', '3.0', '5.0', '4.0', '2.0']
54	['*MainCourtyard*', '0.5', '0.4', '0.6', '0.3', '0.7', '0.5', '0.3', '0.3', '0.4', '0.4', '0.6', '0.6', '0.3', '0.5', '0.5', '0.7', '*SubDivisions*', '4.0', '4.0', '4.0', '4.0', '4.0', '4.0', '4.0', '4.0', '4.0', '4.0', '4.0', '4.0', '4.0', '4.0', '4.0', '4.0', '*Organization*', '6.0', '6.0', '6.0', '6.0', '5.0', '6.0', '6.0', '5.0', '5.0', '6.0', '6.0', '5.0', '6.0', '6.0', '5.0', '5.0', '*Angle*', '2.0', '3.0', '2.0', '0.0', '2.0', '1.0', '3.0', '2.0', '2.0', '1.0', '1.0', '0.0', '3.0', '0.0', '3.0', '0.0', '*Connectivity*', '3.0', '0.0', '1.0', '0.0', '4.0', '3.0', '2.0', '1.0', '2.0', '4.0', '3.0', '0.0', '4.0', '4.0', '3.0', '4.0', '*min_floors*', '5.0', '6.0', '2.0', '5.0', '2.0', '4.0', '5.0', '5.0', '2.0', '6.0', '3.0', '3.0', '5.0', '5.0', '4.0', '4.0', '*max_floors*', '3.0', '4.0', '5.0', '5.0', '3.0', '4.0', '4.0', '5.0', '2.0', '3.0', '6.0', '6.0', '6.0', '5.0', '4.0', '4.0', '*min_extra_floors*', '5.0', '5.0', '5.0', '4.0', '1.0', '4.0', '3.0', '4.0', '2.0', '4.0', '3.0', '5.0', '3.0', '5.0', '3.0', '1.0', '*max_extra_floors*', '4.0', '4.0', '5.0', '4.0', '0.0', '3.0', '2.0', '3.0', '5.0', '1.0', '2.0', '4.0', '5.0', '4.0', '3.0', '0.0']
52	['*MainCourtyard*', '0.5', '0.4', '0.3', '0.3', '0.3', '0.7', '0.6', '0.3', '0.4', '0.5', '0.6', '0.4', '0.7', '0.5', '0.4', '0.6', '*SubDivisions*', '4.0', '4.0', '4.0', '4.0', '4.0', '4.0', '4.0', '4.0', '4.0', '4.0', '4.0', '4.0', '4.0', '4.0', '4.0', '4.0', '*Organization*', '5.0', '6.0', '6.0', '6.0', '5.0', '6.0', '5.0', '6.0', '5.0', '6.0', '6.0', '5.0', '6.0', '6.0', '6.0', '6.0', '*Angle*', '1.0', '2.0', '0.0', '1.0', '1.0', '3.0', '3.0', '1.0', '3.0', '0.0', '2.0', '0.0', '1.0', '1.0', '3.0', '3.0', '*Connectivity*', '3.0', '2.0', '2.0', '4.0', '4.0', '3.0', '1.0', '2.0', '4.0', '4.0', '1.0', '0.0', '4.0', '1.0', '0.0', '1.0', '*min_floors*', '4.0', '5.0', '3.0', '6.0', '4.0', '2.0', '4.0', '5.0', '3.0', '4.0', '6.0', '4.0', '5.0', '2.0', '6.0', '3.0', '*max_floors*', '4.0', '3.0', '5.0', '5.0', '3.0', '3.0', '4.0', '5.0', '2.0', '3.0', '3.0', '3.0', '4.0', '6.0', '6.0', '2.0', '*min_extra_floors*', '0.0', '4.0', '4.0', '5.0', '1.0', '3.0', '5.0', '1.0', '2.0', '3.0', '2.0', '0.0', '0.0', '4.0', '2.0', '2.0', '*max_extra_floors*', '5.0', '5.0', '2.0', '2.0', '5.0', '5.0', '4.0', '3.0', '3.0', '1.0', '1.0', '4.0', '5.0', '2.0', '3.0', '1.0']

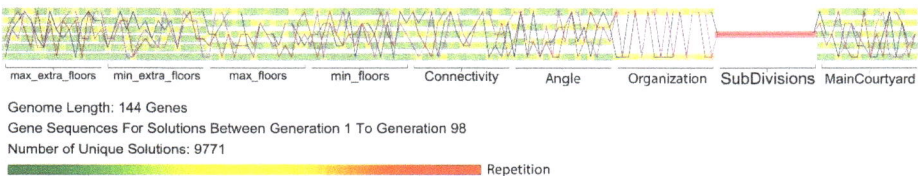

max_extra_floors   min_extra_floors   max_floors   min_floors   Connectivity   Angle   Organization   SubDivisions   MainCourtyard

Genome Length: 144 Genes
Gene Sequences For Solutions Between Generation 1 To Generation 98
Number of Unique Solutions: 9771

Repetition

Figure 11. Geometric representation of each of the three selected phenotypes' genomes. In the graph above, phenotypes are represented through the following polylines: Red—Phenotype 26, Blue—Phenotype 57, and Magenta—Phenotype 81.

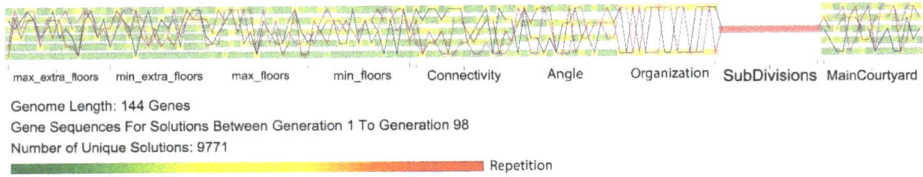

max_extra_floors   min_extra_floors   max_floors   min_floors   Connectivity   Angle   Organization   SubDivisions   MainCourtyard

Genome Length: 144 Genes
Gene Sequences For Solutions Between Generation 1 To Generation 98
Number of Unique Solutions: 9771

Repetition

Figure 12. Geometric representation of each of the three selected phenotypes' genomes. In the graph above, phenotypes are represented through the following polylines: Red—Phenotype 2, Blue—Phenotype 54, and Magenta—Phenotype 52.

The fitness values for each of the six phenotypes presented above are compared to Barcelona's current situation (BCS) and Cerdà original plan (COP) in Tables 3 and 4. Results show that most of the individuals have achieved greater connectivity than the current state while improving density, getting closer to a realistic approach.

Table 3. Selected individuals based on Fitness Average compared to current and original state.

Num. Individual	57	26	81	Barcelona's Current Situation (BCS)	Cerdà Original Plan (COP)
CY—Courtyard exposure (#faces)	1813	5158	6569	5788	5379
B—Building exposure (#faces)	3473	2620	2069	1491	2945
C—Connectivity (value)	43	30	23	24	39
DE—Density (hab/km^2)	9544	17,259	23,596	34,500	10,400

Table 4. Selected individuals based on Relative Difference compared to current and original state.

Num. Individual	2	54	52	BCS	COP
CY—Courtyard exposure (#faces)	3604	3117	3632	5788	5379
B—Building exposure (#faces)	2839	2933	2876	1491	2945
C—Connectivity (value)	29	29	29	24	39
DE—Density (hab/km^2)	14,924	16,146	14,616	34,500	10,400

4. Discussion

Back, Hammel, and Schwefel [15] argue that "the most significant advantage of using evolutionary search lies in the gain of flexibility and adaptability to the task at hand", and while the optimal solution for a single objective problem is clearly defined, multiple objective problems require the "robust and powerful search mechanisms" [16] (p. 13) of evolutionary algorithms to find the fittest solution candidates that take into consideration all of the assigned objectives. The experiments proved successful by breeding a diverse set of individuals across generations that continued to perform better towards their fitness criteria. While the experiment did not provide a single optimized solution, something that is often sought after in design, it did respond to the multiple design objectives of the design model, providing a diverse set of optimal solutions (Figure 10).

Regarding designer strategies in later stages, it has been proven that the fitness average-ranking approach generates an adequate variety that can be helpful in situations with solution uncertainty, especially in complex architectural 3D compositions (Figure 13). On the other hand, in specific scenarios, it could be interesting to add specific external-criteria values to choose individuals in the Pareto front. For instance, a minimum density value or a density range that would greatly reduce options within the front.

The computational environment plays a significant role in the application of an evolutionary model as a design strategy. The experiments carried out were limited to 100 individuals and 100 generations, a limit imposed by the computational load and time required to carry out the experiments. However, a larger population and generation count would generate greater diversity, as well as allow for more optimization of the fitness criteria. As mentioned previously, increasing the mutation rate and probability can help to increase the explorative strength in low-population situations, but will always delay final optimized results.

Figure 13. 3D aerial view render from generation-100.

As a matter of further research, unsupervised-learning data analysis is considered. Nevertheless, the quality and number of data obtained are not yet enough.

It is thought that supervised algorithms should be discarded, as choosing the labeling of the training examples by the authors would be impossible without a minimum grade of subjectivity in the selection. As it is not yet clear what the procedure for weighting different objectives within the Pareto front should be, it is not currently possible to select "the best solution".

In that sense, Machine Learning is thought to be implemented for trying to find and conclude which one of the obtained solutions might be more appropriate in the case of a realistic application to Eixample's urban blocks' future reorganization.

Future data analysis should be based on a higher number of individuals within the population and also on a higher number of iterations. Implementing a clustering algorithm is still to be decided.

Logistic regression Kernel-based analysis would probably be discarded during the first trials stepping into k-means and principal-components analysis (PCA) being self-organizing maps (SOM) within the scope of the analysis.

It could also be considered to introduce Reinforcement Learning in which every individual is considered as an Agent with a reward function towards the optimal solution. In this case, again, the target goal needs to be predefined.

Author Contributions: conceptualization, D.N.-M. and M.M.; methodology, D.N.-M. and M.M.; software, D.N.-M. and M.M.; validation, A.C.-B., M.M., and D.N.-M.; formal analysis, D.N.-M. and M.M.; investigation, D.N.-M., M.M., and A.C.-B.; data curation, D.N.-M.; writing—original draft preparation, D.N.-M., A.C.-B., and M.M.; writing—review and editing, D.N.-M., A.C.-B., and M.M.; visualization, D.N.-M.; supervision, A.C.-B. and M.M.; project administration, D.N.-M.

Funding: This research received no external funding.

Conflicts of Interest: The authors declare no conflict of interest.

References

1. Jacobs, J. *The Death and Life of Great American Cities*; Random House: New York, NY, USA, 1961; ISBN 067974195X.
2. Simon, H.A. *The Sciences of the Artificial*, 3rd ed.; MIT Press: Cambridge, MA, USA, 1997; ISBN 9780262691918.

3. Soddu, C. Recognizability of the Idea: The Evolutionary Process of Argenia. In *Creative Evolutionary Systems*; Corne, D.W., Bentley, P.J., Eds.; Morgan Kaufmann: San Diego, CA, USA, 2002; pp. 109–127. ISBN 978-1-55860-673-9.

4. Koolhaas, R. Whatever Happened to Urbanism? *Des. Q.* **1995**, *164*, 28–31.

5. Mayr, E. Typological Versus Population Thinking. In *Evolution and the Diversity of Life: Selected Essays*; Harvard University Press: Cambridge, MA, USA, 1997; pp. 26–29, ISBN 978-0-674-27105-0.

6. Ajuntament de Barcelona Població per Districtes. 2015–2016. Available online: http://www.bcn.cat/estadistica/catala/dades/anuari/cap02/C020102.htm (accessed on 1 August 2018).

7. Talk Architecture the 3 Drawings of Cerda. Available online: https://talkarchitecture.wordpress.com/2011/02/06/the-3-drawings-of-cerda/ (accessed on 1 August 2018).

8. Deb, K.; Pratap, A.; Agarwal, S.; Meyarivan, T. A fast and elitist multiobjective genetic algorithm: NSGA-II. *IEEE Trans. Evol. Comput.* **2002**, *6*, 182–197. [CrossRef]

9. Busquets, J. *Barcelona: La Construccion Urbanistica de Una Ciudad Compacta*; Ediciones del Serbal: Barcelona, Spain, 2004.

10. Arroyo, F. Barcelona Suspende en Zona Verde | Edición Impresa | EL PAÍS. Available online: https://elpais.com/diario/2009/10/24/catalunya/1256346439_850215.html (accessed on 1 August 2018).

11. Ajuntament de Barcelona la Superilla Pilot de la Maternitat i Sant Ramon Estarà en Marxa al Mes D'abril | Superilles. Available online: http://ajuntament.barcelona.cat/superilles/ca/noticia/la-superilla-pilot-de-la-maternitat-i-sant-ramon-estarza-en-marxa-al-mes-dabril (accessed on 1 August 2018).

12. Luke, S. *Essentials of Metaheuristics*; Lulu Press, Inc.: Morrisville, NC, USA, 2013; ISBN 9781300549628.

13. Zitzler, E.; Laumanns, M.; Thiele, L. SPEA2: Improving the Strength Pareto Evolutionary Algorithm. *Evol. Methods Des. Optim. Control Appl. Ind. Probl.* **2001**.

14. Makki, M.; Farzaneh, A.; Navarro, D. The Evolutionary Adaptation of Urban Tissues through Computational Analysis. In Proceedings of the 33rd eCAADe Conference, Vienna University of Technology, Vienna, Austria, 16–18 September 2015; Martens, B., Wurzer, G., Grasl, T., Lorenz, W.E., Schaffranek, R., Eds.; Volume 2, pp. 563–571.

15. Back, T.; Hammel, U.; Schwefel, H.P. Evolutionary computation: Comments on the history and current state. *IEEE Trans. Evol. Comput.* **1997**, *1*, 3–17. [CrossRef]

16. Zitzler, E. Evolutionary Algorithms for Multiobjective Optimization: Methods and Applications. *TIK-Schriftenreihe* **1999**.

![mathematics logo] *mathematics*

MDPI

Article

A Developed Artificial Bee Colony Algorithm Based on Cloud Model

Ye Jin [1], Yuehong Sun [1,2,*] and Hongjiao Ma [1]

[1] School of Mathematical Sciences, Nanjing Normal University, Nanjing 210023, China;
    160902016@stu.njnu.edu.cn (Y.J.); xuminxi525@163.com (H.M.)
[2] Jiangsu Key Laboratory for NSLSCS, Nanjing Normal University, Nanjing 210023, China
* Correspondence: 05234@njnu.edu.cn; Tel.: +86-13451825524

Received: 11 March 2018; Accepted: 10 April 2018; Published: 18 April 2018

Abstract: The Artificial Bee Colony (ABC) algorithm is a bionic intelligent optimization method. The cloud model is a kind of uncertainty conversion model between a qualitative concept $\widetilde{T}$ that is presented by nature language and its quantitative expression, which integrates probability theory and the fuzzy mathematics. A developed ABC algorithm based on cloud model is proposed to enhance accuracy of the basic ABC algorithm and avoid getting trapped into local optima by introducing a new select mechanism, replacing the onlooker bees' search formula and changing the scout bees' updating formula. Experiments on CEC15 show that the new algorithm has a faster convergence speed and higher accuracy than the basic ABC and some cloud model based ABC variants.

Keywords: artificial bee colony algorithm (ABC); cloud model; normal cloud model; Y conditional cloud generator; global optimum

1. Introduction

In recent years, the development of metaheuristics [1–3] has advanced. Many scholars have made a lot of contributions in this area. The Artificial Bee Colony algorithm (ABC) [4] is a novel swarm intelligence algorithm among the sets of metaheuristics. The ABC algorithm is an optimization algorithm which mimics the foraging behavior of the honey bee. It provides a population-based search procedure in which individuals called food positions are modified by the artificial bees with the increasing of iterations. The bee's aim is to discover the places of food sources with high nectar amount.

From 2007 to 2009, Karaboga et al. [5–7] presented a comparative study on optimizing a large set of numerical test functions. They compared the performance of the ABC algorithm with the genetic algorithm (GA), particle swarm optimization (PSO), differential evolution (DE) and evolution strategy (ES). The simulation results show that ABC can be efficiently employed to solve engineering problems with high dimensions. It can produce very good and effective results at a low computational cost by using only three control parameters (population size, maximum number of fitness evaluations, limit). Akay et al. [8] studied the parameter tuning of the ABC algorithm and investigated the effect of control parameters. Afterwards, two modified versions of the ABC algorithm were proposed successively by Akay et al. [9] and Zhang et al. [10] for efficiently solving real-parameter numerical optimization problems. Aderhold et al. [11] studied the influence of the population size about the optimization behavior of ABC and also proposed two variants of ABC which used the new position update of the artificial bees.

However, ABC was unsuccessful on some complex unimodal and multimodal functions [7]. So, some modified artificial colony algorithms were put forward in order to improve the performance of the basic ABC. Zhu et al. [12] proposed an improved ABC algorithm called gbest-guided ABC (GABC) by incorporating the information of global best solution into the solution search equation to guide the search of candidate solutions. The experimental results tested on six benchmark functions showed that GABC outperformed basic ABC. Wu et al. [13] described an improved ABC algorithm to enhance the global search ability of basic ABC. Guo et al. [14] presented a novel search strategy and the improved algorithm is called global ABC which has great advantages of convergence property and solution quality. In 2013, Yu et al. [15] proposed a modified artificial bee colony algorithm in which global best is introduced to modify the update equation of employed and onlooker bees. Simulation results on the problem of peak-to-average power ratio reduction in orthogonal frequency division multiplexing signal and multi-level image segmentation showed that the new algorithm had better performance than the basic ABC algorithm with the same computational complexity. Rajasekhar et al. [16] proposed a simple and effective variant of the ABC algorithm based on the improved self-adaptive mechanism of Rechenbergs 1/5th success rule to enhance the exploitation capabilities of the basic ABC. Yaghoobi [17] proposed an improved artificial bee colony algorithm for global numerical optimization in 2017 from three aspects: initialising the population based on chaos theory; utilising multiple searches in employee and onlooker bee phases; controlling the frequency of perturbation by a modification rate.

Multi-objective evolutionary algorithms (MOEAs) gained wide attention to solve various optimization problems in fields of science and engineering. In 2011, Zou et al. [18] presented a novel algorithm based on ABC to deal with multi-objective optimization problems. The concept of Pareto dominance was used to determine the flight direction of a bee, and the nondominated solution vectors which had been found in an external archive were maintained in the proposed algorithm. The proposed approach was highly competitive and was a viable alternative to solve multi-objective optimization problems.

The performances of Pareto-dominance based MOEAs degrade if the number of objective functions is greater than three. Amarjeet et al. [19] proposed a Fuzzy-Pareto dominance driven Artificial Bee Colony (FP-ABC) to solve the many-objective software module clustering problems (MaSMCPs) effectively and efficiently in 2017. The contribution of the article was as follows: the selection process of candidate solution was improved by fuzzy-Pareto dominance; two external archives had been integrated into the ABC algorithm to balance the convergence and diversity. A comparative study validated the supremacy of the proposed approach compared to the existing many-objective optimization algorithms.

A decomposition-based ABC algorithm [20] was also proposed to handle many-objective optimization problems (MaOPs). In the proposed algorithm, an MaOP was converted into a number of subproblems which were simultaneously optimized by a modified ABC algorithm. With the help of a set of weight vectors, a good diversity among solutions is maintained by the decomposition-based algorithm. The ABC algorithm is highly effective when solving a scalar optimization problem with a fast convergence speed. Therefore, the new algorithm can balance the convergence and diversity well. Moreover, subproblems in the proposed algorithm were handled unequally, and computational resources were dynamically allocated through specially designed onlooker bees and scout bees, which indeed contributed to performance improvements of the algorithm. The proposed algorithm could approximate a set of well-converged and properly distributed nondominated solutions for MaOPs with the high quality of solutions and the rapid running speed.

The basic ABC algorithms is often combined with other algorithms and techniques. In 2016, an additional update equation [21] for all ABC-based optimization algorithms was developed to speed up the convergence utilizing Bollinger bands [22] which is a technical analysis tool to predict maximum or minimum future stock prices. Wang et al. [23] proposed a hybridization method based on krill herd [24] and ABC (KHABC) in 2017. A neighbor food source for onlooker bees in ABC was obtained from the global optimal solutions, which was found by the KHABC algorithm. During the information exchange process, the globally best solutions were shared by the krill and bees. The effectiveness of the proposed methodology was tested for the continuous and discrete optimization.

In this paper, another technique called cloud model [25] will be embedded into the ABC algorithm. Cloud model is an uncertainty conversion model between a qualitative concept and its quantitative expression. In 1999, an uncertainty reasoning mechanism of the cloud model was presented and the cloud model theory was be expanded after that. In addition, Li successfully applied cloud model to inverted pendulum [26]. Some scholars combined cloud model with ABC because cloud model has the characteristics of stable tendentiousness and randomness [27–29].

We propose a new algorithm which inherits the excellent exploration ability of the basic ABC algorithm and the stability and randomness of the cloud model by modifying the selection mechanism of onlookers, the search formula of onlookers and scout bees' update formula. The innovation points of the new algorithm are:

1. The population becomes more diverse in the whole search process by using a different selection mechanism of onlookers, in which the worse individual will have a larger selection probability than in basic ABC;
2. Local search ability of the algorithm can be improved by applying the normal cloud generator as the search formula of onlookers to control the search of onlookers in a suitable range;
3. Historical optimal solutions can be used by Y-conditional cloud generator as scout bees's update formula to ensure that the algorithm not only jumps out of local optimum but also avoids a blind random search.

The remainder of the paper is structured as follows. Section 2 provides the description of the basic ABC algorithm, followed by the details and framework of the developed ABC algorithm based on cloud model, as shown in Section 3. Subsequently, Section 4 gives the experiment results on CEC15 of the comparison among the proposed DCABC, the basic ABC, and other ABC variants based on cloud model. Then in Section 5, the current work is summarized, and the acknowledgements are given in the end.

2. The Basic ABC Algorithm

There are three kinds of bees, namely, employed bees, onlooker bees and scout bees in the ABC algorithm. The total population number is N_s; the number of employed bees is N_e and onlookers is N_u (General define $N_e = N_u = \frac{N_s}{2}$). In the initialization phase, food sources in the population are randomly generated and assigned to employed bees as

$$X_i^j = X_{min}^j + rand(0,1)(X_{max}^j - X_{min}^j),\tag{1}$$

where $j \in \{1, 2, \ldots, D\}$, X_{max}, X_{min} are the upper and lower bounds of the solution vectors, D is the dimension of the decision variable.

Each employed bee X_i generates a new food source V_i in the neighborhood of its present position:

$$V_i^j = X_i^j + \phi_i^j(X_i^j - X_k^j),\tag{2}$$

where $j \in \{1, 2, \ldots, D\}$, $k \in \{1, 2, \ldots, N_e\}$, k must be different from i, k and j are random generating indexes, ϕ_i^j is a random number between $[-1, 1]$. At the same time, we should guarantee V_i in the field of definition domain. V_i will be compared to X_i and the employed bee exploits the better food source by greedy selection mechanism in terms of fitness value fit_i in Equation (3):

$$fit_i = \begin{cases} \dfrac{1}{1 + f_i} & f_i \geq 0 \\ 1 + abs(f_i) & f_i < 0 \end{cases} \tag{3}$$

where f_i is the objective value of solution X_i or V_i. Equation (3) is used to calculate fitness values for a minimization problem, while for maximization problems the objective function can be directly used as a fitness function.

An onlooker bee evaluates the fitness value of all employed bees, and uses the roulette wheel method to select a food source X_i updated the same as employed bees according to its probability value P calculated by the following expression:

$$P = \frac{0.9 fit(X_i)}{\max\limits_{m=1}^{N_e} fit(X_m)} + 0.1, \tag{4}$$

If a food source X_i cannot be improved beyond a predetermined number (*limit*) of trial counters, it will be abandoned and the corresponding employed bee will become a scout bee randomly produced by Equation (1). The algorithm will be terminated after repeating a predefined maximum number of cycles, denoted as *Max_Cycles*. The flow chart of ABC algorithm is shown in Algorithm 1.

Algorithm 1: The basic ABC algorithm.

Initialization phase
    Initialize the food sources using Equation (1).
    Evaluate the fitness value of the food sources using Equation (3), set the current generation
    $t = 0$.
While $t \leq Max\_Cycles$ **do**
    Employed bees phase
        Send employed bees to produce new solutions via Equation (2).
        Apply greedy selection to evaluate the new solutions.
        Calculate the probability using Equation (4).
    Onlooker bees phase
        Send onlooker bees to produce new solutions via Equation (2).
        Apply greedy selection to evaluate the new solutions.
    Scout bee phase
        Send one scout bee produced by Equation (1) into the search area for discovering a new
    food source.
        Memorize the best solution found so far.
    $t = t + 1$.
end while
Return the best solution.

---

3. A Developed Artificial Bee Colony Algorithm Based on Cloud Model (DCABC)

The ABC algorithm is a relatively new and mature swarm intelligence optimization algorithm. Compared to GA and PSO, ABC has a higher robustness [6]. More and more scholars want to improve the performance of the ABC algorithm. Zhang [27] put forward an algorithm named PABC with the new select scheme based on cloud model. For the individual with a better fitness characteristic, the value of probability was likely relatively high, and vice versa. Lin et al. [29] proposed an improved ABC algorithm based on cloud model (cmABC) to solve the problem that the basic ABC algorithm suffered from slow convergence and easy stagnation in local optima by calculating food source through the normal cloud particle operator and reducing the radius of the local search space. In cmABC, the author also introduced a new selection strategy that made the inferior individual have more chances to be selected for maintaining diversity. In addition, the best solution found over time was used to explore a new position in the algorithm. A number of experiments on composition functions showed that the proposed algorithm had been improved in terms of convergence speed and solution quality. In this section, we propose a developed ABC algorithm named DCABC which is based on cloud model with a new choice mechanism of onlookers and new search strategies of onlooker bees and scouts.

3.1. Cloud Model

Professor Li presented an uncertainty conversion model between a qualitative concept $\widetilde{T}$ [30] presented by nature language and its quantitative expression which is called cloud model on the basis of traditional fuzzy set theory and probability statistics. He developed and improved a complete set of cloud theory [31] which consists of cloud model, virtual cloud, cloud operations, cloud transform, uncertain reasoning and so on.

Suppose U is a quantitative domain of discourse that are represented by precise values (one-dimensional, two-dimensional or multi-dimensional), and $\widetilde{T}$ is a qualitative concept in U. X is an arbitrary element in U and a random implementation of qualitative concept $\widetilde{T}$. The degree of certainty of X to $\widetilde{T}$ expressed as $\mu(X)$ is an random number that has stable tendency. The distribution of X on the domain of discourse U is called cloud model, or simply 'cloud' for short. Each pare $(X, \mu(X))$ is called a cloud droplet, and cloud model can be formulated as follows:

$$\forall X \in U \longrightarrow \mu(X) \in [0,1] \tag{5}$$

The normal cloud is scattered point cloud model based on normal distribution or half-normal distribution. Normal cloud model uses a set of independent parameters to work together in order to express digital characteristics of a qualitative concept and reflect the uncertainty of the concept. Based on the normal distribution function and membership function, this group of parameters are represented by three digital characteristics including expectation Ex, entropy En, and hyper entropy He.

Expectation Ex is a point which can best represent the qualitative concept in domain of discourse space. It can be considered as the center of gravity of all cloud drop, which can best represent the coordinates of the qualitative concept on number field. Entropy En stands for the measurable granularity of the qualitative concept. Entropy also reflects the uncertainty and fuzzy degree of the qualitative concept. Fuzzy degree means value range that can be accepted by the qualitative concept in domain of discourse. Hyper entropy He is the measure of entropy's uncertainty, namely entropy of En. It reflects randomness of samples which represent qualitative concept values, and reveals the relevance of fuzziness and randomness. Hyper entropy can also reflect the aggregation extent of cloud droplets.

Given three digital characteristics Ex, En and He, forward cloud generator in Equations (6)–(8) can produce N cloud droplets of the normal cloud model (Algorithm 2), which are two-dimensional points (x_i, μ_i) ($i \in \{1, 2, \ldots, N\}$).

$$En'_i = N(En, He^2), \tag{6}$$

$$x_i = N(Ex, (En'_i)^2), \tag{7}$$

$$\mu_i = \exp\{-\frac{(x_i - Ex)^2}{2(En'_i)^2}\}, \tag{8}$$

Algorithm 2: Forward cloud generator algorithm.

Input: Ex, En, He and N.

Output: quantitative value x_i of ith cloud droplet and its degree of certainty μ_i.

Forward cloud generator

  Generate a normal random number En_i' with expectation En and hyper entropy He by Equation (6);

  Generate a normal random number x_i with expectation Ex and hyper entropy En_i' by Equation (7).

Drop (x_i, μ_i)

  Calculate μ_i by Equation (8);

  A cloud droplet(x_i, μ_i) is get.

Repeat

  Repeat the above step until N cloud droplets have come into being. (Figure 1)

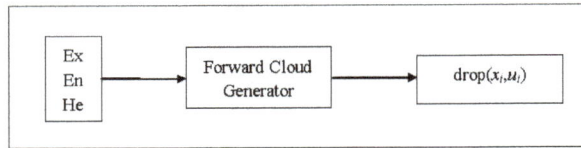

Figure 1. Forward Cloud Generator.

Given three digital characteristics (Ex, En, He) and a specific degree of certainty μ, cloud generator refers to Y-conditional cloud generator based on uncertainty reasoning of cloud model. In other words, every cloud droplet (x_i, μ) has the same degree of certainty which belongs to concept $\widetilde{T}$. The formula of Y-conditional cloud generator (Algorithm 3) is:

$$x_i = Ex \pm \sqrt{-2ln(\mu)} * En_i', \tag{9}$$

Algorithm 3: Y-conditional cloud generator algorithm.

Input: Ex, En, He, N and μ.

Output: Quantitative values x_i of ith cloud droplet and its degree of certainty μ.

Y-conditional cloud generator

  Get a normal random number En_i' with expectation En and hyper entropy He by Equation (6);

  Calculate x_i with Ex, En_i' and μ by Equation (9).

Drop (x_i, μ)

  A Y-conditional cloud droplet (x_i, μ) is get.

Repeat

  Repeat the above step until get N cloud droplets. (Figure 2)

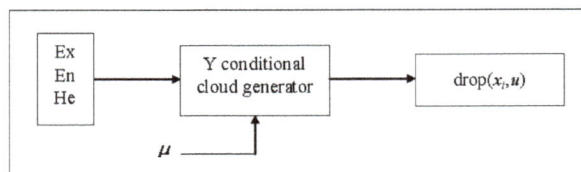

Figure 2. Y-Conditional cloud generator.

3.2. New Choice Mechanism for Onlookers

3.2.1. New Choice Mechanism

In the basic ABC algorithm, onlooker bees choose the good-quality nectars by employing the roulette wheel selection scheme. That is to say, the bigger the nectar's fitness value, the higher the probability it will be chosen by onlookers. The selection mechanism contains three parts: calculating the selection probability of each solution in population according to its fitness value; selecting the candidate solution using the roulette wheel selection method; starting the local search of onlooker bees around the candidate solution. However, the selection scheme is so greedy that it is easy to lead to the rapid decrease of population diversity and fall into local optimum. We hope to obtain a reasonable selection scheme.

Zhang et al. [27] improved the selection strategy based on cloud model with three digital characteristics Ex, En and He in Equation (10):

$$\begin{cases} Ex = \max\limits_{i=1}^{Ne} fit_i \\ En = \frac{Ex - fit_i}{12} \\ He = \frac{En}{3} \end{cases} \tag{10}$$

The possibility of the current individual which is the best can be regarded as the choice probabilities and can be produced by the positive cloud generator. Thinking differently, it will be found that the worst individual also contains useful information after several loop iterations. So, we ensure that the worst individual has larger selection probability. Equation (10) pays more attention to the inferior individuals. Detailed positive cloud generator operations can be described as follows:

$$\begin{cases} Ex = \min\limits_{i=1}^{Ne} fit_i \\ En = \frac{fit_i - Ex}{12} \\ He = \frac{En}{3} \end{cases} \tag{11}$$

The selective probability of the corresponding individual is adjusted as follows:

$$P = \exp\{-\frac{(x - Ex)^2}{2(En')^2}\}, \tag{12}$$

where, $En' = N(En, He^2)$, $x = N(Ex, (En')^2)$, N is a normal random number generator.

We find that the individuals closer to Ex(inferior individuals) will get the higher possibility, namely, selection probability.

3.2.2. Efficiency Analysis

In our proposed algorithm DCABC, Equation (4) is used as the probability selection formula for onlookers when a random number *rand* between 0 and 1 are less than or equal to 0.5; otherwise the selection formula is set by the new choice mechanism in Equation (12). The goal of processing selection probability in two cases is to avoid the algorithm plunging into local optimum.

To test the effectiveness of the current selection mechanism, the modified and the basic ABC run independently on CEC15 [32] with dimensions(D) 10, 30 and 50, respectively. We set the initial population size $N_s = 40$. The number of employed bees equals to the number of onlookers, which is $N_e = N_u = \frac{N_s}{2}$. The value of '*limit*' equals to $N_e * D$ [33]. Every experiment is repeated 30 times. The maximum number of function evaluations (*MaxFES*) is set as $D * 10,000$ for all functions [34]. The simulation results is recorded in Table 1. It can be easily observed that the ABC with new choice mechanism is superior to the basic ABC on most functions. This implies that the new choice mechanism improves the performance of the basic ABC.

Table 1. Experimental Results between ABC with the new choice mechanism (NCMABC) and the basic ABC.

Functions	Criteria	ABC (10D)	NCMABC (10D)	ABC (30D)	NCMABC (30D)	ABC (50D)	NCMABC (50D)
f_1	Mean	1.36e+06	**1.11e+06**	3.73e+06	**3.48e+06**	1.20e+07	**1.05e+07**
	Std	1.12e+06	6.92e+05	1.46e+06	1.42e+06	3.69e+06	3.49e+06
	Rank	2	1	2	1	2	1
f_2	Mean	8.57e+02	**7.23e+02**	7.26e+02	**5.84e+02**	1.43e+03	**1.31e+03**
	Std	7.52e+02	8.35e+02	6.06e+02	6.77e+02	1.20e+03	1.07e+03
	Rank	2	1	2	1	2	1
f_3	Mean	2.02e+01	**1.95e+01**	**2.01e+01**	2.02e+01	**2.02e+01**	2.02e+01
	Std	3.94e-02	3.12e+00	4.10e-02	4.14e-02	4.52e-02	4.13e-02
	Rank	2	1	1	2	2	1
f_4	Mean	1.26e+01	**1.18e+01**	9.77e+01	**9.56e+01**	2.34e+02	**2.32e+02**
	Std	4.18e+01	3.67e+00	1.80e+01	1.78e+01	2.93e+01	2.89e+01
	Rank	2	1	2	1	2	1
f_5	Mean	4.31e+02	**3.80e+02**	2.42e+03	**2.37e+03**	4.24e+03	**4.19e+03**
	Std	1.49e+02	1.51e+02	2.86e+02	2.71e+02	4.77e+02	3.75e+02
	Rank	2	1	2	1	2	1
f_6	Mean	5.03e+03	**4.47e+03**	1.38e+06	**1.24e+06**	2.20e+06	**1.95e+06**
	Std	4.41e+03	3.10e+03	7.15e+02	6.35e+02	8.30e+02	7.53e+05
	Rank	2	1	2	1	2	1
f_7	Mean	8.72e+01	**7.90e-01**	9.39e+00	**9.18e+00**	1.85e+01	**1.59e+01**
	Std	2.58e+01	2.87e-01	1.25e+00	1.13e+00	8.52e+00	2.02e+00
	Rank	2	1	2	1	2	1
f_8	Mean	**1.32e+04**	1.81e+04	4.11e+05	**3.90e+05**	2.89e+06	**2.15e+06**
	Std	2.14e+04	3.02e+04	3.06e+05	1.96e+05	9.87e+05	8.05e+05
	Rank	1	2	2	1	2	1
f_9	Mean	9.48e+01	**9.32e+01**	1.21e+02	**1.19e+02**	1.62e+02	**1.33e+02**
	Std	2.20e+01	2.35e+01	4.44e+01	4.41e+01	1.13e+02	7.43e+01
	Rank	2	1	2	1	2	1
f_{10}	Mean	**4.37e+03**	4.60e+03	6.88e+05	**5.07e+05**	**8.07e+05**	9.57e+05
	Std	4.24e+03	7.02e+01	3.57e+05	2.90e+05	4.68e+05	4.39e+05
	Rank	1	2	2	1	1	2
f_{11}	Mean	3.01e+02	**2.82e+02**	3.22e+02	**3.21e+02**	**3.58e+02**	3.62e+02
	Std	4.70e-01	4.01e-13	7.47e+00	7.40e+00	1.69e+02	1.65e+02
	Rank	2	1	2	1	1	2
f_{12}	Mean	**1.04e+02**	1.04e+02	1.07e+02	1.07e+02	1.10e+02	1.10e+02
	Std	7.72e-01	8.03e-01	8.07e-01	6.08e-01	7.41e-01	7.99e-01
	Rank	1	1	1	1	1	1
f_{13}	Mean	**3.13e+01**	3.14e+01	1.02e+02	1.04e+02	**1.89e+02**	1.89e+02
	Std	2.04e+00	2.21e+00	4.35e+00	3.58e+00	4.91e+00	4.25e+02
	Rank	1	2	1	2	1	1
f_{14}	Mean	1.86e+03	**1.81e+03**	3.06e+04	3.06e+04	5.00e+04	**4.95e+04**
	Std	1.37e+03	1.31e+03	4.51e+03	5.57e+03	1.77e+03	1.42e+01
	Rank	2	1	1	1	2	1
f_{15}	Mean	1.00e+02	1.00e+02	1.00e+02	1.00e+02	1.02e+02	1.01e+02
	Std	2.27e-11	5.54e-12	6.56e-02	6.99e-03	1.23e+00	1.14e+00
	Rank	1	1	1	1	2	1
Mean rank		1.67	1.2	1.67	1.13	1.73	1.13
Overall rank		2	1	2	1	2	1

3.3. The New Search Strategy of Onlooker Bees

Lin et al. [29] proposed an improved ABC algorithm based on cloud model (cmABC). By calculating a candidate food source through the normal cloud operator and reducing the radius of the local search, the cmABC algorithm was proved to enhance the convergence speed, exploitation capability and solution quality on the experiments of composition functions. In cmABC, three digital characteristics of cloud model (Ex, En, He) are given as:

$$\begin{cases} Ex = X_i^j \\ En = ex \\ He = \frac{En}{10} \end{cases} \tag{13}$$

where X_i is the current food sources position, $j \in \{1, 2, \ldots, D\}$, ex is variable. The forward cloud generator can produce a normal random number V_i^j, which will correspond to the new food sources position of jth dimension. Detailed operations were described as:

$$\begin{cases} En' = N(En, He^2) \\ V_i^j = N(Ex, En'^2) \end{cases} \tag{14}$$

The greater the value of entropy En, the wider the distribution of cloud droplets and vice versa. When the search iteration reached a certain number of times, the population was closer and closer to the optimal solution. A nonlinear decrease strategy to self-adaptive adjust the value of ex was used in cmABC for the sake of improving the precision of solution and controlling the bees' search range:

$$ex = -(E_{max} - E_{min})(t/T_{max})^2 + E_{max} \tag{15}$$

where $t \in \{1, 2, 3, \ldots, T_{max}\}$ was the current number of iterations, T_{max} was the maximum number of cycles. The values of parameters E_{max} and E_{min} were set to 5 and 10^{-4}, respectively. In order not to specify too many parameters, in this paper, three digital characteristics of cloud model (Ex, En, He) are given as

$$\begin{cases} Ex = X_i^j \\ En = \frac{2}{3}|X_i^j - X_k^j| \\ He = \frac{En}{10} \end{cases} \tag{16}$$

where $j \in \{1, 2, \ldots, D\}$, $k \in \{1, 2, \ldots, N_e\}$, k must be different from i, k, j are random generating indexes. This amendment is based on the stable tendency and randomness of normal cloud model. The entropy En is selected by '3σ' principle of normal cloud model, which can control the onlooker bees to search in a suitable range.

3.4. Search Strategy of Scouts Combined with Y Conditional Cloud Generator

Employed and onlooker bees look for a better food source around their neighborhoods in each cycle of the search. If the fitness value of a food source is not improved by a predetermined number of trials that is equal to the value of '*Limit*', then that food source is abandoned by its employed bee and the employed bee associated with that food source becomes a scout bee. In the basic ABC, the scout randomly finds a new food source to replace the abandoned one by Equation (2), which makes the convergence rate of the basic ABC slow for not taking full advantage of the historical optimal solution information. In this section we make the scout bee search a candidate position around the historical optimal value fit_{best} (corresponding to $Globalmin$) by Y-conditional cloud operator. Search strategy of scouts combined with Y-conditional cloud generator is described in Algorithm 4. The purpose of setting $\mu \in (0, 0.5)$ in Step 3 is to guarantee population diversity. Cloud droplets which have smaller membership degrees are farther from center Ex, that is to say the new food source is farther from historical optimum $Globalmin$. However, the historical optimum information is used to generate a scout, therefore aimless searching of scout bees in the basic ABC algorithm can be avoided to a certain degree.

Algorithm 4: Search strategy of scouts combined with Y-conditional cloud generator.

Step 1   Set expectation Ex as $GlobalParams$, which is the position parameters of $Globalmin$.

Step 2   Entropy $En = (X_{max} - X_{min})/N_e$.

Step 3   Hyper entropy $He = En/c2$, where $c2 = 10$.

Step 4   Randomly generate membership degrees $\mu^j \in (0, 0.5)$, where $j \in \{1, 2, \ldots, D\}$.

Step 5   Obtain the new food resource X_i according to Equation (9).

3.5. DCABC Algorithm

Pseudo code of DCABC algorithm proposed for solving unconstrained optimization problems is given in Algorithm 5. *MaxFES* represents the maximum number of function evaluations. *FES* represents the number of function evaluations.

Algorithm 5: Pseudo code of DCABC algorithm.

Initialization phase

    Using Equation (1) initialize the population of solutions X_i^j, $i = 1, 2, \ldots, N_e$, $j=1, 2, \ldots, D$.

    Evaluate the fitness of the population by Equation (3), set the current $FES = Ne$.

While $FES \leq MaxFES$ **do**

  Employed bees phase

    Send employed bees to produce new solutions via Equation (2).

    Apply greedy selection to evaluate the new solutions.

    If *rand* less than or equal to 0.5, Calculate the selective probability using Equation (4);

    Otherwise calculate the probability using Equations (11) and (12).

  Onlooker bees phase

    Send onlooker bees to produce new solutions via Equations (14) and (16).

    Apply greedy selection to evaluate the new solutions.

  Scout bee phase

    Send one scout bee generated by Algorithm 4 into the search area for discovering a new food source.

    Memorize the best solution found so far.

end while

Return the best solution.

---

4. Experimental Study of DCABC

4.1. Evaluation Functions

Comparing the proposed DCABC with the basic ABC and the other ABC variants, such as GABC [12], cmABC [29] and PABC [27], the experimental results of benchmark functions with 10, 30 and 50 decision variables in CEC15 [32] are given under the same machine with an Intel 3.20 GHz CPU, 8GB memory, and the operating system is Windows 7 with MATLAB 9.0 (R2016a). All functions in CEC15 have different optimal values $f(x^*)$.

4.2. Parameters Settings

For all compared algorithms including DCABC, the size of initial population is 40, an equal split of employed bees and onlookers. *'limit'* equals to $N_e * D$ [33]; The dimension is set as 10, 30 and 50 in turn. In Equation (2) of GABC [12], $C = 1.5$. In cmABC [29], $Emax = 5$, $Emin = 10^{-4}$. The *MaxFES* is $D * 10,000$, which is used as the terminal criterion of five algorithms. Every experiment is repeated 30 times each starting from a random population with different random seeds, the mean results (Mean) and the standard deviation (Std) of each algorithm are recorded with the format of $f(x) - f(x^*)$ in Tables 2–4. The best results are highlighted in boldface. Rank records the performance-rank of five algorithms for dealing with each benchmark function according to their mean results. The overall rank for each algorithm is defined according to their mean rank values over 15 benchmark problems. The number of (Best/2ndBest/Worst) is counted for each algorithm.

4.3. Experiments Analysis

DCABC algorithm is better than four other compared algorithms on dimension 10. It can be seen from Table 2 that DCABC has the best performance on 10 of 15 test problems. DCABC is only worse than ABC, PABC on one and two functions (f_9, f_3 and f_9). It is worth noting that GABC and cmABC surpass DCABC only on functions f_3, f_5, and f_{14}. GABC and cmABC generate the best results only on functions f_4 and f_9, respectively.

From Table 3, DCABC ranks NO.1 on 11 of 15 functions with dimension 30. Actually, DCABC is superior to ABC and GABC on all the functions. In contrast, DCABC is inferior to cmABC and PABC on functions f_3 and f_9, and cmABC shows the best performance on functions f_3, f_5, f_9, and f_{14}.

Table 2. Experimental Results about ABC and other ABC variants (10D).

Functions	Criteria	ABC	GABC	cmABC	PABC	DCABC
f_1	Mean	1.37e+06	5.88e+05	1.39e+06	1.43e+06	**8.06e+02**
	Std	1.12e+06	4.73e+05	8.96e+05	9.72e+05	2.31e+03
	Rank	3	2	4	5	1
f_2	Mean	8.57e+02	2.32e+03	4.23e+02	8.51e+02	**2.88e+02**
	Std	7.52e+02	2.45e+03	3.52e+02	5.30e+02	1.20e+03
	Rank	4	5	2	3	1
f_3	Mean	2.01e+01	1.88e+01	**1.87e+01**	1.98e+01	2.00e+01
	Std	3.9e-02	5.07e+00	4.77e+00	1.81e+00	7.88e-03
	Rank	5	2	1	3	4
f_4	Mean	1.26e+01	**5.04e+00**	9.30e+00	1.23e+01	7.89e+00
	Std	4.18e+00	1.42e+00	3.28e+00	3.94e+00	2.99e+00
	Rank	5	1	3	4	2
f_5	Mean	4.31e+02	2.31e+02	**1.82e+02**	3.66e+02	2.61e+02
	Std	1.49e+02	1.16e+02	9.83e+01	1.24e+02	1.51e+02
	Rank	5	2	1	4	3
f_6	Mean	5.03e+03	3.16e+03	3.20e+03	4.13e+03	**1.68e+02**
	Std	4.40e+03	2.36e+03	2.96e+03	4.95e+03	1.97e+02
	Rank	5	2	3	4	1
f_7	Mean	8.72e-01	4.02e-01	5.07e-01	8.65e-01	**3.99e-01**
	Std	2.58e-01	2.37e-01	3.29e-01	2.68e-01	4.24e-01
	Rank	4	2	3	5	1
f_8	Mean	1.32e+04	5.04e+03	7.38e+03	1.26e+04	**4.25e+02**
	Std	2.14e+04	4.33e+03	6.80e+03	2.43e+04	1.20e+03
	Rank	5	2	3	4	1
f_9	Mean	9.48e+01	1.00e+02	**6.48e+01**	9.59e+01	1.00e+02
	Std	2.20e+01	5.66e-02	4.79e+01	1.76e+01	4.12e-02
	Rank	2	4	1	3	4
f_{10}	Mean	4.37e+03	1.84e+03	3.28e+03	4.69e+03	**4.41e+02**
	Std	4.24e+03	9.70e+02	5.22e+03	4.73e+03	1.88e+02
	Rank	4	2	3	5	1
f_{11}	Mean	3.00e+2	2.45e+02	2.49e+02	2.90e+02	**1.92e+02**
	Std	4.69e-1	1.14e+02	1.07e+02	4.81e+01	1.44e+02
	Rank	5	2	3	4	1
f_{12}	Mean	1.04e+02	**1.03e+02**	1.04e+02	1.04e+02	**1.03e+02**
	Std	7.72e-01	4.95e-01	6.33e-01	6.29e-01	8.95e-01
	Rank	2	1	2	2	1
f_{13}	Mean	3.13e+01	2.89e+01	2.95e+01	3.11e+01	**2.77e+01**
	Std	2.04e+00	2.70e+00	2.30e+00	2.01e+00	2.74e+00
	Rank	5	2	3	4	1
f_{14}	Mean	1.86e+03	4.76e+02	**4.47e+02**	1.41e+03	1.14e+03
	Std	1.38e+03	8.49e+02	9.03e+02	1.32e+03	1.39e+03
	Rank	5	2	1	4	3
f_{15}	Mean	**1.00e+02**	**1.00e+02**	**1.00e+02**	**1.00e+02**	**1.00e+02**
	Std	2.28e-11	7.75e-06	1.25e-09	1.59e-11	5.42e-04
	Rank	1	1	1	1	1
Mean rank		4.00	2.13	2.26	3.67	1.73
Overall rank		4	2	3	4	1
Best/2nd Best/Worst		1/1/8	3/10/1	5/2/0	1/1/3	10/1/0

Table 3. Experimental Results about ABC and other ABC variants (30D).

Functions	Criteria	ABC	GABC	cmABC	PABC	DCABC
f_1	Mean	3.73e+06	3.11e+06	1.39e+06	1.43e+06	**8.06e+02**
	Std	1.46e+06	2.05e+06	8.96e+05	9.72e+05	2.31e+03
	Rank	5	4	2	3	1
f_2	Mean	7.26e+02	2.85e+03	4.23e+02	8.51e+02	**2.88e+02**
	Std	6.06e+02	3.77e+03	3.52e+02	5.30e+02	1.20e+03
	Rank	3	5	2	4	1
f_3	Mean	2.01e+01	2.02e+01	**1.87e+01**	1.98e+01	2.00e+01
	Std	4.10e-02	8.32e-02	4.77e+00	1.81e+00	7.88e-03
	Rank	4	5	1	2	3
f_4	Mean	9.77e+01	5.50e+01	9.31e+00	1.23e+01	**7.89e+00**
	Std	1.80e+01	9.75e+00	3.28e+00	3.94e+00	2.99e+00
	Rank	5	4	2	3	1
f_5	Mean	2.42e+03	1.92e+03	**1.82e+02**	3.66e+02	2.61e+02
	Std	2.86e+02	2.96e+02	9.83e+01	1.24e+02	1.51e+02
	Rank	5	4	1	3	2
f_6	Mean	1.38e+06	1.45e+06	3.20e+03	4.13e+03	**1.68e+02**
	Std	7.15e+05	7.51e+05	2.96e+03	4.95e+03	1.97e+02
	Rank	4	5	2	3	1
f_7	Mean	9.39e+00	7.04e+00	5.07e-01	8.65e-01	**3.99e-01**
	Std	1.25e+00	1.81e+00	3.29e-01	2.68e-01	4.24e-01
	Rank	5	4	2	3	1
f_8	Mean	4.11e+05	3.31e+05	7.38e+03	1.26e+04	**4.25e+02**
	Std	3.06e+05	1.91e+05	6.80e+03	2.43e+04	1.20e+03
	Rank	5	4	2	3	1
f_9	Mean	1.21e+02	1.05e+02	**6.48e+01**	9.59e+01	1.00e+02
	Std	4.44e+01	4.89e-01	4.79e+01	1.76e+01	4.12e-02
	Rank	5	4	1	2	3
f_{10}	Mean	6.88e+05	6.84e+05	3.28e+03	4.69e+03	**4.41e+02**
	Std	3.57e+05	5.28e+05	5.22e+03	4.73e+03	1.88e+02
	Rank	5	4	2	3	1
f_{11}	Mean	3.22e+02	3.49e+02	2.49e+02	2.90e+02	**1.92e+02**
	Std	7.47e+00	1.11e+02	1.07e+02	4.81e+01	1.44e+02
	Rank	4	5	2	3	1
f_{12}	Mean	1.07e+02	1.07e+02	1.04e+02	1.04e+02	**1.03e+02**
	Std	8.07e-01	5.78e-01	6.33e-01	6.29e-01	8.95e-01
	Rank	3	3	2	2	1
f_{13}	Mean	1.03e+02	9.91e+01	2.95e+01	3.11e+01	**2.77e+01**
	Std	4.35e+00	2.65e+00	2.30e+00	2.01e+00	2.74e+00
	Rank	5	4	2	3	1
f_{14}	Mean	3.06e+04	3.14e+04	**4.47e+02**	1.41e+03	1.14e+03
	Std	4.51e+02	6.51e+02	9.03e+02	1.32e+03	1.39e+03
	Rank	4	5	1	3	2
f_{15}	Mean	**1.00e+02**	**1.00e+02**	**1.00e+02**	**1.00e+02**	**1.00e+02**
	Std	6.56e+02	7.75e-06	1.25e+00	1.59-11	5.42e-04
	Rank	1	1	1	1	1
Mean rank		4.20	4.07	1.67	2.73	1.40
Overall rank		5	4	2	3	1
Best/2nd Best/Worst		1/0/8	1/0/5	5/10/0	1/3/0	11/2/0

In Table 4, DCABC outperforms all compared algorithms with dimension 50 on functions f_1, f_3, f_4, f_5, f_6, f_8, f_9, f_{10}, f_{12}, and f_{15}. DCABC cannot beat ABC, GABC, cmABC and PABC on f_7 and f_{11}. PABC shows the best performance on f_2 and f_{14}. cmABC is superior to all other algorithms on f_3 and f_7. GABC is competitive on function f_{13}. ABC has the best results on function f_{11}, f_{12} and f_{14}. It is worth noting that the overall performance of DCABC is the best.

Table 4. Experimental Results about ABC and other ABC variants (50D).

Functions	Criteria	ABC	GABC	cmABC	PABC	DCABC
f_1	Mean	1.20e+07	1.05e+07	1.00e+07	1.28e+07	**3.92e+02**
	Std	3.69e+06	4.49e+00	2.63e+06	4.65e+06	1.36e+03
	Rank	4	3	2	5	1
f_2	Mean	1.43e+03	7.88e+03	1.52e+03	**1.37e+03**	3.74e+03
	Std	1.20e+03	7.18e+03	1.04e+03	1.32e+03	7.84e+03
	Rank	2	5	3	1	4
f_3	Mean	2.02e+01	2.02e+01	**2.00e+01**	2.01e+01	**2.00e+01**
	Std	4.12e-02	6.70e-02	6.64e-03	3.41e-02	1.56e-02
	Rank	3	3	1	2	1
f_4	Mean	2.34e+02	2.30e+02	2.71e+02	2.19e+02	**1.53e+02**
	Std	2.93e+01	2.90e+01	3.35e+01	3.04+01	2.42e+01
	Rank	4	3	5	2	1
f_5	Mean	4.24e+03	4.76e+03	4.14e+03	4.10e+03	**3.93e+03**
	Std	4.77e+02	4.03e+02	4.09e+02	3.34e+02	4.31e+02
	Rank	5	2	4	3	1
f_6	Mean	2.20e+06	2.45e+06	2.11e+06	2.30e+06	**2.25e+03**
	Std	7.53e+05	1.26e+06	6.63e+05	8.94e+05	1.10e+03
	Rank	3	5	2	4	1
f_7	Mean	1.85e+01	1.91e+01	**1.57e+01**	1.72e+01	3.04e+01
	Std	8.52e+00	9.91e+00	1.48e+00	6.53e+00	1.56e+01
	Rank	3	4	1	2	5
f_8	Mean	2.15e+06	2.97e+06	2.35e+06	3.25e+06	**3.48e+03**
	Std	8.87e+05	1.65e+06	6.22e+05	1.02e+06	1.03e+04
	Rank	2	4	3	5	1
f_9	Mean	1.62e+02	1.08e+02	1.08e+02	1.52e+02	**1.07e+02**
	Std	1.13e+02	5.21e-01	8.44e-01	9.20e+01	5.74e-01
	Rank	4	2	2	3	1
f_{10}	Mean	8.07e+05	1.10e+06	9.27e+05	9.86e+05	**3.59e+03**
	Std	4.68e+05	6.62e+05	3.07e+05	4.57e+05	7.45e+02
	Rank	2	5	3	4	1
f_{11}	Mean	**3.58e+02**	6.68e+02	3.99e+02	4.27e+02	8.16e+02
	Std	1.69e+02	4.05e+02	2.72e+02	2.89e+02	4.05e+02
	Rank	1	4	2	3	5
f_{12}	Mean	**1.10e+02**	1.19e+02	**1.10e+02**	**1.10e+02**	**1.10e+02**
	Std	7.41e-01	4.97e-01	8.35e-01	6.39e-01	9.15e-01
	Rank	1	2	1	1	1
f_{13}	Mean	1.89e+02	**1.85e+02**	1.93e+02	1.88e+02	1.87e+02
	Std	4.91e+00	4.43e+00	5.34e+00	6.20e+00	5.88e+00
	Rank	4	1	5	3	2
f_{14}	Mean	**4.99e+04**	5.51e+04	5.02e+04	**4.99e+04**	5.41e+04
	Std	1.77e+03	6.10+03	2.45e+03	1.76e+03	4.57e+03
	Rank	1	4	2	1	3
f_{15}	Mean	1.02e+02	**1.00e+02**	**1.00e+02**	**1.00e+02**	**1.00e+02**
	Std	1.23e+00	2.00e-07	3.03e-01	1.59-11	3.89e-04
	Rank	2	1	1	1	1
Mean rank		2.73	3.20	2.47	2.67	1.93
Overall rank		4	5	2	3	1
Best/2nd Best/Worst		3/4/1	2/3/3	4/5/2	4/3/2	10/1/2

Unimodal function f_1, hybrid function f_8 and composition functions f_{10} are chosen to exhibit the convergence precision of all compared algorithms. Figures 3–8 are the convergence graphs of five algorithms. The horizontal axis is the number of function evaluations (FES), and the vertical axis is the function values over one independent run. In all the figures, DCABC is represented by the black line with circles, and it has larger descend gradient and gets the minimal error values among the five algorithms. The convergence speed of DCABC is also obviously superior to the other four algorithms.

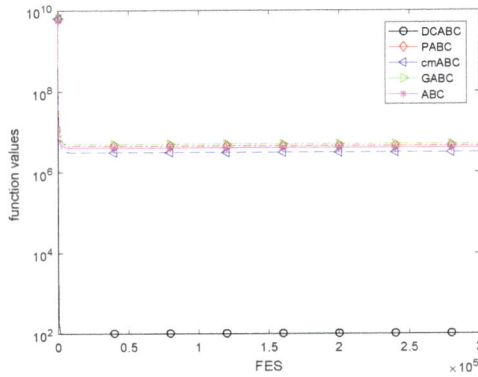

Figure 3. Convergence curves of five algorithms for f_1 with D = 30. (The optimal value of f_1 is 100).

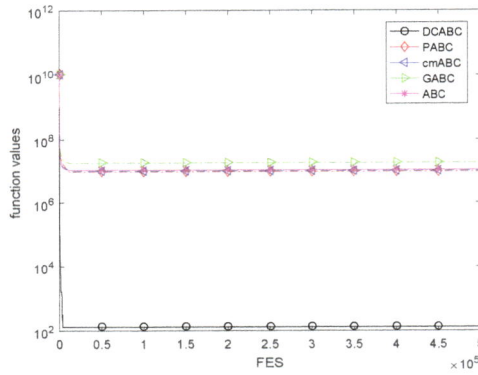

Figure 4. Convergence curves of five algorithms for f_1 with D = 50. (The optimal value of f_1 is 100).

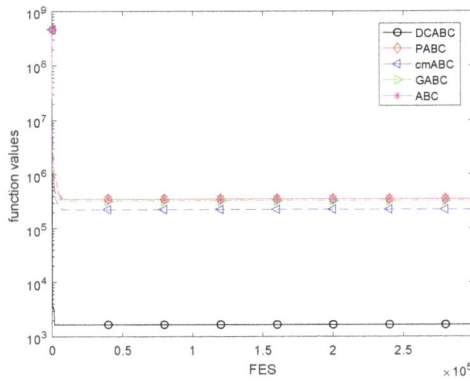

Figure 5. Convergence curves of five algorithms for f_8 with D = 30. (The optimal value of f_8 is 800).

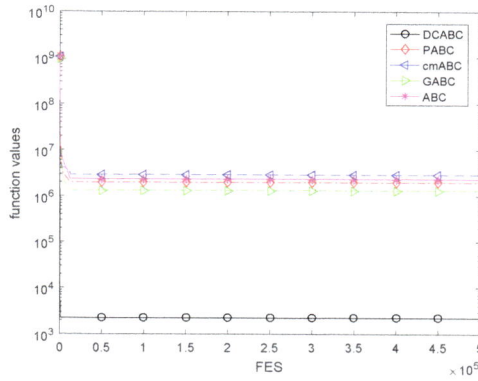

Figure 6. Convergence curves of five algorithms for f_8 with D = 50. (The optimal value of f_8 is 800).

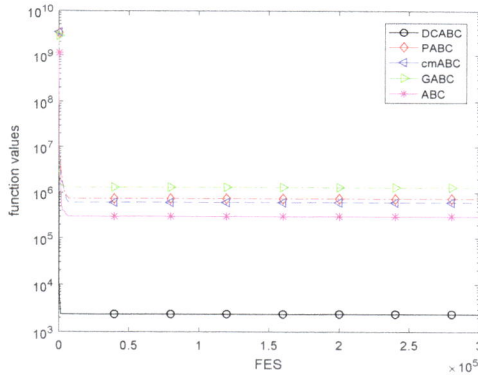

Figure 7. Convergence curves of five algorithms for f_{10} with D = 30. (The optimal value of f_{10} is 1000).

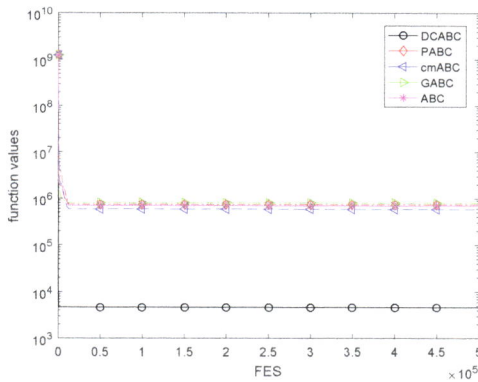

Figure 8. Convergence curves of five algorithms for f_{10} with D = 50. (The optimal value of f_{10} is 1000).

On the whole, compared to the other three modified ABC algorithms, DCABC can show the best performance on most of the functions, that is to say this new algorithm is more stable and the solutions obtained by it have higher precision than other algorithms.

5. Conclusions

In the present study, a developed artificial bee colony algorithm based on cloud model, namely DCABC, is proposed for the continuous optimization. By using a new selection mechanism, the worse individual in DCABC has a larger probability to be selected than in basic ABC. DCABC also improves the local search ability by applying the normal cloud generator as onlookers bees' formula to control the search of onlookers in a suitable range. Moreover, historical optimal solutions are used by Y conditional cloud generator when updating the scout bee to ensure the algorithm jump out of the local optimal. The effectiveness of the proposed method is tested on CEC15. The results clearly show the superiority of DCABC over ABC, GABC, cmABC and PABC.

However, there are quite a few issues that merit further investigation such as the diversity of DCABC. In addition, we hope to show the performance of DCABC by Null Hypothesis Significance Testing (NHST) [35,36] in our future work. We only test the new algorithm on classical benchmark functions and have not used it to solve practical problems, such as fault diagnosis [37], path plan [38], Knapsack [39–41], multi-objective optimization [42], gesture segmentation [43], unit commitment problem [44], and so on. There is an increasing interest in prompting the performance of DCABC, which will be our future research direction.

Acknowledgments: This research is partly supported by Humanity and Social Science Youth foundation of Ministry of Education of China (Grant No. 12YJCZH179), the Natural Science Foundation of the Jiangsu Higher Education Institutions of China (Grant No. 16KJA110001), the National Natural Science Foundation of China (Grant No. 11371197), the Foundation of Jiangsu Key Laboratory for NSLSCS (Grant No. 201601).

Author Contributions: These authors contributed equally to this paper.

Conflicts of Interest: No conflict of interest exists in the submission of this article, and it is approved by all authors for publication.

References

1. Sörensen, K.; Sevaux, M.; Glover, F. A History of Metaheuristics. In *Handbook of Heuristics*; Springer: Berlin/Heidelberg, Germany, 2018.
2. Sörensen, K. Metaheuristics-the metaphor exposed. *Int. Trans. Oper. Res.* **2015**, *22*, 3–18.
3. Črepinšek, M.; Liu, S.; Mernik, M. Exploration and exploitation in evolutionary algorithms: A survey. *ACM Comput. Surv.* **2013**, *45*, 1–33.
4. Karaboga, D. *An Idea Based on Honey bee Swarm for Numerical Optimization*; Technical Report-tr06; Engineering Faculty, Computer Engineering Department, Erciyes University: Kayseri, Turkey, 2005.
5. Karaboga, D.; Basturk, B. Artificial Bee Colony (ABC) optimization algorithm for solving constrained optimization problems. *Found. Fuzzy Log. Soft Comput.* **2007**, *4529*, 789–798.
6. Karaboga, D.; Basturk, B. On the performance of artificial bee colony (ABC) algorithm. *Appl. Soft Comput.* **2008**, *8*, 687–697.
7. Karaboga, D.; Akay, B. A comparative study of artificial bee colony algorithm. *Appl. Math. Comput.* **2009**, *214*, 108–132.
8. Akay, B.; Karaboga, D. Parameter tuning for the artificial bee colony algorithm. *Comput. Collect. Intell.* **2009**, *5796*, 608–619.
9. Akay, B.; Karaboga, D. A modified artificial bee colony algorithm for real-parameter optimization. *Swarm Intell. Appl.* **2012**, *192*, 120–142.
10. Zhang, D.; Guan, X.; Tang, Y.; Tang, Y. Modified artificial bee colony algorithms for numerical optimization. In Proceedings of the 2011 3rd International Workshop on Intelligent Systems and Applications (ISA), Wuhan, China, 28–29 May 2011; pp. 1–4.
11. Aderhold, A.; Diwold, K.; Scheidler, A.; Middendorf, M. Artificial bee colony optimization: A new selection scheme and its performance. *Nat. Inspired Coop. Strateg. Optim. (NICSO 2010)* **2010**, *284*, 283–294.

12. Zhu, G.; Kwong, S. Gbest-guided artificial bee colony algorithm for numerical function optimization. *Appl. Math. Comput.* **2010**, *217*, 3166–3173.

13. Wu, X.; Hao, D.; Xu, C. An improved method of artificial bee colony algorithm. *Appl. Mech. Mater.* **2012**, *101–102*, 315–319.

14. Guo, P.; Cheng, W.; Liang, J. Global artificial bee colony search algorithm for numerical function optimization. In Proceedings of the 2011 Seventh International Conference on Natural Computation (ICNC), Shanghai, China, 26–28 July 2011; Volume 3, pp. 1280–1283.

15. Yu, X.; Zhu, Z. A modified artificial bee colony algorithm with its applications in signal processing. *Int. J. Comput. Appl. Technol.* **2013**, *47*, 297–303.

16. Rajasekhar, A.; Pant, M. An improved self-adaptive artificial bee colony algorithm for global optimisation. *Int. J. Swarm Intell.* **2014**, *1*, 115–132.

17. Yaghoobi, T.; Esmaeili, E. An improved artificial bee colony algorithm for global numerical optimisation. *Int. J. Bio-Inspired Comput.* **2017**, *9*, 251–258.

18. Zou, W.; Zhu, Y.; Chen, H.; Zhang, B. Solving multiobjective optimization problems using artificial bee colony algorithm. *Discret. Dyn. Nat. Soc.* **2011**, *2*, 1–37.

19. Amarjeet; Chhabra, J.K. FP-ABC: Fuzzy Pareto-Dominance Driven Artificial Bee Colony Algorithm for Many-Objective Software Module Clustering. *Comput. Lang. Syst. Struct.* **2018**, *51*, 1–21.

20. Xiang, Y.; Zhou, Y.; Tang, L.; Chen, Z. A Decomposition-Based Many-Objective Artificial Bee Colony Algorithm. *IEEE Trans. Cybern.* **2017**, *99*, 1–14.

21. Koçer, B. Bollinger bands approach on boosting ABC algorithm and its variants. *Appl. Soft Comput.* **2016**, *49*, 292–312.

22. Bollinger Bands—Trademark Details. 2011. Available online: Justia.com (accessed on 1 April 2018) .

23. Wang, H.; Yi, J. An improved optimization method based on krill herd and artificial bee colony with information exchange. *Memet. Comput.* **2017**, *2*, 1–22.

24. Gandomi, A. H.; Alavi, A. H. Krill herd: A new bio-inspired optimization algorithm. *Commun. Nonlinear Sci. Numer. Simul.* **2012**, *17*, 4831–4845.

25. Li, D.; Liu, C.; Du, Y.; Han, X. Artificial Intelligence with Uncertainty. *J. Softw.* **2004**, *15*, 1583–1594.

26. Chen, H.; Li, D.; Shen, D.; Zhang, F. A clouds model applied to controlling inverted pendulum. *J. Comput. Res. Dev.* **1999**, *36*, 1180–1187.

27. Zhang, C.; Pang, Y. Sequential blind signal extraction adopting an artificial bee colony Algorithm algorithm. *J. Inf. Comput. Sci.* **2012**, *9*, 5551–5559.

28. He, D.; Jia, R. Cloud model-based Artificial Bee Colony algorithm's application in the logistics location problem. In Proceedings of the International Conference on Information Management, Innovation Management and Industrial Engineering, Sanya, China, 20–21 October 2012.

29. Lin, X.; Ye, D. Artificial Bee Colony algorithm based on cloud mutation. *J. Comput. Appl.* **2012**, *32*, 2538–2541.

30. Li, D.; Meng, H.; Shi, X. Membership clouds and membership clouds generators. *Comput. Res. Dev.* **1995**, *42*, 32–41.

31. Di, K.; Li, D.; Li, D. Cloud theory and its applications in spatial data mining and knowledge discovery. *J. Image Graph.* **1999**, *4*, 930–935.

32. Chen, Q.; Liu, B.; Zhang, Q.; Liang, J.; Suganthan, P.N.; Qu, B. *Problem Definition and Evaluation Criteria for CEC 2015 Special Session and Competition on Bound Constrained Single-Objective Computationally Expensive Numerical Optimization*; Technical Report; Computational Intelligence Laboratory, Zhengzhou University: Zhengzhou, China; Nanyang Technological University: Singapore, 2014.

33. Veček, N.; Liu, S.; Črepinšek, M.; Mernik, M. On the Importance of the Artificial Bee Colony Control Parameter Limit. *Inf. Technol. Control* **2017**, *46*, 566–604.

34. Mernik, M.; Liu, S.; Karaboga, D. On clarifying misconceptions when comparing variants of the Artificial Bee Colony Algorithm by offering a new implementation. *Inf. Sci.* **2015**, *291*, 115–127.

35. Derrac, J.; García, S.; Molina, D.; Herrera, F. A practical tutorial on the use of nonparametric statistical tests as a methodology for comparing evolutionary and swarm intelligence algorithms. *Swarm Evol. Comput.* **2011**, *1*, 3–18.

36. Veček, N.; Mernik, M.; Črepinšek, M. A chess rating system for evolutionary algorithms: A new method for the comparison and ranking of evolutionary algorithms. *Inf. Sci.* **2014**, *277*, 656–679.

37. Yi, J.; Wang, J.; Wang, G. Improved probabilistic neural networks with self-adaptive strategies for transformer fault diagnosis problem. *Adv. Mech. Eng.* **2016**, *8*, 1–13.

38. Wang, G.; Chu, H.; Mirjalili, S. Three-dimensional path planning for UCAV using an improved bat algorithm. *Aerosp. Sci. Technol.* **2016**, *49*, 231–238.

39. Feng, Y.; Wang, G.; Deb, S.; Lu, M.; Zhao, X. Solving 0-1 knapsack problem by a novel binary monarch butterfly optimization. *Neural Comput. Appl.* **2017**, *28*, 1619–1634.

40. Feng, Y.; Wang, G.; Li, W.; Li, N. Multi-strategy monarch butterfly optimization algorithm for discounted 0-1 knapsack problem. *Neural Comput. Appl.* **2017**, doi:10.1007/s00521-017-2903-1.

41. Feng, Y.; Wang, G.; Dong, J.; Wang, L. Opposition-based learning monarch butterfly optimization with Gaussian perturbation for large-scale 0-1 knapsack problem. *Comput. Electr. Eng.* **2017**, doi:10.1016/j.compeleceng.2017.12.014.

42. Rizk-Allah, R.M.; El-Sehiemy, R.A.; Deb, S.; Wang, G. A novel fruit fly framework for multi-objective shape design of tubular linear synchronous motor. *J. Supercomput.* **2017**, *73*, 1235–1256.

43. Liu, K.; Gong, D.; Meng, F.; Chen, H.; Wang, G. Gesture segmentation based on a two-phase estimation of distribution algorithm. *Inf. Sci.* **2017**, *394–395*, 88–105.

44. Srikanth, K.; Panwar, L.K.; Panigrahi, B.K.; Herrera-Viedma, E.; Sangaiah, A.K.; Wang, G. Meta-heuristic framework: Quantum inspired binary grey wolf optimizer for unit commitment problem. *Comput. Electr. Eng.* **2017**, doi:10.1016/j.compeleceng.2017.07.023.

MDPI

St. Alban-Anlage 66

4052 Basel

Switzerland

Tel. +41 61 683 77 34

Fax +41 61 302 89 18

www.mdpi.com

Mathematics Editorial Office

E-mail: mathematics@mdpi.com

www.mdpi.com/journal/mathematics